Agricultural Economics

Agricultural Economics

Edited by **Adler Bryant**

SYRAWOOD
PUBLISHING HOUSE
New York

Published by Syrawood Publishing House,
750 Third Avenue, 9th Floor,
New York, NY 10017, USA
www.syrawoodpublishinghouse.com

Agricultural Economics
Edited by Adler Bryant

International Standard Book Number: 978-1-68286-365-7 (Hardback)

Printed in China.

Contents

Preface

Agricultural economics is the application of concepts and tools from economics to evaluate the production, distribution and consumption of agricultural products. Agricultural economics plays a vital role in preparing agricultural, food and environmental policies and programmes that are crucial to the economic development of a nation. The significant topics included in this book are analyses of consumer behaviour, supply and demand, government regulations and policies, macroeconomics of agriculture, and market equilibrium, etc. In this book, using case studies and examples, constant effort has been made to make the understanding of the difficult concepts of agricultural economics as easy and informative as possible, for the readers.

After months of intensive research and writing, this book is the end result of all who devoted their time and efforts in the initiation and progress of this book. It will surely be a source of reference in enhancing the required knowledge of the new developments in the area. During the course of developing this book, certain measures such as accuracy, authenticity and research focused analytical studies were given preference in order to produce a comprehensive book in the area of study.

This book would not have been possible without the efforts of the authors and the publisher. I extend my sincere thanks to them. Secondly, I express my gratitude to my family and well-wishers. And most importantly, I thank my students for constantly expressing their willingness and curiosity in enhancing their knowledge in the field, which encourages me to take up further research projects for the advancement of the area.

Editor

Grain quality and determinants of farmers' preference for rice varietal traits in three districts of Ghana: Implications for research and policy

Maxwell Darko Asante[1,2], Bright Owusu Asante[2,3], George Kwasi Acheampong[2], Alexander Nimo Wiredu[4], Samuel Kwame Offei[1], Vernon Gracen[1], Hans Adu-Dapaah[2] and Eric Yirenkyi Danquah[1]

[1]West African Centre for Crop Improvement, College of Agriculture and Consumer Sciences, University of Ghana, P. M. B. LG 30, Legon, Ghana.
[2]CSIR-Crops Research Institute, P. O. Box 3785, Kumasi, Ghana.
[3]Deparment of Agricultural and Resource Economics, Faculty of the Professions, University of New England, Armidale, NSW, 2350, Australia.
[4]CSIR-Savanna Agricultural Research Institute, P. O. Box TL 52, Tamale, Ghana.

Rice has become a popular staple in Africa because it is a convenience food for the rapidly increasing urban population. In Ghana about 70% of the rice consumed is imported. Domestic rice is not competitive because of its perceived poor grain quality. This study assessed the influence of grain quality and other factors on farmers' preference for improved rice varieties. It also investigated how specific grain quality attributes influence farmers' preference for improved rice varieties. The problt regression model was used to assess the influence of grain quality attributes and other factors on farmers' preference for improved rice varieties. The results showed that, age, sex, tolerance to pest and diseases, farm size, and good grain quality had a positive influence on farmers' preference. Experience in rice cultivation however negatively influenced farmers' preference for rice varieties. Furthermore, specific grain quality attributes such as, grain length and shape, fragrance, cooking quality, grain color, and absence of foreign matter also positively influenced farmers' preference for rice varietal traits while chalkiness had a negative influence. Improving grain quality will increase consumer demand for locally produced rice and farmer preferences for improved rice varieties. This has the potential of increasing local rice production and ensuring self-sufficiency in rice production. By implication, there is the need to invest into the development and deployment of rice varieties that have grains that meet the expectations of farmers to propel wide spread adoption and utilization of domestic rice.

Key words: Ghana, grain quality, preference, rice varietal traits.

INTRODUCTION

Rice is a staple food for nearly half of the world population. In Africa, rice has increasingly become popular and it is reported to be the fastest growing food source (Nwanze et al., 2006). Besides being an important food staple for both rural and urban communities across Ghana, it is the most important cash crop in the communities in which it is produced (Asuming–Brempong and Osei-Asare, 2007). According to the 2010 budget statement of Ghana, the country imports about 70% of the domestic rice requirement, accounting for about USD 600 million per annum (Duffuor, 2009). To bridge this gap, successive governments over the years have been promoting domestic rice production and consumption. However, due to the poor grain quality of the locally produced rice, consumers usually prefer imported rice. Availability of improved rice varieties with good grain quality is therefore necessary in order to boost domestic rice production and consumption.

Although many varieties of rice have been developed, few have been adopted possibly because researchers have not considered farmers' preferences and perceptions of varieties during the development process (Efisue et al., 2008). Adoption of improved rice varieties may differ depending upon the preferences of the farmers. Farmers assess a new technology such as crop variety, in terms of a range of attributes, including grain quality, straw yield, and input requirements in addition to grain yield (Joshi and Bauer, 2006).

Previous studies on rice grain quality preference in Ghana and other parts of the world have focused on consumers (Bam et al., 1998; Diako et al., 2010; Abansi et al., 1992; Choudhury et al., 1992; Juliano and Duff, 1991). However, little has been done to assess grain quality in relation to the farmer in order to infer how grain quality could influence farmers' preference for improved rice varieties.

This study assessed the influence of grain quality and other factors affecting farmers' preference for improved rice varieties. It further investigates how grain quality attributes such as colour, grain length and shape, chalkiness, percent broken grains (related to milling yield), cooking quality, fragrance, absence of foreign materials and grain expansion ratio influences farmers' preference for improved rice varieties and their implications for research and policy. This study will serve as a guide to policy making by pointing to areas of intervention for effective promotion of domestic rice production and consumption in Ghana. It provides useful information to researchers especially breeders to develop varieties that are consistent with the needs of both farmers, processors, traders and consumers especially in terms of rice grain quality. The remaining sections of this paper are presented as follows. In the next section, we present description of the methodology and the study area; it includes data collection process and the empirical framework for the study. The third and fourth sections present the results and analysis of the empirical findings and their implications. We end with some concluding remarks and policy recommendations.

METHODOLOGY

The study area

The study was conducted in the Ejisu-Juaben, Atwima Nwabiagya and the Ejura-Sekyedumase districts of the Ashanti region. The Ashanti region is located in the deciduous forest agro-ecological zone. The zone is characterized by a bi-modal rainfall pattern, the major season beginning from April and ending in July and the minor one beginning in September and ending in October. The annual rainfall ranges between 1500 mm and 1600 mm with an average of about 1300 mm per annum. Temperatures are generally low throughout most of the year with the highest of 28°C in March and April. Lower temperatures are experienced during the major season in June and July. Rice, roots and tuber crops such as cassava, yam, cocoyam and sweet potatoes as well as plantain are the major food crops cultivated in the zone. Rice cultivation is basically in inland valleys. Cocoa (*Theobroma cacao*) and oil palm (*Elaeisguineensis*) are the common tree crops that form an integral part of the people's livelihood.

The study was conducted in 6 communities from the three districts of the Ashanti region of Ghana. The communities comprise of, Nobewam and Besease of the Ejisu-Juaben district, Kyeiase and Nkawie of the Atwima Nwabiagya district, and Aframso and Ejura in the Ejura-Sekyedumase district. These are the major rice growing communities in the region and are among the major rice growing communities in southern Ghana. Farming and rice cultivation is the major occupation in the selected communities as they are endowed with inland valleys suitable for rice cultivation (Figure 1). In all, six rice communities (two communities per district) were randomly selected from a list of major rice producing communities in Ashanti region. A random sample of 30 farmers was further obtained from each of the communities. Although the study intended to obtain a sample of 180, due to discrepancies such as non-response and irrelevant data, consequently, a total of 165 rice farmers were involved in the analysis.

Both formal and informal approaches were employed to collect data for the study. Participatory rural appraisal (PRA) tools were used to generate information on farmer perceptions at the community level. Probing, iteration, observation and preference rating (Obiri, 2003; Grandstaff and Grandstaff, 1987) were the PRA techniques used to retrieve information from farmers. In addition, there was triangulation to test the robustness of the responses from the groups. The respondents were thus encouraged to respond more completely and precisely. In addition, it helped to organize the responses and ensured that all necessary issues were covered.

This was followed by a formal survey with the use of semi-structured questionnaires. This enabled individual farmers to express their own views without any community influence. During the interview, useful demonstrations and drawings were frequently used to illustrate difficult points. Data collected included demographic characteristics, factors influencing varietal preference, and the effect of grain quality on preference for improved varieties.

Analytical framework

In the first part of the analyses, the study assessed specific varietal characteristics that are preferred by farmers. In the second part, factors influencing farmers' preference for rice varietal traits were assessed. To assess specific rice varietal characteristics that are most preferred by farmers in a new rice variety, varietal attributes were identified and farmers were made to rank them. The ranking was evaluated using the Kendall's coefficient of concordance (W) to identify the varietal attributes that is most preferred by the farmers. The Kendall's coefficient of concordance (W) is a measure of degree of agreement/concordance among m set of n ranks. It is an index that measures the ratio of the observed variance of the sum of ranks to the maximum possible variance of sum of ranks. The essence of this index is to find the sum of the ranks for each attribute/factor being ranked and to examine the variability of this sum. If the rankings are in perfect agreement the variability among these sums is said to be a minimum (Mattson, 1986).

The analysis is a statistical technique that is used to identify and

Figure 1. A map of the Ashanti Region Ghana.

rank a given set of factors into the most pressing and then measures the degree of agreements among the judges. Usually the identified factors are ranked according to the most pressing using numerals such as 1, 2, 3...,n, in that order. The factor with least total score is ranked as the most pressing, while the one with the highest score is ranked as least pressing. The computed total rank score is then used to calculate the W. The value of W is positive in sign and ranges from 0 to 1. It is 1 when the values assigned by one farmer are exactly the same as those assigned by other farmers, and is 0 when there is maximum disagreement among the farmers.

Given that T represents the sum of ranks of each factor being ranked, the variance of the sum is given by

$$Var_T = \frac{\sum T^2 - (\sum T^2)/n}{n} \quad (1)$$

The maximum variance of T is then given by

$$m^2(n^2-1)/12 \quad (2)$$

Where m = number of sets of rankings (by the farmers) and n = the number of specific varietal attributes being ranked.
The Coefficient of Concordance (W) is therefore given as:

$$W = \frac{\left[\sum T^2 - (\sum T^2)/n\right]/n}{m^2(n^2-1)/12} \quad (3)$$

Equation (3) is further simplifies to the computational formula as:

$$W = \frac{12\left[\sum T^2 - (\sum T^2)/n\right]/n}{nm^2(n^2-1)} \quad (4)$$

The hypothesis and the significance of the rankings are further assessed using the F-test as follows:

The null (H0) and the alternative (H1) hypotheses are stated as follows:
H_0: there is no agreement among the rankings of the varietal characteristics preferred by farmers in a new rice variety, versus and
H_1: there is agreement among the rankings of varietal characteristics preferred by farmers in a new rice variety.
The coefficient of Concordance (W) may be tested for significance in terms of the F distribution. It is given by:

$$F = \frac{[(m-1)W_c]}{(1-W_c)} \quad (5)$$

Where W_c is the calculated Coefficient of Concordance (W).
The F statistic has $v_1 = (n-1) - 2/m$ degrees of freedom for the numerator, and
$v_2 = (m-1)[(n-1) - 2/m]$ degrees of freedom for the denominator.
If the F is greater than the relevant critical F* from Fisher's F statistical distribution table, the null hypothesis is rejected.
The probit model was used to assess the influence of grain quality and other factors on farmers' preference for improved rice varieties. The study used the probit model (Maddala, 2005) specified as:

$$P_i = P(y_i^* < y_i)$$
$$P_i = P(y_i^* < \beta_0 + \beta_j X_{ji}) = F(y_i)$$
$$P_i = F(y_i) = \frac{1}{\sqrt{2\pi}} \int_{-\infty}^{Z_i} e^{s^2/2}\, ds \quad (6)$$

where P_i is the probability that an individual will make a certain choice (prefer an improved rice variety or not), s is a random variable normally distributed with mean zero and a unit variance. y_i is the dependent variable (observed preference for an improved rice variety) and y_i^* is the threshold value of the dependent variable.

To obtain an estimate of the index Z_i, the inverse of the cumulative normal function is used:

$$yi = F^{-1}(P_i) = \beta_0 + \sum \beta_{ki} x_{ki} + u_i \quad (7)$$

The parameter estimates β_s of the probit apart from the information on the effect of changes in the explanatory variable on the probability of adoption also provide information on the relative effect of each explanatory variable on the likelihood that a farmer will prefer an improved variety. The latter can be obtained as follows:

$$\frac{\partial P_i}{\partial X_{ji}} = \beta_{ij} * f(Z_i) \quad (8)$$

Where P_i is the mean of estimated probability that a farmer prefers a variety and

$$f(Z_i) = F^{-1}(P_i) \quad (9)$$

The elasticity of the predicted probability is then computed as:

$$\frac{\partial P_i}{\partial X_{ji}} = \beta_{ij} * f(Z_i) * \frac{\overline{X}}{P_i} \quad (10)$$

Guided by related studies in the past, (Asrat et al., 2010; Ayamga et al., 2006; Gockowski and Ndoumbe, 2004; Nzomoi et al., 2007), socio-economic attributes as well as varietal traits were identified and hypothesis constructed regarding farmers' preference for improved rice varieties. These factors either had a bearing on the individual (farmer) or on the improved variety and therefore policy. The variables include age, gender (sex), level of education, farm size, experience in rice production and other attributes of the variety such as high yield, good grain quality, tolerance to biotic stress and tolerance to pest and diseases. A dummy variable was used to specify sex of farmers (SEX). Rice cultivation is labour intensive in southern Ghana. As a result more males are involved in rice cultivation than females. In view of the above, a value of 1 was assigned to males and 0 to females. If this is true, the sex variable is expected to be significant. A person with formal education is expected to have adequate knowledge on the importance of planting improved varieties. Most researchers have found a positive relationship between education and preference for a new technology (Gamba et al., 2002; Nzomoi et al., 2007; Ayamga et al., 2006; Katungi, 2006; Mussei et al., 2001). The variable education (EDUC) is therefore expected to be positive.

Age is represented by AGE and AGE^2. Many studies have found a negative relationship between age and preference or adoption (Gockowski and Ndoumbe, 2004; McBride and Daberkow, 2003; Gamba et al., 2002). The hypothesis here is that farming requires strength hence, as one ages, his or her strength begin to decline after a peak age hence beyond this age, farmers' desire to explore new technologies begin to decrease. In addition, past success of older farmers could make them complacent, affecting their preference for new varieties. Young farmers are therefore more likely to prefer improved varieties. Hence the variable AGE is expected to be negative. The variable FRMSZE is used to represent farm size and it defines the effective land area (in hectares) under rice cultivation and not total land area owned by farmers. Most researchers have found a positive relationship between farm size and farmers preference for a given technology (Asrat et al., 2010; Adimado, 2001; Kheralla et al., 2001; Langyintuo and Mekuria, 2005). The implication here is that farmers with larger farm sizes for a particular crop tend to require more support in terms of marketing, pricing and inputs than their counterparts with smaller farm sizes hence will prefer an improved variety that will enable them to increase their productivity and income. This variable is thus expected to be positive.

The variable HIGYLD is expected to have a positive influence on farmers' preference for a new rice variety. Grain quality is an important attribute that farmers consider in selecting a particular rice variety. This is because it has a direct effect on the demand for the produce, its market price and resultant income that accrue to the farmer. A study by Ramasamy et al. (2003) has shown a positive effect of grain quality on the willingness of drop outs to continue cultivating hybrid rice varieties in Tamil Nadu. The variable GGQAP was binary, and was assigned a value of 1 if farmers prefer good grain quality and 0 if otherwise. This variable is expected to be positive.

Biotic and abiotic stresses such as drought, iron toxicity, nitrogen use efficiency, diseases and pests, including weeds have debilitating effects on grain yield, grain quality, market price and farmer income. Farmers will thus prefer improved rice varieties with high tolerance to these stresses. Odendo et al. (2001) found a positive relationship between tolerances to stresses especially striga, drought, diseases and insect pests and farmers' preference for improved maize varieties in Western Kenya. The variables TBIOSTRSS and TPED representing tolerance to biotic stress and tolerance to pest and diseases, respectively are thus expected to be positive.

The empirical model is specified as:

$$\begin{aligned} y_b &= \beta_0 + \beta_1 AGE + \beta_2 AGE^2 + \beta_3 FRMSIZE + \beta_4 GGQAP + \beta_5 HIGHYLD \\ &+ \beta_6 EXP + \beta_7 \text{SEX} + \beta_8 \text{TBIOSTRSS} + \beta_9 TPED + \beta_{10} EDUCL + \vartheta_i \end{aligned} \quad (11)$$

The dependent variable y_i is the farmers' preference for an improved rice variety or otherwise and it takes the value of 1 if the farmer prefers an improved variety and 0 if he/she does not. The probit model employed to estimate the effect of factors influencing farmer preference for an improved rice variety included nine explanatory variables. Table 1 presents a description of the explanatory variables used in the model with their means or frequencies.

Further analysis on grain quality attributes

The above model will estimate the influence of grain quality on farmers' preference for an improved rice variety in Ghana. However, the specific quality attributes that significantly bring about this influence were not clear. In view of this, a second analysis which investigates the influence of specific grain quality attributes on farmers' preference for an improved rice variety was imperative. In this analysis, several grain quality attributes were explored to assess their influence on farmers' preference for an improved rice variety. The explanatory variables included in the second model include grain colour, cooking quality, expansion ratio, absence of foreign material, fragrance, grain length and shape, percent brokenness and chalkiness.

Colour, cooking quality, absence of foreign materials, fragrance and chalkiness were represented by dummy variables. All these factors except chalkiness are expected to have a positive effect on farmers' preference. Chalkiness is not preferred among consumers hence is expected to have a negative influence on farmers preferences. The model is specified as:

Table 1. Summary description of explanatory variables used in the model.

Variable	Unit of measure	Frequency/mean
Age of respondents (AGE)	Years	36.84
Sex of respondents (SEX)	Binary variable	0 = 54; 1 = 113
Level of education (Basic level) (EDUC)	Binary variable	0 = 90; 1 = 75
Farm size (FRMSIZE)	Hectares	1.28
Good grain quality and appearance (GGQAP)	Binary variable	0 = 25; 1 = 140
Tolerance to biotic stress (TBIOSTRSS)	Binary variable	0 = 74; 1 = 91
Tolerance to pest and diseases (TPED)	Binary variable	0 = 65; 1 = 100
High yield (HIGYLD)	Binary variable	0 = 25; 1 = 140

$$y_b = \beta_0 + \beta_1 COLOR + \beta_2 \mathrm{COOKQLITY} + \beta_3 \mathrm{EXPRATIO} + \beta_4 \mathrm{FOREIGNM} + \beta_5 \mathrm{FRAGANCE} + \beta_6 \mathrm{GLNS} + \beta_7 \mathrm{PERCENTBGS} + \beta_8 \mathrm{CHKN} + \varepsilon_i \quad (12)$$

where, COLOR represents colour of the grain, COOKQLITY represents cooking quality, EXPRATIO represents expansion ration, FOREIGNM represents absence of foreign material, FRAGANCE represents the presence of fragrance, GLNS represents grain length and shape, PERCENTBGS represents percent brokenness and CHKN represents chalkiness.

RESULTS

Characteristics of the sampled rice producers

The sampled rice producers included 68% male farmers. However, the male proportion is relatively lower in the Ejura-Sekyedumase (58%) district. On the average a farmer is aged 37 years and was responsible for an average of a five member household with an almost equal gender distribution. In general a little over 70% of the sampled rice producers had formal education. The distribution however varied greatly across the district with Ejura-Sekyeredumase having about 69%, Atwima Nwabiagya exactly 50% and about 89% of farmers with formal education in the Ejisu-Juaben District. On the average, Over 59% of the respondents cultivate rice once a year with 38.8% of them cultivating at most twice a year. This was however consistent across all three districts. Overall, about 70% of the rice producers were beneficiaries of extension services. On the whole, majority of the respondents (84.1%) were married. On the average, about 75% of the sampled households had access to credit and this was similar across the three districts (Table 2).

Ranking of rice varietal characteristics preferred by farmers

The results of the rankings are presented in Table 3. High yield potential was the most preferred varietal characteristic. However, this was ranked second among households in Atwima Nwabiagya. The next most preferred varietal characteristic was good grain quality. Conversely, it was ranked as the most preferred varietal characteristic by households in Atwima Nwabiagya. Pests and diseases was the third most preferred varietal characteristic and was followed by resistance to abiotic stress the least preferred varietal characteristic. Overall, the Kendall's W of 0.650 was significant at 1% level suggesting that 65% of the farmers agree on the outcome of the ranking. However, this was quite variable across the districts with the highest agreement among households in Ejisu-Juaben (72.6**%**), followed by Atwima Nwabiagya with about 70% agreement and the least agreement in Ejura-Sekyedumase (58.5%) districts (Table 3).

Estimated effects of factors influencing farmers' preference for a new rice variety

The results of the probit model are summarized in Table 4. From the results, a likelihood ratio (LR) statistic of 122.65 with a chi-square distribution at 10 degrees of freedom was significant at 1%. This means that at least one of the variables in the model has a significant effect on farmer preference for an improved rice variety and that the explanatory variables jointly influence the preference. With the exception of education, all the explanatory variables were found to be in conformity with the prior expectations. The coefficient of age (AGE), sex (SEX) and tolerance to pest and diseases (TPED) were significant at 10% probability level; the coefficient of the square of age (AGE^2), farm size (FRMSZE), tolerance to biotic stress (TBIOSTRSS) and good grain quality and appearance (GGQAP) were significant at 5% probability. The coefficient of the remaining variable level of education (EDUCL) was not significant.

Because age is quadratic in nature, squaring the age was necessary to assess the effect of age after the threshold (turning point) on the preference for improved rice variety. The age of the farmer had a positive significant effect on preference for improved rice variety.

Table 2. Characteristics of rice producing households by Districts.

Characteristics	Ejura- Sekyedumase (N=64)	Atwima Nwabiagya (N=32)	Ejisu- Juaben (N=69)	Overall (N=165)
Gender (%)				
Male	59.4	78.1	72.5	68.5
Female	40.6	21.9	27.5	31.5
Education (%)				
Tertiary	0.0	6.3	2.9	2.4
Secondary	17.2	0.0	4.4	8.5
Basic	28.1	40.6	69.6	47.9
Primary	21.9	3.1	13.0	14.6
No formal education	32.8	50.0	10.1	26.7
Cultivations per year (%)				
Once	51.6	40.6	75.4	59.4
Twice	46.9	59.4	21.7	38.8
Three times	1.6	0.0	2.9	1.8
Household size (N)	4.3	4.0	5.1	4.6
Married (%)	81.3	93.8	82.6	84.2
Average age (years)	39.3	35.0	35.3	36.8
Access to extension (%)	78.1	46.9	69.6	68.5
Access to credit (%)	75.0	78.1	73.9	75.2

Table 3. Ranks of farmers' preferred rice varietal traits by districts.

Varietal traits	Ejura-Sekyedumase		Atwima Nwabiagya		Ejisu-Juaben		Overall	
	Mean rank	Rank	Mean rank	Rank	Mean rank	Rank	Mean rank	Rank
Good grain quality – appearance, cooking and fragrance quality	1.57	2	1.59	1	1.88	2	1.71	2
High yields	1.81	1	1.72	2	1.36	1	1.61	1
Resistance to diseases and pest including weeds	2.94	3	2.78	3	2.94	3	2.91	3
Resistance to abiotic stresses – drought, iron toxicity	3.68	4	3.91	4	3.81	4	3.78	4
Kendall's W (Significance)	0.585 (0.000)		0.698 (0.000)		0.726 (0.000)		0.650 (0.000)	

When age was squared, the effect became negative and significant. There is a high probability for older farmers to prefer an improved rice variety and this probability of their preference increases with increase in age. Increasing the age of the farmer by 1 year increases the probability of preference for an improved rice variety by 0.0076% (Table 5). However, beyond some point, increasing age decreases the probability of preference for an improved rice variety. To investigate further the particular age at which the preference for an improved variety begins to decrease, we differentiated the estimated probit equation with respect to age and set the resultant derivative to zero as shown below:

$$P_r = -14.11072 + 0.45041 AGE - 0.00644 AGE^2 + 1.000514\, FRMSIZE + 3.886725\, GGQAP + 2.473187\, HIGHYLD + 1.84187\, SEX + 1.924135\, TBIOSTRSS + 1.3638\, TPED + 1.785202\, EDUCL + \vartheta_i \quad (13)$$

$$0.45041 - 0.01288\, AGE = 0 \Rightarrow AGE = 34.97 \cong 35 \text{ years}$$

By doing this and assuming all other things constant, the estimated optimal age is 35 years. This means that below

Table 4. Probit estimates of factors influencing farmers' preference for improved rice varieties.

Variable	Coefficient	Std. error	z-Statistic	Prob.
AGE	0.450410*	0.232917	1.933780	0.0531
AGE^2	-0.006440**	0.002970	-2.168635	0.0301
FRMSIZE	1.000514**	0.404030	2.476335	0.0133
GGQAP	3.886725**	1.614356	2.407601	0.0161
HIGYLD	3.473187**	1.519006	2.286487	0.0222
SEX	1.841870*	1.072536	1.717304	0.0859
TBIOSTRSS	1.924135**	0.905446	2.125070	0.0336
TPED	1.363800*	0.786746	1.733468	0.0830
EDUCL	-1.785202ns	1.114081	-1.602399	0.1091
C	-14.11072	5.761233	-2.449254	0.0143
Mean dependent var	0.842424	LR statistic (10 df)		122.6532
S.E. of regression	0.150466	McFadden R-squared		0.853197
Log likelihood	-10.55200	. Probability(LR stat)		0.000000
Obs with Dep=0	26	Total obs		165
Obs with Dep=1	139			

*, ** , and ***: Significant at 10%, 5% and 1% probabilities respectively; ns: not significant.

Table 5. Marginal effects of significant variables in the model.

Variable	Coefficient	Marginal Effects
AGE	0.45041	0.0000758
AGE2	-0.00644	-0.000108
FRMSIZE	1.000514	0.0001683
GGQAP	3.886725	0.2596016
HIGYLD	3.473187	0.1597558
EXP	-0.091444	-0.0000154
SEX	1.84187	0.0036672
TBIOSTRSS	1.924135	0.0003237
TPED	1.3638	0.0009133

35 years farmers have a stronger preference for improved rice varieties than older ones.

Farm size of the respondents had the expected positive sign and was significant (Table 5). Increasing farm size by 1 ha increases the probability of farmers' preference for improved rice variety by 0.00017. The sex of the farmers positively influenced their preference for new rice varieties and was significant.

The probability that a male farmer prefers improved rice varieties is 0.37. Higher yields had a significant positive influence on farmers' preference for an improved rice variety. Increase in yield by 1 kg increases the probability of farmers' preference for an improved rice variety by 15.98%. Tolerance to biotic stress had significant positive influences on farmers' preference for an improved rice variety. This increases the probability of farmers' preference for an improved rice variety by 0.032 (Table 5).

The quality and appearance of the grain had the expected significant positive effect on farmers' preference for an improved rice variety. Increasing the quality and appearance of the grains increases the probability of farmers' preference for an improved rice variety by 26.0%.

Results of further analysis on grain quality attributes

This was done to further investigate which of the grain quality attributes significantly influence farmers preference for improved rice varieties. The various grain quality attributes that were investigated included colour, grain length and size, fragrance, cooking quality, grain color, absence of foreign materials, percent broken grains and grain expansion ratio and chalkiness. The probit

Table 6. Probit estimates of grain quality attributes influencing farmers' preference for improved rice varieties.

Variable	Coefficient	Std. error	z-Statistic	Prob.
COLOR	1.995244**	0.785746	2.539299	0.0111
COOKQLITY	1.619973***	0.609496	2.657892	0.0079
EXPRATIO	0.200719[ns]	0.682215	0.294216	0.7686
FOREIGNM	2.393931***	0.633938	3.776287	0.0002
FRAGANCE	2.507448*	1.426872	1.757305	0.0789
GLNS	2.130330**	0.748136	2.847518	0.0044
PERCENTBGS	0.307811[ns]	0.584799	0.526354	0.5986
CHKN	-3.436279**	1.172298	-2.931233	0.0034
C	-5.382626	1.530020	-3.518010	0.0004
Mean dependent var.	0.842424	McFadden R-squared		0.728446
Log likelihood	-19.51895	Probability(LR stat)		0.000000
LR statistic (8 df)	104.7193	Total obs		165
Obs with Dep=1	139	Obs with Dep=0		26

*, ** and ***: Significant at 10%, 5% and 1% probabilities respectively; [ns]: not significant.

Table 7. Marginal effects of significant variables of grain quality attributes.

Variable	Coefficient	Marginal effects
COLOR	1.995244	0.4464842
COOKQLITY	1.619973	0.3207381
FOREIGNM	2.393931	0.5651178
FRAGANCE	2.507448	0.6634943
GLNS	2.130330	0.5048045
CHKN	-3.436279	-0.0993324
C	-5.382626	0.4464842

result of the grain quality model is presented in Table 6.

From the results in Table 6, a likelihood ratio (LR) statistic of 104.7193 with a chi-square distribution at 8 degrees of freedom was significant at less than 1% level of significance. This means that at least one of the grain quality attributes in the model has a significant effect on farmers' preference for an improved rice variety and that these attributes jointly influence the farmers' preference for an improved rice variety. With the exception of expansion ratio and percent broken grains, all the explanatory variables were found to be in conformity with the apriori expectations. The coefficient of cooking quality (COOKQLITY) and absence of foreign material (FOREIGNM) were significant at 1% probability level; the coefficient of grain length and shape (GLNS), chalkiness (CHKN) and colour (COLOR) were significant at 5% probability and the coefficient of fragrance (FRAGANCE) was significant at 10% probability level. The coefficient of the remaining variable expansion ratio (EXPRATIO) and percent broken grains (PERCENTBGS) were not significant at 10% probability. Hence, we do not reject the null hypotheses. The marginal effects of the grain quality attributes used in the model are presented in Table 7.

The cooking quality of the grain had a positive significant effect on farmers' preference for improved rice variety. An improvement in cooking quality increases the probability of farmers' preference for an improved variety by 32.07**%,** whilst the presence of fragrance increases preference for a variety by as much as 66.34%. Grain length and shape increased farmers' preference for an improved variety by over 50.0%. The colour of the grain also had a positive significant effect on farmers' preference for improved rice variety. White grain colour increases the probability of farmers' preference by 44.65%. Absence of foreign material had a positive significant influence on farmers' preference for improved rice variety and it increases the probability of farmers' preference for improved rice variety by 56.51% (Table 7).

DISCUSSION

Characteristics of rice producers and production system

The bi-modal nature of the rainfall in the study area accounts for some of the farmers cultivating rice twice a year. Production in the study area and Ghana as a whole is perceived to be very tedious. This may explain why production is dominated by males. The high gender imbalance in rice production can however limit the full potential the rice sector. Contrary to earlier studies, rice production in the study area includes farmers in their middle ages (Asuming-Brempong et al., 2011; Wiredu et al., 2010; Asante et al., 2011). This observation is interesting and can boost efforts to improve rice production in the country. These are farmers who are targeted by various interventions and with their zeal they are more likely to invest in new technologies.

Most studies have shown relatively low level of education among farm households in Ghana; this situation is predominant in northern Ghana (Akudugu et al., 2012; Wiredu et al., 2011; Wiredu et al., 2010). In this study however, the large majority of the heads of rice producing families had at least basic education. This result is interesting because the farmers in question have the ability to assess and evaluate the potential benefits of the improved technologies and make informed decisions. The good farmer contact with extension officers will enhance access to information on new technologies. It is important for development interventions to provide adequate support for institutions such extension and research to ensure effective promotion of agricultural technologies such as improved rice varieties.

Factors influencing preference for improved rice varieties

Age was used as a proxy for experience in rice cultivation. There was a positive relationship between age and preference. However, this relationship was not stable but reversed after age 35. This shows that farmers below this age are energetic, hence have the physical capability of expanding their rice farms and so are interested in new rice technologies than older farmers. These findings are quite consistent with that of previous research (McBride and Daberkow, 2003; Ayamga et al., 2006; Gamba et al., 2002). Asante et al. (2011) also found a positive relationship between age and farmers' decision to join farmer based organizations in Ghana. It is therefore important for the country to develop policies that would create incentives to attract and maintain young farmers in rice production. It is necessary to identify the preferences of this age group for effective targeting of such policies.

Some researchers have found a positive relationship between farm size and farmers preference for a given technology (Asrat et al., 2010; Adimado, 2001; Kheralla et al., 2001; Langyintuo and Mekuria, 2005; Asante et al., 2011). The larger the size of the farm, the higher the investments made with the corresponding expected income and higher risks hence farmers will prefer an improved variety that will enable them to increase their productivity and income and at the same time reduce risks. Promotional strategies for improved rice varieties should also target farmers with relatively large farm sizes with an average of about 1.28 ha. However, expansion of rice area should also come with strategies to increase yields through training in good agricultural practices. More importantly, the use of recommended fertilizer rates and other complementary technologies should be encouraged to help increase yields. Farmers must also be sensitized and encouraged to take advantage of the on-going fertilizer and seed subsidy program in order to obtain the best from increased area of the preferred improved variety. Increase in yields have been a major driving determinant of farmers' preference for improved varieties (Langyintuo and Mekuria 2005; Ramasamy et al., 2003; Odendo et al., 2001). Farmers with higher yield are motivated to further improve their yields in the subsequent year and this will require them to adopt improved varieties with high yield potential coupled with other factors (Asrat et al., 2010).

Abiotic stresses such as drought, iron toxicity, nitrogen deficiency as well as biotic stresses (pests and diseases) could have adverse effects on yield even when a variety with high yield potential is cultivated. Tolerance to these stresses motivates farmers to prefer improved rice varieties positively. Similar finding was obtained by Odendo et al. (2001) when assessing farmers' preferences and constraints to maize production in Western Kenya. Kshirsagar et al. (2002) also found tolerance to biotic and abiotic stresses as having positive influence in determining varietal preference of farmers in Orissa. Another important factor that influences farmers' preference for an improved variety is superior grain quality. Superior grain quality increased farmers' preference for improved rice varieties by as much 26.0%. Unlike other cereals, rice is mostly consumed as whole grains. The physical appearance of the rice grain is therefore an important attribute when it comes to marketing of the produce. Poor appearance and cooking qualities will reduce demand and hence the price of domestic rice and this will mean low incomes for farmers. Against this background, farmers in recent times consider attributes that interests consumers in their production decision-making. Researchers and policy makers should consider grain quality as one of the major criteria in developing varieties to boost domestic rice production and consumption in Ghana.

Grain quality and farmer preference for improved varieties

Rice grain quality includes physical appearance of the grain as well as cooking and eating quality. In the present study, appearance related traits including colour, grain length and shape had positive effect on farmers' preference for improved varieties. Chalkiness on the other had a negative effect on farmers' preference for a new variety. These findings are not surprising because most rice consumers in Ghana prefer white coloured, long grain rice that is not chalky but translucent (Bam et al., 1998). Consequently, farmers would need varieties with these consumer preferred attributes to be able to sell their rice. Another consumer preferred attribute is low percentage of broken grains but this attribute did not significantly affect farmers' preference for rice varieties. This could be because the farmers saw percent broken rice as a trait which is wholly controlled by post-harvest handling and the effectiveness a milling machine. There is however, some genetic control for milling yield (Aluko et al., 2004) and it can therefore be bred for by breeders. Breeders can also breed for preferred cooking and eating attributes of rice including fragrance. The presence of fragrance and the desired cooking quality had significant influence on farmers' preference for improved varieties. Such varieties will be easier to market because taste, aroma and texture are reported to be the three most critical attributes that Ghanaian consumers look for in cooked rice (Diaka et al., 2010). Another cooking attribute, expansion ratio, did not influence farmer-preference for improved rice varieties possibly because consumers put more emphasis on taste rather than volume expansion for cooked rice (Diaka et al., 2010).

Conclusions

This study assessed the influence of grain quality and other factors in farmers' preference for improved rice varieties. Key variables that significantly influenced farmers' preference for a new rice variety were; age, farm size, high yield, experience, sex, tolerance to abiotic and biotic stresses and good grain quality. Grain quality was ranked second to high yield as the most important varietal characteristics that farmers require. Grain quality attributes that significantly influenced farmers' preference for a new rice variety were grain colour, grain length and shape, chalkiness, fragrance, cooking behaviour and absence of foreign material.

Implication for research

Factors such as age, sex, farm size, experience, tolerance to pest and diseases, tolerance to abiotic stress and good grain quality need to be considered by researchers in the development and dissemination of new rice varieties. It is imperative to assess farmers' perception and opinion from the beginning of the research process and incorporate them into research programmes. This will enhance the acceptability of research output such as improved varieties.

Grain quality attributes are of keen interest to farmers and are likely to influence their preference for an improved rice variety. Hence researchers should consider grain quality attributes particularly, white coloured, long, slender and translucent grains (low chalkiness), fragrance and preferred cooking quality in their rice breeding efforts. Breeders of other crops should also consider quality issues to increase adoptability of varieties. This is because like any other economic agent, production decisions of small scale farmers are profit oriented. Hence farmers will choose varieties that are not only high yielding but have grain qualities that will make it easily marketable. Because rice is consumed as whole grains, the quality of the grain greatly influence its demand. Research into superior grain quality would therefore improve consumer acceptability and farmer adoption of improved rice varieties.

Policy implications

Generally, rice farmers respond to demand signals among other factors to produce. If the demand is high, all things being equal, farmers are likely to produce to meet the high demand given the available resources. Additionally, since the demand for the grain is affected by its quality, efforts towards addressing the quality issues are likely to boost local rice production and consumption. In view of this, efforts should be targeted at enhancing the development of rice varieties with inherent good grain quality attributes. In addition, government could partner with the private sector to invest in the establishment of rice post-harvest centers. Such centers should be equipped with the necessary post-harvest and processing facilities to ensure the production of good quality rice grains. The production of good quality rice will increase consumer and trader acceptability for locally produced rice and result in higher incomes of small scale rice farmers.

ACKNOWLEDGEMENTS

This work was sponsored by the Alliance for Green Revolution in Africa. Many thanks go to the agricultural extension officers in the study areas for the enormous role they played in organizing farmers for the study. All technicians of the CSIR-Crops Research Institute, Ghana who worked on this study are appreciated for their assistance.

REFERENCES

Abansi CL, Duff B, Lantican FA, Juliano BO (1992). Consumer demand for rice grain quality in selected rural and urban markets in the Philippine. *in* Consumer demand for rice grain quality. International Rice Research Institute, P. O. Box 933, Manila, Philippines. pp. 37-57.

Adimado S (2001). Willingness to Pay for Research Findings: A Case Study of Pineapple Farmers in Ghana. MPhil Thesis Submitted to the Department of Agricultural Economics and Agribusiness, University of Ghana, Legon. pp. 38-50.

Akudugu MA, Guo E, Dadzie SK (2012). Adoption of Modern

Agricultural Production Technologies by Farm Households in Ghana: What Factors Influence their Decisions. J. Biol. Agric. Health Care 2(3).
Aluko G, Martinez C, Tohme J, Castano C, Bergman C, Oard JH (2004). QTL mapping of grain quality traits from the interspecific cross Oryza sativa x O. glaberrima. Theor. Appl. Genet. 109:630-639.
Asante BO, Afari-Sefa V, Sarpong DB (2011). Determinants of small-scale farmers' decision to join farmer based organizations in Ghana. Afr. J. Agric. Res. 6(10):2273-2279.
Asrat S, Yesuf M, Carlsson F, Wale E (2010). Farmers' preferences for crop variety traits: Lessons for on-farm conservation and technology adoption. J. Ecol. Econ. 69:2394-2401.
Asuming-Brempong S, Gyasi KO, Marfo KA, Diagne A, Wiredu AN, Asuming-Boakye A, Haleegoah J, Frimpong BN (2011). The exposure and adoption of New Rice for Africa (NERICAs) among Ghanaian rice farmers: What is the evidence. Afr. J. Agric. Res. 6(27):5911-5917.
Asuming-Brempong S, Osei-Asare YB (2007). Has imported rice crowded-out domestic rice production in Ghana? What has been the role of policy? African Association of Agricultural Economists (AAAE) Conference proceedings: pp. 91-97.
Ayamga M (2006). Factors influencing the decision to participate in microcredit Programmes: An illustration for Northern Ghana. Ghana J. Dev. Stud. 3(2):57-65.
Bam RK, Anchirinah VM, Manful JT, Ansere-Bioh F, Agyeman A (1998). A preliminary study on consumer preference and price quality relationship in Ghana.Improving the competitiveness of local rice in Ghana, NRI, UK.
Choudhury N, Kabir KA, Biswas SK, Islam R (1992). Influence of rice grain properties on market price in Bangladesh. in Consumer demand for rice grain quality. International Rice Research Institute, P. O. Box 933, Manila, Philippines. pp. 117-133.
Diako C, Sakyi-Dawson E, Bediako-Amoa B, Saalia FK, Manful JT (2010). Consumer perceptions, knowledge and preferences for aromatic rice types in Ghana. Nat. Sci. 8(12):12-19.
Duffuor K (2009). Budget Statement and Economic Policy of the Government of Ghana for the 2010 Fiscal Year.
Efisue A, Tongoona P, Derera J, Langyintuo A, Laing M, Ubi B (2008). Farmers' Perceptions on Rice Varieties in Sikasso of Mali and their Implications for Rice Breeding. J. Agron. Crop Sci. 194:393-400.
Gamba P, Ngugi C, Verkuijl H, Mwangi W, Kiriswa F (2002). Wheat Farmers' Seed Management and Varietal Adoption in Kenya. Mexico, D.F.: CIMMYT, Egerton University, Njoro, Kenya, and KARI, Nairobi. P. 22.
Gockowski J, Ndoumbe M (2004). The Adoption of Intensive Monocrop Horticulture in Southern Cameroon. J. Agric. Econ. 30:195-202.
Grandstaff TB, Grandstaff SW (1987). A conceptual basis for methodological development in rapid rural appraisal. In: Proceeding of the 1985 Institutional Conference on Rapid Rural Appraisal. Rural Systems Research Project and Farming Systems Research Project, 69-88. KhonKaen University, KhonKaen, Thailand.
Joshi G, Bauer S (2006). Farmers' Choice of the Modern Rice Varieties in the Rainfed Ecosystem of Nepal. J. Agric. Rural Dev. Trop. Subtropics 107(2):129-138.
Juliano BO, Duff B (1991). Rice grain quality as an emerging priority in national rice breeding programs.Pages 55-64 in Rice grain marketing and quality issues.International Rice Research Institute, Manila, Philippines.
Katungi EM (2006). Social Capital and Technology Adoption on Small Farms: The Case of Banana Production Technology in Uganda. Doctor of Philosophy Thesis, Department of Agricultural Economics, Extension and Rural Development, Faculty of Natural and Agricultural Sciences, University of Pretoria, South Africa.
Kheralla M, Minet N, Kachule R, Souce BG, Berry P (2001). Impact of Agricultural Market Reforms on Smallholder Farmers in Benin and Malawi, Research Report, P. 2, IFPRI.
Kshirsagar KG, Pandey S, BellonSource MR (2002). Farmer Perceptions, Varietal Characteristics and Technology Adoption: A Rainfed Rice Villagein Orissa, Economic and Political Weekly, 37(13):1239-1246.
Langyintuo AS, Mekuria M (2005). Accounting for neighborhood influence in estimating factors determining the adoption of improved agricultural technologies. A paper at American Agricultural Economics Association annual meeting, providence, Rode Island: pp. 1-28
Maddala GS (2005). Introduction to Econometrics.3rdEdition, John Wiley & Sons Ltd, The Atrium, Southern Gate, England: pp. 318-323.
Mattson DE (1986). Statistics-Difficult Concepts Understandable Explanations. Bolchazy-Carducci Pub. Inc. pp. 281-283.
McBride WW, Daberkow SG (2003). Information and the Adoption of Precision Farming Technologies. J. Agribus. 21:21-28.
Mussei A, Mwanga J, Mwangi W, Verkuijl H, Mungi R, Elang A (2001). Adoption of Improved Wheat Technologies by Small-Scale Farmers in Mbeya District, Southern Highlands, Tanzania. Mexico D.F.: International Maize and Wheat Improvement Centre (CIMMYT) and the United Republic of Tanzania.
Nwanze KF, Mohapatra S, Kormawa P, Keya S, Bruce-Oliver S (2006). Rice development in sub-Saharan Africa. J. Sci. Food. Agric. 86(5):675-677.
Nzomoi JN, Byaruhanga JK, Maritim HK, Omboto PI (2007). Determinants of technology adoption in the production of horticultural export produce in Kenya. Afr. J. Bus. Manage. 1(5):129-135.
Obiri BD (2003). Developing and Testing Technologies with Farmers.Natural Resources Systems Programme. Project report, DFID Project No. R7446. 30 pp. http://www.nrsp.org.uk./database/documentation/2950.pdf [accessed 25 May 2007].
Odendo M, De Groote H, Odongo OM (2001). Assessment Of Farmers' Preferences And Constraints To Maize Production In Moist Mid-altitude Zone Of Western Kenya. Paper presented at the 5th International Conference of the African Crop Science Society, Lagos, Nigeria October 21-26, 2001.
Ramasamy C., Aldas J, Selvaraj KN, Hossain M (2003). Hybrid Rice in Tamil Nadu: Evaluation of Farmers' Experience. Econ. Polit. Weekly 38(25):2509-2512.
Wiredu AN, Mensah-Bonsu A, Andah EK, Fosu KY (2011). Hybrid Cocoa and Land Productivity of Cocoa Farmers in Ashanti Region of Ghana. World J. Agric. Sci. 7(2):172-178.
Wiredu NA, Gyasi KO, Abdoulaye T, Sanogo D, Langyintuo A (2010). Characterization of maize producing households in the Northern Region of Ghana. Country Report – Ghana: CSRI/SARI – IITA, Ibadan, Nigeria.

2

Economic impacts of cassava research and extension in Malawi and Zambia

Arega D. Alene[1], R. Khataza[2], C. Chibwana[3], P. Ntawuruhunga[1] and C. Moyo[1]

[1]International Institute of Tropical Agriculture (IITA), Lilongwe, Malawi.
[2]Bunda College of Agriculture, University of Malawi, Lilongwe, Malawi.
[3]United States Agency for International Development (USAID), Lilongwe, Malawi.

This paper estimates the economic impacts of cassava research and extension in Malawi and Zambia over the period 1990-2008. The data come from sample household surveys, planting material production records, and a series of cassava improvement experiments conducted in the two countries. Past investments in cassava improvement have led to the development and release of a good number of high-yielding and cassava mosaic virus disease (CMD)-tolerant cassava varieties. The results show relatively higher adoption rates for the CMD-free local varieties compared to CMD-tolerant varieties that have been released in the two countries. The adoption of new varieties has been low and slow largely due to the fact that most of these varieties lacked the consumption attributes highly valued by farmers. The multiplication and distribution of CMD-free planting materials of the recommended local varieties led to greater adoption, but infection with CMD three to four years after adoption meant that the yield gains and economic benefits could not be sustained. Nevertheless, the multiplication and distribution of clean cassava planting materials generated a modest rate of return of 24%, which is actually consistent with an earlier rate of return estimate of 9 to 22% for cassava improvement in developing countries. Analysis of the ex ante impacts of current and future investments in cassava improvement shows that cassava improvement research that focuses on the development and dissemination of varieties with highly preferred consumption and industrial attributes would yield a greater rate of return of 40%.

Key words: Adoption, cassava, economic surplus, impact, Malawi, Zambia.

INTRODUCTION

Cassava is Africa's second most important food staple after maize and provides more than half of the dietary calories for over 200 million people (Nweke et al., 2002). In Malawi, cassava is a staple food for more than 30% of the population and occupies 60% of the area under roots and tubers and nearly 50% of the total production. Cassava has wide agro-ecological adaptation, but the main growing areas are the northern belt along the lakeshore (Karonga, Rumphi, Nkhatabay, and Nkhotakota), the southern cassava belt (Mangochi, Machinga, Zomba, and the southern Shire Highlands), and the central belt of Dedza and Lilongwe. The marketed surplus of cassava increased from 11% in 2002 to 75% in the central region and 60% in the southern region (Mataya et al., 2001; Phiri, 2001; Haggblade and Zulu, 2003) The fresh market takes up about 80% whereas the remainder is absorbed in the manufacturing and confectionary industries. Similarly, in Zambia, cassava accounts for roughly 15% of national calorie consumption (Dorosh et al., 2007) and is mostly grown in the five provinces of Luapula, Northern, North-Western, Copperbelt, and Western provinces where the crop is regarded as a staple

(Soenarjo, 1992). The importance of the crop is fast increasing such that in the recent years cassava production has expanded to the Southern and Eastern parts of the country (Chitundu, 1999).

The expansion of cassava production in Africa in the face of longstanding as well as emerging threats to cassava production and productivity is largely attributed to sustained investments in research and extension aimed at addressing a wide range of biotic and abiotic constraints (Nweke et al., 2002). One of the major biotic constraints to cassava production is cassava mosaic virus disease (CMD) and is transmitted by the whiteflies and infected cuttings. Since recently, cassava brown streak virus disease (CBSD) has become yet another major constraint to cassava production. As part of a major long term crop improvement effort since its creation in 1967, the International Institute of Tropical Agriculture (IITA) initiated cassava research in the early 1970s with a focus on developing varieties with resistance to major diseases such as CMD. Cassava breeding was initiated using breeding materials from Moor plantation near Ibadan and a limited number of east African landraces with resistance to CMD developed through interspecific hybridization in the 1930s (Haggblade and Zulu, 2003). This work resulted in several elite genotypes that had resistance

to CMD as well as high and stable yields and good consumer acceptability. The development of these resistant varieties, and their delivery to national programs for testing under specific local conditions during the late 1970s and 1980s, has led to the successful deployment of CMD-resistant cassava in Sub-Saharan Africa (Nweke et al., 2002). In addition to their resistance to CMD, the improved varieties combine enhanced postharvest qualities, multiple pest and disease resistance, wide agro-ecological adaptation, and greatly improved yield potential where yield increases of 50-100% without the use of fertilizer were demonstrated in many African countries.

The national cassava improvement programs in Malawi and Zambia have developed and released varieties that outperform the local varieties using breeding materials received from IITA. The improved genetic materials from IITA, referred to as the Tropical Manihot Selections (TMS), were distributed to several countries in the cassava-growing belt of eastern and southern Africa during the late 1980s at a time when governments were dismantling large-scale maize subsidy programs. Over the period 1990–2011, IITA and the respective national programs released a total of 12 improved varieties in Malawi and 8 improved varieties in Zambia. The improved cassava varieties coupled with the declining profitability of maize due to the withdrawal of subsidies contributed to a surge in cassava production in Malawi and Zambia beginning in the early to mid-1990s (Haggblade and Zulu, 2003). The increased availability of improved cassava varieties opened up a range of profitable commercial opportunities for production of cassava-based foods, feeds, and industrial products. At the same time, improved disease tolerance and higher productivity as well as a flexible harvesting calendar offered prospects for improving household food security.

Despite major efforts to develop and disseminate a growing number of improved varieties, there is lack of comprehensive evidence on the adoption and economic impacts of improved cassava varieties. This paper used household survey data as well as planting material production estimates for measuring variety adoption and on-farm experimental data for estimating yield gains.

Cassava research in Malawi and Zambia

Cassava improvement dates back to the 1930s in Malawi and 1940s in Zambia. Cassava was then regarded as an important famine reserve crop and each household was encouraged to have a piece of land under cassava as a fallback plan. The research focus then was on agronomic practices as well plant health in order to generate information on local planting conditions and select mosaic disease tolerant varieties. In Malawi, some 22 local varieties were characterized and put under mosaic observations at Mulanje and other stations together with a number of new introductions from Amani in north-east Tanzania. Cassava production expanded following the removal of fertilizer subsidies in the late 1980s and the droughts in the early 1990s which required an emergency response involving accelerated multiplication and distribution of planting materials of the best local varieties.

In Malawi, the Root and Tuber Crops Research Program was established in 1978, whereas the Zambia Root and Tuber Improvement Program (RTIP) was established in 1979. National cassava research activities initially focused on identification of best local varieties, cleaning, and distribution of planting materials. The national programs in both countries adopted IITA's breeding scheme in order to speed up selection, evaluation, and release of new varieties. The varieties released by the root and tuber improvement programs have IITA parent material introduced directly in the form of tissue culture or seed population (Tables 1 and 2). The cassava breeding program in Malawi released three waves of improved cassava varieties: first the best local varieties (Chitembwere, Gomani, Mbundumali, and Nyasungwi) were released in the 1980s and recommended to farmers on the basis of mosaic virus tolerance and early bulking. In 1992, IITA through SARRNET launched a three-year drought recovery program of accelerated multiplication and distribution of cassava planting materials. This program targeted to provide planting materials to 300,000 smallholder farmers throughout the country. Under the same program in 1995 there was massive planting material multiplication and

Table 1. Improved cassava varieties released in Malawi, 1980–2002.

Variety	Category	Release year	IITA material used	Major attributes			
				Yield (t/ha)	Maturity (MAP)	Taste	Disease tolerance
Nyasungwi	Local selection	1980s	None	12-21	12-15	Semisweet	
Chitembwere	Local selection	1980s	None	20-23	15-18	Sweet	
Manyokola	Local selection	1980s	None	25	9-15	Sweet	Tolerates CGM, CBSD but is susceptible to CMD
Gomani	Local selection	1980s	None	25	9-12	Bitter	Susceptible to CGM, CBSD, CMD
Mkondezi (MK91/478)	Improved	1999	Seed population	40	9-15	Bitter	Tolerates CMD and CM
Maunjili (TMS 91934)	Improved	1999	IITA introduction	35	9-12	Bitter	Tolerates CMD, CM and CGM
Silira (TMS 60142)	Improved	1999	IITA introduction	25	12-15	Bitter	Tolerates CMD and CM but is susceptible to CGM
Sauti (CH92/077)	Improved	2002	IITA seed population	25	12-15	Bitter	Tolerates CMD, CM and CGM
Yizaso (CH92/112)	Improved	2002	IITA seed population	25	12-15	Bitter	Tolerates CMD, CM and CGM
Phoso (LCN 8010)	Improved	2008	IITA introduction	35	9-15	Bitter	Tolerates CMD and CBSD
Mulola (TMS 83350)	Improved	2008	IITA introduction	40	9-15	Bitter	Tolerates CMD, CM, and CGM
Sagonja (CH92/082)	Improved	2009	IITA seed population	25-35	9-15	Bitter	Tolerates CMD, CBSD, CM, and CGM
Chiombola (TME 6)	Improved	2009	IITA introduction	45	9-15	Bitter	Tolerates CMD and CGM

distribution of existing cassava clones, including Gomani and Mbundumali (Haggblade and Zulu, 2003). The second phase of the accelerated multiplication and distribution program targeted the establishment of 15,000 hectares of cassava nurseries to be eventually distributed to 75,000 farmers. The second series of varieties came out of hybridization and screening trials which started in 1992 from which process three new clones were identified and released in May 1999 (Mkondezi, a bitter variety; Silira, categorized as semi-sweet, and Maunjili, a bitter variety). In 2002 a further two bitter varieties (Sauti and Yizaso) were released. These new varieties increased yield by about 54% from the already high 13 tons per hectare for the best CMD-free local varieties (Gomani and Mbundumali) to 20 tons per hectare (Benesi et al., 1999).

In Zambia, the breeding program by the Root and Tuber Improvement Program has led to two waves of varietal releases, the first was in 1993 and the second in 2000. In 1993, three varieties, namely Bangweulu (LUC55), Kapumba (LUC327), and Nalumino (LUC304), were released. These varieties have higher yield ability and possess superior attributes compared to other traditional cassava varieties (Table 3). Historically the heavy fertilizer subsidies provided a strong incentive for maize production in the country. However, these recommended varieties coincided with a policy shift towards cassava production following the removal of fertilizer subsidies (Haggblade and Zulu, 2003). Starting from 1988/89 there were a series of multiplication and distribution of cassava planting materials to respond to the increased demand for cassava material in the country. Through efforts of the Zambian Root and Tuber Improvement Program and a consortium under Program Against Malnutrition (PAM) engaged in the distribution of cassava planting materials, a total of 552,000 cuttings were distributed in three consecutive seasons (1989-1992) to individual farmers looking for planting materials Soenarjo, 1992). Most of the cassava materials were susceptible to cassava mosaic, but a local clone called Nalumino was identified as being resistant and has been used in breeding program as a source of resistance (Soenarjo, 1992).

Table 2. Improved cassava varieties released in Zambia, 1990–2000.

Variety	Category	IITA material	Year of release	Yield (t/ha)	Maturity (MAP)	Taste
Bangweulu	Local selection	None	1993	31	12-16	Bitter
Kapumba	Local selection	None	1993	22	16-24	Sweet
Nalumino	Local selection	None	1993	29	16-24	Bitter
Mweru	Improved	IITA male x Nalumino	2000	41	16	Sweet
Chila	Improved	IITA male x Nalumino	2000	35	16	Bitter
Tanganyika	Improved	IITA male x Nalumino	2000	36	16	Sweet
Kampolombo	Improved	IITA male x Nalumino	2000	39	16	Sweet

Source: Haggblade and Nyembe (2008).

Table 3. Average yields of cassava varieties across districts in Malawi.

District	Yield (tons/ha)		
	Local CMD-infected	Local CMD-free	Improved CMD-resistant
Nkhatabay	12	17	17
Mzimba	12	12	9
Nkhotakota	11	16	17
Lilongwe	10	11	11
Zomba	7	11	12
Mulanje	6	9	20
All	9	12	14
Yield gain (%)	-	**33**	**55**

The conventional breeding program was also started in 1988/89 in Mansa with 15,077 (seedlings from twelve different crosses (Soenarjo, 1992). By 1992, preliminary evaluation identified 15 clones as being tolerant to CMD. Further evaluations led to the release of four new varieties of Mweru, Chila, Tanganyika, and Kampolombo in 2000 (Soenarjo, 1992). A total of four out of the seven or 57% of the released varieties had IITA parent material crossed with best local variety in order to enhance local adaptation and variety attributes.

MATERIALS AND METHODS

The economic surplus method for ex-post impact analysis

Several impact studies of agricultural technologies have estimated aggregate economic benefits through extrapolation of farm-level yield or income gains using partial equilibrium simulation models such as the economic surplus model (Alston et al., 1995). The economic surplus method is the most widely used procedure for economic evaluation of benefits and costs of a technological change. Technological change due to research in agriculture increases the yield or reduces the cost of production once the new technology is adopted. If the new technology is yield increasing, the producer sells more of the good in the market and if demand is downward-sloping the price decreases. Technology adoption reduces the per-unit cost of production and hence shifts the supply function of the commodity down and to the right. If the market for the commodity is perfectly competitive, this will lead to an increase in the quantity exchanged in the market and a fall in price. As a result, consumers benefit from the price reduction and producers may benefit from selling a greater quantity.

The basic model of research benefits in a closed economy is shown in Figure 1. The demand for the commodity is denoted by D, whereas the pre-research supply curve is S_0 and the post-research supply curve following technological change is S_1. The initial equilibrium is denoted as (P_0, Q_0), while the post-research equilibrium is (P_1, Q_1). That is, the initial equilibrium price and quantity are P_0 and Q_0, whereas after the supply shift they are P_1 and Q_1. The total benefit from the research-induced supply shift is equal to the area beneath the demand curve and between the two supply curves (ΔTS=area *abce*). The total benefit comprises the sum of benefits to consumers (ΔCS=area P_0bcP_1) and the benefits to producers in the form of the change in producer surplus (ΔPS=area P_1ce minus area P_0ba). Under the assumption of a parallel shift (so that the vertical difference between the two curves is constant) area I_0de equals area P_0ba.

This allows estimation of the economic surplus in a closed economy as follows: (1) Economic Surplus $\Delta TS=P_0Q_0K_t(1-0.5Z_t\eta)$ for ex-post analysis and $\Delta TS=P_0Q_0K_t(1+0.5Z_t\eta)$ for ex-ante analysis of potential impacts; (2) consumer surplus $\Delta CS=P_0Q_0Z_t(1-0.5Z_t\eta)$ for ex-post analysis and $\Delta CS=P_0Q_0Z_t(1+0.5Z_t\eta)$ for ex-ante assessment of potential benefits to consumers; and producer surplus $\Delta PS=(K_t-Z_t)P_0Q_0(1-0.5Z\eta)$ for ex-post analysis and $\Delta PS=(K_t-Z_t)P_0Q_0(1+0.5Z\eta)$ for ex-ante assessment of potential benefits to producers. In this model, K_t is the supply shift representing per-unit cost reduction due to technological change and derives from net yield gains due to research and technology adoption rates, whereas $Z = K_t\varepsilon/(\varepsilon+\eta)$ represents the percentage

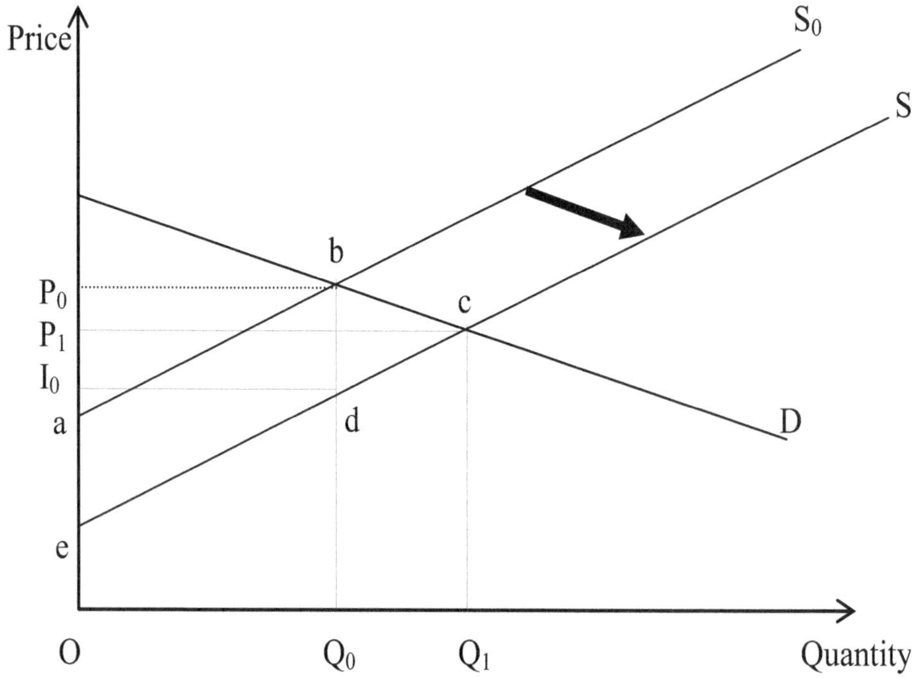

Figure 1. Effects of technological change on producer and consumer surplus.

reduction in price due to the supply shift and ε and η are price elasticity of supply and demand, respectively. Similarly, Alston et al. (1995) show that in a small open economy, change in economic surplus is equal to change in producer surplus and can be calculated as $\Delta TS=\Delta PS=P_wQ_0K_t(1+0.5K_t\varepsilon)$ for ex-post analysis and $\Delta TS=\Delta PS=P_wQ_0K_t(1+0.5K_t\varepsilon)$ for ex-ante analysis of potential impacts of research, where P_w is the real world price.

The research-induced supply shift parameter K_t is the single most important parameter influencing total economic surplus resulting from unit cost reductions. Following Alston et al. (1995), the supply shift was derived as:

$$K_t=[(\Delta Y/Y)/\varepsilon-(\Delta PC/PC)/(1+\Delta Y/Y)]A_t,$$

Where, ΔY/Y is the proportional yield increase per hectare, given that research is successful and the resulting innovation fully adopted; $\Delta PC/PC$ is the proportional increase or decrease in the variable production costs required to achieve the yield increase; and A_t is the rate of adoption of the innovation at time *t*. For improved performance, the adoption of improved verities may require some investment in new inputs like improved seeds or planting materials, chemical fertilizer, pesticide and more labor in operations. Such investments constitute adoption costs required to achieve the necessary yield advantage that improved varieties have over the traditional varieties. However, cassava is famous for its ease of cultivation and does not require more extensive use of labor than is required for the traditional varieties. In view of this, the supply shift equation reduces to $K_t=[(\Delta Y/Y)/\varepsilon]A_t$.

Data sources

Adoption of improved varieties

Adoption rates of improved cassava varieties over the years were estimated based on data coming from household surveys and planting material distribution efforts. In Malawi, a survey of adoption of improved cassava varieties was conducted in 2007. In Zambia, on the other hand, variety adoption data for 2007 were obtained from the Central Statistical Office. The adoption profiles of improved cassava varieties over time were derived using the S-shaped logistic function (Griliches, 1957), which has been used widely to analyze adoption patterns over time (Maredia et al., 2000; Feder et al. 1985; CIMMYT, 1993; Bantilan et al. 2005). Specifically, the diffusion of improved cassava varieties was assumed to follow a logistic curve given as a sigmoid function of time *t*, $Y_t = K[1-e^{-(a+bt)}]^{-1}$, where *K* is the long-run upper limit on adoption; *b* is the slope coefficient measuring the rate of acceptance of the new technology; and *a* is the intercept reflecting aggregate adoption at the start of the estimation period (Feder et al. 1985; CIMMYT, 1993).

The results of the adoption survey in Malawi showed generally high adoption of the local selection *Manyokola,* which is highly preferred by farmers but is susceptible to CMD. Only 7% of the farmers adopted improved varieties (that is, *Mkondezi*, *Silira*, *Maunjili*, *Sauti,* or *Yizaso*) that are tolerant to CMD but are less preferred by farmers due to lack of good consumption attributes. Improved cassava varieties like *Mkondezi* were mostly reported in the north by 7% of the sampled farmers, whereas *Silira* was found to be popular (10%) in the central region. Similarly, in Zambia, local varieties are still popular among cassava growers and thus the largest share of cassava area (over 70%) is still under local varieties. Eight years after their release, the improved varieties (Chila, Mweru, Tanganyika, and Kampolombo) have not been widely adopted. The results showed adoption rates of about 15% in 2007 for improved cassava varieties in Zambia. Nearly 75% of the households surveyed in Malawi cited lack of planting materials and information regarding their availability as the major constraint to the adoption of improved cassava varieties. Awareness, access to planting materials, and farmer perception are important factors in the adoption of improved varieties. Variety adoption will not take place unless farmers are aware of the varieties that exist and have access to planting materials.

Table 4. Average yields of cassava varieties across stations in Malawi, 1990–1997.

Varieties	Station					
	Mkondezi	Chitala	Chitedze	Bvumbwe	Makoka	National
Improved	Yields (tons/ha)					
Mkondezi	25	17	16	18	19	20
Maunjili	20	21	21	17	25	21
Yizaso	17	-	-	5	13	13
Silira	15	17	11	11	16	14
Sauti	-	17	-	-	-	17
Mean (Y_1)	19	19	16	15	20	18
Local						
Manyokola	7	16	16	11	21	13
Gomani	15	17	9	6	13	13
Chitembwere	12	12	13	3	12	11
Nyasungwi	10	11	-	4	17	11
Mean (Y_0)	12	15	12	9	17	13
$(Y_1-Y_0)/Y_0$ (%)	58	31	27	80	20	42

Source: Calculated from various SARRNET reports.

Table 5. Average yields of cassava varieties in Zambia, 2002-2004.

Variety	Lusaka province		Mansa	Mean	Yield gain (%)
	Yield 15 MAP	Yield 16 MAP	Yield 16 MAP		
Bangweulu (local, CMD-free)	21.30	22	31	24.8	21
Chila	18.20			18.2	
Mweru (Improved, CMD-resistant)	19.60	26	41	28.9	41
Muganga (local check)	20.50			20.5	
Manyokola	11.00			11.0	

Source: Calculated from various SARRNET reports.

Recognizing the high preference of farmers for the local selection *Manyokola* but also the high susceptibility of this variety to CMD, IITA and national program partners focused early efforts on the multiplication and distribution of CMD-free planting materials using tissue culture-based cleaning technology. The ex-post impact analysis in this study focuses first on this aspect of the IITA-led cassava improvement effort in Malawi and Zambia. The logistic function was used to estimate the adoption pattern of the CMD-free planting materials from 1993 to 2010 (Fig. 2). IITA and the national programs participated in the multiplication and distribution of clean cassava planting materials. As a result of such multiplication and distribution efforts, the area under improved and CMD-free local varieties reached an estimated 13% of the cassava area in 2003.

Yield gains due to research

Table 3 presents average yields of local and improved cassava varieties based on the household survey of adoption. The results show significant yield differences between the improved and traditional cultivars in the country. The average yield for CMD-free and newly released improved varieties were 12 tons/ha and 14 tons/ha, respectively, compared to 9 tons/ha for CMD-infected local varieties. This translates to a corresponding 33 and 55% yield gains through disease-free local varieties and newly bred varieties, over the generic local varieties. Manyong et al. (2000) reported that the improved cassava varieties have a yield advantage of up to 63% over local varieties grown under similar farmer-managed field conditions. The effects of cassava improvement program (either through cleaning or breeding) were dominant in all ecological zones as evidenced by significant yield gains for the CMD-free cassava varieties as well as improved cassava varieties over local check.

For purposes of comparison, the experimental yields recorded at five research stations in Malawi from 1990 to 1997 are presented in Table 4. The experimental results show that the yield gains for new varieties over local varieties range from 30 to 58% depending on the region but the national average was estimated at 42%.

In Zambia, experiments were conducted at Mansa research station and in Lusaka province from 2002 to 2004. These results are presented in Table 5. The experimental results show that variety Bangweulu gave 25 tons/ha compared to 21 tons/ha for Muganga, whereas Mweru, one of the improved varieties bred by the Zambian root and tuber improvement program, produced 29 tons/ha. Despite data limitations, this still demonstrates the marked effects of using CMD-free local varieties as well as improved varieties, with respective yield gains of 21 and 41% over the localcheck Muganga.

Supply and demand elasticity estimates for cassava

Supply elasticity estimates for cassava in Malawi and Zambia were not readily available. However, Masters et al. (1996) generalize estimates for supply elasticity to be within the range of 0.2-1.2 but are usually low for major crops with little expansion potential because they already take up large share of available resources. Alston et al. (1995) proposed unitary supply elasticity in the absence of exact measures indicating that a one percent increase in cassava prices would lead to an increase in cassava supply by the same margin. The relevant estimates for demand elasticity range from 0.4-10 and are lower for food crops in a small market and higher for export crops or import substitutes whose sales can grow quickly (Masters et al., 1996). Other studies in the recent past like Deaton (1989), Tsegai and Kormawa (2002), Alene et al. (2009) and Dorosh et al. (2009) reported price elasticity of demand of 0.33, 0.46, 0.38, and 0.20, respectively. Dorosh et al. (2009) found that a 10% price increase would reduce the demand for cassava products by 2%. On the basis of past empirical work and given the unique features of cassava as a major staple, this study adopted unitary supply elasticity and demand elasticity of 0.2.

Cassava multiplication and distribution costs

Massive cassava multiplication and distribution program started in the 1992/93 season as a joint response to the 1991/92 drought season. Initially, the Government of Malawi and NGOs established cassava and sweet potato multiplication scheme on a small scale. In the 1992/93 season, the Government of Malawi through IITA/SARRNET (1993/94) and NGOs launched the first phase of the accelerated multiplication and distribution of cassava and sweet potato planting materials. This project was worth US$700,000 in which US$250,000 came from USAID/Malawi and US$450,000 was from United States Department of Agriculture/Overseas Famine Disaster Administration (USDA/OFDA). The funded activities were planned for two years from September 1992 to September 1994 but were granted two-year no-cost extension to September 1995 and later to March 1996. The project was quite successful in terms of increasing area under cassava production and raising cassava productivity hence a second phase of the program was initiated in 1998 in order to intensify and sustain the project achievements realized. The second phase of the accelerated multiplication and distribution program was planned for two years from December 1998 to May 2001. The implementation of activities in the second phase was also made possible through US$382,334 financial support from USAID-Malawi.

The accelerated multiplication and distribution projects were running concurrently with other cassava improvement activities like breeding which also had a component of seed multiplication and distribution. The first phase of SARRNET was a US$7 million regional project launched in 1994 where Malawi, Mozambique, and Zambia benefited US$130,000, US$145,000, and US$100,000, respectively, for five years up to 1998. In 1996 the United States Foreign Disaster Assistance/Bureau for Humanitarian Response (OFDA/BHR) committed US$4.6 million in the SADC multiplication and distribution activities. From this, Malawi and Zambia were allocated US$492,000 and US$651,000. In the 1997/98 season, IITA/SARRNET received another funding from USAID/OFDA under the project framework of Strategic Action Plan for Root and Tuber Crops for El Nino Southern Oscillation (SAP-RT) mitigation in Eastern and Southern Africa. Malawi benefited with US$21,500 and Zambia got US$26,000. Zambia also got extra resources from USAID-Zambia in 1998 to the tune of US$781,700 for purposes of distributing disease free planting materials to the farmers.

The SARRNET phase II operating with a budget of US$3.5 million started in 1999 and ended in September 2002. At the end of the project period, US$895,460 was still available and was committed for a one year no-extra cost extension to September 2003. Between September 2003 and August 2004, a US$257,000 USAID/RCSA crop diversification and enhanced productivity was implemented under the umbrella project of improving rural livelihoods in Southern Africa where Malawi got US$48,000 and Zambia was allocated US$53,500.

From the year 1990 to 2008, a total of US$12 million is estimated to have been used for purposes of root and tuber crops research, multiplication and distribution of disease free planting materials in the SARRNET member countries. These expenses were incurred in the process of promoting planting materials for cassava and sweet potatoes in all the SARRNET countries. Based on the level of activities for the two crops, the costs were equally distributed such that 50% of the total costs accrued to cassava multiplication and distribution. This gives an average expenditure of US$0.32 million every year and a total of US$6 million as the investment in cassava research, multiplication and distribution program over the period of the analysis.

Cassava prices in Malawi and Zambia

The benefits for cassava multiplication and distribution (and the ex-ante benefits) are based on average domestic market prices prevailing in the two countries. In Zambia, the average cassava price was estimated at US$110 per ton dry weight after netting out the effect of inflation and was based on Otterdijk (1999), Haggblade and Zulu (2003), and Haggblade and Nyembe (2008). Otterdijk (1999) reported that farmers in 1994/95 season received an average price within the range of US$0.05 to US$0.08 per kilogram of fresh cassava root—equivalent to US$167–US$267 per ton dry weight using a conversion factor of 0.3 from fresh to dry weight. In another study, Haggblade and Zulu (2003) estimated that in 1985 producers got US$375/ha for a production of 6 tons/ha and in 2002 they realized US$675/ha for cassava yield of 12 tons/ha. This implied that the average cassava producer price was at US$63/ton in 1985 and US$56/ton in 2002. In a marketing margin analysis, Haggblade and Nyembe (2008) reported farm gate price of ZK10,000–ZK15,000 per 50 kg bag of dried cassava chips for the 2006 season. With the exchange rate in 2006 of US$1 to ZK3,500, this gives an average producer price of US$71/ton. For Malawi, cassava prices for the period 1990-2006 were readily available from FAOSTAT online publication. Therefore, the average price of about US$70 per ton of cassava was used in the ex-post analysis, whereas a four-year average price of US$138/ton dry weight (2003-2006) was used in the ex-ante analysis discussed below.

RESULTS AND DISCUSSION

Ex-post impacts of cassava research and extension

The estimated benefits from cassava research and extension involving multiplication and distribution of improved and CMD-free planting materials are presented in Table 6. In the base model, the stream of benefits and costs were compounded at a 5% rate to their respective 2008 values. The benefits were accumulated annually at a rate in tandem with that of cassava variety adoption and supply shift. The results show gross economic benefits of over US$17 million for Malawi and Zambia, which is equivalent to US$1 million per year for the two

Table 6. Benefits and costs of cassava improvement in Malawi and Zambia (US$ million).

Year	Malawi			Zambia			Total		NPV
	ΔTS	ΔCS	ΔPS	ΔTS	ΔCS	ΔPS	ΔTS	Costs	
1990	0.00	0.00	0.00	0.00	0.00	0.00	0.00	0.29	-0.12
1991	0.00	0.00	0.00	0.00	0.00	0.00	0.00	0.12	-0.05
1992	0.00	0.00	0.00	0.00	0.00	0.00	0.00	0.12	-0.05
1993	0.11	0.09	0.02	0.08	0.06	0.01	0.19	0.11	0.04
1994	0.05	0.04	0.01	0.10	0.09	0.02	0.15	0.96	-0.41
1995	0.04	0.04	0.01	0.14	0.12	0.02	0.18	0.73	-0.29
1996	0.72	0.60	0.12	0.18	0.15	0.03	0.90	0.69	0.12
1997	1.02	0.85	0.17	0.23	0.19	0.04	1.24	0.75	0.28
1998	1.11	0.93	0.19	0.33	0.28	0.06	1.45	0.50	0.58
1999	0.31	0.26	0.05	0.49	0.40	0.08	0.80	0.50	0.19
2000	0.81	0.67	0.13	0.49	0.41	0.08	1.29	0.49	0.55
2001	0.93	0.78	0.16	0.66	0.55	0.11	1.60	0.48	0.80
2002	0.53	0.44	0.09	0.75	0.62	0.13	1.28	0.47	0.60
2003	0.42	0.35	0.07	0.83	0.69	0.14	1.25	0.12	0.89
2004	0.51	0.42	0.08	0.90	0.75	0.15	1.40	0.12	1.06
2005	0.44	0.37	0.07	1.05	0.87	0.17	1.48	0.11	1.18
2006	0.56	0.46	0.09	0.98	0.82	0.16	1.54	0.11	1.30
2007	0.50	0.42	0.08	1.00	0.84	0.17	1.51	0.17	1.28
2008	0.42	0.35	0.07	0.84	0.70	0.14	1.26	0.16	1.10
Total	8.48	7.06	1.41	9.05	7.54	1.51	17.52	7.00	9.03
Annual	0.47	0.39	0.08	0.48	0.40	0.08	0.92	0.37	0.48

countries combined. The results provide further insights into the distribution of research benefits where some 83% of the benefits accrued to consumers, whereas only 17% of the benefits were captured by the producers. Since the cassava markets in the two countries are not well integrated and developed, there is limited commoditytrading or movement outside production zones. This implies that the producer households are largely the same as the consumer households hence the total welfare gains accrue largely to the same cassava producing households.

The cassava multiplication and distribution program was quite successful and worthwhile when evaluated on the basis of benefit-cost ratio and internal rate of return (IRR). Using a discount rate of 5%, the benefit-cost ratio was estimated at 3:1 and the rate of return for the program was found to be 24%. A benefit-cost ratio criterion recommends as viable any investment plan with a ratio-value equal to or greater than one. If the ratio is greater than one, the project is returning more benefits than it costs. The rate of return is the rate that equates NPV to zero and the higher the rate above the opportunity cost of capital the better the investment plan. The benefit-cost ratio of 3:1 suggests that the returns were three times higher than the research investment costs incurred. In other words, every dollar invested in the multiplication and distribution activities paid back three times through production gains associated with the use of CMD-free planting materials. The accrued benefits are taken as a measure of success and it can be concluded that the IITA/SARRNET cassava multiplication and distribution program has achieved reasonable success in Malawi and Zambia.

Potential economic impacts of cassava research and extension

New variety development

Early efforts to control the effects of CMD in Malawi and Zambia through the IITA-led SARRNET project focused on breeding cassava for disease tolerance. Several varietal trails conducted from early 1990s showed successes achieved in terms of developing high yielding and CMD tolerant varieties. So far, all the five improved varieties released in Malawi have a bitter taste and are only suitable for traditional cassava consuming areas such as along the lakeshore. In addition, it has been observed that some of the varieties such as *Mkondezi* were rejected by many farmers because of poor *Nsima* quality. Given the low and slow adoption of the first generation improved varieties due largely to lack of attributes valued by farmers, the focus for future work on

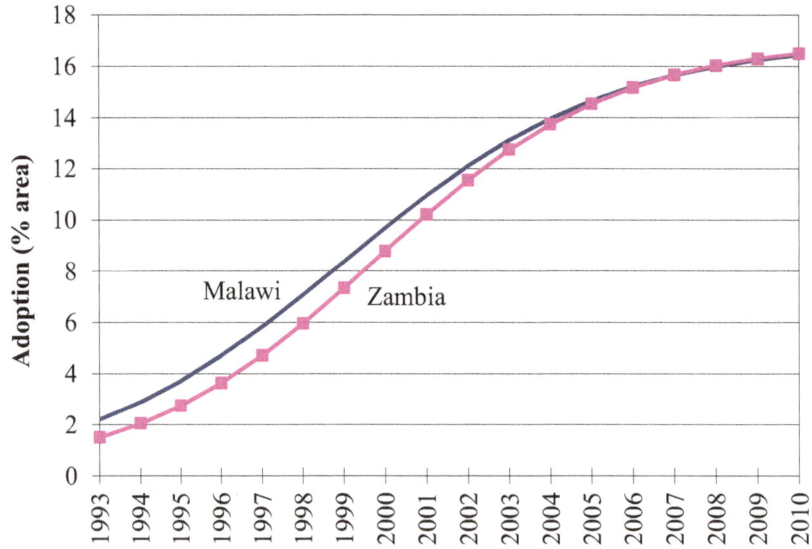

Figure 2. Adoption of CMD-free local cassava varieties in Malawi and Zambia.

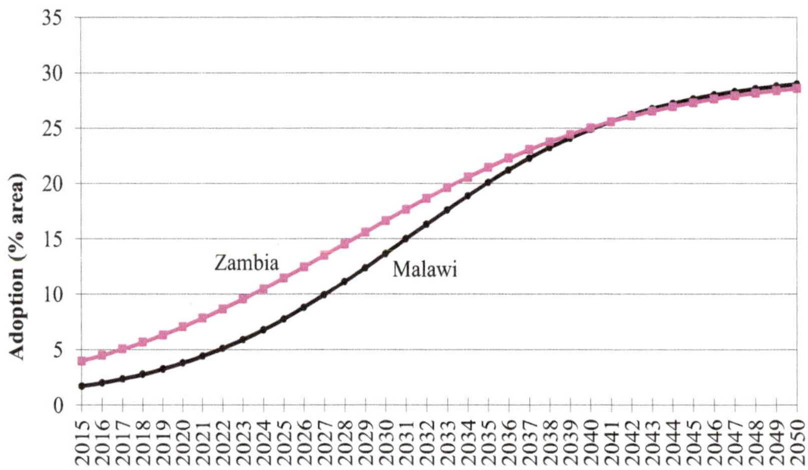

Figure 3. Projected adoption profile for new cassava varieties in Malawi and Zambia.

cassava research in the two countries should be on developing new varieties that are not only high yielding and CMD resistant but should also have consumption attributes highly valued by farmers who are also the major consumers of their own production. The other focus would be to enhance industrial characteristics in order to fast track industrial use and successfully catalyze the cassava commercialization process.

Expected adoption

If such line of cassava research succeeds in developing varieties with desired production and consumption characteristics, the adoption of the new cassava varieties is expected to be higher and faster in Malawi and Zambia, with an estimated adoption ceiling of 30%. On the basis of planting material production records and historical adoption rates of varieties preferred by farmers, expected adoption patterns for new varieties were projected using the logistic function for the period 2015–2050 (Figure 3). The economic surplus model for ex-ante analysis described earlier was used for the projection of potential impacts of cassava research that develops varieties with a range of characteristics preferred by farmers and other actors along the value chain.

Costs of variety development and dissemination

Estimation of net returns to cassava improvement research requires a comparison of the stream of benefits and the corresponding costs of the development and dissemination of improved cassava varieties. The research costs used in this analysis were adapted from previous and current cassava improvement activity budgets but were adjusted to cater for additional extension costs necessary to disseminate the new varieties upon release. The initial research costs were extracted from SARRNET project documents and are summarized in Table 2. Historical expenditures show that variety development accounts for 70% of the total cost, whereas variety dissemination accounts for the remaining 30%. The average annual expenditure for cassava research and extension in the SARRNET member countries was estimated at US$390,000 for twenty-three years until maximum adoption is attained. A total of US$8 million is estimated to be devoted to the development of the new varieties and dissemination of information to the farming families regarding their availability and potential benefits associated with the new varieties. As noted by Johnson et al. (2003), it is worth mentioning that when computing cassava research costs it is challenging to isolate breeding costs from other components of cassava improvement research investment.

Table 7 presents the net present value of potential economic benefits from cassava research and the corresponding estimates of the benefit–cost ratios and rates of return. The results show that, over the period 2015–2050, cassava research and extension can generate an estimated net benefit of US$97 million. This is equivalent to annual gross benefits of US$6 million for Malawi and US$3 million for Zambia following adoption of the new cassava varieties. The results also show nearly 85% of the benefits would accrue to the consumers and the remaining 15% go to the producers. Cassava research benefits are estimated to be higher in Malawi than in Zambia. International and national cassava research in Malawi and Zambia can have an impressive potential benefit–cost ratio of 21:1, indicating that each dollar invested in cassava improvement research generates US$21 worth of additional cassava.

Consistent with the payoffs implied by the estimated benefit–cost ratio, cassava research in Malawi and Zambia has the potential to generate a rate of return of 40%. The 40% internal rate of return to cassava improvement estimated for the two countries is much higher than the estimated rate of return to cassava research focusing on the multiplication and distribution of disease-free planting materials of the available farmer-preferred varieties such as Manyokola in Malawi and Muganga in Zambia. Furthermore, the estimated rate of return is much higher than the prevailing market interest rates and confirms that cassava research holds promise for generating a stream of benefits in excess of the expenditures. By all summary measures such as rate of return and benefit–cost ratios, the results suggest that benefits from cassava improvement are in excess of all cassava research costs in the region.

Sensitivity analysis

In an effort to gain confidence in the results, we evaluated the sensitivity of the base model estimates to variations in the values of some key parameters. Recognizing that the supply shift parameter is the major determinant of research benefits, the model was estimated with the proportional yield gains attributable to cassava research assumed to be half of the base yield gains. Given that the supply shift parameter is also a function of expected adoption of improved varieties, the model was estimated with the maximum adoption assumed to be 40%, up from the base value of 30%. The sensitivity of the estimated rate of return was also evaluated by estimating the model with research and extension costs assumed to be double the base value. The results of the sensitivity analysis presented in Table 8 show that, as a consequence of changes in supply shift, the present value of benefits is sensitive to changes in yield gains. Halving yield gains to about 20% has a proportional effect of halving research benefits to about US$46 million, but the rate of return drops only by 9 percentage points to 31%. On the other hand, raising the ceiling of expected varietal adoption from 30 to 40% has no proportional effect on net benefits and rate of return and has no effect on the benefit-cost ratio. More specifically, the net present value of benefits increases only by US$4 million and rate of return increases marginally from 40 to 42%, whereas the benefit-cost ratio remains the same as the base value of 21. Doubling of research and extension costs reduces net benefits by only 5% and the rate of return (to research and extension) also drops to 28%.

Overall, the summary measures suggest that the scenario with lower yield gains is the most conservative because it has a proportional effect of halving the net benefits to US$46 million as well as the benefit-cost ratios to 10. Similarly, the scenario with the research and extension costs double the base value has a comparable effect of halving aggregate benefit–cost ratio from 21 to 10 with a lower rate of return of 28%. The scenario with greater adoption of new varieties which reaches a maximum adoption rate of 40% before it stabilizes gives the highest rate of return of 42%. In general, the sensitivity analysis demonstrated that total net benefits from cassava research and extension in Malawi and Zambia vary between US$46 and 100 million and the benefit-cost ratio ranges from 10 to 21, with the rate of return varying from 28 to 42%. Although the sensitivity analysis lends credence to the main results, the minimum net benefits implied by the alternative scenarios are still an impressive US$46 million with a modest rate of return of 28% and benefit-cost ratio of 10.

CONCLUSION AND IMPLICATIONS

The national cassava improvement programs in Malawi

Table 7. Potential benefits of cassava research and extension in Malawi and Zambia.

Year	Malawi (US$ million)	Zambia (US$ million)	All (US$ million)	Net present value (US$ million)
2015	0.78	0.09	0.87	0.22
2016	0.92	0.15	1.07	0.35
2017	1.08	0.26	1.34	(0.28)
2018	1.27	0.43	1.70	1.06
2019	1.49	0.68	2.17	1.14
2020	1.74	1.05	2.79	1.48
2021	2.02	1.52	3.54	1.84
2022	2.35	2.05	4.40	2.15
2023	2.71	2.58	5.29	2.28
2024	3.12	3.03	6.14	2.93
2025	3.56	3.37	6.93	3.15
2026	4.05	3.60	7.65	3.26
2027	4.57	3.76	8.33	3.44
2028	5.13	3.85	8.98	3.52
2029	5.71	3.91	9.61	3.60
2030	6.31	3.94	10.25	3.66
2031	6.92	3.96	10.88	3.70
2032	7.53	3.97	11.50	3.71
2033	8.13	3.98	12.11	3.74
2034	8.72	3.98	12.70	3.74
2035	9.28	3.98	13.26	3.72
2036	9.81	3.98	13.79	3.68
2037	10.30	3.99	14.28	3.63
2038	10.75	3.99	14.74	3.57
2039	11.16	3.99	15.15	3.49
2040	11.53	3.99	15.51	3.41
2041	11.86	3.99	15.84	3.31
2042	12.15	3.99	16.13	3.21
2043	12.40	3.99	16.39	3.11
2044	12.62	3.99	16.61	3.00
2045	12.82	3.99	16.80	2.89
2046	12.98	3.99	16.97	2.78
2047	13.13	3.99	17.11	2.67
2048	13.25	3.99	17.23	2.56
2049	13.35	3.99	17.34	2.46
2050	13.44	3.99	17.43	2.35
Total	268.9	114.0	382.86	97.17
Annual	6.40	2.71	9.12	2.31
NPV (US$ million) = 97				
B:C ratio = 21				
IRR (%) = 40				

and Zambia have developed and released varieties that outperform the local varieties using breeding materials received from IITA. Past investments in cassava improvement have led to the development and release of a good number of high-yielding and CMD-tolerant cassava varieties. Over the period 1990–2011, IITA and the respective national programs released a total of 12 improved varieties in Malawi and 8 improved varieties in Zambia. The increased availability of improved cassava varieties opened up a range of profitable commercial opportunities for production of cassava-based foods, feeds, and industrial products. At the same time, improved

Table 8. Sensitivity analysis of the economic benefits of cassava research in Malawi and Zambia.

Parameter	Parameter value			NPV (US$ million)		IRR (%)		B:C ratio	
	Base	New	Δ (%)	New	Δ (%)	New	Δ (%)	New	Δ (%)
Yield gains (%)	0.40	0.20	-50	46	-47	31	-25	10	-50
Adoption ceiling (%)	30	40	33	101	4	42	5	21	0
Res. & Ext. costs (US$ m)	5	10	100	92	-5	28	-30	10	-50

disease tolerance and higher productivity as well as a flexible harvesting calendar offered prospects for improving household food security.

Despite major efforts to develop and disseminate a growing number of improved varieties, however, there is lack of comprehensive evidence on the adoption and economic impacts of improved cassava varieties. Using household survey as well as planting material production data for estimating variety adoption and on-farm experimental data for yield gains, this paper estimates the economic impacts of IITA-led cassava improvement research in Malawi and Zambia over the period 1990-2008. Historical as well as future adoption patterns were estimated using the logistic function with the minimum variety adoption data assembled from various sources. Consistent with the need for a gradual transformation of the scientific capacity of national programs, the content of earlier varietal releases points to the predominance of IITA germplasm supplied for direct release to farmers, whereas the content of recent releases shows that national programs are developing varieties using IITA material as a parent.

The results show relatively higher adoption rates for the CMD-free local varieties compared to CMD-tolerant varieties that have been released in the two countries. The adoption of new varieties has been low and slow largely due to the fact that most of these varieties lacked the consumption attributes highly valued by farmers. The multiplication and distribution of CMD-free planting materials of the recommended local varieties led to greater adoption, but infection with CMD three to four years after adoption meant that the yield gains and economic benefits could not be sustained. Nevertheless, the multiplication and distribution of clean cassava planting materials generated a modest rate of return of 24%, which is actually consistent with an earlier rate of return estimate of 9 to 22% for cassava improvement in developing countries. Analysis of the ex ante impacts of current and future investments in cassava research and extension shows that cassava research that focuses on the development and dissemination of varieties with highly preferred production, consumption, and industrial attributes would yield a greater rate of return of 40%.

Finally, it is worth noting that high rates of return to agricultural research are difficult to sustain in an environment where inputs are not accessible or affordable to farmers. A critical input for achieving greater adoption of improved cassava varieties is an efficient seed system for the production and distribution of highquality and disease-free planting material. Improved varieties can disseminate only with the help of an effective national seed industry, but this is still lacking in many countries in Africa especially for vegetatively propagated crops such as cassava.

REFERENCES

Alene AD, Manyong VM, Tollens E, Abele S (2009). Efficiency–equity tradeoffs and the scope for resource reallocation in agricultural research: Evidence from Nigeria. Agric. Econ. 40:1–14.

Alston JM, Norton GW, Pardey PG (1995). Science under Scarcity: Principles and Practice for Agricultural Evaluation and Priority Setting. Cornell University Press, New York, USA.

Benesi IRM, Moyo CC, Mahungu NM, Mkumbira JM, Soko MM, Sauti RFN, Sandifolo VS (1999). Progress in developing improved cassava varieties in Malawi. In Akoroda, M.O. and Teri, J.M. (Eds.) (1999) Food Security and Crop Diversification in SADC Countries: The Role of Cassava and Sweet Potato. Proceedings of the scientific workshop of the Southern Africa Root Crops Research Network (SARRNET), Lusaka, Zambia, 17 – 19 August 1998.

Chitundu DC (1999). Accelerated multiplication and distribution of cassava and sweetpotato planting materials in Luapula Province of Zambia. In Akoroda MO, Teri, JM. (Eds.) (1999). Food Security and Crop Diversification in SADC Countries: the Role of Cassava and Sweet Potato. Proceedings of the scientific workshop of the Southern Africa Root Crops Research Network (SARRNET), Lusaka, Zambia, 17 – 19 August 1998.

CIMMYT (1993). The Adoption of Agricultural Technology: A Guide for Survey Design. Mexico, D.F., CIMMYT.

Deaton A (1989). Household Survey Data and Pricing Policies in Developing Countries. World Bank Econ. Rev. 3(1):183 -210

Dorosh PA, Dradri S, Haggblade S (2007). Alternative Instruments for Ensuring Food Security and Price Stability in Zambia. Food Security Research Project Working P. 29 Lusaka, Zambia.

Dorosh PA, Dradri S, Haggblade S (2009). Regional trade, government policy and food security: Recent evidence from Zambia. Food Policy 34:350–366.

Feder G, Just RE, Zilberman D (1985). Adoption of Agricultural Innovations in Developing Countries: A Survey. Econ. Develop. Cultural Change 33(2):255-298.

Griliches Z (1957). Hybrid corn: an exploration in the economics of technological change. Econometrica 25(4):501–522.

Haggblade S, Nyembe M (2008). Commercial dynamics in Zambia's cassava value chain. Food Security Research Project Working Lusaka, Zambia P. 32.

Haggblade S, Zulu B (2003). The recent cassava surge in Zambia and Malawi. InWEnt, IFPRI, NEPAD, CTA Conference. P. 9.

Johnson NL, Manyong VM, Dixon AGO, Pachico D (2003). The Impact of IARC Genetic Improvement Programmes on Cassava. In R.E. Evenson and D. Gollin (Eds.) (2003), Crop Variety Improvement and its Effect on Productivity: The Impact of International Agricultural Research. CABI International, Wallingford, UK.

Manyong VM, Dixon AGO, Makinde KO, Bokanga M, White J (2000). The contribution of IITA-improved cassava to food security in sub-Saharan Africa: an impact study. IITA, Ibadan, Nigeria.

Maredia M, Byerlee D, Anderson J (2000). Ex-post evaluation of economic impacts of agricultural research programs: a tour of good practice. Paper presented to the Workshop on "The Future of Impact Assessment in CGIAR: Needs, Constraints, and Options", Standing Panel on Impact Assessment (SPIA) of the Technical Advisory Committee, Rome, 3-5 May, 2000.

Masters WA, Coulibaly B, Sanogo D, Sidibe M, Williams A (1996). The Economic Impact of Agricultural Research: A Practical Guide. Department of Agricultural Economics, Purdue University, West Lafayette, IN.

Mataya C, Sichinga K, Kachule R, Tchale H (2001). A Policy review of cassava processing, marketing and distribution in Malawi. In Mahungu N.M., Banda J.W. and Mataya C (Eds.) (2001) Cassava Commercialisation for Economic Development in Malawi. Proceedings of the Symposium held at Kwacha International Conference Centre 21-23 May 2001, Blantyre, Malawi.

Nweke FI, Spencer DSC, Lynam JK (2002). The cassava transformation: Africa's best-kept secret, Michigan State University Press, East Lansing.

Otterdijk RLM (1999). The importance and potential of cassava and sweet potato for household food security in rural Zambia. In Akoroda, M.O. and Teri, J.M. (Eds.) (1999) Food Security and Crop Diversification in SADC Countries: the Role of Cassava and Sweet Potato. Proceedings of the scientific workshop of the Southern Africa Root Crops Research Network (SARRNET), Lusaka, Zambia, 17-19 August 1998.

Phiri MAR (2001). A Policy review of cassava processing, marketing, and distribution in Malawi. In: Mahungu N.M., Banda J.W. and Mataya C. (Eds.) (2001) Cassava Commercialization for Economic Development in Malawi. Proceedings of the Symposium held at Kwacha International Conference Centre 21-23 May 2001, Blantyre, Malawi.

Soenarjo R (1992). Research activities to combat major pests and diseases of cassava in Zambia. In Allard GB, Skoglund LG, Neuenschwander P and Murphy RJ (eds) 1992. Root and Tuber Pest Management in East and Southern Africa. Proceedings of a regional workshop held in Mombasa, Kenya, 10-14 August 1992. International Institute of Biological Control, Kenya.

Tsegai D, Kormawa P (2002). Determinants of urban households' demand for cassava and cassava products in Kaduna, northern Nigeria: The application of AIDS model. Paper prepared for Conference on International Agricultural Research for Development, Witzenhausen, 9-11 October 2002.

3

Assessment of operational aspects of the input supply chain under national agriculture input voucher scheme (NAIVS) in Tanzania

Aloyce G. M.[1], Gabagambi D. M.[2] and Hella J. P.[2]

[1]Mikocheni Agricultural Research Institute, P. O. BOX 6226, Dar es Salaam, Tanzania.
[2]Department of Agricultural Economics and Agribusiness, Sokoine University of Agriculture, P. O. Box 3007, Chuo Kikuu, Morogoro, Tanzania.

This study examines the operational aspects of the National Agriculture Input Voucher Scheme in Tanzania from 2009/2010 to 2010/2011 based on secondary and primary data. Primary data were collected from a random sample of 300 households in four regions namely: Rukwa, Mbeya, Morogoro and Shinyanga. Secondary data were collected from the Ministry of Agriculture Food Security and Cooperatives (MAFC), Agro dealers and Local Government Authorities. Results indicate that 88% of farmers reported delayed subsidized inputs significant at p = 0.05. The inputs become available during planting season when most of the household food stocks and income is exhausted and this makes top up price unaffordable. It was also observed that the top up price is more than stipulated cost sharing of 50% between farmers and the government. Other pitfalls reported in the system include input adulteration and violation of NAIVS guidelines for input distribution. It is recommended that inputs and crop calenders be established that would ensure inputs arrived to beneficiaries ahead of the planting season In addition, efficient monitoring and evaluation system should be put in place to minimize inefficiency emanating from violation of NAIVS guidelines.

Key words: Tanzania, input vouchers, farmers, institutional framework, subsidy programme.

....

INTRODUCTION

Many African countries resumed fertilizer subsidy in early 2000 in an attempt to enable smallholder resource poor farmers use inputs to boost production and reduce poverty (Ricker-Gilbert and Jayne, 2009; Chibwana et al., 2010; Danning et al., 2009b). The new system of subsidy is market "Smart" and concurrent with Abuja declaration which resolved to increase timely access and raise fertilizer use by farmers in African Union (AU) member states to an average of 50 kg/ha by 2015 (Danning et al., 2009a; Tiba, 2010; Yawson et al., 2010). It is market smart as pointed out in Minde et al. (2008) and Baltzer and Hansen (2011) because it has a specific target, measurable impacts, achievable goals, results orientation and timely duration of implementation. Additionally, the declaration aims to eliminate barriers to fertilizer access such as tariffs on fertilizers and fertilizer raw materials in order to increase productivity, reduce food insecurity and poverty levels among smallholders. The new scheme originated from Malawi as a small starter pack in 1998 (Dorward and Chirwa, 2011; Druilhe and Barreiro-Hurlé, 2012). Other African countries such as Nigeria, Zambia, Tanzania, Kenya, and Ghana adopted the initiative at different time (Druilhe and Barreiro-Hurlé, 2012). The Government of United

Republic of Tanzania resumed subsidy programme in 2003/2004 which was implemented through subsidising transport for companies that were involved in the distribution process of the inputs. The idea was to reduce the input cost below the market price to all farmers. Constraints in the distribution system of agricultural input under this arrangement were frequently reported. Some of the complaints raised include subsidy fertilizer ending up in the shops of input suppliers and thus being sold at the market price, inputs being smuggled to neighbouring countries, delayed inputs delivery, re-bagging fertilizer in warehouses and inputs not being effective due to quality deterioration. Additionally, many targeted farmers could not access the inputs under the 2003/2004 programme modalities and it was difficult to identify beneficiaries and non beneficiaries.

The National Agriculture Input Voucher Scheme was introduced in 2008/2009 as a reform of the previous subsidy policy programmes. Under NAIVS beneficiary, farmers are selected based on eligibility criteria. Some of eligibility criteria require farmers be residing in the village, and be willing to apply the subsidy inputs in the target crops in the area of a ½ ha (Pan and Christiaensen, 2012). Also, it targets farmers who have not afforded to apply inputs in the previous five years and able to pay the cash top up. Eligible farmers are provided vouchers that entitle them to buy inputs from agro-input dealers at a subsidized price. The voucher has a face value of inputs that government supports the farmer. It is contrary to previous National agricultural input subsidy programmes where the subsidy inputs were sold at a lower price than the market price of inputs to all farmers. NAIVS intends to increase the existing 9 kg/ha of fertilizer use, which is below Africa average of 21 kg/ha, and the world average of 100 kg/ha (Ricker-Gilbert and Jayne, 2009; Eboh et al., 2006; Baltzer and Hansen, 2011). It is also a government response to escalating food and input prices in the world aimed at increasing productivity and food security.

Allocation and distribution of inputs under NAIVS involves a chain of actors with established committees from national to village level. Selection of eligible farmers and committee members is guided by set guidelines. Each committee is assigned responsibilities for implementation to ensure targeting and timely delivery (MAFC, 2012). A pilot study by Pan and Christiaensen (2012) pointed out poor targeting performance whereby 60% of vouchers were captured by village elites. Moreover, the increase in number of vouchers available for distribution was found to enhance the targeting performance. Baltzer and Hansen (2011) have pointed out that, study by Pan and Christiaensen (2012) focused mainly on targeting and did not discuss the performance of the input voucher delivery system. Consequently, there is limited knowledge on how the framework operates in delivering inputs to beneficiaries. It is uncertain whether or not the distribution process abides to stipulated national guidelines.

Also, it is not well known whether the programme has managed to correct the inefficiencies of previous subsidy programmes. Effective implementation of subsidy programmes requires a well functioning institutional framework in the supply chain. Polski and Ostrom (1999) defined institution as a broadly known rule, custom or strategy that creates incentives for behaviour in a repetitive circumstance. Institutions are distinguished into three pillars of regulative, normative and cultural-cognitive institutions. Regulative institutions encompass incentive schemes, public procurement policies that are referred to as the formal institutions, while the remaining pillars are referred to as informal institutions (Truffer et al., 2009). The success of subsidy policy reform depends on the institutional arrangements available for translation of intentions into actions and outcomes. Evaluation, design or policy reform requires a systematic way of analysing existing arrangements, generation and comparison of alternatives. The analysis should contain well organised survey of how stakeholders act and reason for acting in a certain way rather than another (Polski and Ostrom, 1999).

The objective of this study therefore is to examine the operational framework of the input supply chain under the National Agricultural Input Voucher scheme in Tanzania. Specifically, the study to assess subsidy inputs distribution system in term of availability of vouchers, timely delivery, input quality and shared price. The knowledge from this study will provide insights on the existing strengths and weaknesses in the delivery system and form the basis for policy reform recommendations.

MATERIALS AND METHODS

Sampling design

Research was carried out in four regions namely: Mbeya, Morogoro, Rukwa and Shinyanga, whereby purposive and random sampling methods were used. Mbeya and Rukwa regions were chosen because these were pilot areas and main food crop producers. Recently, Morogoro and Shinyanga are new comers into the programme, moreover, Morogoro has been identified by the government as an emerging grain basket for the nation, and Shinyanga is among the major cash crop producing regions such as cotton that have benefited from NAIVS. From each participating village, farmers registers were used as sampling frames. Simple random sampling technique was used to obtain 5% of farmers for interview from respective registers. Boyd et al. (1981) contends that, a sample of 5% of the total population is statistically adequate in sampling. The distribution of respondents in the villages involved in this study is shown in Table 1.

Data collection procedure

The study used both primary and secondary data. Primary data were collected in a survey using a semi structured questionnaire administered to 300 households. Some of the collected data were on constraints and challenges on NAIVS input supply chain, access to inputs, distribution procedures, selection of the end users and agro dealers as well as participation in service delivery. Focus group discussions were conducted with key informants such as stockists, District Agriculture and Livestock Development officers

Table 1. Names of region, districts and villages involved in the survey.

Region	District	Village	Criteria of selecting	Non beneficiaries	Beneficiaries	Sample of farmers	Major crop(s) for input voucher
Rukwa	Sumbawanga	Chitete	Accessible	21	14	35	Maize
		Katete	Inaccessible	16	12	28	Maize
Mbeya	Mbozi	Isangu	Accessible	16	27	43	Maize/paddy
		Itaka	Inaccessible	7	36	43	Maize/paddy
Shinyanga	Bariadi	Ibulyu	Accessible	23	29	52	Cotton
		Nguliyati	Inaccessible	17	32	49	Cotton
Morogoro	Mvomero	Lusanga	Accessible	21	13	34	Maize/paddy
		Kinda	Inaccessible	10	6	16	Maize

(DALDOs) and Ginnery managers.

The main crops under subsidy that were considered in this study were maize and rice as food crops and cotton as a traditional cash crop. From each district, two villages were selected based on accessibility criterion; one village was accessible in terms of road network and other village inaccessible. Secondary data were gathered from public and private institutions and internet sources. Pretesting of the questionnaire was undertaken in "Wami-Luhindo" village located in Mvomero district. This village received input subsidy and has similar characteristic to villages under the subsidy programme.

Analytical framework

Institutional Analysis and Development (IAD) framework (Figure 1) was applied in the assessment of NAIVS operational aspects. In IAD framework, policy issue or objective is defined clearly, followed by analysis of physical and material conditions as they influence policy action and situation and constrain institutional arrangements. Physical and material conditions refer to physical and human resource capabilities related to providing goods and services. These include capital, labour, technology, financial resources, storage and distribution channels which play a significant role in policy design and implementation (Polski and Ostrom, 1999). Determination of physical and material condition of goods or services requires answering important questions that are focused in economic nature of activity.

Also, it considers the way a good or service is provided, produced, as well as the physical and human resources required for producing goods or services, and other important aspects in the distribution channel of goods and services. It analyses the community attributes such as demographic features, norms, values, beliefs, degree of general understanding and preferences on policy oriented strategies and outcome.

NAIVS operational framework

Six main actors were involved in the NAIVS implementation framework. Every actor assigned specific responsibilities stipulated in the National Voucher Guidelines (MAFC, 2012). Efficient operations were expected to result into timely delivery of quality inputs to target farmers and increased crop production. IAD framework is applied to evaluate the activities and performance of involved actors in the NAIV framework presented in Figure 2.

Data analysis

Data were analysed based on descriptive statistics using frequencies and cross tabulation with chi-square test. The statistical package for social sciences (SPSS) version 18 and Excel programmes were used.

RESULTS AND DISCUSSION

Government budget on input subsidy

Results revealed continued increase in government budgets for input subsidies from Tsh. 2.0 billion in 2003/2004 to Tsh. 128.7 billion in 2010/2011 fiscal year. Also the quantity of subsidized fertilizer and the number of beneficiaries has substantially increased as depicted in Table 2. In earlier years, the subsidy was supported from government budget, although from 2009 World Bank allocated a total of US$ 299 million to support the subsidy programme (World Bank, 2012). It was difficult to account the number of beneficiaries in the previous subsidy programs although under NAIVS the number of beneficiaries is determined. In 2008/2009, the numbers of beneficiaries were 740,000 and increased to 2,011,000 in 2010/2011. However, the number decreased to 1,800,000 in 2011/2012 as farmers received the subsidy in 2008/2009 graduated from the program and were expected to be self dependent. These results are promising although it is unclear whether or not graduated farmers are self dependent. Baltzer and Hansen (2011) have pointed out that, sustainability of smart subsidies are expected if farmers are able to accumulate finance and productive assets to overcome market barriers after programme termination.

Input subsidies reduce fertilizer cost to farmers who otherwise would be excluded from fertilizer use due to limited ability to afford its cost. Increases in subsidy budget are associated with increase in demand for agricultural inputs which is likely to increase productivity.

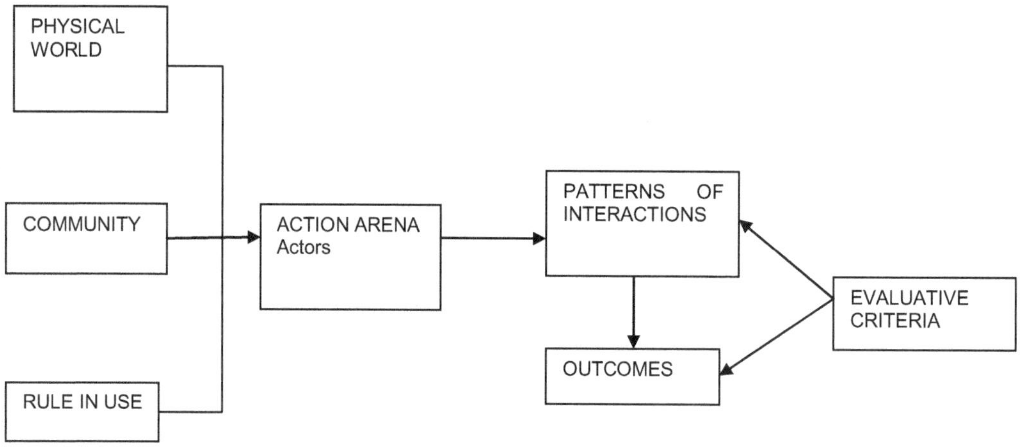

Figure 1. Institutional analysis and development (IAD) framework (Polski and Ostrom, 1999).

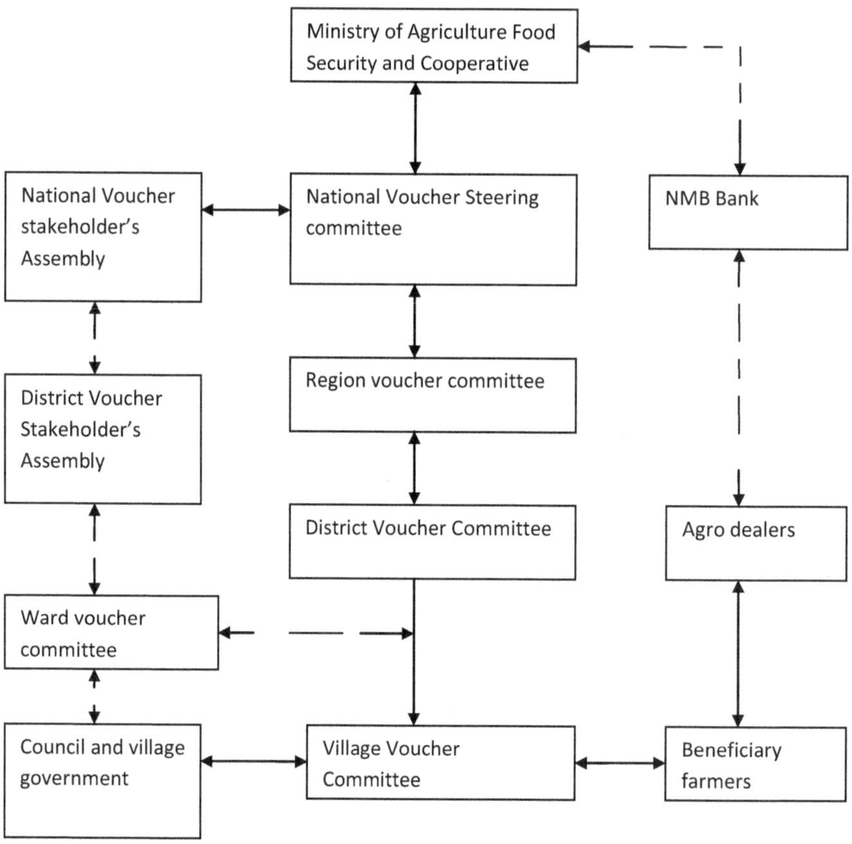

Figure 2. NAIVS implementation framework (MAFC, 2012).

However, increased demand for agriculture inputs might inflate input prices in the long run depending on the supply side of the inputs. Also, increased budgets on subsidies drive resources away from other public goods (Wiggins and Brooks, 2010; Nagy and Edun, 2002).

Cost sharing / cash top up

Fifty percent (50%) cost sharing between the government and eligible farmers was not practical in some districts as could be noted from Table 3. In Sumbawanga district,

Table 2. Budget for fertiliser subsidies to farmers (MAFC, 2013).

Year	Quantity of fertiliser subsidised (tons)	Amount of money spent on fertilizer (billions)	Subsidy as percentage of total agricultural budget	Number of beneficiaries
2003/2004	39,387	2.0	3.7	
2004/2005	81,766	7.2	11.3	
2005/2006	63,000	7.5	6.0	
2006/2007	90,755	21.0	17.3	
2007/2008	83,076	19.5	14.8	
2008/2009	130,000	31.9	28.0	740,000
2009/2010	142,000	69.2	30.3	1,511,900
2010/2011	201,015	128.7	50.8	2,011,000
2011/2012	195,959	118.6	45.9	1,800,000

Source: MAFC (2013).

Table 3. Variation in input costs at free market price and subsidy price.

Location	Input type	Free market price (Tsh)	Voucher value (Tsh)	Farmers contribution (Tsh)	Percentage of farmers contribution
	OPV maize seeds	39000	20000	19000	49
Chitete	DAP/NPK	82500	30000	52500	64
	Urea	72167	22000	50167	70
	OPV maize seeds	39700	20000	19700	50
Katete	DAP/NPK	82900	30000	52900	64
	UREA	72400	22000	50400	70

there was variation in farmer contribution depending on the location. The district established transaction costs to be charged by an agro dealer per bag of input including the profit margin depending on distance from main input supplier. There was variation in the amount of cash top up charged to farmers depending on input type and this cost exceeded 50% stipulated in the national input subsidy guideline. In Mbozi, prices were under the control of market forces of supply and demand. Farmers were free to use the voucher to buy fertilizer at any agro dealer in the district and there were many agro dealers. In Bariadi, the cost sharing was as stipulated in the national voucher guidelines, whereas Mvomero revealed mixed results. In Lusanga, cost sharing was 50% while in Kinda farmers received seeds free of charge with no fertilizers disbursed in 2010/2011 season. It was not clear whether the fertilizer was not in the distributed package or fertilizer was used to compensate the cost of seeds. However, report from the Ministry of Agriculture Food Security and Cooperatives (MAFC) revealed distribution of subsidies as a package suggesting malpractice in Kinda.

Total cost of top up contribution per subsidy package per eligible household in Rukwa was 123,000 TSh; and this was considered 41% higher than anticipated for interviewed households. Top up was considered afforded by large scale farmers with more than 35% households undecided on who is able to afford the top up. This implies that, rich farmers benefit more from subsidies than poor farmers. It was noted that input distribution is done during planting season when most farmers do not have food stock left for sale to purchase inputs. Although, some farmers expressed the need for loans to purchase inputs, there were limited alternatives for credit services. To cope with top up, some farmers shared the cost for the input package and share the inputs. In case, farmers offered the voucher to their relatives who were able to pay top up although those relatives were not in the list of beneficiaries. By implication, farmers who share the subsidy package do not attain technical efficiency in production. Also, wealthier farmers benefit more from subsidy than poorer making economic efficiency of the programme doubtiful.

Number of beneficiaries during 2008/2009 to 2009/2010 period

Generally, there has been increased trend in the number of beneficiaries (Figure 3). In absolute terms, Mbeya region leads in the number of beneficiaries followed by

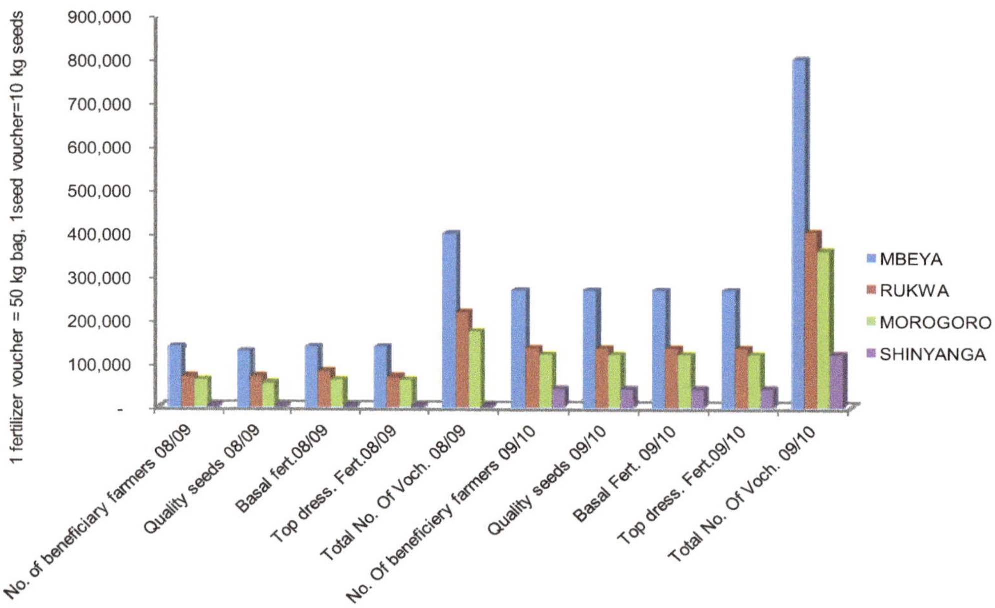

Figure 3. Number of input vouchers distributed to beneficiary farmers in the study regions.

Table 4. Percent response on subsidy received by household through voucher from 2008/2009 to 2011/2012.

Districts	One year	Two years	Three years	More than three years
Sumbawanga	37.04	37.04	25.93	0.00
Mvomero	73.68	26.32	0.00	0.00
Mbozi	17.74	30.65	51.61	0.00
Bariadi	38.98	40.68	18.64	1.69

Rukwa region. Morogoro and Shinyanga regions have lowest number of beneficiaries and input vouchers. Shinyanga region did not receive vouchers in 2008/2009. This is because cash crops were not targeted during the first year of NAIVS although in the subsequent years it was introduced. Increase in the number of beneficiaries and vouchers imply more fertilizer use and likely increase in productivity. However, the impact of these changes on productivity of target crops is a subject for another paper.

Targeting delivery

In Mvomero, 73.68% of households reported to have benefited from the subsidy for one year, and there was no household that reported to receive vouchers for three years (Table 4). In Mbozi and Sumbawanga, some smallholder farmers benefited from subsidy for up to three years. Even with some farmers receiving subsidy for three years, it was not provided consecutively. No farmers were found to benefit from subsidy for four years in the three districts. However, in Bariadi district, smallholder farmers received vouchers for up to four years. In Bariadi district, provision of subsidy through vouchers system was changed during 2011/2012 season by introducing contract farming. Subsidy through contract farming does not limit the number of years the smallholder farmers should benefit from the programme. Under contract farming, farmers borrow inputs from a ginnery company and are required to sell cotton to the company at harvest time. Deductions for the input costs are made from sales of cotton and the balance remitted to the farmers. This is achieved trough farmer groups. This implies that farmers who do not belong to groups do not benefit from the programme. Farmer groups are used for insurance purpose as social cohesion act as collateral. However, it is unclear how farmers will respond at harvesting time as the contract farming is still new.

Evaluative criteria and outcome

Eligibility criteria and farmers selection

Selected farmers were eligible to receive one bag of fertilizer (di ammonium phosphate) DAP (50 kg), one bag

Table 5. The likelihood of poor and rich farmers getting subsidy.

District	Response (%)			
	Sumbawanga	Mvomero	Mbozi	Bariadi
Selection in favour of relatively rich people	62.26	43.24	47.50	41.18
Selection in favour of relatively poor people	18.87	13.51	15.00	11.76
Equal chances of selection for poor and rich people	9.43	43.24	28.75	43.53
Poor chances of selection for rich people	0.00	0.00	0.00	1.18
Poor chances of selection for poor people	9.43	0.00	8.75	2.35
Total	100	100	100	100

Table 6. Voucher delivery time in the four districts (2008/2009 to 2011/2012).

Districts	Responses (%)	
	Late delivery	On - time delivery
Sumbawanga	30(91)	3(9)
Mvomero	32(88)	3(12)
Mbozi	61(95)	3(5)
Bariadi	50(79)	13(21)
Total	163(88)	22(12)

X^2 (3, N = 185) = 8.01, p = 0.05.

of urea (50 kg) and 10 kg of maize seeds or 15 kg of rice seeds. Cotton farmers were eligible for cotton seeds and one acre pack of pesticides. The number of vouchers that were distributed to districts and villages did not meet demand of eligible farmers. As a result vouchers were given to farmers alternatingly; farmers who received vouchers in one year did not get them the following year. Inadequate vouchers created corruption and social conflicts among politicians, village leaders and farmers. On the other hand, non beneficiaries started refusing to contribute in community development projects in the village. They claimed to be discriminated and argued that subsidy input beneficiaries should be responsible to pay back in terms of contribution to community development projects. Farmers in the list of beneficiaries were forced to accomplish pending contributions in community development projects before they were given the vouchers. This was associated with inadequate awareness of the eligibility criteria. It was noted that farmers did not graduate after three years, which is contrary to exiting strategy in smart subsidies. Smart subsidy requires farmers to be subsidized for three consecutive years and graduate from the programme. After graduation, farmers are expected to become self sufficient and able to support themselves. Results presented in Table 5 show the likelihood of poor and rich farmer's selection in the programme.

In Sumbawanga and Mbozi, selection favoured relatively rich people likely related to the targeting criteria for farmer's ability to contribute the cash top up. Similar findings have been reported in Malawi where subsidized fertilizer was often provided to wealthier households with community and political connections (Ricker-Gilbert, 2011).

Input subsidy delivery

The government prepares vouchers and deliver them to selected farmers through a series of of committees from the national to the village level. Farmers submit the vouchers to selected agro dealers to redeem the inputs. The voucher has the value the government contributes and should be approved at district level. Agro dealers submit vouchers to a selected bank, which has been contracted by government; in this case, the National Microfinance Bank (NMB) to redeem money. Distribution is supposed to follow the National voucher guidelines stipulating the procedures to be followed by every actor in the channel. For example; recruitment of village vouchers committee and eligible farmers required farmer's participation through village assembly. Also village assembly were to select names of agro dealers to enter competition at district level where qualified dealers were approved. Observations indicate that village committees existed in all villages were gender balanced.

Timely delivery

Results on whether or not vouchers were delivered on time reveals that; 88% of households did not receive the inputs on time significant at X^2 (3, N = 185) = 8.01, p = 0.05 (Table 6). Also, 68% of vouchers were delayed for more than seven weeks. In some situations where inputs were delayed, farmers had already planted maize using saved grains from the previous season. In such situation, they required only top dressing fertilizer, but instead they were forced to take the whole input package including basal fertilizer and asked to save the seeds and basal fertilizer for use in next cropping season. This was difficult as farmers resources are limited and have other priorities to allocate their money. Delayed inputs had consequences on continued reliance on poor quality

Table 7. Percent response on quality of distributed inputs.

Full name of village	Not quality inputs	Quality inputs
Chitete	1(5.9)	16(94.1)
Katete	0(0.0)	16(100.0)
Isangu	14(46.7)	16(53.3)
Itaka	8(22.2)	28(77.8)
Ibulyu	15(50.0)	15(50.0)
Nguliyati	13(38.2)	21(61.8)
Lusanga	1(6.3)	15(93.8)
Kinda	5(62.5)	3(37.5)

X^2 (7, N = 187) = 31.4, p = 0.00. Note: Number in the parenthesis is in percentages.

seeds and low adoption of production technologies. The farmer's decision of timely planting was interfered and in some situations late planting was practised, thereby increasing the probability of getting lower yields as they were totally dependent on rain fed agriculture. These findings imply failure to attain the NAIVS objective of timely delivery of inputs to farmers at reduced costs. Also, it is contrary to NAIVS guideline which requires vouchers to be distributed prior to planting season.

In some situations, special forms (paper certificates) were used to substitute vouchers to reduce the delayed time. However, the forms were useful when the MAFC informed Local Government Authorities to print the documents in advance. Even with these government efforts, about half of households who received the paper certificates were not able to purchase subsidized inputs as the certificates were rejected by agro-dealers (Malhotra, 2013). The implication is that, printed paper certificates were not the best option to overcome delayed vouchers. Also, a delay in disbursements of funds for used vouchers was noted leading to agro dealers default in loan repayment. The consequence was limited loan access from the National Microfinance Bank (NMB) to supply inputs in the following season, and failure of small agro dealers to remain in the input supply business. Smart subsidies such as NAIVS are aimed to promote private sector involvement in input business. Agro-dealers aim to generate profit from input business. Profit limiting environment is likely to encourage agro-dealers to exit from the input business lowering market competition which is not in favour of farmers.

Quality of distributed inputs

Results on whether or not the distributed inputs are of the right quality are significant X^2 (7, N = 187) = 31.4, p = 0.00 (Table 7). In Kinda, 62.5% of farmers reported poor quality of distributed maize seeds because they had low germination rate. Households in Isangu (46%), reported poor quality of distributed fertilizers. They complained of DAP being mixed with Minjingu which is relatively cheap; and table salt was sold as calcium ammonium nitrate (CAN). Despite distribution of inputs through ginneries under contract farming in Bariadi, poor quality inputs were also prevalent. In Ibulyu village, 50% of farming households reported poor quality of pesticides , and argued that cotton seeds did not germinate. Quality inputs have standard criteria according to manufacturer specifications. Quality fertilizer has ability to release desired nutrients in the applied site where quality seeds have higher germination, plant vigour and high productivity. Seed germination, plant vigour and physical characteristic of fertilizers such as texture were determining factors of input quality.

Efficacy of pesticide was the main quality factor. Use of quality seeds, fertilizers and pesticides are crucial in agriculture productivity. Poor quality of delivered inputs implies that farmers have to incur extra costs of inputs and labour for replanting. Application of poor quality inputs is also a moral hazard especially under unpredictable weather conditions. Farmers in the study area depend entirely on rain fed agriculture and missing the first rains or late planting/replanting is likely to cause low production or even crop failures.

Voucher committee selection

Leadership reputation was stipulated as a requirement of members of village voucher committee. Some village leaders recruited weak representatives in the village voucher committee to protect their dishonesty interests. Reports from key informants revealed that, in some situations farmers were involved in selection of committee members, but were not courageous to refuse the names suggested by their leaders. Farmers were not well informed about their role stipulated in the national voucher guideline. They felt that by rejecting the appointed candidates they would face problems in their community because they would be considered to have acted against the system. In addition, more than 73% of farmers did not know where to report problems associated with the programme. In the guideline there is

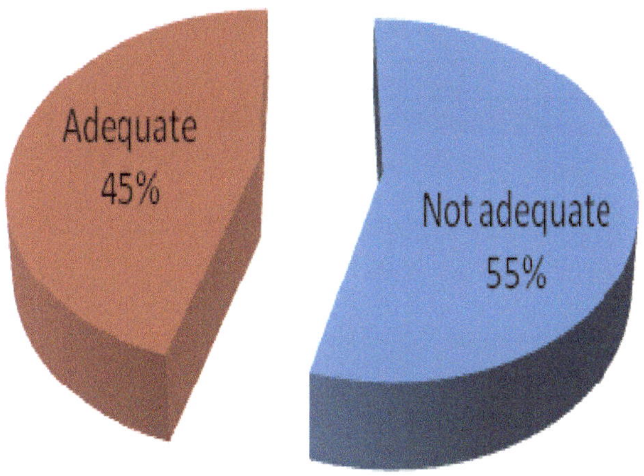

Figure 4. Availability of extension service under NAIVS.

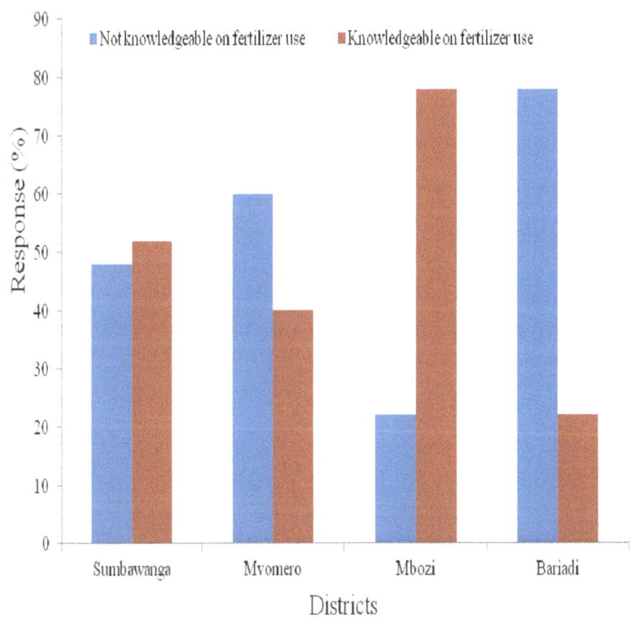

Figure 5. Household knowledge on chemical fertilizer use in percentage.

a provision of an opportunity to appeal to higher levels. Village complaints needs to be reported at ward level, ward to district and district complaints reported to region level, but farmers were not aware. The only option was to report problems to village leaders but in situations where village leaders were the source of problem, farmers were in dilemma.

Selection of agro dealers

Criteria for selection of agro dealers included possession of capital enough to run smoothly the service, short distance from village, experience in input business and possession of Taxpayer Identification number (TIN). The application procedure was to send application to village government where village assembly chose the names of preferred stockists. Stockists with business based in the village were given priority. The chosen stockists were screened at district level. Research investigations revealed violation of procedures for some districts. In such districts, selected agro dealers were not from the list suggested by the village assembly and had no input shop in the district. Chosen agro dealers had limited capital to be able to distribute subsidy inputs on time and throughout the season.

Nevertheless, key informants reported that selection of committees and agro dealers was interrupted by some politicians.

Extension and farmer's knowledge on input use

Extension service

Inadequate extension service under the voucher scheme was reported by 55% of farmers (Figure 4). Sixty five percent of respondents claimed that extension service was mainly based on verbal communication and only 20% mentioned use of leaflets. Key informants reported existence of demonstration plots in Bariadi contrary to other study locations. It was unlikely to deliver adequate message to farmers due to limited capacity of extension staff. Farmers in Bariadi were of the opinion that extension service provided by Techno-Serve could be better if assigned to the district council as they posses strong extension capacity. It was also noted that contract farming is more likely to improve the extension service revealed by use of demonstration plots.

Knowledge on chemical fertilizer use

Figure 5 presents variations in the number of farmers knowledgeable about level of chemical fertilizer application from one study location to another. Results show that 78, 52, 40 and 22% of farmers in Mbozi, Sumbawanga, Mvomero and Bariadi, respectively were knowledgeable to use chemical fertilizer. High knowledge of farmers on chemical fertilizers use in Mbozi and Sumbawanga is associated with earlier introduction of subsidy in these locations. Subsidies are known to stimulate the adoption of improved technologies (Lee, 2005). The long term objective of the government support of subsidy programme was to promote adoption and efficiency application of essential productivity enhancing inputs (URT, 2012). Farmers in these locations might have adopted the use of fertilizers and improved seeds associated with early introduction of subsidy in these locations. This is expected to increase agriculture productivity and food security.

Table 8. Percent response on quality of distributed inputs

Full name of village	Not quality inputs	Quality inputs
Chitete	1(5.9)	16(94.1)
Katete	0(0.0)	16(100.0)
Isangu	14(46.7)	16(53.3)
Itaka	8(22.2)	28(77.8)
Ibulyu	15(50.0)	15(50.0)
Nguliyati	13(38.2)	21(61.8)
Lusanga	1(6.3)	15(93.8)
Kinda	5(62.5)	3(37.5)

Household education and awareness on input subsidy

The level of awareness on input subsidy distribution slightly varied among educated and low educated household heads [X^2 (3, N = 300) = 7.7, p = 0.05]. Awareness of household heads with no formal education was 87.2%, compared to primary education 95.2% and secondary and above 100% (Table 8). High education exposure is associated with more ability to process relevant information and increases the chances of adoption of improved technologies (Morris et al., 1999, Kaliba et al., 2000).

Influence of socio-economic attributes in receipt of subsidy inputs

Some of the socio-economic characteristics such as age, gender and marital status were tested to find out whether or not they have relationship with vouchers received by household. A chi square statistic test results reveals no significant relationship between these variables and subsidy input received by household. However, when same variables were tested for awareness of subsidy distributed to farmers, only age of household head was significant X^2 (3, N = 300) = 8.32, p = 0.04. Household heads with age below 30 years were less aware than household heads with age 30 and above. Age has been reported to be used as indicator of farming experience. Such experience makes certain information search and cost easier (Mpogole and Kadigi, 2012). In addition, aged male households have been associated with strong networks and better connection to community and village leaders. These findings imply that, sometimes farmers were equally treated in the voucher scheme. However, women and disadvantaged groups were not given priority as stipulated in the guideline.

Monitoring the distribution of vouchers

Strong security was maintained in distribution of voucher from national level to district level. Observation fom focus group discussions and key informants revealed that after submitting the vouchers to villages, there was no security force to take care of vouchers. This was a risk to village leaders who were responsible for handling the vouchers. Furthermore, voucher committees were not paid anything despite workload involved in voucher distribution. This could be one of the sources of temptation to cheat when bribed by unfaithful stockists. Furthermore, there were limited funds and facilities for monitoring the programme. Also, the training of committee members on their role in the scheme was not adequate and some were overpowered by village leaders. Additionally, key informants reported political interference in the system in which politicians especially councillors have more power in subsidy inputs than technical staff. They demanded equal share of vouchers regardless of land attributes in order to impress their voters.

Follow-ups interviews from Prevention and Combating of Corruption Bureau (PCCB) caused annoyance to Local Government Authority workers and threatened their work security during voucher distribution.

Leakage and elite capture

Leakages were reported to happen in various ways in the NAIVS. Agro dealers colluded with farmers and village committee members in some locations to cheat. Farmers were paid 10,000 Tsh and village committee 20,000 Tsh to sign the voucher without receiving the inputs. This was also reported by the politician in one of the study location (Luhwago, 2011). Such situation benefited agro dealers than intended farmers and hampered the intended programme objectives. Other reported strategies used for stealing the vouchers were: inclusion of names of children and dead people in the list of beneficiaries. Unfaithful village executive officers (VEOs) and village committees colluded with agro dealers and forged vouchers using fake signatures as if farmers had received payments. Hiding subsidy inputs during distribution and selling them later at full price was also reported. These problems were reported in Mbozi and

Bariadi. Also, unfaithful agro dealers mixed inputs with poor quality /cheap material and sold them as good quality inputs. Existence of ineffective pesticides and poor quality seeds was associated with bureaucracy in the framework. Agro dealers were claimed to arrange with input manufacturers/stockists to provide substandard inputs.

CONCLUSION AND RECOMMENDATIONS

This study has assessed the operational framework of NAIVS in Tanzania. While in some locations the system operated well, challenges were encountered in other areas. The system reduced the cost of inputs to farmers. However, the top up contribution was still unaffordable by farmers. Distributions of inputs were done in lean season when farmers have no alternative source of income. In most cases the inputs were not distributed on time. Delayed input delivery was caused by lack of capital by agro dealers and long chain involved in the distribution of voucher. Bureaucracy existed in selection of agro dealers and independent monitoring and evaluation committee did not exist in the scheme. Generally, there were violations of guidelines in NAIVS framework. Therefore, we recommends establishment of credit institutions or system in the villages where farmers can obtain loans to purchase the inputs. Bureaucracy should be eliminated through provision of full mandates of agro dealer's selection to farmers. Ministry of Agriculture Food Security and Cooperatives should reimburse the vouchers in time to sustain agro dealers in input business. There should be established independent monitoring and evaluation system particularly at village level in order to control leakages and adulterated inputs. Programme awareness need to be raised to all farmers to enable them understand their rights and disciplinary measures to undertake for unfaithful leaders. Due to limited number of extension staff at village level, Ward Agricultural Resource Centres (WARC) should be strengthened by providing enough information regarding NAIVS.

Extension material such as leaflets, posters and magazines needs to be introduced into districts and villages to supplement the extension service. Public and private partnership collaboration in service delivery seems to be a promising option in extension service delivery which should be promoted. Generally, there is a need to establish and experiment a new short chain electronic vouchers delivery system to overcome the NAIVS inefficiencies.

REFERENCES

Baltzer KT, Hansen H (2011). Agricultural input subsidies in Sub-Saharan Africa. Ministry of Foreign Affairs of Denmark. Danida, pp. 8-36.

Boyd HW, Westfall R, Stasch SF (1981). Marketing Research Text and Cases, P. 813.

Chibwana C, Fisher M, Jumbe C, Masters W, Shively G (2010). Measuring the Impacts of Malawi's farm input subsidy program, P. 2.

Danning G, Kabambe P, Sanchez P (2009a). The African Green Revolution moves forward. Springer Science + Business Media B.V. and International Society for Plant Pathology 2009.

Danning G, Kabambe P, Sanchez P, Malik A, Flor R, Harawa R, Nkhoma P, Zamba C, Banda C, Magombo C, Kaeating M, Wangila J, Sachs J (2009b). Input subsidies to improve smallholder maize productivity in Malawi: Toward an African Green Revolution PLoS Biol. P. 7.

Dorward A, Chirwa E (2011). The Malawi agricultural input subsidy programme: 2005/06 to 2008/09. Int. J. Agric. Sustain. 9:232-247.

Druilhe Z, Barreiro-Hurlé J (2012). Fertilizer subsidies in sub-Saharan Africa. Food and Agriculture Organization of the United Nations, pp. 2.

Eboh EC, Ujah OC, Amaechina EC (2006). Do Government Fertiliser Subsidies Benefit Rural Poor Farmers in Nigeria? Making Sense out of Existing Data. PMMA Network Session Paper. Available online at: http://132.203, 59.

Kaliba AR, Verkuijl H, Mwangi W (2000). Factors affecting adoption of improved maize seeds and use of inorganic fertilizer for maize production in the intermediate and lowland zones of Tanzania. J. Agric. Appl. Econ. 32:35-48.

Lee DR (2005). Agricultural sustainability and technology adoption: Issues and policies for developing countries. Am. J. Agric. Econ. 87: 1325-1334.

Luhwago R (2011). How govt loses billions through the Farm Input Voucher System. The Guardian. Available at: http://ippmedia.com/frontend/index.php?l=31035

MAFC (2013). Budget Speeches [Online]. [Accessed Available online at: http://www.kilimo.go.tz/speeches/budget%20speeches/budget%20speeches.htm

Malhotra K (2013). National Agricultural Input Voucher Scheme (NAIVS 2009–2012), Tanzania: Opportunities for Improvement, P. 3.

Morris ML, Tripp R, Dankyi A (1999). Adoption and Impacts of Improved maize production technology. A case study of the Ghana Grains Development Project. Economic Programme paper, pp. 99-101.

Mpogole H, Kadigi RMJ (2012). Round potato (*Solanum tuberosum*) profitability and implications for variety selections in the Southern Highlands of Tanzania. J. Dev. Agric. Econ. 4:258-267.

Nagy JG, Edun O (2002). Assessment of Nigerian Government Fertilizer Policy and Suggested Alternative Market-Friendly Policies. IFDC, P. 9.

Pan L, Christiaensen L (2012). Who Is Vouching for the Input Voucher? Decentralized Targeting and Elite Capture in Tanzania World Development. 40:1619–1633.

Polskl MM, Ostrom E (1999). An Institutional Framework for Policy Analysis and Design. Indiana University, Department of Political Science.

Ricker-Gilbert J (2011). Household-level impacts of fertilizer subsidies in Malawi. Doctor of Philosophy, Michigan State University, pp. 19.

Ricker-Gilbert J, Jayne TS (2009). Do Fertilizer Subsidies Affect the Demand for Commercial Fertilizer? An Example from Malawi. Inter national Association of Agricultural Economists, Beijing.

Tiba Z (2010). The role of input subsidies: Operational guidelines on implementation. FAO Trade and Markets Division.

Truffer B, Rohracher H, Markard J (2009). The Analysis of Institutions in Technological Innovation Systems-A conceptual framework applied to biogas development in Austria. Copenhagen: Copenhagen Business School, P. 7.

URT (2012). Agricultural Public Expenditure Review (PER) for Inputs Support Programme, for FY 2012/13.

Wiggins S, Brooks J (2010). The use of input subsidies in developing countries. Paris: OECD Headquaters, P. 13.

World Bank (2012). Tanzania - Accelerated Food Security Project.

Yawson DO, Armah F, Afrifa, EKA, Dadzie SKN (2010). Ghana's Fertilizer Subsidy Policy: Early Field Lessons From Farmers in the Central Region. J. Sustain. Dev. Afr. P.12.

4

Econometric analysis of suitability and marginal value productivity of farmlands for cassava production in Imo State, Nigeria

Ehirim, N. C.[1], R. A. Okere[2], S. U. O. Onyeagocha[1], M. A. Y Rahji[3], T. T. Awoyemi[3], B. I. Oluwatayo[3] and K. K. Salman[3]

[1]Department of Agricultural Economics, Federal University of Technology Owerri, Imo State, Nigeria.
[2]Nigerian Institute for Oil Palm Research, Benin City, Edo State, Nigeria.
[3]Department of Agricultural Economics, University of Ibadan, Ibadan Oyo State, Nigeria.

Indiscriminate allocation of farm farmlands beyond efficient level affect farmers return, hence suitability and standard value of lands for cassava production for optimal use were investigated. Data were obtained using a multi-stage sampling technique, from a sample of 203 cassava farmers in the state and were analysed using descriptive statistics, land productivity ratio and econonometric tools. Most farmers are female of 55.7 6.8 years of age and post primary education status. Nutrient ammendment is at the rate of 442.36 ± 102.73 Kg/Ha and at mean cost of N36401.77 ± 28575.84/Ha. The performance rate of 0.96 tons/person/Ha was obtained. The suitability index ranges from 0.139 to 0.908 with 46.3% cultivating on non-suitable land of 0.0-0.339 while only 20.7% cultivated on suitable lands of above 0.723 suitability index with a smaller mean area of 1.48Ha. The performance rate across these classes of land ranges from 0.44 tons/Ha/person to 2.11 tons/Ha/person for non-suitable land and suitable lands respectively. Suitable lands had the highest land productivity of 5.71 while moderate and non suitable lands had only 4.00 and 3.72 respectively, hence about 53.4% and only 7.5% increase in production is achieved as a piece of land is improved from non-suitable and moderately suiatble lands respectively to suitable land in the area. The MVP ranges from N302429.76/Ha for non-suitable lands to N718535.2/Ha for suitable lands. The study noted that suitable lands have higher opportunity cost than others for cassava production, hence recommentds that opportunity cost of land must be based on higher MVP for optimal use.

Key words: Farmland, suitability, marginal, value, productivity, cassava.

INTRODUCTION

Land is a portion of the earth surface that houses the biosphere, soil with its geological properties (which include the hydrological portion) and the atmosphere (FAO, 1976). Each of these portions is subject to some natural and human factors that have advertently, accounted for its development and value. The framework of farmland comprises of economic and social attribute that enhances its value (Anyigo, 1982). Farmland has a derived demand and its demand is for the minerals and nutrient it has for crop production (Olayide and Heady, 1982). Hence, its value (opprotunity cost) is therefore, the amount (of money, goods or services) that is considered to be a fair equivalent for its use and or what the land can produce. This means that land must be suitable having possessed a high potential for maximum output. Hence suitable farmland produces a relatively increased quantity of crop per unit area than other parcels of land within the same area.

Suitable farmlands are linked with increased opportunity cost. Such lands are productive, hence attract more opportunity cost than other areas (Okere, 2013). Land productivity is the ratio of output per unit piece of land used in crop production (Olayide and Heady, 1982), hence a suitable piece of land is at a high productivity potential. Land with high marginal value products (MVP) must have equivalent high marginal factor cost (MFC) to attract optimum allocation (Olayinde and Heady, 1982). Al-Kaisi (2012) noted that soil conservation practices can play a significant role in sustaining soil quality and suitability even in adverse himan activities and climate change. This implies that with good soil conservation and management practices, productivity of agricultural land will be sustainably high, thus making such land more valuable for crop production. Agricultural Lands are valued based on their suitability for crop production and its productivity.

Agricultural lands are no doubt scarce with high opportunity cost when project and infrastructural development are present. However, the value is to a greater extent, determined by the MVP of the farmland (Bassey, 2008). Although climate change and environmental degredation have been fingered for the low productive potentials of most arable land used for cassava production in Nigeria, the mining of sand and graver deposits (of solid and liquid minerals) as well as excavation of topsoils for urban development, have left extensive tracts of exposed subsoils with adverse soil chemical and physical properties that do not support plant growth (Hornick and Parr, 2010). Such degraded and marginal soils properties or destructions result to infertile (low in organic matter), often acidic and are subject to severe erosion and surface runoff.

Soil ammendment and conservation practices increase the suitability of soil for crop production and consequently the value system (Al-Kaisi, 2012). Research has shown that proper soil management practices such as liming and timely use of organic amendments such as animal manures and sewage sludge compost on lands can restore increased land productivity (Oyekale, 2008; Hornick and Parr, 2010; Al-Kaisi, 2012). Hence, the relative suitability of land is mainly dependent on the available soil management practices, which farmers adop t to gradually return improvised farm lands to a suitable land for crop production. This presupposes that lands can be classified into different value systems based on its degree of suitability to arable crop production (Oyekale, 2008).

Cassava adapt to marginal soils though, its productivity varies within soils in the same location and between locations in geographical space depending on the constituent minerals and management practices. Projects Coordinating Unit (2003), noted that though Enugu and Imo dominated the production of cassava in the South East with an output of 0.56 tonnes/per person in 2002 the output is not commensurate with the area of land allocated to it and farmers returns has not improved their consumption and welfare need. As a food security component, its ability to close food gap is declning. The dietry need of man with his livestock and industrial needs for cassava products have intensified (without careful management), the use of farmland, thus making lands vulnerable to farming farming risk and low output that fall short of its demand (FDALR, 1982; Ehirim et al., 2006).

To bridge this rising food gap, efforts need to be made by farmers to improve land productivity so that output can be raised to meet the food consumption needs (Adetunji and Adeyemo, 2012). Low level of farm size, technical and economic inefficiency of food and primitive technology may be a draw back to the effort to achieve the progress in food production (Adetunji and Adeyemo, 2012). The arguement that increased crop production requires increased use of farm inputs, which of cause does not exclude expanded use of land, may not translate to production efficiency (Olayide and Heady, 1982). Therefore, an optimal allocation of land base on its MVP placed cassava enterprise at a ventage position for maximum profit and economic sustainability. This study investigated the the productivity of different suitabilty levels of land for cassava production and the expected opportunity cost so that allocation of resources especially for land will generate optimum returns for cassava farmers in Imo state.

MATERIALS AND METHODS

The study used a multi-stage sampling technique to draw 203 cassava farmers for the study. First was a purposive selection of the three agricultural zones for proper representation of the state. Second was a purposive selection of three Local Government Areas (LGA) with evidence of erosion, high and moderate topography, and traditional farming systems such as bush burning, continuous cropping and excessive grazing form each zone, thus giving a total of nine (9) LGA's. The LGA's selected were Ngor Okpala, Owerri North and Owerri West LGA's from Owerri zone while Orsu, Isu and Nwangele LGAs were from Orlu zone and Isiala Mbano, Obowu and Okigwe North from Okigwe Zone. Third is random selection of cassava farmers from the list of cassava farmers with ADP's in each of the selected LGA. The study sampled about 316 (70%) of the total registered farmers from these LGAs and only 203 of the responses were found useful for data analysis.

Data on socio-economic features of the farmers, the type, quatity and prices of farm inputs especially the opportunity cost of land as well as quantity and prices of cassava output and the various land management and soil conservation practices used were collected. Data were analysed using both descriptive, partial productivity of land and econometric tools.

Land suitability index was used to classify land into suitable, moderately suitable and non suitable for cassava production. The suitability index was obtained from the number of 'yes' response (as it affect farmers) on the improved land managementand and or soil conservation practices as well as "no" responses to bad management farm and or soil conservation practices on his land. The practices are application of fertilizers or organic manure to the soil, the use of lime or wood ash for soil acidity, the use of mulch materials to protect the soil, use of bush fallow or shifting cultivation, the use of disease resistant varieties, construction of drainages and trench to remove excess or water logging, crop rotation system in case of mixed crop farming as well as ensuring adequate spacing distance and early planting. Others are bad management practices of the farmers in which a "no" response are expected. They include bush burning, continuous cropping, excessive grazing, making ridges along the slopes, deforestation and the use of heavy machines for tilling the soil. A total of 8 yes response and 6 no response gave a total of 14 responses that suitabily sustain the marginal value productivity of the farmland (FAO, 1991; Oyekale, 2008).

Suitability index is therefore, the ratio of the actual aggregate number good management practices and bad practices avoided to the total score. This is expressed as;

$$SI = \frac{V}{N} \quad \text{and} \quad 0 \leq SI \leq 1 \qquad (1)$$

Where SI = Suitability index, V = Actual number aggregate good land management practices and bad practices avoided by an ith farmer in the area and N = The Total number of both good land management activities that farmers responded yes and the bad practices avoidable as used in this study. As the SI approaches 1, then the land is very suitable for arable crop production but if it approaches 0, then it is non-suitable inbetween these extreems lies moderately suiatble land for cassava production. This is statistically classified using normal distribution approach adopted by Olowu and Oladeji (2004). The estimated mean and the standard deviation of the distribution were used to classify the land into three classes; Suitable, moderately suitable and non-suitable.

$$SC = \bar{SI} \pm SD \qquad (2)$$

Where; SC = Suitability class, $\bar{SI}$ = Mean suitability index and SD= Standard deviation

Partial productivity of land is the ratio of total output of an ith class of land to a unit area of that land cultivated by cassava. This is expressed as:

$$A = \frac{Y}{L} \qquad (3)$$

Where, A = Productivity of land; Y = Output of cassava tubers from an ith class of Land in kg, and L = Area of an ith class of land in hecatres.

The value placed on each class of farmland is based on the MVP of such farmland in the area. The suitability potential of farmlands determines its MVP, hence allocation of farmland for cassava production is efficienct where the ratio of MVP and the MFC is equal to unity.

$$Optimum\ Allocation\ of\ Land = \frac{Marginal\ Value\ productivity}{Marginal\ factor\ Cost} = 1 \qquad (4)$$

$$\text{Marginal Value Productivity (MVP)} = MPP_L\ X\ P_Y \qquad (5)$$

Where MPP_L is the marginal physical product of cassava from an ith class of land and P_Y is the unit market price of cassava produced in the area. The marginal physical product of cassava from an ith class of land is simply an additional unit of cassava from an additional unit ith class of land (Olayide and Heady, 1982). This is obtained as a partial derivative from a linear production function of cassava from an ith class of land in the area. The model is implicitly expressed as;

$$Q_c = f(Labour + Farmsize + Pl_{mat} + capital + \varepsilon) \qquad (6)$$

Where; Q_c = Output of cassava in kg, Pl_{mat} = Plant Materials in naira, Labour = man-days, Farm size = hectares and Capital = sum of depreciation of fixed inputs, rent and interest in naira. ε = the error term.

RESULTS AND DISCUSSION

Socio-economic features of cassava farmers in Imo State

Table 1 shows that majority (40.4%) of cassava farmers falls within 50 to 59 years while only few (3.0%) of them are between 30 to 39 years. The mean age of the cassava farmers is 55.7 years. This implies that cassava farmers in the State are relatively old. This finding is supported by Alfred (2001) that farming is now in the hands of old farmers whose risk absorption, adoption of innovation and productive effort per unit of labour may be declining. Again farming may go extinct with older men dominating the occupation. The result also showed that majority (40.9%) of cassava farmers in the State had post primary education and 17.7% of them had up to tertiary education.

However, only 9.4% of the farmers had relatively no formal education in the area and the mean formal education attainment is 9.5 years. This implies that at average, the farmers most have acquired post primary education. Williams (1984) noted that secondary education can equip farmers with some managerial skills for agri-business and may help in understanding innovations. The result showed mean farming experience of 26.4 years with 48.8% of the cassava farmers having between 11 and 20 years. Only less than 1% has planted cassava in less than 10years. This showed that cassava farming is an old enterprise and can increase mastering of different technologies. This finding is consistent with Ehirim et.al. (2006) who observed that changes is expected over time due to high farming experiences. It could be deduced from the result that extension contact to cassava farmers is small with a mean extension visit 8 visits per farmer in a planting season. Majority (75.9%) of cassava farmers have extension visit of only less than 10 times and 10.8% of them had less than 15 times visit per farmer in a farming season. The result showed that 19.2% of the cassava farmers have less than 4 members per household and 34.5% of them have less than 8 members per household. The mean household size per farmer is about 7 persons per household. This shows that

Table 1. Socio-economic characteristics of cassava farmers in Imo State.

Variable	Frequency	Percentage	Mean ± SD
Age			
30-39	6	3.0	
40-49	43	21.2	
50-59	82	40.4	
60-69	65	32.0	
70-79	6	3.0	55.7 ± 6.8
≥ 80	1	0.5	
Total	203	100	
Education			
0	19	9.4	
Less than 2 Years	8	3.9	
Not more than 6 years	57	28.1	
7 – 12 Years	83	40.9	9.5 ± 2.3
13 – 18 Years	36	17.7	
Total	203	100.0	
Experience			
Less than 10	2	1.0	
11-20	99	48.8	
21-30	31	15.3	
31-40	34	16.7	
41-50	24	11.8	26.4 ± 5.8
≥ 51	13	6.4	
Total	203	100.0	
Extension contact			
Less than 5	25	12.3	
5-10	154	75.9	
11-15	22	10.8	
≥ 16	2	1.0	8.0 ± 2.0
Total	203	100.0	
Household size			
Less than 4	39	19.2	
4-8	70	34.5	
9-12	62	30.5	
≥ 13	32	15.8	7.0 ± 3.0
Total	203	100.0	
Gender			
Male	95	46.8	
Female	108	53.2	
Total	203	100.0	
Marital status			
Single	72	35.5	
Married	81	39.9	
Divorced	35	17.2	
Widowed	15	7.4	
Separated	0	0.0	
Total	203	100.0	

Source: Field Survey, 2010.

Table 2. Descriptive statistics of input and output of cassava production in imo state.

Variable	Units	Mean	Standard deviation
Farm size	Ha	1.62	1.38
Rent	₦	9459.36	6016.25
Labour	Manday	25.00	9.81
Wage		15957.88	14470.42
Plant material	₦	13759.36	1397.50
Cost of improving land	₦	36401.77	28575.84
Soil nutrients	₦	442.36	1002.73
Depreciation	Kg/Ha	1257.07	1246.63
Output	₦	9584.19	2016.56

Source : Field Survey, 2010.

cassava farmers in the area have a relatively large household size to supply the family labour neede for cassava production in the area. This finding is supported by Nweke et al. (2002) about the significant population of cassava growers in Nigeria. The enterprise is female dominated with 53.2% of them as females and 46.8% as males. This finding is supported by the works of Ugwoke et al. (2004) who observed that agricultural activities are female dominated in Imo State.

Decriptive statistics of inpus and output of cassava in Imo State

The reuslt in Table 2 showed a mean area of land cultivated is 1.62 ± 1.38Ha. This implies that cassava production is still within small scale production as a relatively small fragment of land is allocated to its production in the area. A mean labour size of 25 ± 9.81 man-days and a mean wage of ₦15957.88 ± ₦14470.42 per labour per hectare is spent per farmer. This finding implies that the quantity of labour allocated to cassava production in the area is very small as a farmer can hardly make more than one visit to his farm in every 2 weeks within a cropping season. The mean depreciation of all fixed inputs used in the production of cassava in the area is ₦1257.07 ± ₦1246.63. Soil nutrient is applied at the rate of 442.36 ± 1002.73 kg/ha. The wide standard deviation for soil nutrient could be as a result of extensive application of the nutrient in the areas where arable crop lands are not suitable. The mean cost of improving land (which include construction of irrigation and drainage facilities etc) for cassava production is ₦36401.77 + 28575.84/Ha is very high. There is a slight increase in rate of performance of 0.56 tonn/person in 2002 (PCU, 2003), to 0.96 tons/person in the area during the study.

Land suitability, productivity and mariginal value productivity for cassava production in Imo State

The suitability index ranges from 0.139 to 0.908. Majority (46.3%) of the farmers cultivated on non-suitable lands with suitability limit of between 0 to 0.339 while 33.0% of them cultivated cassava on moderately suitable lands of 0.340 to 0.722 (Table 3). Only a few of them (20.7) had cultivated their cassava on suitable lands of between 0.723 to 1.00 suitability index. Non Suitable land is relatively larger in area of about 2.08 ha, it has a relatively smaller output performance rate of 0.44 tons/Ha/person than suitable land that has about 28.8% lower land area but demonstarted a high performance rate of 2.11 tons/ha/person. The increased performance could be due to suitability of land cassava production. This finding is consistent with Oyekale (2008) that improved land use system can ensure a high performance rate than degraded lands.

Similarly, suitable land had the highest land productivity of 5.71. This is greater than the land productivity of moderatel and non suitable lands of 4.00 and 3.72 respectively. There is about 53.4% and only 7.5% increase in production as a piece of land is improved from non-suitable and moderately suiatble lands respectively, to suitable land in the area. This shows that non-suitable lands have higher potential productivity with intensified sustainable land management practices. This will make the non-suitable lands most suitable for crop production in the area. This can be achived by applying suitable soil ammendments and soil conservation practices.

In a similar way, the marginall physical productivity estimates from cassava production function showed the changes in the quantity of cassava produced as farmsize increased by 1 unit. The estimated model for the three different classes of land is shown in Table 4. It could be deduced from the result that an increase in farmsize by 1 hectare will increase output of cassava tubers by 2351.42 kg in non-suitable farmlands, 2087.45 kg in moderately suitable farmlands and 4959.52 kg in suitable farmlands.

Again, the MVP of land in Imo State is very high for all classes of farmland. The value ranges from ₦302429.76 in non-suitable land to ₦718535.2 in suitable lands. This implies that areas with suitable farmlands must attract

Table 3. Land suitability classes and rate of performance in cassava production in Imo State.

Variable	Non suitable land	Marginal suitable land	Suitable land
Suitability index class limit	0 to 0.339	0.340 to 0.722	0.723 to 1.000
Frequency	94	67	42
Relative frequency (%)	46.0	33.0	20.7
Area cultivated mean (std error) (ha)	2.08 (0.92)	1.32 (0.74)	1.48 (0.92)
Mean output (std error)	9327.78 (6087.89)	10365.31 (7083.60)	31191.95 11087.89)
Rate of performance (tons/ha)	0.444	0.79	2.11
Mean Land productivity (std error)	3.72 (2.32)	4.00 (2.47)	5.71(3.22)
Percentage change in productivity	-	7.5	53.5
Marginal productivity of land (std error) (kg)	2351.42 (650.70)***	2087.45 (823.03)***	4959.52 (2000.03)***
Unit price of cassava	₦144.88/kg	₦144.88/kg	₦144.88/kg
Value marginal product of land	₦302429.76/ha	₦340673.73/ha	₦718535.20/ha

The mean suitability is 0.5304 and standard deviation is 0.1913); Source: Field Survey, 2010.

Table 4. Linear production model showing the marginal productivity of the various classes of lands in Imo State.

Variable	Suitable lands		Moderately suitable lands		Non-suitable lands	
	Co-Efficient	t-Value	Co-efficient	t-Value	Co-efficient	t-Value
Constant	-3972.52	1.011	5176.54**	2.352	13380.83***	7.087
Labour	269.70**	2.376	-35.76	0.46	-09.18**	2.390
Farm size	4959.52**	2.48	2087.45***	2.536	2351.42***	3.614
Plant Materials	30.51	1.35	45.92	1.140	8.32	0.389
Capital	-0.974	0.355	0.516	1.175	-0.75	0.749
R2	0.505		0.437		0.611	
Adj R^2	0.451		0.304		0.483	
F-value	9.420***		12.275***		36.13***	
No. of observation	42		67		94	

Source: Field Survey, 2010.

higher opportunity cost than others. Again, there is only a marginal difference in MVP of Moderately suitable land and non suitable land. This could be due the slight differences in their marginal productivities. The study suggests an equal opportunity cost with MVP on suitable farmlands for efficient resource allocation in the area. It is therefore suggested that land use intensification through soil ammendements and conservative practices will make no only make non-suitable lands sustainable but increases its value and returns.

CONCLUSION AND RECOMMENDATION

There is a strong evidence that land allocation for agriculture does not follow its productivity. Standard value of land are estimated from the MVP, hence the comparism with the opportunity cost for oprimal land use system. The study disintegaterd arable farmlands in Imo State based on their suitability for cassava production, hence estimated the MVP for each so as to establish a standard value for their optimal allocation. The study revealed that productivity follows suitability level and to increase the suitability farmers need to employ soil conservative and good management practices as listed in the study to increase the MVP. The study recommends that opprotunity cost of different suitability level of farmlands should based on their corresponding MVP for oprimal allocation.

REFERENCES

Adetunji MO, Adeyemo KE (2012). Economic Efficiency of Pig Production inOyo State, Nigeria: A Stochastic Production Frontier Approach. Am. J. Exp. Agric. 2(3):382-394. SCIENCEDOMAIN international. www.sciencedomain.org

Alfred SDY (2001) Socio-economic and Psychological Factors Affecting Adoption of Agricultural Innovation by Farm Household in Kogi State. Unpublished Ph.D. Thesis, Federal University of Technology. Akure. P. 58.

Al-Kaisi M (2012). Conservation Practices Saves Soils and Sustains Agriculture.*Integrated Crop* Management News Department of Agronomy, Iowa State University, USA.

Anyigo N (1982). The need for Land Resource services in Agricultural Production in Anambra State and Nigeria.In Proceedings of the First National Seminar on Agricultural Land Resources,Kaduna, Federal

Department of Agricultural Land Resources. P. 225.
Bassey EE (2008). The Effect of Land tenure on Natural Resource conservation in Nigerian Rainforest Ecosystem. www.environsecurity.org/espa/PDF/ESA-DRC.
Ehirim NC, Odurukwe SN, Ajaero, J, Emenyonu CA (2006). Socio-economic Analysis of Farming Risk in Cassava Production in Owerri, Nigeria. Int. J. Natural.Appl. Sci. 2(2):89-95.
F.A.O (1991). Nigerian Land Resources management study. Main Report and Annex 1. Rome FAO 1:17-24.
FAO (1976); A framework for Land Evaluation. Report on the Agro-ecological zones. Soils Bull. Fao Rome 32:72.
Federal Department of Agricultural Land Resources (FDALR) (1982). The Need for Land Resource Services to State. In proceedings of the First National Seminar on Agricultural Land Resources, Kaduna, Federal Department of Agricultural Land Resources. P. 273.
Hornick SB, Parr JF (2010). Restoring the productivity of marginal soils with organic amendments Ecological Agriculture Projects, McGill University (Macdonald Campus) Ste-Anne-de-Bellevue, QC, H9X 3V9 Canada Am. J. Alternat. Agric., 2(2): 64-68.
Nweke FI, Spender DSC, Lynam J (2002). The Cassava Transformation: Africa's Best Kept Secret. Michigan State University Press, East Lansing, Michigan, USA.
Olayide SO, Heady EO (1982); Introduction to Agricultural Production Economics. Department of Agricultural Economics University of Ibadan. University Press Ibadan. pp. 9-247.
Olowu AO, Oladeji JO (2004). Research Method in Agricultural Extension. Agricultural Extension Society of Nigeria (AESON). Agricultural and Rural Management Training Institute (ARMTI) Ilorin.
Oyekale SA (2008). Suitability of land, Journal of sustainable development. Department of Agricultural Economics, University of Ibadan.
Project Coordinating Unit (PCU) (2003). Crop Area Yield Survey. Abuja, Nigeria: Federal Ministry of Agriculture and Rural Development: Project Coordinating Unit.
Ugwoke FO, Odurukwe SN, Oguamah .C (2004). Gender Roles and Impacts of Oil Palm Commercialization on Intra-Household Dynamics in Imo State, Nigeria. Int. J. Agric. Rural Develop. 1(9);12-18.
Williams SKT (1984). Identification of Local Leaders. In A Manual of Agricultural Extension Workers in Nigeria. Les Shyraden, Ibadan, Nigeria. pp. 41-45.

5

The role of the rural labor market in reducing poverty in Western Ethiopia

Amsalu Bedemo[1], Kindie Getnet[2], Belay Kassa[3] and S. P. R. Chaurasia[3]

[1]Department of Economics, Faculty of Business and Economics, Assosa University, P. O. Box 018, Assosa, Ethiopia.
[2]International Water Management Institute (IWMI), P. O. Box, 5689, Addis Ababa, Ethiopia.
[3]School of Agricultural Economics and Agribusiness, Haramaya University, P. O. Box 138, Ethiopia.

This study examines the role of the rural labor market in reducing poverty and improving the well-being of smallholder farmers in rural Ethiopia. Propensity score matching technique is used to estimate the effect of labor market participation on poverty, consumption expenditure and income of smallholder farmers. The overall result indicates that the rural labor market contributes significantly to income growth, consumption expenditure and poverty reduction among smallholder farmers. Particularly, participation in off-farm wage activity has a positive and significant effect on household consumption expenditure and income but a negative and significant effect on the likelihood of a household being poor.

Key words: Poverty, rural, labor market, household, smallholder, propensity score.

....

INTRODUCTION

The rural poverty and the living conditions of the rural people in developing countries are highly heterogeneous problems. The problems are much sever and diverse in rural Africa, as a result of which farm households in such countries adopt different livelihood strategies (David, 2010). Although, many different efforts were made by governments of developing countries to fight rural poverty, the heterogeneous nature of the problem along with diversified livelihood strategies adopted by households made the efforts more difficult; and therefore poverty reduction remained to be the major policy challenge facing almost all countries in the developing world. In fact, poverty reduction requires that individuals be engaged in productive employments and economic activities that could help them generate adequate income to secure better living; and that development endeavors need to be targeted to the sectors where most of the poor employed and live (Ellis, 2001; Haggblade et al., 2010; Bernardin, 2012). Obviously, in the context of developing countries, the appropriate area is the rural and agricultural sector. Because three out of every four poor people in developing countries live in rural areas; most of them depend directly or indirectly on agriculture for their livelihoods and daily consumption (World Bank, 2008). Thus, agriculture remains to be the main source of livelihood for the majority of households in developing countries and thus expected to make significant contribution to poverty reduction efforts. However, the potential of the sector to contribute to poverty reduction efforts depend to a large extent on broad-based productivity growth. But such growth require improving the asset position of the rural poor, creating an access to technological innovation, making smallholder farming more competitive and sustainable, diversifying income sources towards the labor market and the rural non-farm economy (David, 2010), which could of course not be actually realized in sub-Sahar African countries.

The importance of agriculture to poverty reduction efforts were clearly demonstrated during the economic transformation of Asian countries where rapid growth of productivity in the farm sector helped drive this process (Awuor, 2007). However, this is not the case in most parts of sub-Saharan Africa, where farm households failed to achieve rapid growth in agricultural productivity (Jayne et al., 2010; Kwadwo and Samson, 2012). Therefore, it is evidenced that agriculture on its own is unable to provide sufficient means of survival and the escape out of poverty for the majority of poor households in Africa (Awuor, 2007; Emmanuel, 2011). As a result, households in rural areas participate in multiple economic activities and diversify income sources to minimize the effect of low farm income.

In view of this and as in the case of most African countries, agriculture in Ethiopia is dominated by smallholder farming. In such farming system with low agricultural productivity the engagement of farmers solely in to agriculture may not be adequate to successfully fight poverty because of poor access of smallholder farmers to key agricultural inputs. Thus, rural households look for additional employment opportunities to supplement subsistence farming. One such opportunity available for most smallholder farmers in rural western Ethiopia is participation in off-farm labor market. Even though there are vast literatures showing the contribution of off-farm activities to rural households in developing countries of Asia and Latin America (Verner, 2006; Otusu and Yamano, 2006; Micevska and Rahut, 2008), there is still little empirical evidence on the role of the rural labor market in reducing poverty and improving household welfare in rural Africa, particularly in Ethiopia. Most of the previous studies in Ethiopia had been largely limited to analyzing the determinants of participation in off-farm work, with no evidences on the importance of participation on the welfare of households. Generally, empirical studies that analyzed the role of off-farm labor markets on the farm household income and poverty status are scarce and they are almost absent in case of Ethiopia.

Therefore, the objective of this study is to empirically examine the role of off-farm labor markets in reducing poverty and improving welfare of smallholder farmers in rural Ethiopia. To this end, the effects of labor market participation on rural poverty, household income and consumption expenditure were examined, separately for participation in off-farm wage and self-employment and overall off-farm employment activities. Considering the self-selection nature of participation in the labor market, propensity score matching technique was employed to account for selection bias that may arise when participation is not randomly assigned.

Labor markets and rural poverty: Theoretical and empirical overview

Poverty and labor markets are strongly related in that earnings from participation in the labor market are among the main sources of income for participants. The implication of this is that the likelihood of participation in the labor market and the ability to earn income from such activities could be considered to be important in affecting the earning potential and level of poverty. However, this ability and the amount of income earned from such participation may depend to a large extent on the functioning of labor markets, the nature of off-farm activities that the poor engage in, and labor protections the markets accord (Barbara et al., 2012).

The rural labor market consists of employment opportunities both in the farm and off-farm activities. The farm labor market may be of large-scale mechanized and large family farms that depend heavily on hired farm labor and small-scale sub-sector characterized by smallholder farming where labor is mostly obtained from family sources with limited use of hired workers in varying proportions. The labor market in the later case is usually characterized by hiring farm labor on a casual, day-to-day basis (Bardhan and Udry, 1999; Mazumdar, 1989). But until recently, the academic and institutional literature on the labor market in rural Africa revealed a picture in which rural labor markets are either absent or very thin reflecting the situation that they are relatively neglected areas in research (Bardhan and Udry, 1999; Leavy and White, 2003). As pointed out by Carlos (2010), despite the relative neglect, the labor market in rural Africa is alive and operational; even though it is imperfect and fails to operate efficiently.

Recently, off-farm labor markets have become important components of livelihood strategies among rural households in most developing countries. A number of studies have documented substantial contribution of such markets to household income, consumption expenditure and level of poverty. Some of these empirical studies are discussed subsequently. Sosina et al. (2012) explored whether non-farm employment leads to higher consumption expenditure growth in Ethiopia. Their findings indicated that the household consumption expenditure growth is positively correlated with the initial share of non-farm income and that the growth elasticity of non-farm income share is higher for wealthier households. Similarly, in Ghana, Victor and Awudu (2009) investigated the impact of non-farm employment on farm household income and way out of poverty using propensity score matching method. Their finding indicated that non-farm employment had a positive and robust effect on farm household income and consumption expenditure but a negative and significant effect on the likelihood of being poor.

In rural Argentina, Verner (2006) conducted empirical study on the rural labor market and its income generation ability. His findings revealed that the vast majority of the rural employees were engaged in the off-farm activity. Similarly in Asia, Otusu and Yamano (2006) examined the role of the rural labor market in long-term process of poverty reduction in comparison with the current situation in East Africa. Their findings indicated that the reliance on agricultural labor market alone would not reduce poverty to a significant extent in view of the declining share of agricultural wage income in Asia and its negligibly low level in East Africa.

Finally, in Ethiopian, Mulat et al. (2006) has documented that employment and labor market variables had a significant impact on poverty. Since employment was identified as one of the critical avenues for poverty reduction, it is important to examine the structure of the labor market to identify the areas where the poor or vulnerable groups are concentrated for intervention in reducing poverty.

MATERIALS AND METHODS

The data set and situation of off-farm labor market in the study area

Description of data set

The data set used in this study was obtained from survey of 324 sample households in rural western Ethiopia. The sample households were selected from three districts of western Ethiopia (namely Guto Gida, Gida Ayana and Jima Arjo). They were selected purposively based on their diversity in terms of access to off-farm work, experience and exposure to labor market participation and variations in the nature and extent of participation. Moreover, they represent broad climatic condition reflecting high land and low land area, variations in markets and socio-economic infrastructure. Sample households were selected randomly and proportionately. They were interviewed using structured questionnaires that require short recall period. The data collection process took almost one year (June, 2010 to April, 2011) and is conducted in three rounds following main agricultural seasons in the study area. The first round representing ploughing and weeding seasons (May, 2010 to October, 2010), the second representing harvesting and threshing seasons (November, 2010 to February, 2011) and the final round for off-agricultural season (February, 2011 to April, 2011). Round surveys were used in order to capture variations in household time allocation and prices that change following agricultural seasons. But the data representing a variable in each of the three surveys were summed to arrive at annual figures.

The off-farm labor markets in the study area[1]

Farm households in the study area participated in different types of off-farm activities that include both wage employment and self-employment. About 73.5% of sample households reported that they participated in off-farm activities (both in wage employment and self-employment) out of which 77% were participants in wage employment and the remaining in off-farm self-employment. As the data indicates, out of the total sample size, about 52.2% reported that they participated in off-farm wage employment and 21.3% in off-farm self-employment activity. The fact that farmers participated more in wage employment during busy agricultural season but in off-farm self-employment activity during slack agricultural season reflects the importance of agricultural as main source of wage employment in the study area.

The most important types of wage employment activities in terms of participation were causal agricultural employment (39.4%) followed by employment in government sector (20.2%), unskilled wage worker (16%) and private sector employment(14.7 percent). Similarly, there are different types of non-farm self-employment activities in the study area. Among the major ones are production and sale of local food and drink (28.9%), trade in food grain, manufactured goods, livestock and livestock products (24.6%), collecting and selling firewood, water, grass, straw and charcoal (14.4%), handicraft, including weaving, making and selling equipment and pottery (13.4%) and others.

The two major reasons for participating in off-farm activities as identified by respondents were limited farm income to support livelihood and inadequate land to cultivate. Most sample farm households (about 76%) reported that they were engaged in off-farm activities because farm work is not able to generate adequate income for their livelihoods. About 65% of the participants reported that they participated in such activities because they do not have adequate land to cultivate.

Another important aspect of the rural labor market in the study area is participation in different labor market regimes. Out of 324 total sample households, 35% participated as sellers only, 21% as buyers only, 17% as simultaneous buyers and sellers and the remaining 27% are self-sufficient households.

Estimation strategy

As stated previously, the main purpose of this study was to analyze the role of the rural labor market in reducing poverty and improving wellbeing of households in the study area. Even though simple regression analysis may serve the purpose, it is not an appropriate approach to capture the effect of market participation as it may generate biased estimates. This is because as indicated by Caliendo and Kopeinig (2008) such regression assumes exogenous determination of participation while it may be potentially endogenous. The difficulty arises since researchers want to know the difference between the outcomes of treated and non-treated groups at the same time, which of course could not be observed at the same time due to the problem of self-selection bias. Thus, linear regression will not be appropriate and thus the use of the non-parametric approach called Propensity Score Matching (PSM) was preferred.

Even though the PSM technique was initially used to evaluate the impact of a project/program by considering the implementation of the programs as treatment, recently the approach has been extended to problems that self-select due to individual decisions. There are a number of similar empirical studies that dealt with the problems of self-selection due to individual decisions using PSM technique. For instance, Mariapia (2007) analyzed the impact of agricultural technology adoption (considering participation to adopt as a treatment variable) on poverty alleviation strategies in rural Bangladesh. In the same way, Fydess et al. (2011) measured the effect of participation in charcoal production on household income and poverty in three districts of western Uganda using household survey data and PSM techniques. They took participation in

[1] The study area is located in western part of Ethiopia, 330 km from Addis Ababa.

charcoal production as a treatment and tried to find the average effect of participation on household income, poverty. Similarly, Menale et al. (2010) analyzed the impact of adopting improved varieties on crop income and rural poverty in rural Uganda using cross-sectional farm household data and PSM method. Tanguy et al. (2007) had also assessed the impact of marketing cooperatives on the behaviour and welfare of their members based on detail household data in rural Ethiopia employing PSM techniques. Finally, Victor and Awudu (2009) investigated the impact of non-farm employment on farm household income and way out of poverty using farm household data from Brong-Ahafo region of Ghana and employing the PSM technique. All these empirical studies justify the use of PSM technique under situations that require self-selection due to individual decision as in the case of rural off-farm labor market participation on poverty.

In the context of this study, propensity score matching constructs a statistical comparison group by matching every individual observation of labor market participants with an observation of similar characteristics from the group of non-participants. It is a two-step procedure. First, a probability model for participation decision is estimated to obtain propensity scores of participation for each observation. In the second step, each participant is matched to a non-participant with similar propensity score values to estimate the average treatment effects. As defined by Rosenbaum and Rubin (1983), the propensity score P(X) is the conditional probability of receiving a treatment given pre-treatment characteristics. It is given as;

$$P(X) = P_r(T_i=1/X) = E\ (T/X) \tag{1}$$

Where T_i = (0, 1) is the indicator of an exposure to off-farm labor market given the covariates (X's) which represent the vector of pre-treatment characteristics.

In order to estimate ATT, the potential outcome framework was adopted (Wooldridge, 2002) where each household is viewed as facing two potential outcomes: One arising from participation in labor market and the other from non-participation. Therefore,

$$\begin{aligned} ATT &= E\,(Y_i^1 - Y_i^0 \backslash T_i = 1) \\ &= E\,[E\,(Y_i^1 \backslash T_i = 1, P(X_i) - E(Y_i^0 \backslash T_i = 0, P(X_i) \backslash T_i = 1] \end{aligned} \tag{2}$$

Where $P(X_i)$ is the propensity score, Y_i^1 is the potential outcome in the situation of participation and Y_1^0 is the potential outcome in the situation of non-participation.

As outlined by Caliendo and Kopeinig (2008), important properties (conditions) for the implementation of propensity score matching technique are the balancing property, conditional independence assumption (CIA) and common support condition. The balancing property indicates the condition that each participant is required to be matched with a non-participant of similar propensity score values. Testing for balancing property is important to make sure that household behaviour within each group is actually similar. The second is the Conditional Independence Assumption. This states that, once the set of all observable characteristics are controlled for, participation in off-farm work is random and uncorrelated with outcome indicators. That means systematic differences in outcome indicators between participant and non-participant individuals with the same values for covariates are attributable to treatment. The third requirement is the common support condition which requires that sample households with the same values of covariates X have positive probabilities of being both participants and non-participants. Therefore, all the individuals in the common support region actually participate in all states.

The actual implementation of the propensity score matching technique in this study was carried out through the following simple model. Second, the matching algorithm was chosen. In this case both Nearest Neighbour (NN) matching with calliper and Kernel-based matching were used just for the purpose of comparison (Imbens, 2004; Abadie and Imbens, 2006). Third, the issues of overlaing and common support region was check for to ensure that any combination of characteristics observed in the labor market participants can also be observed among non-participants (Bryson et al., 2002). Fourth, the matching quality was tested to ensure that the distribution of the relevant variables in both groups is balanced. The standardized bias, mean difference of covariates, Pseudo-R^2 was used for this purpose (Rosenbaum and Rubin, 1983; Sianesi, 2004). Existence of differences after matching may suggest a fundamental lack of comparability between the two groups (Blundell et al., 2005) which indicate that the technique was not successful requiring some remedial measures.

Finally, the technique is applicable for estimating average treatment effects provided that the assumption of conditional independence is satisfied. For this purpose it is important to check the sensitivity of the estimated results as it helps to know whether unobservable factors have an effect strong enough to undermine the implications of the matching analysis. If there are unobserved variables that affect participation and the outcome variable simultaneously, a hidden bias might arise to which matching estimators are not robust. The sensitivity analysis test is conducted using bounding approach proposed by Rosenbaum (2002) and applied by Caliendo and Kopeinig (2008). The probability of participation, π_i, is not only determined by observable factors (x_i) but also by an unobservable component (u_i): $\pi_i = \Pr(D_i=1 \backslash x_i) = F(\beta x_i + \alpha u_i)$, where the parameter α is the effect of u_i on the participation decision. From this, if the analysis is free from hidden bias, α will be zero and the participation probability will solely be determined by x_i. However, if there is hidden bias, two individuals with the same observed covariates x will have differing chances of participation in the labor market. For matched pair of individuals i and j, following logistic distribution, the odds that individuals receive a treatment (participation) are then given by $P_i/(1- P_i)$ and $P_j/(1- P_j)$, and the odds ratio is given by;

$$\frac{P_i/(1-P_i)}{P_j/(1-P_j)} = \frac{P_i(1-P_j)}{P_j(1-P_i)} = \frac{\exp(\beta x_i + \gamma\mu_i)}{\exp(\beta x_j + \gamma\mu_j)} \tag{3}$$

If both participants and non-participants have identical observed covariates as implied by the matching procedure, the x vector cancels out, implying that:

$$\frac{\exp(\beta x_i + \gamma\mu_i)}{\exp(\beta x_j + \gamma\mu_j)} = \exp\{\gamma(\mu_i - \mu_j\} \tag{4}$$

But, still both individuals may differ in their odds of receiving treatment by a factor that involves the parameter γ and the difference in their unobserved covariates u. Sensitivity analysis evaluates how changes in the values of γ and (u_i–u_j) alter inferences about the estimated effect. According to Rosenbaum (2002), Equation (4) above implies the following bounds on the odds ratio that either of the two matched individuals will receive treatment:

$$\frac{1}{e^{\gamma}} \le \frac{P_i(1-P_j)}{P_j(1-P_i)} \le e^{\gamma} \tag{5}$$

So, both matched individuals have the same probability of participating only if e^{γ} =1. Otherwise, if for example e^{γ}= 2, individuals who appear to be similar in terms of covariate x but could differ in their odds of receiving the treatment by as much as a

Table 1. Characteristics of labor market participants and non-participants.

Name of the variables	Participants	Non-participants	Mean dif.
Age of the head	40.52 (11.2)	38.38(10.4)	2.15**
Education level of the head	4.32(3.7)	5.22(3.7)	0.89***
Sex of the head (1= male)	0.98(.00)	0.98(.0)	0.00
Family size	6.01 (0.2)	5.89(0.2)	0.02
Number of dependents	2.35(1.8)	3.05(1.8)	0.36**
Number of adult laborers(aged 15-64)	4.12(1.8)	3.76(1.6)	0.25**
Animal wealth in TLU	5.38(0.3)	5.49(0.3)	0.10
Value of variable farm input	714.95(1003)	1,132.6(1494)	408***
Value of farm implement	384.7(185)	434(181)	49***
Land owned in hectares	2.22(1.38)	3.11(1.70)	0.89***
Land owned in adult equivalent	0.51(0.34)	0.72(0.44)	0.21***
Land cultivated in hectares	2.07(0.10)	2.89(1.44)	0.83***
Value of off-farm equipment owned	364.13(1022)	151.85(748)	212**
Number of draft animals owned	0.20(0.62)	0.39(0.77)	0.18***
Amount of non-labor income in Birr	144.03 (632)	287.3(774)	143*
Amount of credit in Birr	309.59(777)	333.19(834)	23.60
Dummy for Jima Arjo	0.33(0.47)	0.29(0.46)	0.04
Dummy for Guto Gida	0.38(0.49)	0.33(0.47)	0.05
Annual household income	9,095.52(9179.9)	8,408.73(9717)	686.79**
Total annual consumption expenditure	8,140.07(2652)	7,684.42(2638)	455.65***
Household income per adult equivalent	5,780.97(1128)	5,393.24 (1164)	387.73*
Consumption per adult equivalent	5,126.27(673)	4,839.58(712)	286.69***
Poverty status (head count ratio)	0.341	0.389	- 0.048**

Source: Own computation, 2012; ***, **, and * significant at 1, 5, and 10% level of significance. Incidence of poverty for the whole sample is 35.8%. Standard deviations in parentheses. All values are in Birr (Ethiopian currency unit).

factor of 2. In this sense, e^{γ} is a measure of the degree of departure steps. First, the propensity scores were estimated using the logit from a study that is free of hidden bias (Rosenbaum, 2002).

RESULTS AND DISCUSSION

Descriptive analysis of labor market participants and non-participants

The participation in off-farm labor market is measured as a dummy variable taking value 1 if the household participated during the survey year and 0 otherwise. The result of mean comparison test for off-farm wage participants and non-participants is provided in Table 1. The result reveals existence of a number of differences in household endowments, farm characteristics, family composition and socio-economic variables across participants and non-participants. For instance, on average, non-participant households have significantly younger heads (38 compared with 40 years) and are better educated than participants. Key differences were also observed between both groups in terms of family composition. On average, participant households have smaller number of dependents and larger number of adult laborers as expected. Participant and non-participant households also differ significantly in terms of the value of farm assets and variable farm inputs owned. For instance, on average, labor market participant farmers cultivated smaller farm size, owned significantly lower value of farm equipment and farm variable inputs. Moreover, significant differences were observed between the two groups in non-labor income, ownership in off-farm assets such as draft animals and value of off-farm equipment. However, there were no significant differences between participants and non-participants in terms of variables such as sex of the head, family size, animal wealth, the amount of credit, location and distance to the nearest market center.

Finally, differences between labor market participant and non-participant households in terms of the outcome indicators are reported in the same table. The outcome indicators used in the analysis were level of poverty (head count ratio), annual household consumption expenditure and income both in adult equivalent units. In order to classify households as poor and non-poor, the recent consumption-based poverty line which is officially declared by the government of Ethiopia (based on the 2010/2011 Welfare Monitoring Survey) was used.

Accordingly, the total poverty line per adult person per

year was estimated to be Birr 3,781(MoFED, 2012).
The total poverty line of Birr 3,781 and the information on consumption expenditure obtained from sample households in the study area were used to classify households as poor and non-poor. The poverty status of a household is measured as a binary variable indicating 1 if the household is non-poor and 0 otherwise. Based on the data, the incidence of poverty for the whole sample is 35.8%. That is, among 324 sample households included in this study, 115 (about 35.8%) were found to be poor during the survey year. Moreover, as shown in the table, about 38.9% of non-participants and 34.1% of the participant households fell below the poverty line.

The poverty figures are higher than the national average may be because of the method of data collection which was very intensive in this case. The annual household consumption expenditure includes total expenditure on all purchased as well as own produced consumption goods (evaluated at the current local prices in the study areas). Participants are significantly distinguishable from non-participants in terms of annual consumption expenditure and income per adult equivalent. The annual average total and per adult equivalent consumption expenditures during the survey year were Birr 8,140.1 and 5,126.3 respectively for labor market participant households. The similar figures for non-participants were Birr 7,684.42 and 4,839.6 respectively. Likewise, average annual household income is higher for participants than non-participants (Birr 9,095.5 as compared to Birr 8,408.7).

Average effects of off-farm labor market participation on welfare indicators

The result of logit model estimated for predicting the propensity score is provide in Annex Table A. The result shows fairly low pseudo R^2(0.29, 0.20 and 0.37 for off-farm wage work, self-employment and the total off-farm work, respectively). This may indicate that it would be easier to find good match between participant and non-participant households. In general, the estimated result indicates that the probability of participation in wage work is positively and significantly influenced by age and sex of the household head, the number of adult laborers in the family and distance to the nearest market center. The number of dependents and the size of land cultivated significantly reduced the probability of participation in wage work. Moreover, availability of financial resources such as credit obtained had no significant effect on off-farm wage employment, but it significantly increased the probability of participation in off-farm self-employment activity. In all cases, the amount of non-labor income and the animal wealth variables measured in a tropical livestock unit were not significant in influencing the likelihood of participation in any type of off-farm activity. Finally, location differences also affect the probability of participation in off-farm wage employment but not off-farm self-employment.

Before predicting the average effects of participation on outcome indicators, it is essential to test the balancing property and the matching quality of the propensity score. Accordingly, the t-test on covariates, the significance of pseudo R^2 and Likelihood Ratio tests were conducted before and after matching. The test results for participation in wage employment are reported in Annex Tables B and C. It shows that for almost all covariates, the t-ratios are insignificant after matching indicating that there are no systematic differences between the two groups after matching. Similarly, after matching the pseudo-R^2 value is very low as compared to before (0.042 against 0.29 for off-farm wage work) and also the p-value for LR χ^2 is insignificant after matching. The results in general indicate that there are no systematic differences between the two groups after matching. This implies that PSM has created a high degree of covariate balance in this study.

Finally, the goal of propensity score matching in this study is to obtain the average effect of off-farm labor market participation on consumption expenditure, income and poverty. To this end, two alternative matching algorithms, namely the Nearest Neighbor Matching (NNM) and the Kernel Based Matching (KBM) were used based on implementation of common support region with caliper so that the distributions of observations for both groups are located in common support region. As suggested by Rosenbaum and Rubin (1983), a caliper size of one-quarter of the standard deviation of the propensity score was used. The result is presented in Table 2.

In general, the result indicated that labor market participants are better off than non-participants in terms consumption expenditure, household income and poverty. As shown in the table, the NNM result reveals that participating in off-farm wage activity has significantly increased household income per adult equivalent by 39% and consumption expenditure by about 26%. That means, the income and consumption expenditure per adult equivalent family size for households participated in off-farm wage employment is greater than that of non-participants by 39 and 26% respectively. The KBM also provides comparable and significant results. Furthermore, it can be seen from the table that participation in off-farm wage employment has significantly reduced the level of poverty. Accordingly, the coefficient of poverty for NNM (-0.096) indicates that the probability of participant household falling below the poverty line is 9.6 percent less than that of non-participants. The KBM approach also provides significant reduction in poverty.

Similarly, the annual household income per adult equivalent for participants of off-farm self-employment activity is significantly higher than non-participants by about 49% in NNM approach and by 35% in KBM respectively. However, the impact on consumption

Table 2. Average effects of participation in off-farm work on household welfare indicators (PSM-estimation result).

Type of participation	Welfare indicators	NNM matching		KBM matching	
		ATT(S.Dv.)	t-stat	ATT(S.Dv.)	t-stat
Off-farm wage employment	Ln (Income)	0.394(0.13)	3.001***	0.312(0.06)	4.567***
	Ln (Expenditure)	0.262(0.08)	3.150***	0.239(0.04)	5.728***
	Poverty (1/0)	-0.096(0.03)	-1.874*	-0.073(0.03)	-1.763*
\Off-farm self-employment	Ln (Income)	0.493(0.20)	2.457***	0.346(0.11)	3.222***
	Ln (Expenditure)	0.067(0.08)	0.828	0.018(0.05)	0.155
	Poverty (1/0)	-0.022(0.01)	-0.167	-0.008(0.02)	-0.267
Overall off-farm employment (both self and wage)	Ln (Income)	0.581(0.13)	4.482***	0.506(0.08)	6.446***
	Ln (Expenditure)	0.304(0.11)	2.759***	0.292(0.04)	7.055***
	Poverty (1/0)	-0.292(0.08)	-2.102**	-0.172(0.11)	-1.96**

Source: Own computation, 2012. ***, **, and * significant at 1, 5, and 10% level of significance.

expenditure and poverty is very small and insignificant. Moreover, participants of off-farm self-employment are less likely to be poor by about 2.2% using the NNM method which is also not significant indicating that compared to wage employment activity, the importance of self-employment activity is limited in the study area.

The average effect of overall off-farm employment (both wage and self-employment) is also reported in the same table. The result indicates that the income per adult equivalent of off-farm labor market participants is higher than those of non-participants by about 58% in NNM and by about 51% in case of KBM. Similarly, the annual household consumption expenditure per adult equivalent for off-farm participants is 30% higher than that of non-participants (and 29% in case of KBM). Finally, the average effect of participation in off-farm work on poverty reduction is also larger and significant as shown in the table. The coefficient of poverty in NNM (-0.292) indicates that off-farm participants are less likely to be poor by about 29% on average as compared to non-participants.

The average effects of participation in off-farm labor market on outcome indicators identified in this study are generally low as compared to an empirical finding by Victor and Awudu (2009) in Brong-Ahafo region of Ghana which were 62 and 35%, respectively for the impact of participation in wage employment and self-employment on poverty respectively.

Finally, the test result of sensitivity analysis for participation in off-farm wage work reported in annex Table D indicates that the critical values of gamma (e^{γ}), the values at which we would question our conclusion of the effect of labor market participation on consumption expenditure, income and poverty starts from values equal to 1.8, 1.9 and 1.5 respectively. Since these values are large, the effects of participation are not generally sensitive to problem of unobserved variables. For instance, the value of gamma = 1.7 for consumption expenditure shows that the impact of participation on household consumption expenditure is not sensitive to selection bias due to unobserved variables even if participants and non- participants were allowed to differ by as much as 70% in terms of unobserved covariates. Generally, the test results lead us to conclude that the estimated average effects are totally insensitive to hidden bias, and thus are generally pure effects of labor market participation.

SUMMARY AND CONCLUSION

In this study efforts were made to examine the role of the rural labor market in reducing poverty and improving welfare of households by using propensity score matching technique. The effects of participation in off-farm employment on household poverty, consumption expenditure and income were examined. The finding, in general, confirmed that off-farm participant households are better off than non-participants in terms consumption expenditure, overall household income and level of poverty. The income and consumption expenditure of households who participated in off-farm wage employment were significantly greater than that of non-participants. Moreover, on average, participant households are less likely to be poorer by about 29%, as compared to non-participants. Finally, off-farm self-employment activity (mostly non-agricultural activity) is limited in the study area and not significant at influencing consumption expenditure and reducing poverty.

The findings are in harmony with the mounting attention of governments in promoting off-farm activities in rural areas of developing countries as an alternative means to get out of poverty. Therefore, the implication is that policy measures which are directed towards promoting off-farm work opportunities in the study area are relevant to achieve the goals of poverty reduction. Moreover, it is essential to provide incentives and increase the capacity of households to participate in rural non-farm activities to take advantage of such opportunities. Such intervention would help overcome some entry barriers to off-farm employment thereby promoting efficient functioning of the rural labor market in the study area. In addition to being one important source of income for rural farm households, off-farm employment could help smooth incomes, which in turn smoothens consumption over long periods of time.

Conflict of Interests

The author(s) have not declared any conflict of interests.

ACKNOWLEDGEMENTS

The author express their thanks to Haramaya University and Wollega University (both in Ethiopia) for the financial support they provided us to successfully complete the research; East Wollega Zone Office of Agriculture and Rural Development for facilitating the contact with the districts in the study area and with the development agents, and the development agents who helped in conducting the survey.

REFERENCES

Abadie A, Imbens G (2006). Large sample properties of matching estimators for average treatment effects. Econometrica 74(1):235-267.

Awuor Z (2007). Agricultural growth, rural poverty and hunger in Africa. AAAE Conference Proceedings (2007) College of Agriculture and Veterinary Sciences Depart. Agric. Econ. Ghana. 465-468.

Barbara T, Sophia B, Alastair B (2012). Key Issues in Agricultural Labour Markets: A review of major studies and project reports on agriculture and rural labour markets. P. 20, February 2012. ISBN 978-94-6138-191-0.

Bardhan P, Udry C (1999). Development Microeconomics. Oxford University Press Inc., New York.

Bernardin S (2012). Non-farm income diversification in rural Ghana: Patterns and determinants. Afr. Dev. Rev 24(3):233-244.

Blundell R, Dearden L, Sianesi B (2005). Evaluating the impact of education on earnings in the UK: models, methods and results from the NCDS. J. Roy. Stat. Soc. 168(3):473-512.

Bryson A, Dorsett R, Purdon S (2002). The use of propensity score matching in the evaluation of active labour market policies. Working paper number 4, London: Policy studies institute and national centre for social research.

Caliendo M, Kopeinig S (2008). Some practical guidance for implementation of propensity scores matching. J. Econ. Surveys. 22(1):31-72.

Carlos O (2010). Rural inequality, wage employment and labour market formation in Africa: Historical and micro-level evidence. Working Policy Integration Department International Labour Office, Geneva. P. 97.

David S (2010). The rural non-farm economy, livelihood strategies and household welfare. Afr. J. Agri. Res. 4(1):82-109.

Ellis F (2001). Rural livelihoods, diversity and poverty reduction policies: Uganda, Tanzania, Malawi and Kenya. Ladder Working P. 1.

Emmanuel E (2011). Rural livelihood diversification and agricultural household welfare in Ghana. J. Dev. Agric. Econ. 3(7):325-334.

Fydess K, Pamela J, Shively G, Dick S (2011). Income, poverty and charcoal production in Uganda. Forest Policy. Econ. 13(2011):199-205.

Haggblade S, Hazell P, Reardon T (2010). The Rural Non-farm Economy: Prospects for Growth and Poverty Reduction. World Dev. 38(10):1429-1441.

Imbens G (2004). Nonparametric estimation of average treatment effects under exogeneity. Rev. Econ. Stat. 86(1):4-29.

Jayne T, David M, Mghen E (2010). Principal Challenges Confronting Smallholder Agriculture in Sub-Saharan Africa. World Dev. (38)10:1384-1398.

Kwadwo A, Samson J (2012). Increasing agricultural productivity and enhancing food security in Africa: New Challenges and Opportunities. Synopsis of an International Conference. International Food Policy Research Institute, Washington, DC, March 2012.

Leavy J, White H (2003). Rural labor markets and poverty in Sub-Saharan Africa. Working Paper, Brighton, Institute of Development Studies, University of Sussex.

Mariapia M (2007). Agricultural technology adoption and poverty reduction: A propensity-score matching analysis for rural Bangladesh. Food Pol. 32:372-393.

Mazumdar D (1989). Microeconomic issues of labour markets in developing countries; analysis and policy implications. Economic Development Institute of the World Bank.

Menale K, Bekele S, Geoffrey M (2010). Adoption and Impact of Improved Groundnut Varieties on Rural Poverty. Evidence from Rural *Uganda*. Discussion Paper Series May 2010, EfD DP 10-11.

Micevska M, Rahut D (2008). Rural non-farm employment and incomes in the Himalayas. Econ. Dev. Cul. Change 57(1):163-193.

MoFED (Ministry of Finance and Economic Development-Ethiopia) (2012). Ethiopia's Progress towards Eradicating Poverty: An Interim Report on Poverty Analysis Study (2010/2011). March 2012, Addis Ababa, Ethiopia.

Mulat D, Fantu G, Tadele F (2006). Towards a more employment-intensive and pro-poor economic growth in Ethiopia: Issues in Employment and Poverty. Discussion Employment Strategy Department, International Labour Office, Geneva. P. 22.

Otusu K, Yamano T (2006). The role of rural labor markets in poverty reduction: Evidence from Asia and East Africa. FASID discussion series on international development strategies 2006-12-007.

Rosenbaum P (2002). Observational Studies. New York, NY: Springer.

Rosenbaum P, Rubin D (1983). The central role of the propensity score in observational studies for causal effects. Biometrika. 70(1):41-55.

Sianesi B (2004). An evaluation of the Swedish system of active labor market programmes in the 1990s. Rev. Econ. Stat. 86(1):133-155.

Sosina B, Barrett C, Holden S (2012). Does the nonfarm economy offer pathways for upward mobility? Evidence from a panel data study in Ethiopia. World Dev. 40(8):1634-1646.

Tanguy B, Eleni G, Alemayehu S (2007). Smallholders' commercialization through cooperatives: A Diagnostic for Ethiopia. IFPRI Discussion P 00722

Verner D (2006). Labor markets and income generation in rural Argentina. World Bank Policy Research Working P. 4095.

Victor O, Awudu A (2009). Non-farm employment and poverty reduction in rural Ghana: A propensity-score matching analysis. Paper prepared at the International Association of Agricultural Economists conference, Beijing, China, August 16-22.

World Bank (2008). World Development Report 2008: Agriculture for Development Washington, DC.

Wooldridge J (2002). Econometric Analysis of Cross Section and Panel Data. The MIT Press, Cambridge.

ANNEX

Table A. Logit estimation result (for predicting the propensity scores) [Dependent variable: Participation in off-farm work (1/0)].

Explanatory variable	Participation					
	Wage work		Self-employment		Overall off-farm work	
	Coef.	St.Er	Coef.	St.Er	Coef.	St.Er
Age in years	0.024***	0.011	-0.140	0.112	-0.102	0.095
Age square	-0.001***	0.0005	0.002	0.001	0.001	0.001
Education level of the head	0.045	0.055	-0.051	0.064	0.032	0.064
Gender(1=male;0 = female)	0.618***	0.243	0.357**	0.172	0.337***	0.149
Adult laborers(aged 15-64)	0.387**	0.201	0.056	0.036	0.213*	0.130
Elder children(aged 10-14)	0.127	0.259	0.145**	0.075	0.086*	0.044
Dependents	-0.262**	0.126	-0.075	0.153	-0.366**	0.153
Land cultivated in hectares	-1.524***	0.471	-0.101	0.476	-1.007***	0.315
Amount of credit obtained	-0.076	0.052	0.041***	0.008	0.023***	0.007
Animal wealth in TLU	0.250	0.214	-0.065	0.048	-0.012	0.009
Non-labor income	-0.054	0.062	-0.001	0.0007	0.004	0.005
Distance in kms	0.061*	0.035	-0.010	0.041	0.047	0.042
Dummy for Guto Gida	0.343*	0.177	0.083	0.069	0.237**	0.111
Dummy for Jima Arjo	0.169*	0.092	0.120	0.099	0.167*	0.102
Constant	-2.160	2.629	-1.408	0.926	2.285	1.344
Pseudo R^2	0.29		0.20		0.37	
LR chi^2	128.6		37.01		193.44	
Log likelihood	-160.1		-122.4		-114.0	
Prob> chi^2	0.000		0.052		0.000	

Source: Own calculations, 2012. ***, **, and * significant at 1, 5, and 10% level of significance.

Table B. Covariate balancing before and after matching (for participation in wage work).

Variable name	Sample	Mean			Reduction	t-test	
		Treated	Control	% Bias	%|Bias|	t	p>|t|
Age	Unmatched	40.524	38.641	17.8		1.60	0.089
	Matched	39.582	38.563	7.6	57.3	0.75	0.451
Year of schooling of head	Unmatched	4.321	5.218	-24.4		-2.20	0.029
	Matched	4.546	5.214	-18.2	25.5	-1.33	0.183
Average family schooling	Unmatched	0.708	0.732	-9.2		-0.82	0.411
	Matched	0.711	0.731	-7.6	16.8	-0.58	0.565
Gender (1= male)	Unmatched	0.976	0.981	-3.1		-0.28	0.778
	Matched	0.964	0.980	-11.3	-260.1	-0.84	0.400
Number of adult labourers	Unmatched	4.214	3.968	14.5		1.90	0.083
	Matched	4.146	3.945	11.8	18.7	0.90	0.369
Number of young children aged1015	Unmatched	1.191	1.218	-2.6		-0.23	0.817
	Matched	1.273	1.198	7.0	-173.2	0.50	0.620
Number of dependents	Unmatched	2.869	3.237	-20.6		-1.85	0.065
	Matched	3.327	3.208	6.7	67.7	0.49	0.628
Land cultivated in adult equivalent	Unmatched	0.465	0.660	-51.4		-4.65	0.000
	Matched	0.463	0.578	-66.6	-29.6	-1.63	0.094

Table B Contd.

Non-labour income	Unmatched	287.32	144.04	20.3		1.82	0.070
	Matched	258.36	141.06	16.6	18.1	1.35	0.176
Animal wealth in TLU	Unmatched	5.384	5.487	-2.6		-0.24	0.812
	Matched	4.803	5.477	-17.4	-556.1	-1.23	0.218
Credit	Unmatched	309.6	333.20	-2.9		-0.26	0.792
	Matched	132.6	348.12	-26.7	-813.2	-1.86	0.103
Distance to the nearest market in kms.	Unmatched	5.049	4.525	12.3		1.11	0.270
	Matched	4.318	4.466	-3.5	71.8	-0.26	0.794

Source: Own calculation using survey data, 2012.

Table C. Additional indicators of matching quality.

Type of off-farm participation	Pseudo R^2 before matching	Pseudo R^2 after matching	LR χ^2(*p-value*) before matching	LR χ^2(*p-value*) after matching	Total % mean standardize bias reduction
Total off-farm work	0.371	0.057	193.44(0.000)	9.57 (0.793)	87.72
Wage work	0.287	0.042	128.57(0.000)	24.15 (0.256)	84.84
Self-employment	0.197	0.036	37.01(0.052)	13.56(0.631)	84.33

Source: Own calculation using survey data, 2012.

Table D. Sensitivity analysis test for hidden bias (for participation in wage work).

Gamma (e^{γ})	Income	Consumption expenditure	Poverty
	p-value	p-value	p-value
1.0	0.004	0.0001	0.007
1.1	0.008	0.004	0.016
1.2	0.015	0.012	0.041
1.3	0.027	0.022	0.069
1.4	0.036	0.037	0.099
1.5	0.049	0.057	0.134
1.6	0.064	0.078	0.174
1.7	0.081	0.096	0.219
1.8	0.092	0.125	0.268
1.9	0.113	0.154	0.339
2.0	0.168	0.196	0.372
2.1	0.212	0.247	0.429

Source: Own calculation using survey data, 2012.

6

Improved rice variety adoption and its effects on farmers' output in Ghana

Abel Kwaku K. Bruce, Samuel A. Donkoh and Michael Ayamga

Department of Agricultural and Resource Economics, Faculty of Agribusiness and Communication Sciences, University for Development Studies, Tamale, Ghana.

Sub-Sahara Africa and for that matter Ghana, missed out of the first Green revolution. However, with the instrumentality of the former United Nations Secretary, Kofi Annan, through the Alliance for Green revolution in Africa (AGRA) and other bodies, the revolution is being introduced in some parts of Africa, including Ghana. The extent to which the new revolution would work depends on a careful study of the socioeconomic underpinnings of technology adoption. This study sought to investigate the factors that influence the adoption of improved rice varieties and its effects on rice output in Ghana. The method of analysis involved an estimation of treatment effect model comprising a Probit equation and a production function. The empirical results show that the adoption of improved rice variety had a positive effect on farm output. Other inputs that had significant and positive impact on output were farm size, labour and fertilizer. The probability of adopting improved rice variety was high for the following: farmers who had formal education; farmers who had bigger household sizes; and farmers who had smaller farms. Contrary to our a *priori* expectations, however, farmers who had access to extension services had lower probability of adoption. The authors recommend that farmers be supported with more fertilizer subsidization. Farmers should also form farmer groups to support one another on the field. Also, the fundamental problems of illiteracy among farmers must be addressed.

Key words: Ghana, improved rice seeds, technology adoption, treatment effect model.

INTRODUCTION

Agriculture has a direct influence on the attainment of at least five of the Millennium Development Goals (MDGs) (MoFA, 2010a). The first goal of eradicating poverty and extreme hunger can only be achieved through increased agricultural productivity.

Agriculture continues to play a pivotal role in Ghana's economy, contributing to about 30% of gross domestic product (GDP). The agricultural sector also provides employment to about 50.6% of the labour force and in 2010, it was the largest foreign exchange earner (MoFA, 2010a). The overall growth rate of the agricultural sector, vis-à-vis the current annual population growth rate of 2.6% is 2.8%. The small margin between these figures has serious implications for the attainment of food security, employment generation and improvement in rural incomes and national economy (NAPCDD, 2003).

In spite of the significant role agriculture plays in the provision of food, food shortage still persists among many households in Ghana (Carr Jr., 2001). Ghana depends largely on imported rice (400,000 tonnes/annum) to

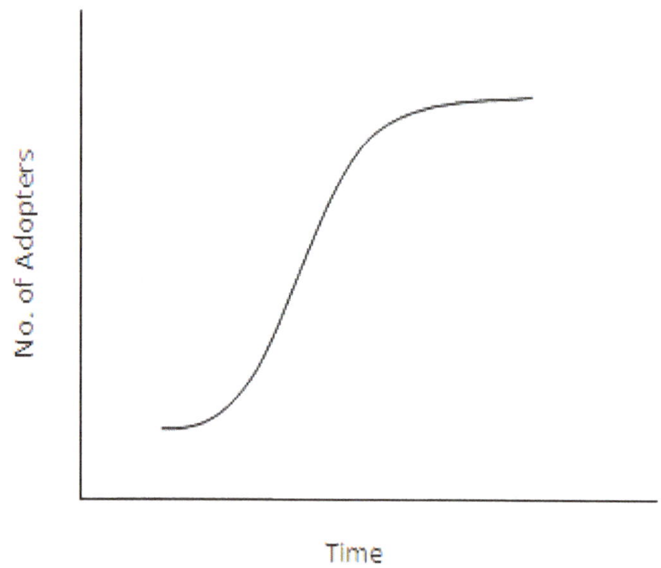

Figure 1. The s-shaped diffusion curve.

make up for the deficit in rice supply. The self-sufficiency ratio of rice in Ghana has declined from 38% in 1999 to 24% in 2006 (MoFA, 2010a).

The Government of Ghana in her drive to increase and promote the quality of locally produced rice in 2003, established and implemented the Nerica Rice Dissemination Project (NRDP). The main goal of the five year project was to contribute to poverty-reduction and food security through the adoption of high yielding NERICA upland rice varieties. Ten districts benefited from the implementation of the project which resulted in a cultivated area of 22,561.40 ha yielding about 56,400 metric tons of paddy Nerica rice (MoFA, 2010b). However, judging from the yields of farmers, there still appears a yawning gap between results of on-farm demonstration plots (6.5 Mt/Ha) and actual yields (2.4 Mt/ha) from the farmers' fields (MoFA, 2010b). Despite the prospects of improved varieties, adoption rates are still low. The objective of this study is to determine the socio-economic, farm characteristics as well as institutional factors that influence the adoption of improved varieties and the effects on rice output in Ghana.

MATERIALS AND METHODS

Literature review-diffusion of innovation

Rogers (1962) defined an innovation as an idea, practice or object that is perceived as new by an individual or other units of adoption. Also, he defined diffusion as the process by which an innovation is communicated through channels over time among members of the social system. The Innovation Diffusion Theory seeks to explain how, why and at what rate new ideas and technologies spread through cultures (Rogers, 1962). The origin of the theory is varied and spans over multiple disciplines. From the literature, the concept of diffusion was first studied by Gabriel Tarde in 1890 and then anthropologists Friedrich Ratzel and Leo Frobenius. Later on, Rogers made extensive studies and came up with four main elements that influence the spread of a new idea, namely; the innovation in question, the communication channels, time and the social system. These elements work in conjunction with one another. This means that for any technology to be adopted, it requires that first, the innovation should be communicated through a channel over a period of time, and this process takes place in a social system. The process relies heavily on human capital and the innovation must be widely adopted in order to be self-sustainable. Within the rate of adoption, there is a point where an innovation reaches mass saturation point; where the largest adoption rate is experienced. This is shown by a logistic curve with S shape as shown in Figure 1.

Rogers categorized the adopters as innovators, early adopters, early majority, late majority and laggards. As a new technology is introduced through a communication channel, innovators are normally the first to adopt the new technology, followed by the early adopters. These two categories are risk takers, who would adopt the technology despite the fact that they may not have full knowledge about its prospects. After some time when some positive benefits have been seen, the early majority joins and then the late majority. When the technology has proven to be good beyond every doubt, the laggards also join. However, despite some promising attributes of the technology, some people may not adopt it. The reasons are varying; some may not be aware of it, others may not have the means to access it, and some would still have some misgivings about the technology. Whatever, the reason, however, they would also have made a choice. It is important for socio-economists to intensify their research into the farmers' socio-economic, farm characteristics as well as some institutional factors that may encourage or discourage the adoption of agricultural innovations.

Rogers (1962) also identified five stages of accepting an innovation, namely; knowledge, persuasion, decision, implementation and confirmation. That is to say that when an innovation is introduced, the potential adopter needs to have knowledge about its benefits and limitations. After gaining the knowledge he/she needs to be persuaded that the benefits far outweigh the costs. If he/she is convinced he/she makes a decision to adopt the technology, otherwise he/she rejects it. Once a positive decision has been made, he/she implements it. Often times, this involves trying the technology on a small portion of the farmer's field. If it is successful, the technology is confirmed, and the farmer can increase the portion of his or her field in the next farming season.

Lastly, Rogers (1962) noted five attributes of a good technology as follows; simplicity, compatibility, trialability, relative advantage, and observability. That is to say that the innovation should not only be seen, it should be easy to understand and adopt. It should also be consistent with the farming practices already adopted by the farmer. Furthermore, the farmer should be able to experiment the new technology to know for him/herself its usefulness and viability.

Theoretical framework

Following Foltz's (2003) framework, technology adopters have a positive net willingness to pay for the technology or adaptation strategy. Farmers have what Foltz calls a reservation price $P_r = (w, k, \eta)$ for the technology that is greater than or equal to the actual market price P_m. He defined the reservation price as the amount that an individual would be willing to pay for the technology given his asset position w; other inputs he uses, k; and the parameter of his preference, η. According to him, P_m is the given price of the technology which is constant for all individual. Thus, for a given individual, the dependent variable A is defined as an index variable for whether or not he/she adopts the new technology. It takes on the value zero and one as follows:

$$A = 1 \quad if \quad P_r(w,k,\eta) - P_m > 0$$
$$A = 0 \quad if \quad P_r(w,k,\eta) - P_m \leq 0 \tag{1}$$

where the variables are as defined. The function $P_r = (w,k,\eta)$ represents a shadow price for an individual adopting an adaptation strategy. The inference problem in terms of econometrics, according to Foltz, then becomes a question of parameterizing the equation that defines the net benefits of the technology to farmers. The standard model in the literature is the random utility model. Researchers are not able to observe the preference parameters of the utility function. It is however, assumed that they are known to decision makers.

Let these parameters be an unobserved variable so that the actual utilities of an individual can be written as:

$$U_i = P_{ri}(w,k,\eta) - P_m = z'\gamma + e_1 \tag{2}$$

where z is a set of characteristics of the decision maker observable to the econometrician, γ is a parameter vector and e_1 is the error term. Hence, $z'\gamma$ becomes an index function that allows us to estimate the probability of adoption $A = 1$ in the following fashion:

$$\text{Pr}ob\, P_r(w,k,\eta) - P_m > 0 = \text{Pr}ob\,(z'\gamma + u_1 > 0) \tag{3}$$

Assuming that the disturbance term is normally distributed, this becomes a standard Probit model. By symmetry of the normal distribution, Equation 4 is obtained as follows:

$$\text{Pr}\ ob\ (P_r - P_m > 0) = \text{Pr}\ ob\ (u_1 < z'\gamma) = F(z'\gamma) \tag{4}$$

Where $F(\cdot)$ is the cumulative density function of the normal distribution.

This is then estimated using the maximum likelihood estimation, in which the likelihood function is as follows:

$$\ln L = \sum_{y_i=0} \ln(1 - \Phi_i) + \sum_{y_i=1} \ln(\Phi_i) \tag{5}$$

Thus, in general the Probit model to be estimated to determine the factors influencing the adoption of improved rice varieties is of the form:

$$A^* = z'\gamma + u_1 \tag{6}$$

where A* can be viewed as an indicator for whether or not this latent variable is positive as depicted in Equation 7.

$$A = \begin{cases} 1 & if\ A^* > 0\ i.e.\ -u_1 < z'\gamma \\ 0 & otherwise \end{cases} \tag{7}$$

The rest of the variables are as defined.

Sample selection bias

According to Barnow et al. (1980), sample selection bias arises when in program evaluations, the treatment (or control) status of the subjects is related to unmeasured characteristics that themselves are related to the program outcome under study. In this case, they define the term bias as the potential mis-estimate of the impact of the treatment (or programme) on the outcome. In this present study, selection bias can arise when improved rice adoption is related to unmeasured characteristics like farmers' innate ability which are also related to their rice output. Sample selection has been well expounded in Heckman (1979) and Smit (2003). One common version of sample selection which is related to this present study is where information on the dependent variable is available for all respondents, but the distribution of respondents over categories of the independent variable of interest has occurred in a selective way.

For example, in this present study, the main objective is to determine the effects of improved rice adoption on output. Thus, there are two main categories of respondents, namely adopters and non-adopters of improved rice varieties. If adoption (normally specified as a dummy variable) is simply regressed on output, the estimate of the adoption effect may be biased because the distribution of respondents over the categories of adopters and non-adopters was not random. Adopters may differ in several (measured and unmeasured) ways from non-adopters. If these characteristics are related to output, the coefficient of the adoption variable may catch up these effects and be biased because of this. In other words, supposed an output equation was estimated and it was found that adopters on a whole had a greater output level than non-adopters, how sure would the researchers be that adopters' relatively high level of output was as a result of the adoption of improved variety and not other positive innate characteristics that adopters possess? The idea behind Heckman's sample selection procedure is to estimate a selection equation (the Probit model in this study) and use the predicted values to form a selection control factor (λ) equivalent to the Inverse Mills Ratio (IMR) which will serve as an additional regressor in the substantive equation. In this case, the pure effect of adoption on output was measured. Besides, the other determinants of output are freed from the effects of the unmeasured characteristics and therefore the coefficients are unbiased (Smit, 2003).

Thus, given a substantive equation of the form:

$$Y_i = X_i'\beta + e_2 \tag{8}$$

Where Y_i is output, X_i is a vector of farm inputs, β is a vector of parameters to be estimated, e_2 is the two sided error term also with mean zero and constant variance.

Equation 6 is estimated and the predicted values of adoption used to construct λ equivalent to the IMR, which is included in equation 8 as an additional regressor. According to Heckman (1979) when this is done the pure effects of adoption can be evaluated and also, the other explanatory variables in equation 8 are freed from the unmeasured characteristics, such as the farmer's innate ability.

Treatment effect model

Treatment effect model is similar to the Heckman's two stage sample selection model. The main difference between the two

Table 1. Definition of variables used in the study.

Variable	Definition	Expected sign
Age	Age of the farmer in years	+/-
Education	Number of years of formal education	+
Household size	Number of people in farmer's house eating in the same bowl	+/-
Farm size	Size of farmer's yam plot in acres	+
Extension	Dummy; 1 if farmer had access to extension service during farm season in question formal education; 0 if otherwise	+
Adoption	Dummy; 1 if farmer adopted improved rice seed; 0 if otherwise	+
y_1	Natural Logarithm of rice output in kilograms	+
x_1	Natural Logarithm of farm size in acres	+
x_2	Natural Logarithm of labour cost in Cedis	+
x_3	Natural Logarithm of seeds in kilograms	+
x_4	Natural Logarithm of fertilizer cost in Cedis	+

however, is that in the case of the former, the treatment condition (adoption in this case) enters the substantive equation to measure the direct effect on output (Maddala, 1983). Thus, the regression equation becomes:

$$Y_i = X_i^{'}\beta + A_i\delta + u_2 \tag{9}$$

Where δ measures the effect of adoption on output and A_i is as defined earlier. Adding the IMR translates into:

$$\ln Y_i = \beta'(\Phi_i \ln X_i) + \delta'(\Phi_i A_i) + \sigma\phi_i + u_3 \tag{10}$$

(Maddala, 1983)

where ϕ_i and Φ_i are the probability density function (PDF) and the cumulative density function (CDF) respectively of the standard normal distribution, and $\Phi_i \equiv \Phi(w_i^{'}\gamma)$. u_3 is two sided error term with $N(0, \sigma_v^2)$. The rest are as defined earlier.

EMPIRICAL MODEL

The empirical models of the study are as follows:

$$A = \gamma_0 + \gamma_1 Age + \gamma_2 Agesqd + \lambda_3 Education + \gamma_4 Extensio \tag{11}$$

(Adoption Model)

$$y_1 = \beta_0 + \beta_1 x_1 + \beta_2 x_2 + \beta_3 x_3 + \beta_4 x_4 + \delta_1 A + u_2$$

(Output model)
The variables are as defined in Table 1.

Data and study area

The data for this study come from the Statistical, Research and Information Directorate (SRID) of the Ministry of Food and Agriculture (MoFA) in conjunction with the Ghana Strategy Support Program (GSSP) of the International Food Policy Research Institute (IFPRI). It must be mentioned that the data was collected as a pilot study and as such the sample size for rice producers was only 414 from rice producing communities in thirteen selected districts in Ghana as follows: Gushiegu and Yendi districts of the Northern region, Bawku municipality and Kassena Nankana East in the Upper East region, Lawra and Sisala East in the Upper West of Ghana, North Tongu in the Volta, Sekyere Afram Plains in Ashanti, Assin North in the Central, Atiwa in the Eastern and Ga East and Ga West both in the Greater Accra Region of Ghana. The final data was sorted to select 406 because 8 of the farmers did not have all the information that are needed.

Definition of variables used in the study

Table 1 shows the definition of variables used in the estimation of the model and their expected signs. From the literature, the effect of the farmer's age is ambiguous; it can be positive or negative depending on the study. The argument is that older farmers may have more experience, resources, or authority that may give them more possibilities for trying a new technology. On the other hand, younger farmers have been found to be more knowledgeable about new practices and may be more willing to bear risk and adopt new technology because of their longer planning horizons. Education is also expected to have a positive effect on adoption because it increases knowledge thereby enhancing the ability to derive, decode and evaluate useful information for technology adoption.

Household size has been identified to have either positive or negative influence on adoption. Larger family size is generally associated with a greater labour force for the timely operation of farm activities. The negative relationship of the variable with adoption has been linked to increased consumption pressure associated with large family which does not permit them to have the means to invest in new technologies for their farms. Normally, farming households with bigger landholdings are supposed to have an enhanced ability to afford improved technologies and a greater capacity to cope with losses if the technologies fail. Furthermore, access to extension gives famers the opportunity to gain knowledge and also obtain some encouragement with respect to the adoption of technologies. Lastly, from neoclassical production economics output is a positive function of land (farm size), labour and capital (seeds and fertilizer) (Koutsoyannis, 1979).

RESULTS AND DISCUSSION

The results and analyses of the model estimation are presented in this section. The descriptive statistics of the variables used in the study are first discussed.

Descriptive statistics of the variables used in the study

Majority of the respondents were males constituting 75.4% of the sampled population. Also, the percentages of respondents who used improved seeds and traditional seeds were 49 and 51, respectively while 54.5% had access to extension services. Furthermore, from Table 2 the mean age of the farmers was 48.3. This age falls within the adult population in Ghana. One of the challenges facing agriculture in Ghana is the ageing population of the farmers. The farming profession appears unattractive to the youth. The average years spent in school by the farmers was 2.15. In Ghana, six years is spent in primary school while an additional three years is spent in the Junior High School (JHS) to complete basic education. The 2.15 average years spent in school by the sample farmers attests to the fact that there was low level of education among the farmers. Specifically, 76.2% of the farmers had no formal education, while 10.9% each finished basic and secondary education. Only 2.5% made it to the tertiary level. The mean farm size of 4.92 compares with the national average of 5 acres. In Ghana, small-scale farmers are about 92% of the farming population (MoFA, 2010c). On the average, it took Gh¢ 11.01 and Gh¢ 6.50 of labour and fertilizer costs respectively to cultivate 4.92 acres of rice plot. The average quantity of seeds for the same plot size was 24.41 kg.

Determinants of improved rice variety adoption

From the results in Table 3, all the variables, except age and age squared were significant. However, household size and education had a positive effect on the probability of adoption, farm size and extension service had a negative effect on adoption. Our findings are consistent with that of Foltz in (2003) who argued that formal education helps farmers to understand the information about a technology which in turn facilitates the adoption of a technology. Similarly, education gives farmers the ability to perceive, interpret and respond to new information much faster (Uaiene et al., 2009; Nzomoi et al., 2007; Salasya et al., 1996). Also, the significance of the household size variable can be attributed to the fact that the large household size served as labor for their farm plot. However, the negative sign of the coefficients of extension service and farm size variables did not meet our *a priori* expectations. Normally, the extension contact should lead to increased probability of adoption, since farmers who have contacts with extension staff have the opportunity of learning about improved varieties. Also, normally, farmers who have large farms size tend to have a higher probability of adoption because they are able to allocate some portions of their field to cultivating the improved seed as a trial, pending their full acceptance of the new technology. In this present study, the negative coefficient means that rather farmers with smaller farms as well as those who did not receive extension services had high probability of adopting improved rice varieties.

The effects of adoption on output

The main objective of this study was to investigate the effects of improved seed adoption on rice output. In other words, the study sought to find out whether the adoption of improved rice seeds leads to increased output as opposed to the adoption of traditional varieties, other things being constant. From Table 4, not only was the adoption significant but it maintained its expected positive sign confirming our *a priori* expectation that the adoption of improved rice seeds leads to increased output. This is consistent with the findings of Wiredu et al. (2010), Uaiene et al. (2009) and Sserunkuuma (2005).

It can also be observed from the table that apart from seeds, all the other variables were significant and maintained their positive sign. The sum of the coefficient of the conventional input is 0.76, implying that there was decreasing returns to scale. A 100% increase in land led to a 26% increase in output while a 100% increase in labour led to a 21% increase in output. Also, while a 100% increase in seeds led to a 5% increase in output, a 100% increase in fertilizer led to a 24% increase in output.

The significance of lambda (λ) in Table 1 implies that selectivity bias was present in the model and that if it was not correct, the estimated coefficients, including the adoption variable, would have been bias, meaning that the pure effects of the explanatory variables on output could not be measured. Thus, the pure effects of the explanatory variables on output would not have been measured. However, the correction of the selectivity problem ensured that the estimated coefficients were freed from the effects of unobserved factors that correlated with the adoption variable.

Policy implication

As indicated earlier, the findings of this study are consistent with that of many studies that evaluated the effects of the Asian Green revolution that took place in the early 1960s (Johnson et al., 2003; Janvry and Sadoulet, 2002; Evenson and Gollin, 2000; Hazell and

Table 2. Descriptive statistics of the variables used in the study.

Variable	Mean	Standard Dev.	Minimum	Maximum
Age	48.63	16.44	4.00	90.00
Education	2.15	4.16	0	15.00
Household size	7.47	5.88	1.00	40.00
Farm size	4.92	5.32	0.50	42.00
Labour cost	110.11	251.24	2.00	2876.00
Seed	24.41	40.62	2.00	250.00
Fertilizer cost	60.56	93.92	0	850.00

Note that the amounts quoted here are in old Ghana Cedis. The equivalence is as follows: 2,000 Old Ghana Cedis=2 New Ghana Cedis=1 US Dollar.

Table 3. Maximum likelihood estimates for the parameters of the Probit adoption model.

Variable	Coefficient	Standard error	Z	P > |Z|	[95% Conf. Interval]	
Adoption						
Age	0.026	0.029	0.90	0.367	-0.031	0.083
Age²	-0.000	0.003	-0.74	0.460	-0.001	0.000
Education	0.518	0.225	2.30	0.021**	0.077	0.959
Extension	-0.330	0.189	-1.74	0.082*	-0.700	0.042
Farm size	-0.031	0.175	-1.79	0.074*	-0.065	0.003
HH size	0.666	0.017	3.96	0.000***	0.034	0.995
Constant	-0.966	0.693	-1.39	0.163	-2.325	0.393
λ	-0.585	0.338	-1.73	0.084*	-1.248	0.780
Rho	-0.461					

Source: Field Survey *** Significant at 1%, ** significant at 5%. Note: Dependent variable: Adoption of improved seeds. Number of observation = 406. Wald chi² (5) = 66.27 and pro > chi² = 0.000.

Table 4. Maximum Likelihood Estimates of treatment effect model-two step estimates.

Variable	Coefficient	Standard error	Z	P > |Z|	[95% Conf. Interval]	
Output						
Farm size	0.255	0.122	2.09	0.036*	0.016	0.494
Labor	0.212	0.628	3.38	0.001***	0.892	0.335
Seeds	0.550	0.801	0.69	0.492	-0.102	0.212
Fertilizer	0.240	0.055	4.38	0.000***	0.132	0.347
Adoption	1.419	0.535	2.65	0.008***	0.370	2.467
Constant	3.237	0.431	7.50	0.000***	2.392	4.083

Source: Field survey ***, significant at 1%, ** significant at 5%, * significant at 10%. Note: Number of observation = 406, Wald chi² (5) = 66.27, Prob > chi² = 0.000.

Ramasamy, 1991). These studies found that with complementary inputs like fertilizer, irrigation and insecticides, the improved varieties of rice and maize did far better than the traditional seeds. The net effects of the Green revolution was that many countries that were hitherto net rice importers became net exporters leading to overall increased world output. Proponents of the Green revolution argued further that with expanded market as a result of exports, farmers had the opportunity to increase their output, leading to increased income, and for that matter poverty reduction.

On the other hand, critics argued that the Green revolution led to income inequalities in favour of large-scale farmers who had access to the complementary inputs. They stressed that since the high yielding varieties of rice and maize could not do well without the complementary inputs, little or no access on the part of poor small-scale farmers meant that they were often out-competed and marginalized by their well-to-do counterparts, leading to a further widening of the gap between them (Cleaver, 1972; Gadgil and Guha, 1995; Todaro and Smith, 2003). The implication then is that for the former to also benefit from the adoption of improved seeds, there should be conscious and affirmative efforts to support them in accessing the complementary inputs.

As indicated earlier, SSA and for that matter, Ghana, missed out of the initial Green revolution. However, with support from AGRA through the instrumentality of Kofi Annan, a former UN Secretary-General, the revolution has been re-introduced into the country. For the revolution to succeed this time, there is the need to correct the mistakes associated with the first one. Currently in Ghana, the fertilizer subsidy programme that was removed some years back has been restored. However, not only is the price of the input the same for all farmers, the mode of sale is such that large-scale farmers can have greater access to the disadvantage of small scale farmers. In the long run, if this is not checked the consequence would be that some small scale-farmers may buy the input at a higher price. Fortunately, in this study the probability of adoption was greater for small-scale farmers. Since they constitute over 90% of the farming population (MoFA, 2010c), they need to be supported. However, the fact that output increased with farm size means that large-scale farmers cannot be relegated to the background. Generally, both groups of farmers must be supported but there should be a conscious effort to ensure that large-scale farmers do not enjoy at the detriment of small scale farmers.

Conflict of Interests

The authors have not declared any conflict of interests.
List of acronyms: AGRA, Alliance for a Green revolution in Africa; CDF, cumulative density function; GDP, gross domestic product; GSSP, Ghana strategy support program; IFPRI, international food policy research institute; IMR, inverse mills ratio; JHS, Junior High School; MDGs, millennium development goals; MoFA, ministry of food and agriculture; NERICA, new rice for Africa; NRDP, rice dissemination project; PDF, probability density function; SRID, statistical, research and information directorate; SSA, Sub Saharan Africa; UN, United Nations.

REFERENCES

Barnow BS, Cain GG, Goldberger AS (1980). Issues in the analysis of selectivity bias. In E. Stromsdorfer and G. Farkas (Eds.), Eval. Stud. Rev. 5:42-59.
Carr Jr, VH (2001). Technology adoption and diffusion. Cornish and J. E Pratley, Melbourne, Australia.
Cleaver H (1972). The contradictions of the Green Revolution (GR). Monthly Rev. 24:2.
Evenson RE, Gollin D (2000). The green revolution: An end of century perspective. [online] Available from: http://web.williams.edu/Economics/wp/Gollin_The_Green_Revolution.pdf. [Accessed 24 June 2011].
Foltz JD (2003). The economics of water-conserving technology adoption in Tunisia: An empirical estimation of farmer technology choice. Econ. Dev. Cult. Change 51(2):359-373.
Gadgil M, Guha R (1995). Ecology and Equity: Use and Abuse of Nature in Contemporary India., New York: Routledge.
Hazell PBR, Ramasamy C (Eds.) (1991). The Green Revolution Re-considered. London: The John Hopkins University Press pp. 1-9.
Heckman JJ (1979). Sample selection as a specification error. Econometrica 47:153-161.
Janvry A, Sadoulet E (2002). World poverty and the role of agricultural technology: Direct and indirect effects. J. Dev. Stud. 38(4):1-26.
Johnson M, Hazell P, Gulati A (2003). The role of intermediate factor markets in Asia's Green Revolution: lessons for Africa? Am. J. Agric. Econ. 85(5):1211-1216.
Koutsoyannis A (1979). Modern Microeconomics. (2nd Edition), London: Macmillan.
Ministry of Food and Agriculture (MoFA) (2010a). Medium term Agricultural Sector Investment Plan (METASIP) National Rice Development Strategy (NRDS), Draft, Ministry of Food and Agriculture, Accra.
Ministry of Food and Agriculture (MoFA) (2010b). NERICA Rice Dissemination Project (NRDP). [Online] Available from: http://mofa.gov.gh/site/?page_id=4626. Access 28th June, 2013.
Ministry of Food and Agriculture (MoFA) (2010c). Agriculture in Ghana. Facts and Figures (2009).
NAPCDD (2003). National Action Program to Combat Drought and Desertification, Environmental Protection Agency, Accra, Ghana, April 2002.
Nzomoi JN, Byaruhanga JK, Maritim HK, Omboto PI (2007). Determinants of technology adoption in the production of horticultural export produce in Kenya. Afr. J. Bus. Stud. 5:129-135. [Online] Available from: http://www.academicjournals.org/ajbm. [Accessed 10 June 2013).
Rogers EM (1962). Diffusion of Innovations Glencoe: Free Press.
Smit J (2003). Estimating the Heckman's two-step procedure to control for selection bias with SPSS.[online] Available from: Http://Home.Planet.Nl/~Smits.Jeroen. [Accessed on 20 May 2011]
Sserunkuuma D (2005). The adoption and impact of improved maize and land management technology in Uganda. Electr. J. Agric. Dev. Econ. 2(1):67-84. [Online] Available from: ftp://ftp.fao.org/docrep/fao/008/ae694e/ae694e00.pdf [Accessed on 20 May 2011] 1.
Salasya B, Mwangi W, Hugo V, Odendo M, Dona J (1996). An assessment of the adoption of seed and fertilizer packages and the role of credit in smallholder maize production in Kakamega and Vihiga districts. CIMMYT Nairobi.
Todaro MP, Smith SC (2003). Economic Development London: Pearson, Addison Wesley.
Uaiene RN, Arndt C, Masters WA (2009). Determinants of agricultural technology adoption in Mozambique. Discussing P. 67E.
Wiredu AN, Gyasi KO, Abdoulaye T (2010). Impact of improved varieties on yield of rice producing households on Ghana. Household Survey, Ghana. Paper presented at the second Africa Rice Congress, Bamako, Mali, 22–26 March 2010: Innovation and Partnerships to Realize Africa's Rice Potential. [Online]: Available: http://www.africarice.org/workshop/ARC/3.6%20Wiredu%20fin.pdf [Accessed on 12 June 2012]

Measurement of technical efficiency of smallholder fish production in Ghana: A stochastic frontier approach

Christian Crentsil[1] and Felix Larry Essilfie[2]

[1]School of Economics, Keynes College, University of Kent, Canterbury, CT2 7NP, UK.
[2]Departments of Agricultural Economics and Agribusiness, College of Agriculture and Consumer Sciences, University of Ghana, Legon, Accra, Ghana.

This paper examines the level and determinants of technical efficiency of smallholder fish production in seven of the ten regions of Ghana. We employ the single-stage stochastic frontier approach in this study. Regional location, feed, fingerlings and labour are found to influence technical efficiency positively and significantly. However, formal education, marital status, membership in fish farmer groups and contact with extension services negatively influence inefficiency. Finally, estimates from the study indicate that the average smallholder fish producer in Ghana is 73.88% technically efficient. We recommend a bottom-up participatory approach to policy formulation which involves grassroots participation as well as the inclusion of aquaculture management in the curriculum of schools.

Key words: Smallholder fish production, technical efficiency, stochastic frontier.

INTRODUCTION

Globally, the contribution of fish farming to national development, both from poverty alleviation and national economic development points of view poses some interesting concerns. These concerns may be attributable to the fact that fish is a significant component of the diet for many people around the world, providing essential nutrition for over one billion people, and fish production also provides livelihood for over 200 million people in developing countries (The WorldFish Centre, 2007). In terms of trade, over 37% (by volume) of world fish production is traded internationally, the value of this being the highest in international trade in all animal proteins (World Bank, 2011). It is therefore not surprising that aquaculture has recently been adopted as a means to increase or supplement other sources to meet the deficit in Ghana's fish supply. In 2003, for instance, Ghana produced about 52% of its fish requirements from its domestic sources; this contribution increased to 68% in 2004 (FAO, 2005).

In Ghana, there are generally six major sources of fish ranging from imports, the open sea or marine fisheries, lagoon fisheries, the Lake Volta, other inland water bodies and pond fish production. In fact, of all these sources most of Ghana's fish supply comes from marine sources (Asmah, 2008), which contributed about 80% of the total quantum of domestic production between 1993 and 2000 (FAO, 2004). According to Braimah (2001) in Asmah (2008) Lake Volta is the single most important source of fish of all inland fisheries, supporting about 140 species of fish. Furthermore, the vast size of the Lake also lends itself to canoe fishing by fisher folk along its banks. It was estimated to have produced over 70,000 tonnes of fish in 2002 which is about 16% of total domestic production and 85% of inland fishery output

(Asmah, 2008). Fishing along the lake is seasonal: the peak season spans the months of July and August, while the lean season is from January to February. It is necessary to mention that apart from the Volta Lake there are many other water bodies from which fish is obtained for consumption, of which include Bosomtwi, Weija, Barekese, Tano, Vea and Kpong; altogether covering approximately 1,000,000 ha, and over 50 lagoons covering 40,000 ha (ibid). It is generally argued that in terms of fish, Ghana as a country has a self-sufficiency ratio of 60% meaning that the demand of fish exceeds supply by some 40%. According to Adutwum (2001) in Asmah (2008) the nation has over the years tried to meet this deficit through the importation of frozen fish. This raises concern for especially the small holder fish farmer who cannot effectively compete on the basis of prices: the imported frozen fish is cheaper and has increasingly become an important part of the diet of low income urban and rural consumers. Per capita fish consumption in Ghana is between 20 and 25 kg, making it one of the highest in Africa. This is an indication of the availability of a market for fish and fish products.

In spite of the high demand pointed out above available data indicates that the contribution of small-scale pond fish production to total quantum of fish produced in Ghana for local consumption is less than 1% (Abban et al., 2009). This is an interesting discovery and one that calls for some consideration and action especially at a time when aquaculture is poised to fill the gap between dwindling supply from major capture fisheries and the ever-increasing demand for fish and fishery products in several peer countries. For aquaculture to succeed in Ghana and to play its role towards food security and livelihood both the government and private sector must work together. Plausible as the above may be, there is the inherent assumption that fish farmers need more inputs to reach their potential. However, it is known that increase in agricultural production, and by inference fish production, may be attained through improvement in productivity, which can be increased through one or a combination of factors namely, technology, the types and quantities of resources used and the efficiency with which the resources are used. Of the various determinants, improvement in the efficiency of the resources already available to the farmers is most important (Goyal et al., 2006); hence the objective of this study was to determine the level of technical efficiency of smallholder fish production. Taking the above into consideration, this paper measures and analyzes the performance of small-scale fish farming households in Ghana. The paper applies a stochastic production frontier model, which measures the relative technical efficiency in a consistent way while also shedding light on the factors associated with these efficiency differences.

Statement of hypotheses

In this study, three hypotheses are tested. These hypotheses are;

1. $H_0 = \beta_{ij} = 0$, The null hypothesis that identifies an appropriate functional form between the restrictive Cobb-Douglas and the translog production function.
2. $H_0 = U_i = 0$, The null hypothesis specifies that each smallholder fish farmer is technically efficient and that variations in actual fish output (harvest) are due to random effects.
3. $H_0: \gamma = \delta_0 \ldots \delta_6 = 0$, The null hypothesis that inefficiencies are absent from the model at every level.

METHODOLOGY

Theoretical framework

The estimation of technical efficiency has been carried out with many different approaches, but the Stochastic Frontier Analysis (SFA) and Data Envelopment Analysis (DEA) have been given the most consideration by researchers. The basic difference between these two approaches lies in the method of analysis: the former employs econometrics while the latter uses mathematical programming. The SFA takes cognizance of the presence of stochastic noise-random shocks affecting the production process-outside the control of the producer, as well as technical error and also permits inferences to be drawn from estimation results (Coelli et al., 2005). In fact, the inclusion of the measurement error makes the frontier stochastic, whence the name stochastic frontier model is derived (Koop, 2003). As a consequence the SFA technique is considered appropriate for this study as such factors are expected to abound in smallholder fish farming in Ghana, a developing economy. Coelli et al. (2005) specified, in this case, a Cobb-Douglas Stochastic frontier model as:

$$Y_i = exp(\boldsymbol{\beta}_0 + \boldsymbol{\beta}_1 \ln x_{it}) \text{x } exp(v_{it}) \text{x } exp(-u_{it}) \quad (1)$$

Where: $exp(\boldsymbol{\beta}_0 + \boldsymbol{\beta}_1 \ln x_{it})$ is the deterministic component; $exp(v_{it})$ is the noise component, and $exp(-u_{it})$ is inefficiency. Some assumptions have been associated with the noise term in the existing literature:

1. It is independently and identically distributed (*i.i.d.*)
2. It is symmetric and
3. It is distributed independently of (u_{it})

In the estimation of the effects of the exogenous variables on the technical efficiency of farms, the two main procedures are the one-step and the two-stage modelling. In the two-stage procedure the production function is first estimated and the estimates of the technical efficiency of each farm are obtained. These are then regressed on farm-specific variables known or hypothesised to influence the efficiency. Critics of the two-stage approach -including Kumbhakar et al. (1991) and Reifschneider and Stevenson (1991) argue that this procedure is inconsistent and some of the assumptions of the error term, such as independent distribution are violated in the second stage (Pascoe and Mardle, 2003), and hence it is biased (Wang and Schmidt, 2002) and not as efficient as the single-stage procedure (Reifschneider and Stevenson, 1991). In contrast to this approach is the single-step, which incorporates all the variables affecting either the production function or contributing to inefficiency. In essence the relation between μ_i and z_i is established by this procedure, using the maximum-likelihood estimation (Wang and Schmidt, 2002); hence in this study the one-step modelling procedure is adopted.

Determination of technical efficiency

The technical efficiency (TE) of a given firm is illustrated as the ratio of the output obtained from that firm in comparison to the output of the best producing (frontier) firm using the same technology, as:\

$$TE = \frac{yi}{exp(x_i \boldsymbol{\beta})} = \frac{exp(xi\boldsymbol{\beta} - u_i)}{exp(x_i \boldsymbol{\beta})} = exp(-u_i) \qquad (2)$$

Empirical models

In order to determine the effects of predetermined variables on the value of pond fish production, as well as the efficiency of resources used, the translog stochastic production function is estimated, being motivated by the fact that this functional form has been widely used in frontier production studies and it is also flexible to use (Onyenweaku and Okoye., 2007; Onumah and Acquah, 2010).

The following translog model is used in this study to arrive at the technical efficiency of the smallholder fish farmers in Ghana:

$$\begin{aligned} \ln Q = {} & \beta_0 + \beta_1 \ln labour + \beta_2 \ln capital + \beta_3 \ln fingerlings + \beta_4 \ln feed + \\ & \frac{1}{2}\beta_5 (\ln labour)^2 + \frac{1}{2}\beta_6 (\ln capital)^2 + \frac{1}{2}\beta_7 (\ln fingerlings)^2 + \frac{1}{2}\beta_8 (\ln feed)^2 + \\ & \beta_9 \ln(labour) * \ln(capital) + \beta_{10} \ln(labour) * \ln(fingerlings) + \\ & \beta_{11} \ln(labour) * \ln(feed) + \beta_{12} \ln(capital) * \ln(fingerlings) + \\ & \beta_{13} \ln(capital) * \ln(feed) + \beta_{14} \ln(capital) * \ln(\ln feed) + \\ & \beta_{15} \ln(fingerling) * \ln(feed) + Reg + v_i - u_i \end{aligned} \qquad (3)$$

Where: Q refers to the total output of fish harvested in kilograms; *labour* is labour (man-hours) employed during the production season *capital* is the area of all ponds used in production of fish in the production season in hectares; *feed* (in kilograms) is feed[1] fed to the fish during the production season; *fingerlings* count) is the number of fingerlings stocked at the start of the production season; *Reg* is a dummy, used as a proxy to capture regional effects on the efficiency of fish production by smallholder fish farmers in the different regions. Seven (7) dummies are constructed from this variable, where for a particular region, say Greater Accra the *Reg* takes on a value of 1, and zero for all other regions. In the estimation of the model, one of the dummy regional variables is excluded to conform to the $n-1$ degree of freedom restriction when using dummy variables, and also to avoid the problem of perfect correlation among the dummies and the constant; β_is are regression parameters to be estimated and are as defined earlier.

The empirical technical inefficiency model

The model for various operational and farm-specific variables hypothesised to influence technical inefficiencies of fish farms in Ghana is defined as:

$$\begin{aligned} Zi = {} & \delta_0 + \delta_1 land_i + \delta_2 sex_i + \delta_3 techad_i + \delta_4 ffa_i + \\ & \delta_5 edu_i + \delta_6 maristat_i + W_i \end{aligned} \qquad (4)$$

In this inefficiency model, *land* is a dummy, capturing the effect of land tenure on the efficiency of individual smallholder fish farmers. It has a value of 1 if freehold, otherwise 0; *sex* is a dummy, and has a value of 1 if decision maker is male, 0 otherwise; *techad* is a dummy variable and has a value of 1 if farmer was visited at least once by an extension officer, 0 otherwise; *ffa* is a dummy variable indicating whether the farmer/farm is a registered member of the local fish farmers' association. It has a value of 1 if yes, 0 otherwise; *edu* is the maximum level of formal schooling of the farm owner/manager; *maristat* is a dummy and is an indication of marital status of respondent; it takes a value of 1 if married, 0 otherwise; W is the 'error term' in the model; $\delta_0 - \delta_6$ are parameters to be estimated along with the variance parameters σ^2 and γ.

It must be noted that while the σ^2 is an indicator of how well the functional form specified fits the data and also the appropriateness of the assumption underlying the distributional form of the composed error term, the γ tests whether or not the dominant sources of errors are outside the deterministic part of the production function (Umoh, 2006).

Data description

The data set for this study is secondary and it comes from primary information on aquaculture development in Ghana obtained via questionnaires in 7 of the 10 regions, namely the Greater Accra, Eastern, Ashanti, Volta, Western, Central and Brong-Ahafo Regions. The list of smallholder fish farmers in each region was obtained from the Fisheries Directorate's regional offices. From the list respondents were then randomly sampled and interviewed with structured questionnaires. To facilitate data collection, questionnaires were completed with the farm owner or manager, whichever was available, at the time of visit. Where none was present, the farmers were traced to their homes where the data was gathered. Collection of data by phone was done only in one instance, which was to the owner of a commercial farm whose manager was not ready to give out any information. The surveys were conducted between June 2006 and December 2006. Primary data from134 fish farms were obtained, 124 of which are purposively selected (based on intensity of production) from seven of the ten regions (Greater Accra, Ashanti, Volta, Brong Ahafo, Central, Eastern and Western) for this study. This was motivated by the fact that the remaining three northern regions had no record of pond fish farms, and that these seven regions had 966 pond fish farms, spanning more than half of the entire country. Ecologically, the seven regions fairly represent the climatic conditions of the country.

A limitation to this secondary data set was the inconsistency in data entry and incomplete records (especially for costs and outputs). This may be due to the fact that the farmers did not know the basic booking-keeping methods or were afraid to release financial information for fear of taxation (Hiheglo, 2008).

RESULTS AND DISCUSSION

Summary statistics

This study was conducted to provide baseline information for subsequent monitoring of smallholder fish production efficiencies to assess the impacts of changes in the agricultural policy environment on selected socio-economic factors in the study area. Table 1 shows the summary statistics for the variables used in the stochastic frontier model. The mean harvest (output) per hectare was 266.10 kg. This was obtained by using: 0.39 ha of

[1]Feed is a composite term for all food items given to the fish during the production season.

Table 1. Summary statistics of quantitative variables used in the Stochastic Frontier Model.

Variable description	Unit	Mean	Standard deviation
Harvest (Output)	Kg/ha	266.10	706.00
Pond area	ha	0.39	1.05
Fingerlings	Num/ha	1018.64	3629.32
Feed	Kg/ha	400.80	557.60
Labour	Man - hours	464.38	302.43
Years of education	Years	9.06	5.17

Source: Authors' Computation from FAO dataset (2005).

pond area, 1018 fingerlings, 400.8 kg of feed and 464.8 man -hours of labour, by fish farmers with an average of about 9 years of formal education.

In this study we adopted the single-stage modeling technique and Table 2 is a descriptive statistics of the demographic variables used in this study. There were more male fish farmers than female fish farmers. The result also indicates that about 71% of the sample were members of FFA, while more (89.52%) of them had contacts with extension agents. Having more males is no indication of male dominance in the industry as workers, but rather as the main decision makers and heads of family businesses. It is interesting to note that membership of FFAs is on the higher side, with more extension contacts. From the perspective of policy intervention, policy makers may have to consider reaching farmers with new innovations and better ways of improving efficiencies through their farmer associations using a participatory approach.

Technical efficiency measurement of smallholder fish production in Ghana

Table 3 shows the estimation of the maximum likelihood estimates for parameters of the general translog stochastic production frontier and technical inefficiency effect models for smallholder fish production in Ghana.

Whereas labour and capital had positive and significant coefficients at 1%, fingerlings had a positive and significant coefficient but at 10%. This is an indication that stocking a pond with fish does not necessarily determine the yield obtained but rather the optimum combination of other relevant input factors. Feed, however, had a negative and significant coefficient at 1%; hence excess feeding regime may have been detrimental to the growth and development of the fish stocked.

From the results in Table 3, the output of fish would increase by about 1.3 kg with every 10% increase in man-hours. A 1% increase in the number of fingerlings and capital will result in 18 and 62.75% respectively in the output of fish produced.

The input variable that should be of greatest interest to policy makers is feed. Optimum amounts of feed and the adoption of the most effective feeding regime during the production cycle would help improve output of fish and hence efficiency. Thus to improve productivity primary interest should be on research to determine the optimum amount of feed and the right combination of feed nutrients. Interaction between variables resulted in some important findings. Feed alone as an input had a negative significant effect on the output, but an interaction between feed and capital had a positive significant effect on output. On the other hand a combination of feed and labour as well as feed and fingerlings were significantly negative. This confirms the previous assertion that the right proportion and composition of feed had a very important role to play in the output of fish and hence the efficiency of fish production

Inefficiency model estimates

The estimates for the inefficiency model are reported in the lower section of Table 3. Estimated coefficients of formal education, gender, membership in ffa, technical advice and the dummies for the regions were all significantly negative. These imply that fish farmers who have formal education were more technically efficient than those who had none; female farmers/farm owners were more efficient; members of ffas were more efficient, and farmers in all the regions under consideration except the Brong Ahafo Region, were relatively more efficient than their counterparts in the Ashanti Region. Crentsil (2009), however, in his study concluded that the Ashanti Region was the best fish producing region in Ghana. The point to note here is that the output of a farm does not necessarily correlate with its efficiency, because technical efficiency simply relates the output to the input used. Onumah and Acquah (2010), however, found regional differences to be insignificant in the variation of technical efficiency among smallholder fish producers in Ghana. We therefore conclude that the right combination of inputs bear much more on the output and hence efficiency rather than the location of the farmer, even though the latter cannot be ignored as a determinant of the variation in efficiencies among respondents in this study.

Battese et al. (1996) also found a positive significant

Table 2. Descriptive statistics on other demographic features of the smallholder fish farmers.

Variable names	Frequency	Percentage
Gender (1 = Male; 0 = female)		
Male	93	75.00
Female	31	25.00
Marital status (1 = married; 0 = single)		
Married	108	82.10
Single	16	12.90
FFA membership (1 = member; 0 = non – member)		
Member	88	70.97
Non - member	36	29.03
Land tenure (1 = owner; 0 = tenant)		
Owner	82	66.13
Tenant	42	33.87
Access to technical advice (1 = yes; 0 = no)		
Yes	111	89.52
No	13	10.48
Western region (1 = yes; 0 = no)		
Yes	12	9.68
No	112	90.32
Eastern region (1 = yes; 0 = no)		
Yes	9	7.26
No	115	92.74
Central region (1 = yes; 0 = no)		
Yes	20	16.13
No	104	83.87
Brong Ahafo (1 = yes; 0 = no)		
Yes	26	20.97
No	98	79.03
Greater accra (1 = yes; 0 = no)		
Yes	3	2.42
No	121	97.58
Volta region (1 = yes; 0 = no)		
Yes	22	17.74
No	102	82.26

Source: Authors' Computation from FAO dataset (2005).

relationship between education and technical efficiencyof farmers. Chiang et al. (2004) and Onumah and Acquah (2010), on the contrary found a negative correlation between education and technical efficiency, but indicated that technical know-how had greater influence on productivity than general formal education. It was not surprising to discover that members of ffas were more efficient, because as indicated elsewhere in this study members of a group learn from each other and get assistance from other members of the team, hence could be expected to be more efficient than the non-member, generally.

Technical advice in Ghana takes various forms. It includes informal meeting with an extension agent by a single farmer or as a group of farmers to discuss issues regarding their operations. In this study farmers who had access to technical advice were generally more efficient than those who did not. This has policy implications, because by this outcome, therefore, it may be suggested that more contacts with extension agents could further increase the efficiency of smallholder fish farmers.

γ is a measure of level of the inefficiency in the variance parameter, it ranges between 0 and 1. For the translog model, γ is estimated at 0.7992. This is an indication that about 80% of the random variation in fish production is attributable to inefficiency and the remaining 20% to stochastic factors. In other words, the variation in the output of fish is attributable to factors under the control of farm units much more than stochastic factors. The implication of these findings is that in formulating policy to help boost productivity of farmers, policy makers should

Table 3. Maximum-likelihood estimates of parameters of the Translog Frontier production function for smallholder fish farmers in Ghana.

Variable	Coefficient	Std. Error	t – Value	P - Value
Stochastic Frontier				
Constant	0.4433	0.0613	7.23	0.000
lnlabour	0.1310***	0.0450	2.91	0.004
lncapital	0.6275***	0.1201	5.22	0.000
lnfingerlings	0.1790*	0.0993	1.80	0.071
lnfeed	-0.5077***	0.1445	-3.51	0.000
½(lnlabour)^2	-0.1848***	0.0392	-4.71	0.000
½(lncapital)^2	0.0966***	0.0278	3.47	0.001
½(lnfingerlings)^2	0.4008	0.4018	1.00	0.319
½(lnfeed)^2	0.1181	1.3679	0.09	0.931
ln(lab)*ln(cap)	0.0614	0.0407	1.51	0.132
ln(lab)*ln(fing)	0.1887*	0.1130	1.66	0.061
ln(lab)* ln(feed)	-0.0513***	0.0105	-4.89	0.000
ln(cap)* ln(fing)	-0.3099**	0.1410	-2.20	0.026
ln(cap)*ln(feed)	3.3906***	1.0899	3.11	0.002
ln(fing)*ln(feed)	-1.9822***	0.7342	-2.70	0.007
Inefficiency Model				
Constant	-4.1868	1.5033	-2.79	0.005
Formal education	-0.1251***	0.0251	-4.98	0.000
Gender	-0.4348***	0.1801	-2.41	0.007
Marital status	0.2688	0.6921	0.39	0.698
FFA membership	-0.8930***	0.3196	-2.79	0.006
Land tenure	0.2730	0.2682	1.02	0.309
Technical advice	-0.3396***	0.1208	-2.81	0.003
Western region	-1.8439**	0.7300	-2.53	0.012
Eastern region	-2.0078*	1.0453	-1.92	0.055
Central region	-2.1514**	1.0227	2.10	0.035
Brong Ahafo Reg.	0.6034	0.9764	0.62	0.537
Greater Accra Reg.	-0.8092*	0.4648	-1.74	0.052
Variance parameter				
Sigma squared	0.1803***	0.0292	6.17	0.000
Gamma	0.7992***	0.2508	3.19	0.000
Lambda	1.9949	0.4768		
Log likelihood	-56.24704			
Mean TE	73.88%			

***Significance at 1%; **significance at 5%; *significance at 10%.

not merely think about increasing inputs or making credit available but that a means should be found to conduct efficiency monitoring and evaluation at the farm level with the view to creating awareness about the causes of farm level inefficiency. This finding is also a major indicator of how future policy interventions should be formulated and implemented: not top-down but bottom-up by employing participatory methods. Thirdly, it suggests the need for a follow-up on qualitative research to seek to understand qualitative underpinnings for inefficiency in Ghanaian smallholder fish production in greater depths. The σ^2 value of 0.18, highly significant at 1% is an indication of quite a good fit of the translog model for the data.

Distribution of the technical efficiency of smallholder fish farmers in Ghana

The overall mean technical efficiency of the sample was 73.88%. Stating this figure alone without further analysis of the performance of individual farms could be misleading, to say the least. It may be seen from Table 4 that most farms (43.5%) had technical efficiency scores

Table 4. Distribution of the overall technical efficiency of smallholder fish farmers in Ghana.

T.E Class	No. of fish farmers	Percentage
≤0.50	43.5	54
0.51 - 0.60	11	8.8
0.61 - 0.70	17	13.7
0.71 - 0.80	14	11.4
0.81 - 0.90	21	16.9
0.91 - 1.00	7	5.7

Source: Author's computation.

Table 5. Regional technical efficiency distribution of smallholder fish farmers in Ghana.

T.E Class	WR (%)	ER (%)	CR (%)	BA (%)	GA (%)	VR (%)
≤0.50	74.0	33.1	55.0	21.9	0.0	50.0
0.51 - 0.60	0.0	0.0	10.0	11.5	0.0	13.5
0.61 - 0.70	9.3	22.2	5.0	7.7	0.0	4.5
0.71 - 0.80	0.0	44.7	10.0	15.4	33.1	9.2
0.81 - 0.90	16.7	0.0	20.0	26.1	66.9	18.2
0.91 - 1.00	0.0	0.0	0.0	17.4	0.0	4.6

Source: Author's computation. Number of Fish Farmers: WR = 12; ER = 9; CR = 20; BA = 26; GA = 3 and VR = 22; Mean TEs: WR = 49.5%; ER = 61.0%; CR = 55.4%; BA = 66.0%; GA = 81.2% and VR = 59.5%; TE ranges: WR = 0.1375279 to 0 .8998032; ER = 0.2428515 to 0.7899342; CR = 0.0911026 to 0.8977706; BA = 0.1799706 to 0.9317483; GA = 0.809598 to 0.8621432 and VR = 0.0777726 to 0 .933502

below 50%. However, about 6% of farmers operated on or very close to the frontier. The results also indicate that the least efficient farm needs to improve its technical efficiency by some 23.9% to attain the mean efficiency score and the average farmer needs to adopt the best technology of the frontier farmers to increase its efficiency score by at least 26%.

The outcome of this study also brings to the fore the fact that only a small percentage of farmers is near the frontier and therefore policies to improve efficiency need to critically identify the factors responsible for the discrepancy in technical efficiency among farmers. Could regional differences contribute to these differences? This is considered in Table 5.

Regional technical efficiency distribution of smallholder fish farmers in Ghana

In Table 4 it was concluded that most farms operated below the mean technical efficiency score. We therefore assessed the technical efficiency of the sample based on the region within which they operated. From Table 5 it may be seen that on average the Greater Accra and the Western Regions were the most and least technically efficient regions respectively. However, considering the frontier most farmers (17.4%) in the Brong-Ahafo for instance, operated closest to or on the frontier, though in the same region majority of farmers (21.9%) operated at efficiency levels below 50%. This is an indication that even within the same region variation in efficiency could be observed among farmers. Consequently, though regional differences could explain some of the differences in technical efficiency, much more importantly the operations of individual farmers are very critical in this distinction. Onumah and Acquah (2010) however, concluded that region of production played no significant role in explaining the differences in technical efficiency.

Technical efficiency distribution of smallholder fish farmers in Ghana by gender

On average female respondents were more technically efficient than their male counterparts; however about 4.8% of males had efficiency scores between 0.91 and 1.00 (Table 6), an indication that males operated closer to the frontier than females. In a similar study to measure technical efficiency of maize farmers in the Mfantseman Municipality in Ghana, Essilfie et al. (2011) discovered that female maize farmers were more technically efficient, stating that males were more likely to be involved primarily with the production of cash crops.

In their study of aquaculture in Southern Ghana, Onumah and Acquah (2010) also concluded that males were generally more technically efficient, citing the strenuous and laborious nature of fish farming as a reason. The implication of these varied findings is that

Table 6. Gender technical efficiency distribution of smallholder fish farmers in Ghana.

T.E Class	Male (%)	Female (%)
≤0.50	44.2	35.0
0.51 - 0.60	9.6	15.0
0.61 - 0.70	10.6	20.0
0.71 - 0.80	12.5	10.0
0.81 - 0.90	18.3	20.0
0.91 - 1.00	4.8	0.0

Source: Authors' Computation from FAO dataset (2005). Mean TE: Male = 56.9% and Female = 60.6%.

Table 7. Likelihood ratio tests.

Variable description	Chi^2	Df	P > | Z |	Decision
$H_0 = \beta_{ij} = 0$	42.01	17	0.000	Reject H_0
$H_0 = U_t = 0$	50.23	10	0.000	Reject H_0
$H_0 \mid \gamma = \delta_0 \ldots \delta_6 = 0$	36.49	13	0.000	Reject H_0

Source: Authors' Computation from FAO dataset (2005).

gender may not be conclusive in explaining the variations in technical efficiency among smallholder fish farmers in Ghana.

Tests of hypotheses

As was indicated earlier in this paper, to ensure that the estimation procedure and thus the results obtained were as reliable as possible, we carried out tests on the hypotheses stated. For the first null hypothesis, a nested hypothesis test was performed to determine whether the Cobb-Douglas specification is an adequate representation of the frontier production function. This test uses the log Likelihood ratio test.

Table 7 outlines the results of the null hypothesis. The null hypothesis $H_0 = \beta_{ij} = 0$ is rejected in favour of the translog production function. The second null hypothesis explores the test that specifies each smallholder fish farmer is operating on the technically efficient frontier and that the systematic and random technical efficiency in the inefficiency effects are zero. This is rejected in favour of the presence of inefficiency effects. The final null hypothesis determines whether the variables included in the inefficiency effects model have no effect on the level of technical inefficiency. This is also rejected confirming that the combined effect of these variables on technical inefficiency is statistically significant.

Conclusion

The main objective of this paper was to determine the levels and the factors affecting the technical efficiency of smallholder fish production in Ghana. We started off on the premise that different farms would have different levels of technical efficiency owing to farm-specific factors such as the level of experience of the farm owner, the tenure of land, among others; hence these were incorporated in the stochastic frontier in a single-stage modeling procedure. The results of the study showed that the labour employed, the number of fingerlings stocked and the quantity of feed used were positive and significant determinants of technical efficiency among smallholder fish farmers.

Furthermore, interaction between some exogenous variables were found to have significant and positive effects on the endogenous variable and hence efficiency. For instance it was demonstrated that if pond area simultaneously increased with number of fingerlings feed and labour, *ceteris paribus*, the total output of fish would increase (Table 2). This indicates a holistic approach is needed to improve efficiency.The effect of geography on the efficiency of production as captured by the coefficients of the regions indicates that the region within which a farmer operates does have an effect on technical efficiency. On the average the most technically efficient region was the Greater Accra Region (81.2%); a further study to find out how farmers in this region attain such technical efficiency scores is recommended, to serve as the basis for improving the efficiencies of the other regions. The overall average technical efficiency among smallholder fish producers was estimated to be 73.88%. This means that there is the possibility of increasing the efficiency level by some 26.12% if the best practices of the frontier farmer could be emulated and the necessary support given by the government. The average efficiency scores are however, not very representative of the sample since the standard deviations are high and therefore the distribution of farmers according to efficiency indices is tabulated, from which it is concluded that inefficiency among the respondents does not lie only in over-utilization, but also underutilization of significant inputs.

These findings have very important policy implications. Since technical advice enhances efficiency, training members of fish farmer associations by extension agents could help reduce over-feeding and improve on the technical efficiency and hence output. Formal education should be encouraged in the study area, and where possible fish production should be included in the agriculture and integrated science syllabi in the primary and secondary schools, since this may help improve efficiency of fish production in the future. Increases in pond areas will result in the reduction of output, but membership in FFAs could improve output, therefore the formation of more fish farmer cooperative societies is hereby recommended so that the more highly educated and the less educated ones will have the opportunity to learn from each other and members should be encouraged to construct smaller ponds for easier management and hence improve efficiency. Furthermore, membership in fish farmers' associations is a very important determinant of technical efficiency and this medium could be used as the platform for discussing important innovations that could improve efficiency.

Involving fish farmers in the drafting of policies is a recommendation worth noting, especially because most of the variations in technical efficiency result from factors directly under the control of farmers rather than from stochastic factors. A participatory bottom-up - rather than the traditional top-down - approach to solicit the view of farmers before formulating policy interventions would help in the adoption of innovations and hence the sustainability of such interventions.

Conflict of Interests

The author(s) have not declared any conflict of interests.

REFERENCES

Abban EK, Asmah R, Awity L, Ofori JK (2009). Review on National policies and programmes on aquaculture in Ghana, SARNISSA Publication, Ghana, P.83.

Adutwum RO (2001). Policies and Strategies for Food Security in Ghana; A paper presented at a workshop organized by Reseau Ghaneen of SADAOC Foundation, ISSER, University of Ghana, Legon.

Asmah R (2008). Development potential and financial viability of fish farming in Ghana, Doctor of Philosophy Thesis, Institute of Aquaculture, University of Stirling, Canada.

Battese GE, Malik SJ, Gill MA (1996). An investigation of technical inefficiencies of production of wheat farmers in four districts of Pakistan. J. Agric. Econ. 47:37-49.

Braimah LI (2001). Volta Lake Fisheries Management Plant. 162 Fisheries Subsector Capacity Building Project of the Department of Fisheries.

Coelli T, Rao PDS, O'Donnell CJ, Battese GE (2005). An introduction to efficiency and productivity analysis (2ed), Springer Science + Business Media, Inc., New York, USA.

Chiang FS, Suna CH, Yu J.M (2004). Technical efficiency analysis of milkfish (*Chanos chanos)* production in Taiwan-an application of the stochastic frontier production function. Aquaculture 230(1-4):99-116.

Crentsil C (2009). Production function analysis of fish production in the Amansie West District, Bachelor of Science dissertation (unpublished), Department of Agricultural Economics, Agribusiness and Extension, Kwame Nkrumah University of Science and Technology, Ghana.

Essilfie FL, Asiamah MT, Nimoh F (2011). Estimation of farm level technical efficiency in small scale maize production in the Mfantseman Municipality in the Central Region of Ghana: A stochastic frontier approach. J. Dev. Agric. Econ. 3(14):645-654.

Food and Agriculture Organization (FAO) of the United Nations (2005). Results of the 2003-2004 baseline survey of Ghanaian fish farmers, FAO Regional Office for Africa, Accra, Ghana.

Food and Agriculture Organization (FAO) of the United Nations FAO. (2004). Information on Fisheries Management in the Republic of Ghana. Rome, Italy. http://www.fao.org/fi/fcp/en/GHA/body.htm. Accessed 08/07/2012.

Goyal SK, Suhag KS, Pandey (2006). An estimation of technical efficiency of paddy farmers in Haryana State of India. Indian J. Agric. Econ. 61(1):108-122.

Hiheglo PK (2008). Aquaculture in Ghana; prospects, challenges, antidotes and future perspectives, Master of Science Thesis, Department of Social Science and Marketing Studies Norwegian College of Fishery Science, University of Tromsø, Norway.

Koop G (2003). Bayesian Econometrics, John Wiley and Sons Ltd, England.

Kumbhakar SC, Ghose S, McGuckin JT (1991). A generalized production frontier approach for estimating determinants of inefficiency in U.S. dairy farms. J. Bus. Econ. Stat. 9:279-286.

Onumah EE, Acquah HD (2010). Frontier analysis of aquaculture farms in the Southern sector of Ghana. World Appl. Sci. J. 9:826-835.

Onyenweaku CE, Okoye BC (2007). Technical efficiency of small-holder cocoyam farmers in Anambra State, Nigeria: A translog stochastic frontier production approach. Int. J. Agric. Rural Dev. 9(1):1-6.

Pascoe S, Mardle S (Eds). (2003). Efficiency analysis in EU fisheries: Production frontiers and Data Envelopment Analysis. CEMARE Report 60, University of Portsmouth, UK.

Reifschneider DR, Stevenson R (1991). Systematic departures from the frontier: a framework for the analysis of firm inefficiency. Int. Econ. Rev. 32:715-723.

The WorldFish Centre (2007). The threat to fisheries and aquaculture from climate change: A Policy Brief, The WorldFish Centre, Malaysia. http://www.worldfishcenter.org/resource_centre/ClimateChange2.pdf Accessed 15/05/2011.

Umoh G (2006). Resource use efficiency in urban farming: An application of stochastic frontier production function. Int. J. Agric. Biol. 8(1):38-44.

Wang HJ, Schmidt P (2002). One-step and two-step estimation of the effects of exogenous variables on technical efficiency levels. J. Product. Anal. 18:129-144.

World Bank (2011). The global program on fisheries strategic plan of action for fisheries and aquaculture, FAO, Rome, Italy.

8

Extent and adoption determinants of floating tray technology by small holder tobacco farmers: A case of Zimbabwe

Dadirayi Manyumwa, Never Mafuse, Munyaradzi Matovore, Joseph Musara, Vincent Tinashe Munyati, Joseph Chimvuramabwe, Godfrey Chagwiza, Emmanuel Zivenge and Violet Dudu

Department of Agricultural Economics, Education and Extension, Faculty of Agriculture and Environmental Sciences, Bindura University of Science Education, Bindura, Zimbabwe.

The study was on the extent and adoption determinants of floating tray technology by small holder tobacco farmers in Bindura District of Mashonaland province. The objectives of the study were to determine the extent of adoption, identify adoption determinants that influence farmer's decision to take up floating trays. The study aimed also to identify challenges and opportunities of the floating trays. Questionnaires, focus group discussion and observations were used to gather data. The enumerator had face to face interviews with the farmers and clarified areas were the farmers need clarity on issues in the questionnaire. The logit regression model was used to determine the factors that influence adoption decision. Descriptive statistics was also used to analyse the challenges and opportunities. The major challenges were identified as high costs of technology and lack of capital with reduced input costs as an opportunity. Factors that influence farmers' decision to adopt floating tray technology were identified and these include training, educational status and off-farm hours. Age, training and off-farm hours were found to have significant influence on the adoption decision. Results show that training of the household head has a significant impact as shown by the marginal effect of 0.041. Since tobacco seedling using floating tray system is too costly and requires too much capital to establish, the researcher's hypothesis that there are challenges and opportunities in using floating tray system was not rejected. Recommendations made were to advocate for intervention in the provision of low cost or subsidize the floating tray system for the small holder tobacco farmers to increase the adoption of floating tray system.

Key words: Adoption, float trays, logit regression model.

INTRODUCTION

Zimbabwe is an agro based developing economy where agriculture is the vehicle to economic development prospects (GOZ, 2009). It provides employment and incomes for 70% of the population, 60% of the raw materials required by the industrial sector and 45% of total export earnings and, except in years of severe drought, sufficient food to feed the nation (FAO, 2010). Sustainable agricultural development should be ecologically sound, economically viable and socially responsible. It is through agricultural transformation that sustainable economic growth can be achieved (Rukuni et al., 2006).

Zimbabwe is one of the major tobacco exporters in the world. Tobacco (*Nicotiana tabaccum*) is of paramount importance to the agriculture sector and the economy of Zimbabwe as it is one of the country biggest foreign currency earners (FAO, 2010). Increases in both planting areas and yields have contributed to a significant increase in output of tobacco over the past decades. Large-scale commercial farmers used to dominate tobacco production but however this trend has changed with small scale farmers now dominating production (Z.T.P., 2010). The Large scale farmers are characterized by large land holdings, use of modern machinery, and permanent wage labour as compared to their small scale counterparts who are resource constrained (FAO, 2010). Over the past decades, methyl bromide has been used to fumigate tobacco seedbeds. However, the use of this conventional method in seedling tobacco production has been observed to have negative effects to the environment (Z.T.P, 2010). There has been a global call to phase out methyl bromide by 2015. Zimbabwe has specific regulations towards attaining this. Various stakeholders including the Tobacco Research Board (TRB), Agricultural Extension and Technical Services (AGRITEX) and United Nations Industrial Organization have advocated for the use of the float tray system in seedling production as an alternative. Tobacco growers received training in the use of float tray system for tobacco fumigation and have on average invested US$353 per hectare for the trays which last up to 7 years (Mazarura, 2004). Float tray system is an option of seedling production that has been tried, tested and used successfully in many tobacco growing countries such as Brazil, Malawi and the USA (Mazarura, 2004). Besides being more eco-friendly, the float tray system produces seedlings with intact root system and result in a more uniform crop. It is also notably labour extensive and cost effective as compared to conventional seedling production.

The contribution of new technology to economic growth can only be realized when and if the new technology is widely diffused and used (Namara et al., 2007). In most settings, diffusion of agricultural innovations has been seen as the cumulative or aggregate result of a series of individual calculations that weigh the incremental benefits of adopting a new technology against the costs of change. This is often in an environment characterized by uncertainty (as to the future evolution of the technology and its benefits) and by limited information (about both the benefits and costs and even about the very existence of the technology). There has been an apparent overall slowness and the wide variations in the rates of acceptance of the float tray system by most tobacco farmers in Zimbabwe (Namara et al., 2007). Thus, understanding the workings of the diffusion process is essential to understanding how technological change actually comes about and why it may be slow at times. Exploring the determinants affecting choice is essential both for economists studying the determinants of growth and for the creators and producers of such technologies (Mansfield, 1968). For this study, this analysis will be done in the context of tobacco farmers in Bindura District.

Problem statement

Unsustainable tobacco seedling production using methyl bromide is currently used in Zimbabwe's farming communities (Mazarura, 2004). Even though the floating tray has been identified as an option to convectional tobacco seed farming to guard against ozone depletion, the uptake of this technology has been marginal in most small farming communities (Mazarura, 2004). It is therefore imperative to understand challenges and opportunities and the dynamics of adoption of floating tray system.

METHODOLOGY

Description of study area

The study was carried out in Bindura District of Mashonaland Central. Bindura is located in the Mazowe Valley about 88 km North-east of Harare. Bindura is in natural region 2b and receives high rainfall. The area experiences warm summers and cool winters. Rainfall ranges between on average 700 and 1000 mm per annum and effective rainfall is 500 to 635 mm (AREX, 2008). This is enough for intensive crop production. The main crops grown include tobacco, maize, soya beans and winter wheat. Intensive livestock production is also practiced.

Data collection techniques

Both qualitative and quantitative data were collected using a number of approaches and tools. These techniques allow the researcher to systematically collect information about objects of the study (people, objects, phenomena) and about the settings in which they occur (De Vaus, 1996). Structured questionnaires, structured interview guide and observations were used for primary data collection. The researcher also conducted one focus group discussion to triangulate data collection. Stakeholders such as AGRITEX and TRB were interviewed to get their views on the topic. The researcher took into account ethical considerations to minimize the issues of bias information by conducting focus group discussion.

Sampling procedure

Purposive sampling was initially used to identify tobacco farmers in ward 3 Bindura district where tobacco production is extensively done. The sampling frame comprised of 115 farmers of which 23 were adopters and 92 non-adopters. Proportionate random sampling was used with adoption being the stratum. From the sampling unit, 47 farmers with 5 and 42 adopters, respectively became a sampling unit. This allowed generation of a representative sample since the proportions of adopters and non-adopters were different. Snow balling sampling was also employed. The random selection meant that there were equal chances of

Table 1. Description of floating tray adoption variables.

Variable	Description	Units	Adopters		Non-adopters	
			Mean	SD	Mean	SD
Age	Age of household head	Year	32		50	
Household size	Number of family members	Number	5.4	1.3	3.8	0.8
Number of years in school	Years of schooling by household age	Number	13.8	1.7	8.1	1.5
Land size	Size of arable land	Hectare	3.3	1.2	1.9	0.7
Off farm activities	Access to off farm activities	percentages	5.1	11.9	6.7	13.8

Source: SPSS.

Table 2. Gender in tobacco farming.

Gender	Percentage
Male	66
Female	34

selection for each farmer into the sample. Farmers were free to participate and were not coerced.

Data analytical tools

Logit model

Since in this study, the dependent variable (adoption) has only two possible values for either adopting or not adopting, the logistic regression model was used. Without any loss in generality, the researcher coded these two outcomes as (Y = 1 for adoption) and (Y = 0 for not adopting) thus giving rise to a binary dependent variable. Since the response was qualitative in nature, the researcher used qualitative response model, (the logit model), which is a nonlinear model since binary dependent model ("dummy Ys") cannot be used freely in linear regression models. The aim was to find the conditional expectation of the dependent variable given more than one conditional variables. The likelihood of observing the dependent variable (P) was tested as a function of variables which include gender of household, age of household and training of the household.

$$Z_i = \text{Log}\ (P_i/1\text{-}P_i) = \alpha_0 + \alpha_1 X_1 + \alpha_2 X_2 + \dots \alpha_n X_n + \varepsilon$$

Where: X_1 = Training, X_2 = number of years in schooling, X_3 = off farm activities, X_4 = land size, X_5 = age, X_6 = household hold size, ε = Error term.

Most of the smallholder farmers received training on how the float tray system works. The researcher was interested in the number of farmers who received training on the use of float tray system in the production of tobacco. The level of education is also a major determinant in the adoption of technology by the smallholder farmers. In theory, it is believed that those who are educated will adopted the technology faster than those farmers who are well educated. The smallholder framers are not only into the production of tobacco, there are other off farm activities which they are involved in, these off farm activities is also a major determinant on the adoption of new technologies by the smallholder farmers in Zimbabwe. Farmers will prefer a technology which will leave them with time to do other activities.

Size of the land available is another determinant of the adoption of a technology by the smallholder farmers. If the size of the land is huge, farmers would want a technology that will save them time and which does not require many labour days. The size of the arable land available will determine the adoption of technology by the smallholder farmers. The major source of labour in the smallholder sector is the family. The size of the household will also determine the level of adoption by the smallholder farmers in Zimbabwe. If the household is large then the family can adopt a labour intensive technology and if the household size is small then the family will not adopt a labour intensive technology. Household size is a major determinant of the adoption of a technology by the smallholder farmers in Zimbabwe.

RESULTS

Demographic information of small holder tobacco farmers

Most smallholder tobacco farmers in Bindura are male and they constitute more than half of the farmers in the district (AREX, 2012). A range of about 40 and above years is dominating smallholder tobacco farming in the district. The results show that most farmers at least attended ordinary level of education (Tables 1 and 2).

The results in Figure 1 indicate that 66% and 34% were trained and untrained on using the floating tray system respectively. Therefore, there has been increased awareness among farmers of this environmentally friendly method of raising tobacco seedlings instead of using methyl bromide.

Despite the training exposed to farmers (Figure 1) to gain knowledge on the use of float tray system Figure 2 clearly show the low uptake of this system of raising tobacco seedlings. Therefore, the research needed to explore the determinants affecting float tray system adoption.

Adoption determinants

From Table 3, variables were captured and among them number of year in school, size of the household, training and farm size have valid marginal effect on the decision

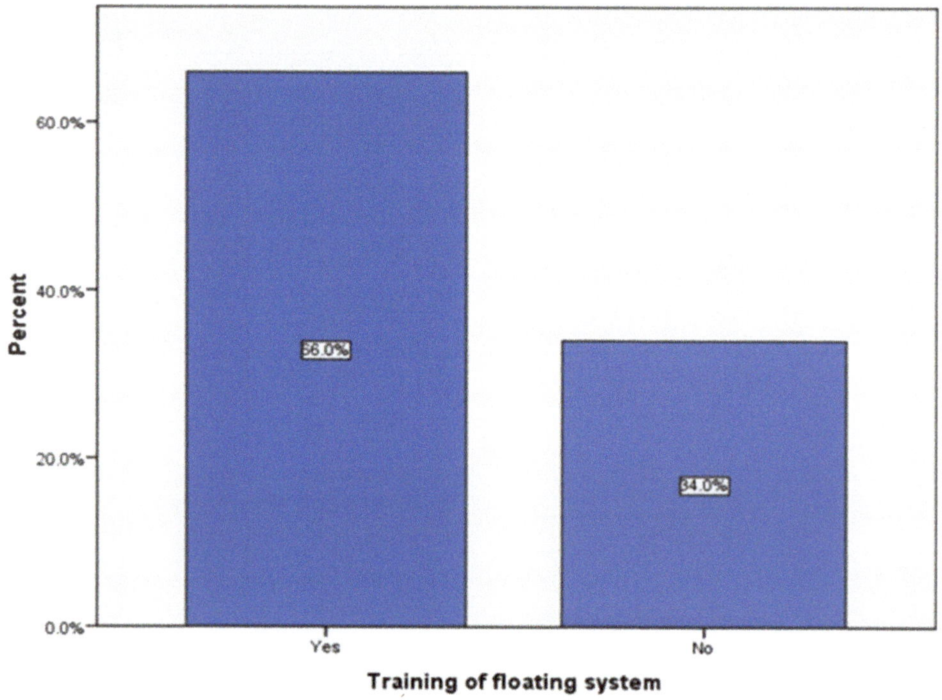

Figure 1. Training of floating trays of farmers in Bindura District, ward 3.

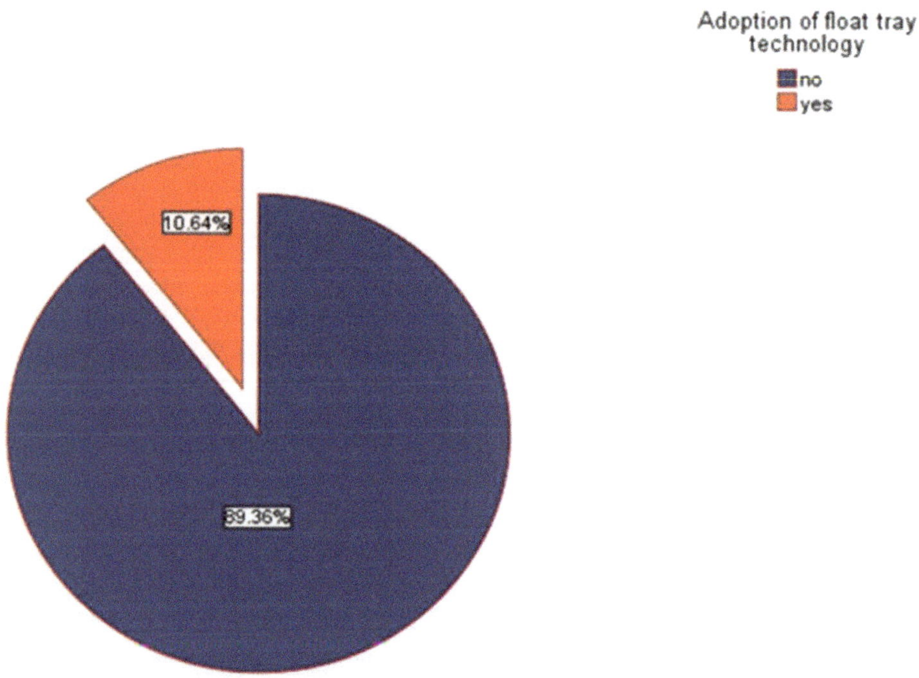

Figure 2. The percentage of adopters and non-adopters of floating tray system in small scale tobacco production.

of a farmer to adopt floating tray system. The t-ratio and the coefficient of variables help to explain and interpret the significance of the variables.

The research (Figure 3) showed that the float tray system uptake by farmers has been quite low because cost of the technology is high therefore the smallholder

Table 3. Effect of factors affecting floating tray system adoption.

Variable	Coefficient	T-ratio	Marginal effect
Age	-0.0034	-0.0029[a]	-0.084
Dependency ratio	0.0091	-0.178	-0.0012
Size of household	0.555	1.562[b]	0.173
Gender	0.026	0.012	-0.0048
Education level	-0.355	1.127[b]	0.041
Training	2.121	1.892	0.4333
Farm size	0.733	1.969[b]	0.0913
Duration	0.0812	1.591[a]	0.0066
Off farm hours	-1.611	0.065[a]	0.0012
Constant	-8.29	1.391	-0.7013
Log likelihood		-60.774	
X^2 (d.f.)	115[b](46)		

Notes: [a] significant at 10%, [b] significant at 5%; The significance is based on p-value, (sig); P-value of the whole model= -2.669; The dependent variable is log (probability of the "adopter"); R^2 = 0.76; Adjusted R^2 = 0.66.

farmers lack initial capital to establish the technology. In addition, the farmers also found the technology to be too technical and complex. Other factors include, lack of knowledge, lack of training, input costs and lack of information.

DISCUSSION

Extent and adoption determinants

Training has shown a valid marginal effect (0.43) in the model meaning that it has great impact in the decision of a farmer to take an innovation. On average, untrained farmers are less likely to participate in float tray seed production as most of them are conservative.

The other factor that affects adoption of the innovation is the level of income a household has. As the farmers' access to income from off farm and non-farm sources increases, the likelihood of participation increases up to some point (Namara et al., 2007). This can mean that an increase in off-farm income, farmers will not willing to adopt new innovations in agricultural productivity as they have the marginal effect of 0.0012. This shows the importance of cash (for leverage in the initial participation decision of farmers). However, at higher levels of off-farm and non-farm income, the farmers are less likely to participate in float tray technology because they have enough money to finance their farming activities and still remain with enough for contingencies. This means that farmers have other sources of income that can be used to cope up with other livelihood option which might be from other enterprises such as soya bean and maize production.

Duration in agricultural activities significantly influences the farmers' decision to use float trays in tobacco seedling. In the context of the studies such as conservation and sustainability, most farmers had on average five seasons under contract. This is constructed in the social dynamics of the communities under review where there are observable trends of dependency on agriculture for survival.

Farmer's age had the expected negative and significant influence on the chances of farmers to use float tray system in tobacco farming. The negative sign for the age variable (-0.0029) could be understood from the commonly observed negative correlation between the age and adoption decision for most technologies in dynamic economic environments. In other words, younger farmers tend to be more willing to adopt than their older counterparts. In most adoption studies, old farmers constitute most of the laggards as much as technology or innovation adoption is concerned. With increase in age farmers tend to shun new farming practices for less demanding procedures and technical skills with low transactional cost associated with them.

Furthermore, older farmers tend to be risk adverse and may avoid innovations in an attempt to avoid risk associated with the initiative. This is because some of the technologies to be used, sunk costs have to be incurred. For example, float tray technology requires more capital to establish as compared to the conventional system, but said to be less costly with time of their use (Mazarura, 2004). So this might discourage older or even some the younger farmers to take up new technologies or innovations .This idea is supported by Rukuni et al. (2006) who argued, that being older creates a conservative feeling among farmers and hence resistance to change. In this study, Baudron (2001) however observed that chances of participation in conservation farming increased with age because youths have little appreciation on the importance of agricultural

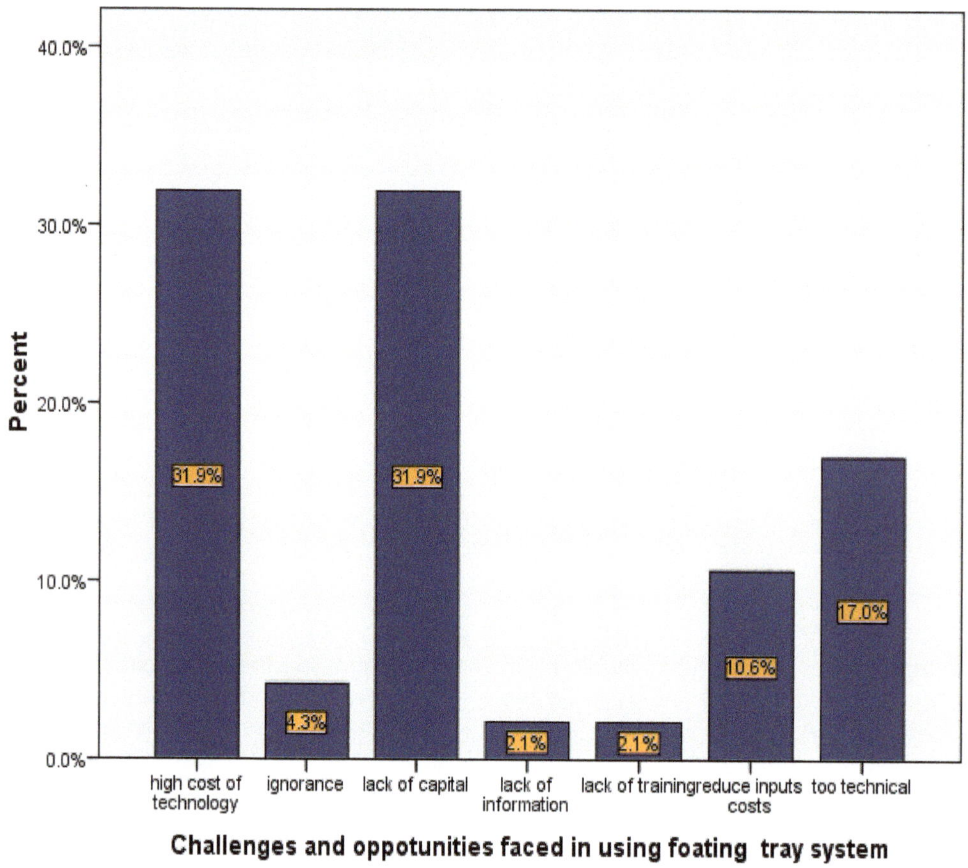

Figure 3. Challenges and opportunities faced by farmers in using the float tray system.

activities in most rural set ups and will take marginal effort to expand these activities. This becomes very relative and similar to those farmers doing tobacco farming.

Education level (as measured by the number of years of schooling by household head) significantly influence farmers' participation but with more years in schooling probability of participation tends to decrease (0.041). A possible explanation to this is that educated people tend to shun agriculture for white colour jobs in Bindura Town and surrounding areas. Some households are more concerned with time value of money and will prefer projects with quick return and profitable like broiler production. However, Thomlow (2007) asserted that education influences household to process information and causes farmers to have better access to understanding and interpretation of information. This tends to differ from what the researcher observed in the field, Ward 3, Bindura district. The researcher observed that farmers who have most number of years in education are not taking up the technology. This may be due to the reason that they have more income generating sources. These farmers tend not to be more concerned about the floating tray system than their little educated counterparts.

Land size significantly influenced farmer decision to use float tray system in tobacco production. A possible explanation to this could be that farmers with large arable land size have the opportunity to spare some sections to try out new practices at less risk. Rogers (1995) supported this by stating that the size of the land is important because the transactional costs are largely fixed cost that are spread across more potential output on large farms. There are also observable indications that highlighted that an increase in participation in tobacco production using floating trays seed is a function of land productivity. Large land size also implies that farmers can use both the conventional and floating tray systems in tobacco seedlings and reduce the inherent risk that is in agricultural tobacco production. This may because some of the farmers want first to try the technology before implementing it as a normal practice in the production. Dependency ratio (that is, the proportion of family members whose ages are less than 14 or more than 65), was introduced into the model as a surrogate for household size to indicate the status of labour availability in the household.

The variable had a positive and insignificant effect on the participation decision. The higher the effective labour available the more likely the household is to participate since chances of labour shortages during peak times are low. This enhances the chances of favourable seedlings. In most households, adoption of new innovations might be due to avoid labour shortages, since innovations are believed to demand few labour requirements. As household size increases, farmers' chances to adopt new innovations will be reduced.

Gender also has a strong bearing on the adoption decision and women who do much of the farming in the communal setup always tend to opt for less labour intensive farming methods. In the study, results show that young female farmers have higher chances of adopting micro irrigation as compared to older males.

Challenges and opportunities faced by tobacco farmers

Major challenges that faced smallholder tobacco farmers are high cost of technology and lack of capital. Most of the farmers did not adopt the float tray systems because of high costs of the innovation. This may be due to very high entry costs of the innovation. Economic theory suggests that a reduction in price of a good or service can result in more of it being demanded. Therefore, adoption can be expected to be dependent on cost of a technology and on whether farmers possess the required resources. The farmers may not understand that floating tray system is associated with sunk costs. This can be contributed by the age of a farmer who, most of them think that new innovations are very expensive before they even tested it. As supported by Khanna (2001). Technologies that are capital-intensive are only affordable by wealthier farmers and hence the adoption of such technologies is limited to larger farmers who have the wealth.

In addition, changes that cost little are adopted more quickly than those requiring large expenditures; hence both extent and rate of adoption may be dependent on the cost of a technology. In the study area, most of the non-adopters did not take up the technology may be due to fear of the technology expenses. Because the technology is capital intensive, farmers are regarding it as the continuous costs of the system in its life span.

One of the major constraint to the adoption of the technology is this study was its complexity. This is the degree to which an innovation is perceived as difficult to understand and use. Most of the non-adopters believe that the floating tray system is too technical. This may be caused by the interaction between age and educational status variables. For those farmers who had few numbers of years in schooling, had low or no adoption of the technology. The reason may be that, they do not want to continue to acquire new skills and knowledge. In addition, old farmers do not want to take new innovations believing the complexity of the floating tray system.

Furthermore, Rogers (1995) suggested that new innovations may be categorized on a complexity-simplicity continuum with a qualification that the meaning (and therefore the relevance) of the innovation may not be clearly understood by potential adopters. When key players perceive innovations as being simple to use the innovations will be more easily adopted.

The small scale tobacco farmers are resource constraint. The indication of the constraint of lack of capital was found on non-adopters. Technologies that are capital-intensive are only affordable by wealthier farmers and hence, the adoption of such technologies is limited to larger farmers who have the wealth (Khanna, 2001). As a major constraint farmers who do not use the float tray system in the study area presented that the technology was capital intensive. This may be due to limited off-farm activities. There are some activities that generate more income than tobacco farming. These activities would allow farmers to have enough capital to invest in new innovation. Capital is therefore a major constraint that has a serious impact on floating tray system adoption.

Information reduces the uncertainty about a technology's performance hence may change individual's assessment from purely subjective to objective over time (Caswell et al., 2001). So, a farmer must have access to various sources of information about an innovation. This information should be complimented with training of the innovation. The combined effect of information and training is a prerequisite to technology adoption. Good training programs and contacts with producers are a key aspect in technology dissemination and adoption. In the case of Bindura District, Ward 3, the researcher found lack of information and training as some of the valid constraints in floating tray system adoption. This may be due to how often the training programs of floating tray system use were being conducted. Farmers presented these constraints; the reason was that when trainings were done they were not yet tobacco growers. They might have tried the technology but failed to use it due to lack of training.

Challenges faced by adopters

Floating tray system may not always be readily adopted by smallholder farmers because it conflicts with convectional seed bed practices which are inherent in farmers (that is, these practices are mutually exclusive). Some of the problems arise from deep socio-cultural beliefs and downgrading of floating tray technology. Below are some of the challenges faced by farmers:

i. There are labour constrains in the first years of floating tray seed farming. Bed construction includes labour intensive tasks and heavy work for those physically challenged,

ii. Lack of self-confidence: Farmers felt that they may not be able to handle the floating tray system. This is because most of them are resource constrained hence they aim lower production level,

iii. Lack of finance to purchase adequate inputs (pine bark, float trays, and black plastic),
iv. Inability to construct durable and suitable beds,
v. Attack of seedlings by crickets at vegetative stage,
vi. High weed and pest infestations in the first year,
vii. Accumulation of algae in the seed beds in the first year,
viii. Unavailability of pine bark,
ix. Long distance from the research station,
x. Very high initial costs to establish the system.

Opportunities lost by non-adopters

The float tray seedling production system has several advantages over the conventional system:

i. Use of fewer chemicals and smaller quantities,
ii. Employs economical integrated management of diseases and pests,
iii. Uses less water and fertilizers,
iv. Produces superior and more uniform drought tolerant seedlings,
v. Facilitates easier field management, arising from a more uniform crop,
vi. Yield and quality from float seedlings are similar or better than conventionally-produced seedlings.

Conclusions

Extent of adoption

The study indicates that there is low adoption of floating tray system in Bindura district, Ward 3. This came as a result from the one sample t-test employed to compare the means. There was no significant difference between the means, hence accepting of the hypothesis that there is low adoption. A number of factors including age, years in schooling and dependency ratio significantly affect the probability of adopting conservation farming technologies. This therefore, means farmers need to be educated or trained more for them to accept the technology of conservation farming.

As the level of education of the household head increases the likelihood of adopting floating tray technology increases. This confirms the fact that floating tray system need special technical and managerial skills for proper utilization. Costs, capital and training issues are the most constraints that affect adoption of float trays. This highlighted the need for institutional support to the resource constraint farmers.

Recommendations

The researcher recommends that smallholder farmers should be educated on sustainable economic technologies in agricultural production such as floating tray technology. This has long term benefits in sustaining production of those who are resource constrained. It is also worth investing in establishing an innovation platform for tobacco seedling technologies. Research institution in this case TRB can review the efficiency of unsubsidized floating tray system to these smallholder tobacco farmers.

The researcher recommends TRB and AGRITEX to mainly focus on senior farmers in training of new innovations since they have a long time in farming tobacco. This is because, in the study, farmers who have more years in tobacco production are non-adopters of the floating tray technology. To reduce transport costs and uncertainty, the researcher recommends TRB to introduce outlets for provision of floating tray equipment in Bindura district. Challenges and opportunities are some of the variables that influence float tray technology adoption decisions of farmers. If Challenges and opportunities alone is the main determinant of adoption, float tray would have dominated the conventional way of seedling. The successful adoption of floating tray system requires, in addition to challenges and opportunities, two additional preconditions were included:

1. The target beneficiaries need to be aware or knowledgeable about the challenges and opportunity superiority of the technology. This may be achieved through extension services in the form of demonstrations, workshops, etc. Farmers' own attributes such as level of education may also augment or complement the public extension services, as educated farmers are active information seekers and experimenters,
2. The technology need to be accessible to the potential users. Awareness or knowledge does not guarantee actual adoption unless the technologies are made accessible to the farmers through institutional support systems.

REFERENCES

Agriculture Research and Extension (AREX) (2008). Annual report for Bindura District.
Baudron F (2001). Challenges for the Adoption of Conservation Farming by Small holders in Semi Arid Zambia. CIRAD, Harare. Zimbabwe.
De Vaus DA (1996). Surveys in Social Research4[th] Edition, London UCL Press and Allen and Umwin.
Food and Agriculture Organization (FAO) (2010). Corporate Document Repository, Issues in the Global Tobacco Economy.
Government of Zimbabwe (GOZ) (2009). Zimbabwe Policy Framework. Harare. Zimbabwe.
Khanna M (2001). Sequential Adoption of Site-Specific Technologies and its Implications for Nitrogen Productivity: A Double Selectivity Model." Am. J. Agric. Econ. 83:35-51.
Mansfield E (1968). Industrial Research and Technological Innovation. New York: Norton.
Mazarura U (2004). A Guide to the Use of the Floating Tray System for Tobacco Seedling Production in Zimbabwe, Harare.
Namara RE, Nagar RK, Upadhyay B (2007). Economics, Adoption Determinants, and Impacts of Micro-Irrigation Technologies: Empirical Results from India. Irrig. Sci. 25:283-297.

Rogers EM (1995). Diffusion of Innovations.3rd Edition. New York: The Free Press, 1983. 4th Edition: The Free Press, New York.
Rukuni M, Tawonezvi P, Eicher C, Munyuki-Hungwe M, Matondi P (2006). Zimbabwe's Agricultural Revolution Revisited, University of Zimbabwe Publications, Harare.
Thomlow S (2007). Assessment of Sustainable Adoption of Conservation Farming in Zimbabwe.
Zimbabwe Tobacco Production (Z.T.P) Records (2008). Tobacco Production in Zimbabwe.

9

Fertilizer adoption in Ethiopia cereal production

Bingxin Yu and Alejandro Nin-Pratt

International Food Policy Research Institute, United States.

This paper was the first to use nationally representative data from the Agricultural Sample Surveys of Ethiopia to examine the factors affecting the adoption of the fertilizer-seed technology "package" promoted by Ethiopia's government. We used a double hurdle model to analyze fertilizer adoption among four major cereal crops (barley, maize, teff, and wheat). This model allowed us to identify factors affecting farmers' access to fertilizer and factors affecting fertilizer demand conditional to input access. Extension was proven to have the biggest impact on fertilizer adoption. We found that knowledge required to adopt new technology represented a high cost for farmers. In addition to extension, other factors that could reduce the cost to access knowledge include farmers' knowledge and skills in cereal production, risk aversion behavior, household wealth and land fragmentation. Substantial yield gain in maize and teff could be achieved from locally tailored extension packages.

Key words: Technology adoption, double hurdle model, Ethiopia, fertilizer, cereal.
JEL Codes: O33, O38, Q16, Q18.

INTRODUCTION

The economic growth strategy formulated by the Ethiopian government in 1991 places very high priority on accelerating agricultural growth to achieve food security and poverty alleviation. A core goal of this strategy is to increase cereal yields by focusing on technological packages that combined credit, fertilizers, improved seeds and better management practices. The participatory demonstration and training extension system (PADETES), started in 1994 to 1995 and in its early stages focused on cereal crops, but expanded to other crops in later years. This technology-package-driven extension approach has been implemented on a large scale and has reached virtually all farming communities in Ethiopia. It represents a significant public investment ($50 million dollars annually or 2% of agricultural GDP in recent years), four to five times the investment in agricultural research. Extensive data from a large number of demonstrations carried out through PADETES, indicates that the adoption of fertilizer-seed technologies could more than double cereal yields and would be profitable to farmers in moisture-reliant areas (Howard et al., 2003).

However, after nearly a decade of implementation the impacts of the program have been mixed, with increased but still limited use of fertilizer (World Bank, 2006). Byerlee et al. (2007) concluded that some of the major factors affecting the results of PADETES program are poor performance of the extension service, promotion of regionally inefficient allocation of fertilizer, low technical efficiency in the use of fertilizer, no emergence of private sector retailers whom were negatively affected by the government's input distribution tied to credit, and the generation of an unleveled playing field in the rural finance sector by the guaranteed loan program with below-market interest rates. Among these issues, low adoption of modern inputs, especially chemical fertilizer, deserves more in-depth study as the fertilizer policy represents a substantial portion of public resources going to agriculture.

Several papers have analyzed the factors affecting adoption of chemical fertilizer in Ethiopia's cereal production. Admassie and Ayele (2004) and Beshir et al. (2012) noted that age of household head, farm land size, education, livestock, non-farm income, gender and access to information are major factors affecting technology adoption. Limited knowledge and education are identified as major constraints for technology adoption by Asfaw and Admassie (2004), while Beshir et al. (2012), Carlsson et al. (2005) and Wubeneh and Sanders (2006) highlighted the positive effect of extension services on fertilizer adoption. Liverpool-Tasie and Winter-Nelson (2012) on the other hand, indicate that technology diffusion in Ethiopia is likely to be enhanced if extension can target intentional networks, rather than spatial clusters; and Abebaw and Haile (2013) show that cooperative membership has a strong positive impact on fertilizer adoption. Dercon and Christiaensen (2011) found an important role of risk on fertilizer adoption, showing that the lack of insurance or alternative means of keeping consumption smooth leaves poor households trapped in low return, lower risk agriculture. Croppenstedt et al. (2003) pointed out supply side factors such as credit constraints and untimely fertilizer supply are some of the most important reasons for non-adoption of fertilizer by farmers. Alternatively, Liverpool-Tasie and Winter-Nelson (2009) find no relationship between participation in microfinance programs and the use of technologies and concluded that participation in micro-finance programs increases the likelihood of technology use for the less poor households only.

A major limitation of these studies is that they use a variety of different data sources that are not nationally representative of Ethiopia's agriculture, with some of them not covering the second half of the 2000s when a substantial increase in fertilizer adoption took place. Some of the sources used include the Ethiopia rural household survey (ERHS), which comprises 1,477 households in 15 Peasant Associations across the four major regions in Ethiopia (as in Dercon and Christiaensen, 2011; Liverpool-Tasie and Winter-Nelson, 2009, 2012); a survey conducted in 1994 covering 6,147 cereal-growing farm households, in Amhara, Tigray, Oromia, and SNNP (Croppenstedt et al., 2003), and a survey focusing only in two districts of south Wollo zone (Beshir et al., 2012). The goal of this paper is to look at the extent and determinants of the adoption of the fertilizer promoted in Ethiopia. The study contributes to the literature of technology adoption in Ethiopia's agriculture in several aspects. First, this is the first attempt to analyze technology adoption in Ethiopia using nationally representative data based on agricultural sample surveys from the central statistical agency (CSA, various years). Second, our approach features the sequential process of decision making in technology adoption by separating the decision to adopt fertilizer and the decision about the quantity of input use for each cereal crop, addressing the endogeneity of extension service to improve our understanding of the effectiveness of PADETES. Third, we estimate average partial effects for determinants of technology adoption, allowing us to examine the unconditional effect of factors that influence the adoption process, which is important when there are observations with zero values for input use. Finally, in addition to traditional social economic indicators, we introduce spatial variables obtained through GIS tools in the analysis. The spatial distribution of biophysical constraints and market accessibility are incorporated in the model to take into account the impact of local agronomic and development conditions on technology adoption.

Cereal production and technology adoption in Ethiopia

Table 1 presents a summary of area, production and yields of cereals in main Ethiopian production regions in 2003 to 2004 and 2007 to 2008. Total cereal production was 13.6 million tons in 2007 to 2008, expanded by 27% from 2003 to 2004. Average cereal yield reached 1.6 ton/ha in 2007 to 2008, exhibiting a 22% growth over 5 years. In 2007 to 2008, the main cereal according to land use was teff (30% of total cereal land), followed by maize (20%), sorghum (18%) and wheat (16%). Ethiopia's yield levels are lower than the average yield in least developed countries, although they are higher than the average yield in Eastern Africa.

Between 2003 to 2004 and 2007 to 2008, the area of

Table 1. Area, production and yields of cereals in Ethiopia, 2003 to 2004 and 2007 to 2008.

Variable	2003 to 2004				2007 to 2008				Growth 2003 to 2004 and 2007 to 2008 (%)			
	Area	Production	Yield	Area share	Area	Production	Yield	Area share	Area	Production	Yield	Area share
Cereal crop	000 ha	000 tons	Tons/ha	%	000 ha	000 tons	Tons/ha	%				
Barley	911	1071	1.2	13.4	985	1355	1.4	11.4	8.1	26.5	17.0	-14.9
Maize	1300	2455	1.9	19.1	1767	3750	2.1	20.4	35.9	52.7	12.3	6.8
Millet	303	304	1.0	4.5	399	538	1.3	4.6	31.7	77.0	34.4	2.2
Sorghum	1242	1695	1.4	18.2	1534	2659	1.7	17.7	23.5	56.9	27.0	-2.7
Teff	1985	1672	0.8	29.1	2565	2993	1.2	29.6	29.2	79.0	38.6	1.7
Wheat	1075	1589	1.5	15.8	1425	2314	1.6	16.4	32.6	45.6	10.0	3.8
Other	35	44	1.3	0.5	55	108	2.0	0.6	57.1	145.5	56.1	20.0
Total cereal	6816	8786	1.3	100	8675	13609	1.6	100	27.3	54.9	21.7	

Source: Author's calculation using CSA Agricultural Sample Survey data (various years).

four of the major cereal crops under the promoted technologies (fertilizer and/or seed) increased at 4% annually (Table 2). The adoption of the promoted package of jointly using fertilizer and improved seed has been very limited, accounting for only 6% of cultivatedarea. Traditional farming practice of using local seed but no chemical fertilizer remains the dominant farming system in barley (73% of land), followed by maize (62%), teff (56%), and wheat (43%) in 2007 to 2008.

More than 50% of the area planted with teff and wheat and 38% of the area under maize used fertilizer during the period, regardless of seed type. Barley shows the lowest levels of fertilizer adoption with only 27% of its area cultivated using fertilizer. However, the information available from CSA allows only identifying the use of improved seed varieties if they are purchased on the year of the survey. Because of this, the survey can only capture the systematic use of improved seed varieties if they are hybrids because they need to be purchased every year, underestimating the adoption of open pollinated varieties. Given the importance of hybrids among improved maize varieties and of open pollinated varieties for other crops, CSA data can only capture adoption of improved seed in maize. For this reason, our analysis focuses on fertilizer adoption rather than on the fertilizer-seed package.

Cereal cultivation is highly concentrated geographically (Figure 1). Almost 80% of total area under cereals is in the Amhara and Oromiya regions to the northwest, west, southwest and south of the capital, Addis Ababa. This area includes a diverse set of conditions for agricultural production. Spatial conditions for production and market access have been discussed in detailed by Diao and Nin Pratt (2005) and Tadesse et al. (2006).

There are substantial regional variations in the adoption of fertilizer (Figure 2). The spatial distribution of fertilizer use varies by crop, although there is also a significant overlap of zones across the different crops. In general, most of the area under fertilizer is concentrated in areas with suitable natural resources for production and roads access.

METHODOLOGY

Conceptual framework

One of the most used methods for modeling technology adoption behavior is the censored regression model, also called the Tobit model (Wubeneh and Sanders, 2006). The key underlying assumption for a Tobit specification is that farmers demanding modern inputs have unconstrained access to the technology. However, in situations where input supply systems are underdeveloped this is often untenable, as farmers wanting to adopt a new technology, fertilizer in this study, often face input access constraints. The Tobit specification has no mechanism to distinguish households with a constrained positive demand for the new technology from those with unconstrained positive demand, and assumes that a household not adopting the technology is making a rational decision. Hence, in the case of access constraints to access inputs, the Tobit model yields inconsistent parameter estimates (Croppenstedt et al., 2003).

Evidence from previous studies shows the critical role that underdeveloped input supply and marketing systems play on input choices and technology adoption in smallholder agriculture (Shiferaw et al., 2008). Smallholder farmers in many rural areas are semi-subsistence producers and consumers who are partially integrated into imperfect rural markets. Factor markets for labor, land, traction power, and credit in rural areas of developing

Table 2. Cereal area under modern and traditional technology.

Crop and technology	Total area (000 ha)				Share in crop area (%)				Annual growth rate (%)
	2003	2004	2006	2007	2003	2004	2006	2007	
Barley									
Fertilizer + improved seed	0.8	1.8	0.9	1.2	0.1	0.3	0.1	0.2	10.7
Fertilizer + local seed	145.6	164.4	173	140.6	25.8	25.6	27.3	26.6	-0.9
No fertilizer + improved seed	1.2	2.1	0.1	0.2	0.2	0.3	0	0	-36.1
No fertilizer + local seed	415.6	474.2	459	386.8	73.8	73.8	72.5	73.1	-1.8
Total	563.1	642.5	632.9	528.9	100	100	100	100	-1.6
Maize									
Fertilizer + improved seed	197.2	158.1	188.9	192.2	23.4	17.7	17.7	21.6	-0.6
Fertilizer + local seed	99.5	124.6	211.2	146.3	11.8	13.9	19.7	16.4	10.1
No fertilizer + improved seed	10.7	9.5	9.9	5	1.3	1.1	0.9	0.6	-17.3
No fertilizer + local seed	536.1	601.6	660.1	547.9	63.6	67.3	61.7	61.5	0.5
Total	843.5	893.8	1070.2	891.3	100	100	100	100	1.4
Teff									
Fertilizer + improved seed	3.7	7.7	8.2	9.7	0.3	0.5	0.5	0.6	27.2
Fertilizer + local seed	634.2	705	902.2	821.4	45.2	47.2	54.4	53.5	6.7
No fertilizer + improved seed	4.7	3.7	2.1	2.2	0.3	0.2	0.1	0.1	-17.3
No fertilizer + local seed	761.4	778.7	745.8	701.7	54.2	52.1	45	45.7	-2.0
Total	1,404	1,495	16,58.3	1,535	100	100	100	100	2.3
Wheat									
Fertilizer + improved seed	24.9	28.3	22.5	14.1	3.7	3.4	2.6	2	-13.3
Fertilizer + local seed	341.6	418.7	533	379.9	50.1	50.4	60.6	53.8	2.7
No fertilizer + improved seed	5.8	5.3	4.2	6.1	0.9	0.6	0.5	0.9	1.3
No fertilizer + local seed	308.9	379	320.3	305.5	45.4	45.6	36.4	43.3	-0.3
Total	681.2	831.3	880	705.7	100	100	100	100	0.9
Major Cereals									
Fertilizer + improved seed	227	196	221	217	6.5	5.1	5.2	5.9	-1.1
Fertilizer + local seed	1,221	1,413	1,819	1,488	35.0	36.6	42.9	40.7	5.1
No fertilizer + improved seed	22	21	16	14	0.6	0.5	0.4	0.4	-11.9
No fertilizer + local seed	2,022	2,234	2,185	1,942	57.9	57.8	51.5	53.0	-1.0
Total	3,492	3,863	4,241	3,661	100.0	100.0	100.0	100.0	1.2

Source: Author's calculation using CSA Agricultural Sample Survey data (various years).

countries are often imperfect and/or even missing in some cases (Holden et al., 2001; Pender and Kerr, 1998). In these cases, access to fertilizer is the key threshold that farmers with positive desired demand for the new technology have to overcome. The double hurdle (DH) model (Cragg, 1971) is a useful and proper approach to analyze technology adoption under constrained access to inputs, as many Ethiopian households face constraints in accessing inputs like fertilizer (Noltze, Schwarze and Qaim, 2012). This paper also adopts the DH model to examine technology adoption in two stages. In the first stage, the farmer decides

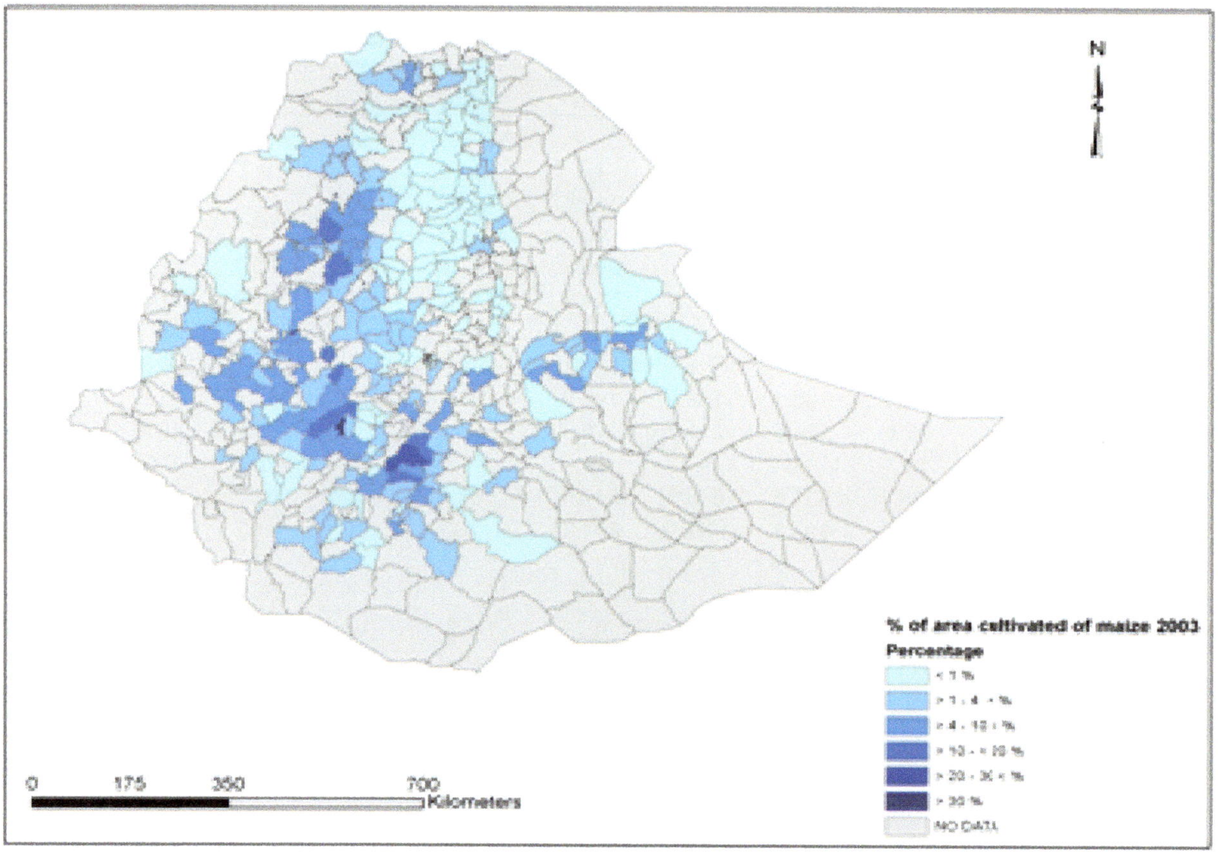
N
% of area cultivated of maize 2003
Percentage
< 1 %
NO DATA
0
175
350
700
Kilometers

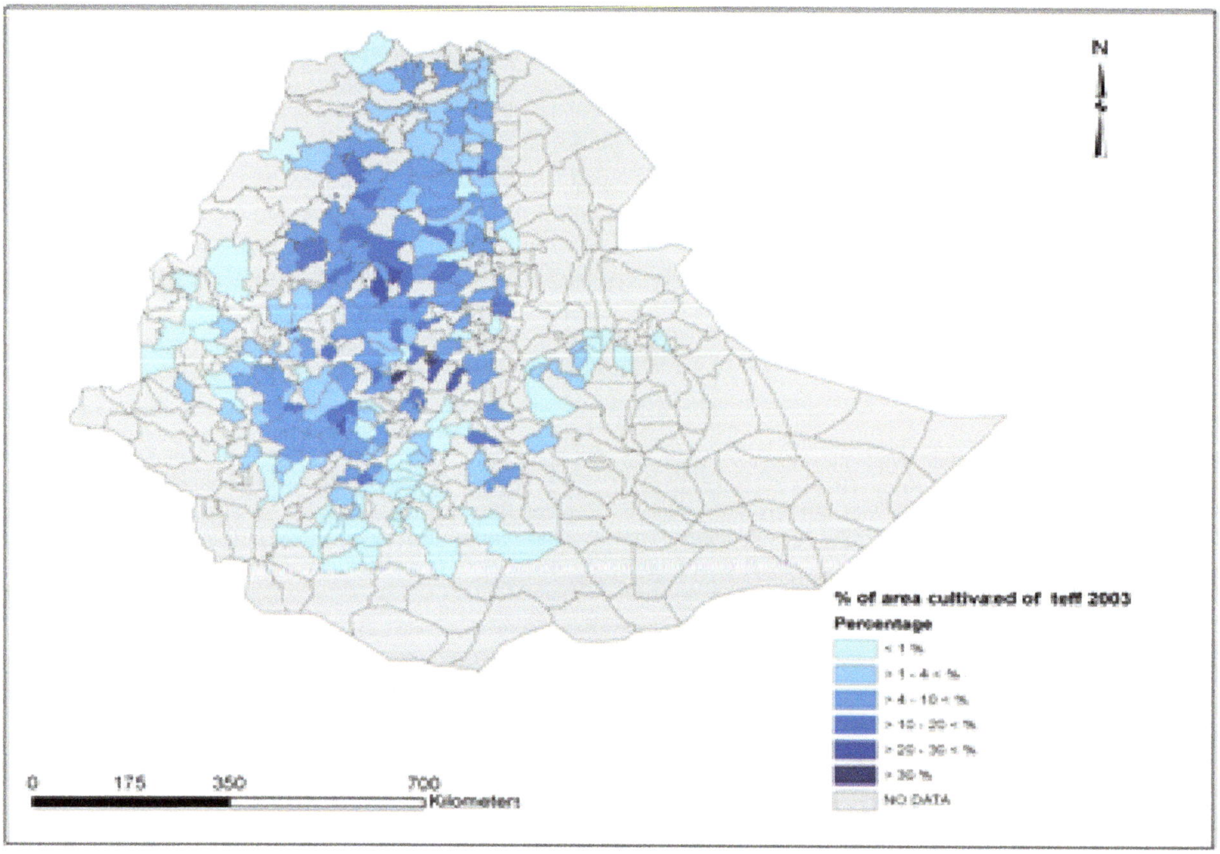
N
% of area cultivared of teff 2003
Percentage
< 1 %
> 4 - 10 < %
NO DATA
0
175
350
700
Kilometers

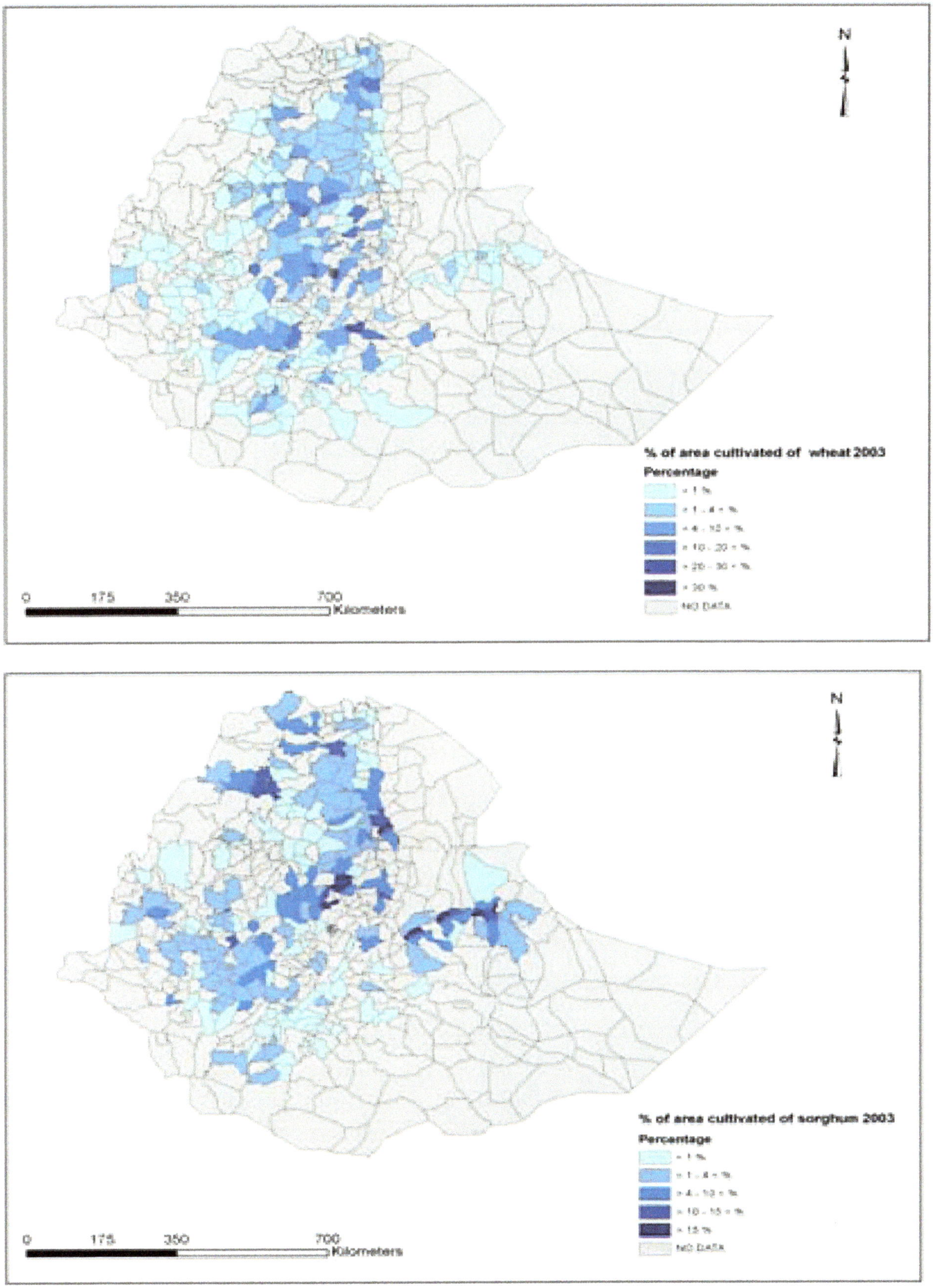

Figure 1. Share of cultivated area in total wereda area for four cereal crops, in percentage; Source: Author's computation using CSA data (various years).

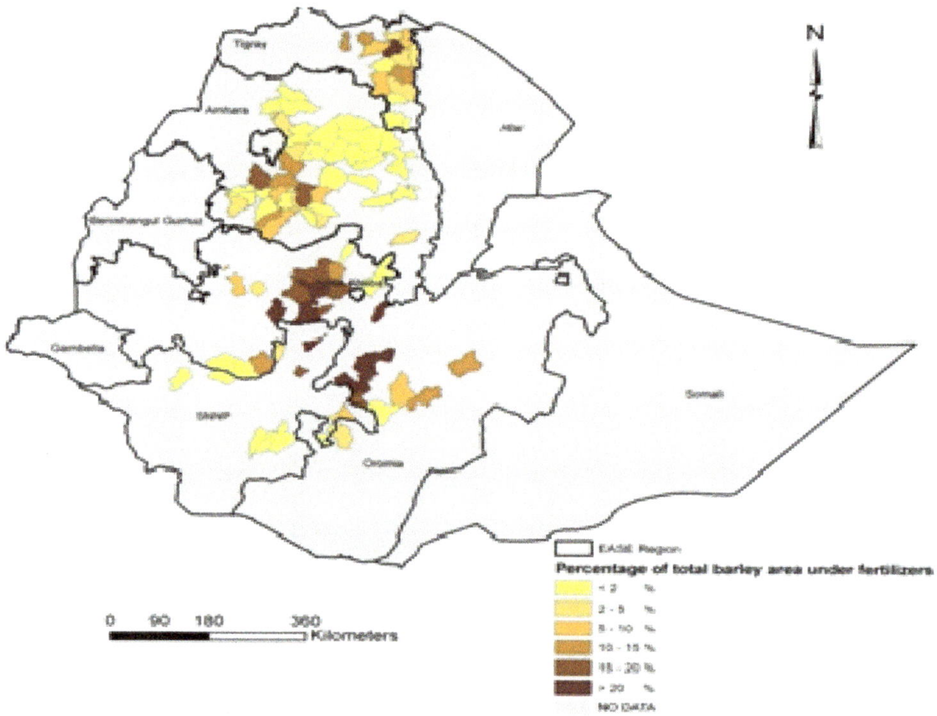
N
Tigray
Amhara
Afar
Benishangul Gumuz
Gambella
SNNP
Oromia
Somali
Percentage of total barley area under fertilizers
< 2 %
2 - 5 %
10 - 15 %
> 20 %
NO DATA
0 90 180 360
Kilometers

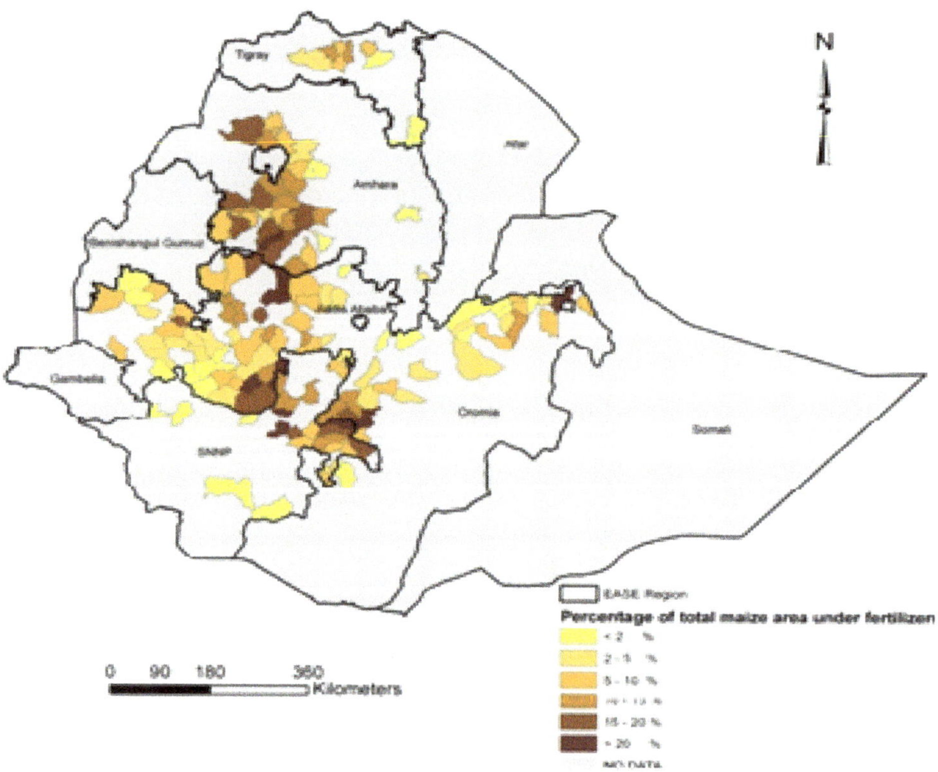
N
Tigray
Amhara
Afar
Benishangul Gumuz
Gambella
SNNP
Oromia
Somali
EASE Region
Percentage of total maize area under fertilizers
< 2 %
5 - 10 %
15 - 20 %
> 20 %
0 90 180 360
Kilometers

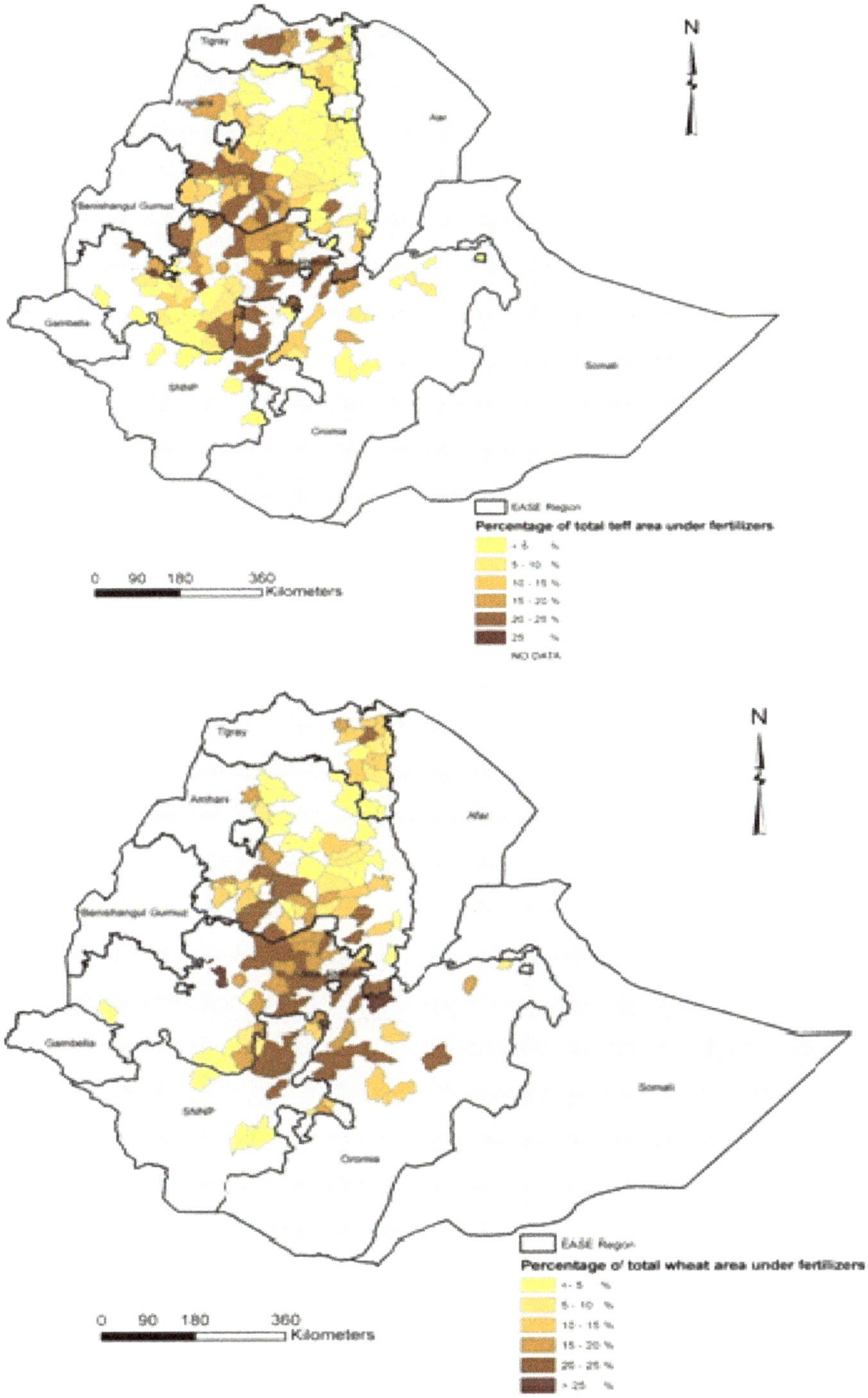

Figure 2. Spatial distribution of fertilizer use in cereal production; Source: Author's calculation using CSA data (various years).

whether or not to participate in the fertilizer market. If he/she chooses to participate, the next step is to decide the quantity to purchase. Through this procedure, the DH model allows separation of the sample of farming households into three groups: households adopting fertilizer, households wanting to adopt but reporting no positive application, and households choosing not to adopt. We incorporate this additional information to the DH model to obtain more efficient and consistent estimates of technology adoption. Similar DH model is adopted by Asfaw et al. (2011) to examine the adoption of chickpea seed variety in Ethiopia.

The DH model used in this study has two equations, one explaining access to fertilizer, and the other one explaining the level of application once access to inputs is granted. First, the latent but unobservable variable underlying an individual farmer's access to fertilizer A^* can be modeled as:

$$A^* = x_1\gamma + e, \tag{1}$$

where x_1 is a vector of variables that affect access, γ is the parameter vector, and *e* is random variable distributed as normal with mean 0 and variance 1. The unobserved demand for fertilizer of farmers (Y^*) can be modeled as:

$$Y^* = x_2\beta + u, \tag{2}$$

where x_2 is a vector of variables that determine the demand function, β is parameter vector, and *u* is normal random variable with mean 0 and variance σ_u^2. The observed input demand (Y) is characterized by the interaction of Equations (1) and (2). A positive use of input like fertilizer is observed only if two thresholds are passed in the decision making process. Hence the farming households are separated into three groups. Group 1 represents the adopters as the farmer has passed the positive demand threshold (Y^* >0) and has access to input (A^* > 0). Group 2 in the sample includes farmers who want input (Y^* >0) but cannot because of some constraints like lack of access (A^* ≤ 0). Group 3 consists of farmers who do not want to use input (Y^* < 0) whether they have access to it or not (A^* > 0 or A^*≤ 0).

We assume that the access and demand equations are independent and that the log likelihood function for the sample-separated data can be expressed as:

$$\ln L = \sum_{G1=1} \ln\left[\Phi(x_1\gamma) \times \left(\frac{1}{\sigma_i}\right) \times \phi\left(\frac{Y - x_2\beta}{\sigma_i}\right)\right] + \sum_{G2=1} \ln\left[\Phi(x_2\beta/\sigma_u) \times (1 - \Phi(x_1\gamma))\right] + \sum_{G3=1} \ln\left[1 - \Phi(x_2\beta/\sigma_u)\right], \tag{3}$$

where φ and Φ are the probability density and cumulative distribution function of the standard normal variable, respectively; 1, 2 and G3 are indicator functions showing whether a given observation belongs to group 1, 2 or 3, respectively. Equation (3) can be estimated using maximum likelihood (ML) techniques, which gives consistent estimates of the parameters. If u_i and e_i are independent, the maximum likelihood function can be separated into a probit and a truncated normal regression model. The model specification of the DH estimator can be tested against the Tobit model using a likelihood ratio (LR) test to determine whether the data supports sequential technology adoption decisions or traditional probit and Tobit approaches are sufficient.

We address potential endogeneity of some of the explanatory variables (in particular the variable representing "extension") using the control function (CF) approach of Rivers and Vuong (1988). According to Imbens and Wooldridge (2008), the CF approach offers some distinct advantages for models that are nonlinear in parameters or endogeneity variables because the CF estimator tackles the endogeneity by adding an additional variable to the regression, generating more precise and efficient estimator than the IV estimator.

After obtaining coefficient estimates for parameters of interest, we can derive the average partial effects (APE) of the explanatory variable across plot and time. The APE is the partial effect averaged across the sample. The first step in obtaining the APE is to derive the partial effect for the explanatory variable of interest x_j for each observation in the sample. The partial effect of a variable x_j on the unconditional expected value of *y* depends on whether x_j is an element of access equation (2) or demand equation (1), or both (Burke, 2009). First, if x_j is an element of both equations, the partial effect is:

$$\frac{dE(y)}{dx_j} = \gamma_j * f(x_1\gamma) * \left[x_2\beta + \sigma \times \lambda\left(\frac{x_2\beta}{\sigma}\right)\right] + \beta_j F(x_1\gamma) \times \left\{1 - \lambda\left(\frac{x_2\beta}{\sigma}\right)\left[\left(\frac{x_2\beta}{\sigma}\right) + \lambda\left(\frac{x_2\beta}{\sigma}\right)\right]\right\}. \tag{4}$$

If x_j is only determining the probability of *Y* > 0 in the access Equation (1), then β_j = 0, and the second term on the right-hand side of Equation (4) disappears. If x_j is only determining the value of *Y* in the demand Equation (2), given that *Y* > 0, then γ_j = 0, and the first term on the right-hand side is dropped.

The APE for a continuous variable is then calculated as the average of the partial effects. The APE of a binary explanatory variable is calculated as the mean difference between unconditional expected value, $E(y)$, valued at the binary variable *D*=0 and *D*=1. The APE is generally of greater interest than the partial effect at the average of the sample mean, particularly in nonlinear models and in the case of discrete variables (Imbens and Wooldridge, 2008). However, the APE obtained from the control function approach outlined above cannot be used for statistical inference. Therefore the bootstrap method is used to obtain the variances of APE and their associated significance levels.

Data and descriptive statistics

The model in Equation (3) is estimated empirically using data from CSA annual agricultural sample surveys conducted in four years: 2003 to 2004, 2004 to 2005, 2006 to 2007, 2007 to 2008, covering all crop production regions of Ethiopia. More recent years are not included in the analysis because administrative border shifts substantially due to government reconstruction. The surveys are nationally representative under a stratified two-stage cluster sample design. First enumeration areas (EAs) were selected using probability proportional to the number of agricultural households from the census and adjusted for the sub-sampling effect (with the caveat that the survey does not cover the non-sedentary population in some zones). At the second stage, 25 agricultural households, households with at least one member engaged in growing crops and/or raising livestock, from each sample EA were systematically selected at the second stage. Each year more than 2,000 enumeration areas (EAs) were selected, resulting in about 50,000 agricultural households growing cereals and other annual and permanent (perennial) crops. The exact number of EAs and agricultural households covered in the survey varies slightly each year due to cost and other considerations. In the selected rural peasant households the agricultural data were collected from sedentary agricultural holders, who operate the land that is used wholly or partly for agricultural production. Unfortunately due to the nature of the survey, it is impossible to build a panel of households for analysis.

Data on crop production and agricultural practice are recorded at plot level. Information on farming practice of irrigation, using agricultural chemicals, growing a single crop (mono-cropping) and land rental access are binary variables. Extension access is defined as whether the plot under government extension programs. At holder level, variables about access to credit and advisory services are binary variables defined as whether the holder benefit from credit service and agricultural advisory services in the locality. Agricultural advisory service is related with extension but not the same because a holder can have plots not under government extension programs but still receiving advisory service from other organizations like NGOs and vice versa. Land fragmentation is represented by the number of plots operated by the holder, and crop rotation refers to whether the holder practices crop rotation or not. Gender, age, and education grade refer to the holder's demographic characteristics.

At household level, household size and total cereal area are included to capture household labor and land resources. At community level, having experience with fertilizer is defined as the share of crop area using fertilizer in total cereal area, while the importance of a particular crop is the proportion of the crop area in total cereal area. Knowledge availability in community is captured by the area share of crop land using fertilizer in the wereda (districts in Ethiopia).

Location affects technology adoption through social and agro-climatic effects, and measures of location and spatially-differentiated variables can explicitly quantify effects of spatial factors on technology uptake and land use. This survey database is complemented by spatial information, incorporating variables that reflect heterogeneity in the quality and availability of natural resources, demographic distribution, and infrastructure and market access.

The spatial variables include market access, population density, road density and land productivity at wereda level. Market access is defined as average travel time in minutes to reach a market of 50,000 or more people (CSA, EDRI and IFPRI, 2007). Road density is the ratio of the length of total road network to land area in the wereda, measured in kilometers per square kilometer (CSA, EDRI and IFPRI, 2007). The road network includes all roads in the country: motorways, highways, main or national roads, and secondary or regional roads. Population density measures the number of persons per square kilometer (CSA, 2010). Crop suitability is calculated based on agro-ecological zones to capture the quality of natural resources for production of the different crops at wereda level. It evaluates land resources and biophysical limitations and potentials for each crop, hence provides the distribution of land, classified into five suitability classes: very suitable, suitable, moderately suitable, marginally suitable and not suitable. Two variables are used to capture suitability for each of the four cereal crops. One variable is area share of highly suitable land, which is the ratio of total very suitable and suitable land in the wereda to total wereda area (EDRI, 2009). The other variable is area share of moderately to marginally suitable land, defined as the ratio of total moderately and marginally suitable land to total wereda area (EDRI, 2009).

Based on Just and Zilberman (1983), we classify explanatory variables for fertilizer access Equation (1) in the DH model and available from CSA data as follows: a) financial constraints -- access to credit; b) fixed costs of adopting the technology; and c) spatial constraints and supply-side effects. Similarly, we group variables affecting the demand of fertilizer (Equation 2) in: a) variables affecting productivity in the use of fertilizer; b) resource availability and risk related variables; and c) spatial variables affecting prices and profitability. Table 3 summarizes the variables used in the analysis.

Variables in the dataset assumed to affect access to inputs (Equation 1) include farmer's access to extension services, farmers' characteristics like gender, age and education, and the level of adoption in the district where the farmer is located (measured as the share of the crop using improved technology in total area of that crop in the district). Most of these variables are related to fixed costs incurred when adopting the new technology and result from the farmer's need to access to knowledge that would allow him/her to implement the new technology We expect a positive relationship between access to fertilizer and access to extension services, education and the level of adoption at the district level. Supply-side effects such as lack of supply, late delivery and inadequate infrastructure are captured by variables representing market access, population and road density and zonal dummies.

Variables explaining demand of fertilizer are irrigation and the use of pesticide and herbicide which we consider respectively as complementary investments and inputs that can increase

Table 3. Factors used to determine fertilizer adoption.

Type	Variable	Plot	Holder	Household	Wereda
Access to fertilizer					
Financial constraints	Access to credit		X		
Fixed costs of adoption	Access to extension	X			
	Access to advisory service		X		
	Gender		X		
	Age		X		
	Education grade		X		
	Area share of crop land using fertilizer			X	X
Spatial constraints, supply-side effects	Market access				X
	Population density				X
	Road density				X
	Zonal dummies				
	Year dummies	X			
Use of fertilizer					
Variables affecting productivity in the use of fertilizer	Irrigation	X			
	Use of pesticide and herbicide	X			
	Mono-crop in the particular plot	X			
	Crop rotation		X		
	Access to extension	X			
	Access to advisory service		X		
	Gender		X		
	Age		X		
	Education grade		X		
	Area share of crop land using fertilizer			X	X
	Area share of highly suitable land				X
	Area share of moderately to marginally suitable land				X
Resource availability and risk related variables	Household size			X	
	Total cereal area			X	
	Area share of the crop in total cereal area			X	
	Number of plots		X		
	Access to land (plot is rented)	X			
Spatial constraints, supply-side effects	Market access				X
	Population density				X
	Road density				X
	Zonal dummies				X
	Year dummies	X			

Source: Variables from CSA Agricultural Sample Survey data (various years).

productivity of fertilizer. Farmer's characteristics like gender, age and education can also affect demand of fertilizer use. Quality of natural resources measured as suitable area in the district where the farmer is located is used as an indicator of expected crop response to fertilizer. Finally, specialization in a particular crop can facilitate use and improve efficiency in the use of fertilizer. Resource availability and risk related variables are also key determinants in the adoption decision and intensity of fertilizer use. According to Coady (1995), a wealthy farmer usually exhibits decreasing absolute risk aversion but increasing relative risk aversion, meaning that the farmer will tend to use higher absolute levels of inputs but less inputs per hectare than less wealthier producers. We expect variables indicating wealth and capital availability as total area and access to additional land (renting land), to be positively related to fertilizer use, with estimated coefficients smaller than 1 if households are relative risk averse. The share of the crop in total area reflects the importance of the crop in the production system, and we expect this variable to be positively related to fertilizer use. The correlation between household size and fertilizer use should be positive for two reasons. First, we assume that fertilizer application is a labor intensive task, and with the cost of family labor being lower than that of hired labor, a positive coefficient for this variable captures this lower cost of applying fertilizer (Coady, 1995). A second explanation for a positive coefficient of household size is related to risk. With labor being a "safe" asset, compared to crop production, more family labor is equivalent to a higher level of non-stochastic assets, allowing for higher use of fertilizer.

Spatial variables like market access, population density and road density affect the level of fertilizer use through marketing and transportation margins affecting the prices that farmers pay for fertilizer and eventually also the price they receive for their products. Zonal dummies capture other specific spatial effects not captured by other variables.

Descriptive statistics

Table 4 presents descriptive statistics of explanatory variables in the DH model by group of fertilizer usage for four major cereal crops (barley, maize, teff, and wheat). The table shows substantial differences between technology adopters and non-adopters. Compared to non-adopters, adopters report larger plot size, higher yields, they are more specialized, they show higher use of pesticides and herbicides, they are younger, more educated, more experienced and wealthier than non-adopters (more oxen, crop fields and larger cereal area), and they have better access to extension, credit, advisory services, larger household size. There are also differences in the spatial location of adopters and non-adopters. Adopters tend to have better market access, improved infrastructure (higher road density), they are located in regions with higher population density, better natural endowments (crop suitability), and live in weredas where technology has disseminated broadly.

RESULTS AND DISCUSSION

Following the framework outlined in the previous section, the endogeneity of extension is addressed by a control function at plot level. The reduced form equation of endogenous explanatory variable, extension access, is regressed over exogenous variables including land rental, farming practice (mono-crop, chemical use, crop rotation, damage control and irrigation), holder characteristics (gender, age, education grade), access to credit and advisory services, household characteristics (household size, farm size, experience in fertilizer, importance of crop) and wereda fertilizer adoption level. Year dummies are used to control time variations and the error term incorporated cluster effect at EA level.

Determinants of fertilizer access

Treating extension as endogenous variable, Table 5 reports results of the econometric estimation of the DH model for fertilizer access at plot level. The first result to be noted is that of the Wald test for independent equations at the bottom of the table indicating that the extension service is endogenous in the decision making process of fertilizer adoption. Compared to the coefficients obtained under the assumption of exogenous extension, the coefficients estimated using the CF approach is smaller in the access function, but larger in the demand function. It suggests that extension service boosts the probability of having fertilizer access but does not affect the amount of fertilizer used among users. We also tested the model by checking the hypothesis that farmers make input decisions simultaneously instead of sequentially as assumed by the DH model. To do this we estimated the Tobit and the DH model separately and compared their log-likelihoods (LR test at the bottom of Table 5). We found that the log-likelihood of the DH model is significantly larger than that of the Tobit model, confirming the relative superiority of the DH specification for this dataset over the Tobit model.

Looking at the main results of the model explaining access to chemical fertilizer we find that the main explanation of access to fertilizer is the possibility of reducing the fixed knowledge cost related to adoption of the new technology, mainly through access to extension services. Also important in explaining access to fertilizer is the share of total cereal land under fertilizer both at household level and at the wereda level where the household is located, suggesting that fertilizer is more likely to be adopted in households who have already used this input in other crops, and in districts with better

Table 4. Descriptive statistics of adopters and non-adopters of fertilizer by crop and input use.

Variable	Barley		Teff		Wheat		Maize	
	Non-adopter	Adopter	Non-adopter	Adopter	Non-adopter	Adopter	Non-adopter	Adopter
Plot level								
Plot area (ha)	0.12	0.16	0.21	0.27	0.14	0.24	0.10	0.18
Plot yield (ton/ha)	1.09	1.27	0.90	1.00	1.25	1.60	1.66	2.05
Extension (yes = 1)	0.08	0.31	0.07	0.27	0.10	0.29	0.05	0.55
Irrigation (yes = 1)	0.01	0.01	0.01	0.00	0.01	0.01	0.03	0.02
Improved seed (yes = 1)	0.00	0.01	0.00	0.01	0.01	0.05	0.01	0.44
Pest. and herbicide (yes = 1)	0.02	0.12	0.06	0.14	0.05	0.18	0.01	0.02
Holder level								
Gender (male = 1)	0.85	0.84	0.88	0.87	0.86	0.85	0.83	0.87
Age	45.5	44.7	43.3	42.9	45.1	43.7	43.3	41.3
Education grade	2.1	2.8	2.2	2.8	2.3	3.0	2.4	2.8
Credit (yes = 1)	0.21	0.41	0.18	0.39	0.21	0.38	0.18	0.37
Advisory service (yes = 1)	0.47	0.51	0.45	0.54	0.47	0.50	0.38	0.58
Number of oxen	1.2	1.3	1.3	1.5	1.2	1.4	1.1	1.2
Household level								
Household size	5.37	5.82	5.31	5.66	5.36	5.76	5.28	5.68
Cereal area (ha)	0.82	1.03	0.93	1.19	0.86	1.14	0.78	0.95
Crop land using fertilizer (%)	15.5	84.0	8.9	76.7	12.7	81.8	18.8	74.6
Wereda level								
Market access (minutes)	258	230	261	239	257	233	263	248
Road density (km/km^2)	30.8	34.8	29.5	31.6	30.5	34.2	29.3	32.4
Population density (persons/km^2)	199	221	177	194	193	223	193	213
Area share of highly suitable land (%)	0.13	0.19	0.29	0.32	0.2	0.2	0.25	0.29
Crop land using fertilizer (%)	20.3	37.2	39.2	51.7	36.3	55.4	22.0	31.2

Source: Author's calculation using CSA agricultural sample survey data (various years).

access to inputs and knowledge on the new technology.

Holder's characteristics also affect household's access to fertilizer. In particular, age has a significant and negative effect on the likelihood of fertilizer adoption in the case of maize, wheat and barley, supporting the hypothesis that older holders are less likely to access the modern technology than younger holders. Accessibility is better in male-headed households than their female-headed counterparts among teff and barley farmers. No relation between access to

Table 5. Double hurdle regression estimates for fertilizer access.[a]

Variable	Maize		Teff		Wheat		Barley	
	Coefficient	P > z	Coefficient	P > z	Coefficient	P > z	Coefficient	P > z
Fertilizer access (yes=1)								
Credit (yes = 1)	-0.094	0.000	0.027	0.185	0.067	0.002	-0.115	0.000
Extension (yes = 1)	2.646	0.000	0.014	0.902	0.231	0.059	1.611	0.000
Advisory service (yes = 1)	-0.299	0.000	0.012	0.677	0.070	0.083	-0.263	0.000
Gender (male = 1)	0.013	0.527	0.093	0.000	-0.000	0.987	0.067	0.024
Age	-0.002	0.000	0.001	0.306	-0.003	0.000	-0.004	0.000
Education grade	0.004	0.173	-0.003	0.440	0.001	0.828	0.000	0.997
Area share of total crop land using fertilizer (household)	0.022	0.000	0.040	0.000	0.035	0.000	0.031	0.000
Area share of total crop land using fertilizer (wereda)	0.010	0.000	0.012	0.000	0.013	0.000	0.013	0.000
Market access (wereda)	-0.000	0.000	0.000	0.001	-0.001	0.000	-0.000	0.000
Population density (wereda)	0.000	0.048	-0.000	0.037	0.000	0.162	-0.000	0.854
Road density (wereda)	-0.001	0.000	-0.001	0.000	-0.001	0.001	-0.001	0.007
Generalized residual	-0.477	0.000	0.497	0.000	0.383	0.000	-0.449	0.000
Constant	-2.652	0.000	-2.450	0.000	-2.322	0.001	-3.827	0.000
Observations	110162		89533		60228		62026	
Log likelihood	-167.6		4820		7412		3635	
P-value of Wald test of indep. eqns. (rho = 0)	0.274	0.000	0.290	0.000	0.257	0.000	0.259	0.000
P-value of LR test of Tobit model	19983	0.000	25783	0.000	21405	0.000	12830	0.000

Note: [a] Extension is treated as an endogenous variable; Source: Author's calculation using CSA data (various years).

fertilizer and education was found.

The spatial variables included to explain access do not appear to have major impact as their coefficients are quite small. In the case of maize, the spatial effects are better captured by the zonal dummies (not reported). Access to fertilizer in maize production is more likely in the south and southwest, around Awasa and Jimma, in West Oromia, and in the zones crossed by the major road going east to Djibouti: West and East Hararge, West Arsi and Harari). Coefficients of the dummy variables in the case of teff show that farmers of some zones in SNNP and in particular in Amhara with high teff production have difficulties to access the technology. In the case of wheat, none of the coefficients of the zonal dummy variables is significant, indicating that only variables related to fix costs of the technology are relevant explaining access to fertilizer.

Determinants of fertilizer demand

Results for the estimation of the model explaining area planted using fertilizer conditional on access to fertilizer at plot level are presented in Table 6. The conditional area under fertilizer is mainly explained by: specialization in the particular crop (mono crop production at the plot level); access to inputs through extension specialist; previous knowledge and experience in cereal production (crop rotation); access to land rental market and land fragmentation; total cereal area; share of the crop in total household cultivated cereal area; and the area under fertilizer in the wereda.

The area under cereal production is positively

Table 6. Double hurdle regression estimates for fertilizer use.

Variable	Maize		Teff		Wheat		Barley	
	Coefficient	P>z	Coefficient	P>z	Coefficient	P>z	Coefficient	P>z
Area under chemical fertilizer								
Irrigation (yes = 1)	-0.049	0.216	-0.208	0.000	-0.136	0.003	-0.044	0.561
Pesticides and herbicides (yes = 1)	0.019	0.547	0.053	0.000	0.055	0.000	0.089	0.000
Mono-crop (yes = 1)	0.243	0.000	0.132	0.000	0.514	0.000	0.157	0.000
Crop rotation (yes = 1)	0.039	0.008	0.034	0.004	0.055	0.000	0.006	0.769
Extension (yes = 1)	0.187	0.005	0.102	0.005	0.071	0.097	0.148	0.017
Advisory service (yes = 1)	-0.015	0.505	-0.024	0.037	-0.014	0.363	-0.026	0.237
Gender (male = 1)	0.076	0.000	0.029	0.000	0.020	0.014	0.020	0.182
Age	0.001	0.000	0.003	0.000	0.003	0.000	0.002	0.000
Education grade	-0.007	0.000	0.001	0.114	-0.000	0.813	0.004	0.046
Area share of total crop land using fertilizer	0.000	0.525	0.000	0.045	-0.000	0.135	-0.000	0.142
Area share of total crop land using fertilizer (wereda)	0.001	0.043	-0.001	0.000	0.001	0.007	0.002	0.000
Share of highly suitable land	0.025	0.160	0.005	0.785	0.018	0.378	-0.062	0.019
Share of moderately to marginally suitable land	-0.018	0.542	-7.185	0.024	0.573	0.000	-0.098	0.082
Household size	0.000	0.783	-0.001	0.217	0.004	0.002	0.002	0.347
Cereal area of household	0.317	0.000	0.182	0.000	0.155	0.000	0.281	0.000
Share of the crop in total cereal area	0.006	0.000	0.004	0.000	0.004	0.000	0.007	0.000
Number of plots under holder	-0.047	0.000	-0.044	0.000	-0.038	0.000	-0.057	0.000
Plot is rent (yes = 1)	0.050	0.001	0.000	0.992	0.038	0.000	0.105	0.000
Market access	0.000	0.000	0.000	0.000	0.000	0.411	0.000	0.449
Population density	-0.000	0.000	-0.000	0.017	-0.000	0.001	-0.000	0.160
Road density	0.000	0.478	-0.000	0.000	-0.001	0.000	-0.001	0.000
Generalized residual	-0.041	0.290	-0.054	0.011	-0.020	0.419	-0.076	0.026
Constant	-0.676	0.000	-0.420	0.000	-0.443	0.551	-1.626	0.005
Log-likelihood	-10040.9		71.4		-30098		-16162	

Note: [a] Extension is treated as an endogenous variable; Source: Author's calculation using CSA data (various years).

associated with area using fertilizer. The coefficients of advisory services are far smaller and can be insignificant, suggesting extension is the dominants source of knowledge for holders. In addition, farmers increase fertilizer application of a particular cereal crop if the crop is important in the production system (captured by the crop's area share). Households that rent land for crop production tend to have larger area under fertilizer than those without access to land. Similar to total land under cereal production, having access to land rental market results in an increase in the area using fertilizer but a reduced share of fertilized area in total cereal land. Coefficients obtained for land rental in different crops support the hypothesis that households compensate forthe additional risk of increasing area of a crop by reducing input intensity for that crop.

In contrast with other studies (e.g. Croppenstedt et al., 2003), family size does not play a significant role determining fertilizer use. As fertilizer is assumed to be a labor intensive technology, it is expected that availability of family labor would result in higher fertilizer use. Only in the case of wheat we find that household size is positively and significantly related to area under fertilizer. A possible explanation for our results is that farmers make their decisions of the area applied fertilizer mainly based on the crop growing in the plot, and the availability of labor is not a constraint yet given the relatively low adoption rate of fertilizer.

Among holder characteristics we find that age has a positive and statistically significant effect on fertilizer demand in all crops suggesting that among adopters, farming experience is related to efficiency and adoption in the use of fertilizer, at both household and wereda level. We also find a significant effect of gender in conditioning fertilizer demand and households with more educated head exhibit higher fertilizer adoption in barley production but not in the case maize. With the exception of market access for maize and teff, the coefficients of spatial variables (crop suitability, population and road density) are negligible, indicating that biophysical and demographic conditions are not the major constraints in fertilizer adoption in Ethiopia.

Average partial effects

As identification and estimation of average effects become more complicated in the case of nonlinear models with discrete variables, as in this study, it is not easy to examine and compare the effect on fertilizer adoption from different influencing factors. The average partial effect (APE) is hence introduced to measure the average change in technology adoption due to a change in the variables of interest. We obtained average partial effects by bootstrapping the estimated model and results after 500 iterations are reported in Table 7. These results show that extension has a significant positive effect on fertilizer adoption. For example, households having access to extension can increase the average maize area under fertilizer by 0.1 ha.

The higher the share of crop area under fertilizer at household and district level the higher fertilizer use. Farmers' own skills and knowledge, represented by mono crop, crop rotation, and uses of chemicals, all contribute to the quantity of fertilizer used. APEs of variables associated with household wealth confirm that households have exhibited decreasing absolute risk aversion but increasing relative risk aversion. Fragmented land plots prevent wide adoption of technology, while on average a plot managed by a male holder tends to show higher fertilizer use, than those managed by younger female holders. APE results also suggest that although infrastructure related factors like market access, and population and road density do have an impact on fertilizer adoption, their effects are small and not comparable with the agroecological constraints defined by crop suitability.

Ranked by APE, the top three factors affecting average change in area using fertilizer are extension, mono-crop and total area for cereal production. Similarly, Extension, cereal area and mono-crop are found major contributors in changes in fertilizer adoption. Our results are consistent with previous studies on fertilizer adoption in Ethiopia, especially in the role of extension and farmers' experience.

Conclusion

Extension services have played a central role in facilitating access to the promoted technology, as it is the instrument to disseminate new technology package including seed, fertilizer and new farming practice. Ethiopia's agricultural extension system is one of the largest in the world, with over 60,000 development agents working in nearly 10,000 farmer training centers throughout the country. This paper aims to understand the extent and determinants of fertilizer adoption in the country. A double hurdle model is chosen, which allows us to analyze separately the factors affecting access of farmers to the new technology and demand for fertilizer conditional to access in a sequential approach, addressing the endogeneity of extension service. Built on the framework of Asfaw et al. (2011), the study is the first to estimate fertilizer adoption using nationally representative data from Ethiopia Agricultural Sample

Table 7. Average partial effects of factors on chemical fertilizer adoption.

Average increase in area under fertilizer (ha)	Maize		Teff		Wheat		Barley	
	APE	t-value	APE	t-value	APE	t-value	APE	t-value
Variables in both demand and access equations								
Extension (yes = 1)	0.1007	7.9	0.0209	2.6	0.0184	2.2	0.0342	5.4
Advisory service (yes = 1)	-0.0056	-4.0	-0.0043	-2.0	-0.0012	-0.4	-0.0046	-3.7
Gender (male = 1)	0.0040	5.5	0.0072	4.7	0.0034	0.9	0.0017	1.9
Age	0.0000	0.9	0.0005	8.4	0.0004	6.9	0.0001	3.3
Education grade	-0.0003	-3.2	0.0002	0.9	-0.0001	-0.2	0.0002	1.6
Area share of total crop land using fertilizer (household)	0.0004	22.6	0.0010	36.4	0.0007	22.5	0.0004	22.1
Area share of total crop land using fertilizer (wereda)	0.0002	11.8	0.0001	4.9	0.0003	10.7	0.0003	16.4
Market access	0.0000	1.1	0.0000	10.8	0.0000	-1.1	0.0000	-1.2
Population density	0.0000	-2.2	0.0000	-3.0	0.0000	-0.4	0.0000	-1.0
Road density	0.0000	-0.7	-0.0001	-5.6	-0.0002	-3.8	-0.0001	-5.7
Variables in demand equation only								
Irrigation (yes = 1)	-0.0024	-1.1	-0.0319	-5.9	-0.0212	-3.4	-0.0019	-0.6
Pesticides & herbicides (yes = 1)	0.0010	0.5	0.0099	6.1	0.0092	5.8	0.0043	5.3
Mono-crop (yes = 1)	0.0135	12.6	0.0243	3.5	0.0928	9.7	0.0074	4.0
Crop rotation (yes = 1)	0.0020	1.8	0.0060	2.2	0.0085	2.4	0.0003	0.3
Share of highly suitable land	0.0013	1.3	0.0009	0.3	0.0035	0.9	-0.0028	-2.1
Share of moderately to marginally suitable land	-0.0009	-0.5	-1.3125	-2.5	0.0876	4.4	-0.0045	-1.6
Household size	0.0000	0.2	-0.0002	-1.1	0.0006	1.5	0.0001	0.8
Cereal area of household	0.0166	18.4	0.0332	33.5	0.0267	30.8	0.0129	24.1
Share of the crop in total cereal area	0.0003	29.7	0.0008	34.5	0.0008	31.8	0.0003	28.1
Number of plots under holder	-0.0025	-16.7	-0.0081	-18.7	-0.0065	-25.1	-0.0026	-19.4
Plot is rent (yes = 1)	0.0027	3.1	0.0000	0.0	0.0067	4.6	0.0052	6.1

Source: Author's calculation using CSA data (various years).

Surveys while considering model endogeneity using a control function approach.

The major findings are presented below, centered on the impact of extension. First, statistical tests allow us to examine several theoretical and methodological assumptions laid out at the start of the study. The studyconfirms the endogeneity of extension service in the decision making process of fertilizer adoption. Log-likelihood test also indicates that farmers make input decisions sequentially, not simultaneously, highlighting the relative superiority of the DH specification for this analysis. Average partial effect from bootstrapping process is appropriate in inspecting the unconditional effects of factors that influence the adoption process because it is especially helpful in cases when there are observations with zero values for input use.

Second, extension service boosts fertilizer use, which corroborates with many other studies on fertilizer adoption in Ethiopia (Beshir et al., 2012; Carlsson et al., 2005; Wubeneh and Sanders, 2006; Noltze et al., 2012). Extension not only increases the probability of adoption fertilizer but

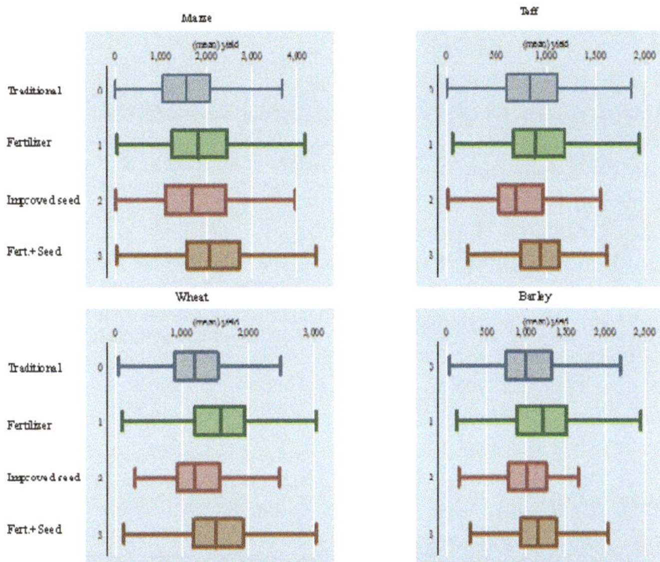

Figure 3. Yield distributions of cereals at the plot level different input combinations (average values 2003-2007 in kilograms per hectare); Source: Author's computation using CSA agricultural sample survey data.

expand the area planted using fertilizer conditional on access to fertilizer. Farmers face a high "knowledge" cost related to adoption of new technology and extension services helps cut the adoption cost (Asfaw and Admassie, 2004). Education and past experience can also effectively lower the adoption barrier and hence promote the diffusion of new technology, which were also identified in earlier studies by Admassie and Ayele (2004) and Beshir et al. (2012). Other factors affects fertilizer use includes household wealth, access to land rental market, land fragmentation and the importance of the crop in the production system.

Third, similar to Dercon and Christiaensen (2011), risk aversion behavior can have a considerable impact on farmer's fertilizer adoption decision. Having access to land rental market can increase the area using fertilizer but reduce the share of fertilized area in total cereal land, suggesting that households compensate for the additional risk of increasing area of a crop by reducing input intensity for that crop. In addition, variables associated with household wealth confirm that households have exhibited decreasing absolute risk aversion but increasing relative risk aversion.

Fourth, spatial variables obtained through GIS tools are introduced in the analysis to capture of biophysical constraints and infrastructure factors like crop suitability and market access. Although we find market access and road density do have an impact on fertilizer adoption, their effects are small and not comparable with the agroecological constraints defined by crop suitability and other local agronomic and development conditions (Diao and Nin-Pratt, 2005; Tadesse et al., 2006). Substantial yield gain in maize and teff can be achieved if technology is provided through a locally tailored extension package. This translates into the need of an extension package focusing on improving access to technological packages that are adapted to local agroecological conditions to fully realize the potential.

Although the results of this analysis highlight the important role and great potential of extension in the development of Ethiopia agriculture, the impacts of the strategy to raise cereal production and yields through extension have been mixed, as fertilizer use increased but access to extension and productivity growth remain low among farmers in the country. Some potential constraints could compromise the effectiveness of the extension system, including service mismatching farmers' need, deficiency of resources, low capabilities and knowledge of extension workers and lack of transparency and motivations. Readers can refer to Davis et al. (2010) for detailed discussion on these constraints.

One problem observed from the data and that it is not necessarily captured by the DH model is the great yield variability among producers using fertilizers (Figure 3). First, the median of the yield distribution obtained using fertilizer + seed in maize and wheat is larger than that of the traditional technology but far from the expectations that the authorities had of doubling yields when the program was launched. Second, the highest yields (the 90th percentile of the distribution) are close to those obtained in trials and experiments during the first face of PADETES (3,700 kg/ha in maize and close to 3,000 kilograms per hectare in wheat). Reducing the high variability observed in yields with fertilizer+seed technology should result in movements of the mean and median of the yield distribution closer to what today are "frontier" values (high yield) resulting in improved conditions and incentives to adopt the technology. Third, median yields obtained in teff and barley using fertilizer technology are low and similar to those obtained using the traditional technology, with frontier values in the improved technology being much lower than those obtained with the traditional technology. This suggests that availability of improved varieties in teff and barley is still a major constrain to increase yields and that the only technical alternative to the traditional technology is the use of chemical fertilizer. The possibilities of increasing yields of these crops using fertilizer only are quite limited.

Conflict of Interests

The author(s) have not declared any conflict of interests.

ACKNOWLEDGEMENTS

The authors acknowledge the significant contribution of José Funes and Sinafikeh Asrat Gemessa to an early version of this paper.

REFERENCES

Abebaw D, Haile MG (2013). The impact of cooperatives on agricultural technology adoption: Empirical evidence from Ethiopia. Food Policy 38:82–91.

Admassie A, Ayele G (2004). Adoption of improved technology in Ethiopia. Addis Ababa, Ethiopia: Ethiopia Development Research Institute. PMid:16895031

Asfaw A, Shiferaw B, Simtowe F, Haile MG (2004). Agricultural technology adoption, seed access constraints and commercialization in Ethiopia. J. Dev. Agric. Econ. 3(9):436-447.

Beshir H, Emana B, Kassa B, Haji J (2012). Determinants of chemical fertilizer technology adoption in North eastern highlands of Ethiopia: the double hurdle approach. J. Res. Econ. Int. Finan. 1(2):39-49.

Burke WJ (2009). Fitting and interpreting Cragg's tobit alternative using Stata. The Stata J. 9(4):584-592.

Byerlee D, Spielman DJ, Alemu D, Gautam M (2007). Policies to promote cereal intensification in Ethiopia: A review of evidence and experience. Discussion Paper 00707. Washington, DC: International Food Policy Research Institute.

Carlsson F, Kohlin G, Mekonnen A, Yesuf M (2005). Are agricultural extension packages what Ethiopian farmers want? A stated preference analysis. Working Papers in Economics. Department of Economics, Goteborg University. P. 172.

Coady DP (1995). An empirical analysis of fertilizer use in Pakistan. Economica 62:213–234.

Cragg JG (1971). Some statistical models for limited dependent variables with applications to the demand for durable goods. Econometrica 39:829–844.

Croppenstedt A, Demeke M, Meschi M (2003). Technology adoption in the presence of constraints: The case of fertilizer demand in Ethiopia. Rev. Dev. Econ. 7(1):58-70.

CSA. Various years. Annual agricultural sample survey (database). Central Statistical Agency. Last accessed September 2011, http://www.csa.gov.et/

CSA, EDRI and IFPRI. 2007. Atlas of Ethiopia Population and Housing Census. Central Statistical Agency, Ethiopian Development Research Institute, and International Food Policy Research Institute, Addis Ababa, Ethiopia.

Davis K, Swanson B, Amudavi D, Mekonnen DA, Flohrs A, Riese J, Lamb C, Zerfu E (2010). In-depth assessment of the public agricultural extension system of Ethiopia and recommendations for improvement. Discussion Paper 01041. Washington, DC: International Food Policy Research Institute.

Dercon S, Christiaensen L (2011). Consumption risk, technology adoption and poverty traps: Evidence from Ethiopia. J. Dev. Econ. 96(2):159-173.

Diao X, Nin -Pratt A (2005). Growth options and poverty reduction in Ethiopia: A spatial, economy-wide model analysis for 2004-15. DSGD Discussion, Washington, DC: International Food Policy Research Institute. P. 20.

EDRI (2009). Crop suitability analysis– agroecological zones. Addis Ababa, Ethiopia: Ethiopian Development Research Institute.

Holden ST, Shiferaw B, Pender J (2001). Market imperfections and land productivity in the Ethiopian highland. J. Agric. Econ. 52(3):53–70.

Howard J, Crawford E, Kelly V, Demeke M, Jeje JJ (2003). Promoting high-input maize technologies in Africa: The Sasakawa-Global 2000 experience in Ethiopia and Mozambique. Food Policy 28:335–348.

Imbens G, Wooldridge JM (2008). Control function and related methods. Applied Microeconometrics Workshop Lecture Notes.

Just RE, Zilberman D (1983). Stochastic structure, farm size, and technology adoption in developing agriculture. Oxford Econ. 35(2):307–328.

Liverpool-Tasie LSO, Winter-Nelson A (2009). Poverty Status and the Impact of Formal Credit on Technology Use and Wellbeing among Ethiopian Smallholders. World Dev. 38(4):541–554.

Liverpool-Tasie LSO, Winter-Nelson A (2012). Social Learning and Farm Technology in Ethiopia: Impacts by Technology, Network Type, and Poverty Status. J. Dev. Stud. 48(10):1505–1521.

Noltze M, Schwarze S, Qaim M (2012). Understanding the adoption of system technologies in smallholder agriculture: The system of rice intensification (SRI) in Timor Leste. Agric. Syst. 108:64-73.

Pender J, Kerr JM (1998). Determinants of farmers' indigenous soil and water conservation investments in semi-arid India. Agri. Econ. 19:113–125.

Rivers D, Vuong QH (1988). Limited information estimators and exogeneity tests for simultaneous probit models. J. Econometr. 39:347-366.

Shiferaw BA, Kebede TA, You L (2008). Technology adoption under seed access constraints and the economic impacts of improved pigeonpea varieties in Tanzania. Agric. Econ. 39:309-323.

Tadesse M, Alemu B, Bekele G, Tebikew T, Chamberlin J, Benson T (2006). Atlas of the Ethiopian Rural Economy. Addis Ababa, Ethiopia: Ethiopian Development Research Institute.

Wubeneh NG, Sanders JH (2006). Farm-level adoption of sorghum technologies in Tigray, Ethiopia. Agri. Systems 91:122-134.

World Bank (2006). Ethiopia: Policies for pro-poor agricultural growth. Washington, DC.

An economic analysis of crude oil pollution effects on crop farms in Rivers State, Nigeria

ThankGod Peter Ojimba[1], Jacob Akintola[2], Sixtus O. Anyanwu[1] and Hubbert A. Manilla[1]

[1]Department of Agricultural Science, Ignatius Ajuru University of Education,Ndele Campus, P.M.B. 5047, Port Harcourt, Nigeria.
[2]Department of Agricultural Economics, Bowen University,Iwo, Osun State, Nigeria.

This study researched on crude oil pollution effects on crop farms in Rivers State, Nigeria using stochastic translog production function. Data were collected in the state, using multi-stage sampling technique. A total of 296 structured questionnaires retrieved from farmers in crude oil polluted and non-polluted areas of the state were used. Stochastic translog production interaction between land and heavy, medium and light oil spillages resulted in crops output reduction by 0.255, 1.257 and 1.027 units, respectively. Interaction across heavy, medium and light oil spillages and fertilizers usage indicated farm crops output decrease by 0.805, 0.586 and 0.729 units, respectively. This study therefore concluded that crude oil pollution on crops farms reduced crops output significantly, hence detrimental to crop production in Rivers State, Nigeria.

Key words: Crude oil, pollution, stochastic translog production function, crops output, Nigeria.

INTRODUCTION

The petroleum industry is the backbone of the Nigeria economy, accounting for over 90% of total foreign exchange revenue. The daily production of crude oil is slightly above two million barrels from more than 240 producing fields, totaling over 5,284 wells drilled. With over half a century of oil and gas exploration, exploitation and production, Nigeria has built up a considerable hydrocarbon infrastructure with over 7,000 km of pipelines linking over 280 producing flow stations all of which are situated in the Niger Delta region of Nigeria (Niger Delta Development Commission, 2006). The Niger Delta region is situated in the southern part of Nigeria and bordered to the south by the Atlantic Ocean, occupying a surface area of about 112, 110 km^2, which represents 12% of Nigeria total surface area with an estimated population of about 28 million inhabitants in 2006 (NDDC, 2006). Within this region, crude oil pollution such as oil spillages and gas flaring regularly occur (Orji, et al., 2011; Nwaichi and Uzazobona, 2011).

Scholars such as Uzoho et al. (2004) evaluated the influence of crude oil on maize growth and soil properties in Ihiagwa, Imo State, Nigeria. The results of their study showed that seed germination, plant height, leaf area and dry matter yield significantly deceased as the levels of crude oil pollution increased. The primary way in which crude oil pollution reduced crop growth and performance according to their study was through reduction of seedling emergence and direct suffocation of plant and oxygen diffusion rates between soil system and the atmosphere.

Dung et al. (2008) explored the spatial variability effects of gas flaring on the growth and development of cassava (*Manihot esculenta*), water leaf (*Talinum triangulare*) and pepper (Piper spp.), which are crops commonly cultivated in the Niger Delta. Their results suggested that a spatial gradient exist in the effect of gas on crop development. Retardation in crop development manifested in decreased dimensions of leaf lengths, and widths of cassava and pepper crops closer to the gas flare points. Their statistical analysis confirmed that cassava yield were higher at locations further away from the flare point. In addition, the amount of starch and ascorbic acid in cassava decreased when plant is grown closer to the gas flare. High temperature around the gas flare appeared to be the most likely cause of the retardation and low yield.

Okonwu et al. (2010) investigation showed that the percentage of germination of maize (*Zea mays*) decreased with increase in concentration of crude oil equilibrated with water. Germination rate decreased significantly with increased time of pre-soaking in crude oils. Crude oil spilled soil immediately after planting increased the length of lag phase preceding germination from 100% in the control to 58% in crude oil contaminated soil. Fernandez-Luqueno et al. (2012) studied the ability of various crops to grow and maintain their yield when they are cultivated in contaminated soils, thereby being able to choose the most appropriate crop when suddenly a gasoline-pipeline collapse on soil of subsistence agricultural systems. Their results showed that gasoline contamination reduced seedling emergence, shoot length, root volume, root dry weight, shoot dry weight and abundance of nodules.

Problem statement

Despite the availability and use of advanced technology in the petroleum industry, various forms of accidents such as blow-outs of production wells, explosions and pipeline ruptures still occur, which are worsened by vandalization of oil installations and pipelines (Otitoloju et al. 2007; Iturbe et al. 2008; Li et al., 2011). Farmers in Rivers State are eventually the most affected judging by the death of marine and terrestrial organisms usually involved in oil spill incidents and the hazardous effect of gas flaring (Saier, 2006; Otitoloju and Dan-Patrick, 2010; Huang, et al., 2011). The rivers and underground water which the inhabitants rely on, for their drinking water have been polluted with crude oil, while buildings and agricultural products had been destroyed (Atakpo and Ayolabi, 2009; Ekpoh and Obia, 2010; Nwaichi and Uzazobona, 2011; Onyenekenwa, 2011). Irregularities had been observed in the major livelihood activities of the people of Rivers State of Nigeria due to crude oil pollution (Okoli, 2006; Adoki and Orugbani, 2007; Orogun, 2009).

Ekunwe and Orewa (2007) (examined the technical efficiency and productivity of yam in Kogi State of Nigeria using stochastic frontier production function. The result indicated that the technical efficiency of farmers varied with a mean of 62%, while only about 23% of the farmers had technical efficiencies exceeding 80%. Erhabor and Emokaro (2007) employed the use of the stochastic frontier production function in the comparative economic analysis of the relative technical efficiency of cassava farmers in the three agro-ecological zones of Edo State, Nigeria. The empirical estimates showed mean technical efficiency of 72, 83 and 91% for Edo South, Edo North and Edo Central agro-ecological zones, respectively. Ajani and Ugwu (2008) used a stochastic production frontier model and obtained the result that gamma which is a measure of variance of output from the frontier attributed to efficiency was 0.114.

Heady et al. (2010) presented multi-output, multi-input total factor productivity (TFP) growth rate in agriculture for 88 countries over the 1970 and 2001 period estimated with both stochastic frontier analysis (SFA) and data envelopment analysis (DEA). They found results with SFA to be more plausible than with DEA, and used them to analyze trends across countries. Large volumes of literature still exist that had used stochastic frontier production analysis in crop production (Ali, 1996; Onyenweaku and Okoye, 2007; Nyagaka et al., 2010; Dlamini et al; 2010). Some scholars (Lachaal et al., 2005; Awoyemi and Adeoti, 2006; Managi et al., 2007) have studied some aspects of stochastic frontier production function but they did not study the economic analysis of crude oil pollution effects on crop farms in Rivers State.

Therefore, there is a dearth of literature on the use of stochastic frontier transcendental logarithmic (traslog) production function for an economic analysis of crude oil polluted and non-polluted crop farms in Rivers State, Nigeria. At this juncture one may seek to understand the economic analysis of crude oil pollution effects on crop farms in Rivers State, Nigeria, using the stochastic frontier transcendental logarithmic (translog) production function as analytical tool to bridge this gap in knowledge.

The objectives of the study

The main objective of this study is to estimate economically crude oil pollution effects on crop farms in Rivers State, Nigeria using stochastic translog production function approach. The specific objectives are to:

1. Determine crude oil pollution effects on crop farms in Rivers State, Nigeria using stochastic translog production function analysis.
2. Make policy statements that could ameliorate the effects of crude oil pollution on crop farms in Rivers State, Nigeria.

MATERIALS AND METHODS

Data collection

This study was conducted in Rivers State of Nigeria in 2003. Data were collected from both the primary and secondary sources. The primary data were collected through personal interviews and observation with the farmers, and structured questionnaires were distributed among farmers in crude oil polluted and non-crude oil polluted areas of an affected community in the state. A multistage sampling technique was used to obtain data for the study. The first stage involved the selection of 17 local government areas (LGAs) out of the existing 23 LGAs in Rivers State. The selected LGAs include: Abua/Odual, Ahoada East, Ahoada West, Andoni, Asaritoru, Degema, Eleme, Emohua, Etche, Gokana, Ikwerre, Khana, Obio/Akpor, Ogba/Egbema/Ndoni, Omuma, Oyigbo and Tai LGAs. These 17 LGAs were selected based on the fact that they were more crop farming inclined than others. The second stage involved the stratification of farmland in a selected LGA into two sampling units namely crude oil polluted and non-crude oil polluted. This stratification of the farmland into two sampling units was based on the fact that information were needed from both crude oil polluted and non-polluted areas.

The third stage involved the random sampling of 10 farmers from crude oil polluted areas in a selected LGA and a corresponding number of 10 farmers from non-crude oil polluted farms (non-polluted) in the same locality (community) in the given area. This gave a total of 20 farmers interviewed per selected LGA in the State, giving a total of 340 questionnaires distributed in the 17 LGAs selected. Out of 340 questionnaires administered, due to difficult terrain, the politicking of oil pollution issues and youth restiveness in the State as at the time of the survey, only 326 questionnaires were retrieved. Furthermore, 30 questionnaires were found inconsistent with the set objectives of the study. Hence, only a total of 296 questionnaires were retained as suitable for analysis. Out of these 296 questionnaires retained as suitable for analysis, 169 questionnaires were retrieved from the crude oil polluted farms and 127 questionnaires from non-polluted farms. The unequal weighting in the data analyzed arose because most of the discarded and unretrieved questionnaires belonged to the non-polluted farms category. Because of the large number of samples retrieved from both polluted and non-polluted crop farms, comparison between the two groups of farms as a measure of efficiency was adequate and not misleading.

Measurement of crude oil pollution and technology indices

To measure the negative effects of crude oil pollution on each farmland polluted, the impact of crude oil pollution index was estimated following the methods specified by Mubana (1978), and Canter and Hill (1979) modified as follows:

$$P = \left(\frac{\overset{n}{\underset{i=1}{E}} \frac{q_{2i}}{q_1}}{n} \cdot X_i \right) \qquad (1)$$

where, P= crude oil pollution index per farmer in the crude oil polluted areas;

q_{2i} = land affected by the crude oil pollution, indicating the farm's degree of crude oil pollution (ha).

q_1 = total land area cultivated (ha)

X_i = percentage of crop yield foregone due to oil pollution (where, i = farmers degrees of pollution, 93 to 100%, 31 to 92% and 0 to 30%).

n = types of crude oil pollution affecting individual farm: n_1 = heavy oil pollution (acquired land); n_2 = medium oil pollution ; n_3 = light oil pollution

X_i was adopted from Udo and Fayemi (1975) and Mubana (1978), which categorized the types of negative effects of oil pollution:

Category A (ni): (i) Heavy oil spillage which leads to 93 to 100% crop yield loss.

(ii) Acquired land for oil well – head sites, flow stations, drilling sites, oil field location, borrow pits, gas flaring sites, pipeline laying operations and other oil related activities which leads to 100% crop yield loss (Mubana, 1978);

Category B (n_2): Medium oil spillage which leads to 31 to 92% crop yield reduction;

Category C (n_3): Light oil spillage which leads to 0 to 30% crop yield reduction.

The level of technology was captured using in a chain index method proposed the Harper (1971) and Mubana (1978). It is mathematically expressed as:

$$T = \frac{\overset{n}{\underset{i=1}{\Gamma}} \left(\frac{\forall_{2i}}{\forall_{1i}} \times 100 \right)}{K} \qquad (2)$$

Where, T = level of technology index, $\%_{2i}$ = quantity of each technology type used in current year t, (2003) measured in bags of fertilizers, packets of improved seeds and dressing of seeds. These inputs were converted into percentages before the summations;

$\%_{1i}$ = quantity of each technology type used in year t - 1, (2002) measured as above, i = 1,2, ... 296,

K = number of types of technology adopted by the farmer in t (2003) and t-1 (year (2002) 2.3

Stochastic translog production function

Christensen et al. (1973) studied translog production function which is general, flexible and allowed analysis of interactions among variables. Ali (1996) used stochastic translog production function to analyse socio-economic determinants of sustainable crop production in Nepal. This study will apply the stochastic translog production function with moderation from Christensen et al. (1973) and Ali (1996) to estimate economically crude oil pollution effects on crop farms in Rivers State, Nigeria. The stochastic frontier translog production function given in equation (3) was estimated for the crude oil polluted crop farms only, while the estimation of non-polluted crop farms did not include the P variables. The general form of the translog stochastic production function used in this study is:

$$\ln Y_j = \forall_0 + \overset{n}{\underset{i=1}{\Gamma}} a_i \ln X_{ij} + \tfrac{1}{2} \overset{n}{\underset{i=1}{E}} \overset{n}{\underset{g=1}{\Gamma}} b_{ig} (\ln X_{ij} \ln X_{ij})$$

$$+ \overset{m}{\underset{i=1}{E}} c_k \ln P_{kj} + \overset{p}{\underset{t=1}{\Gamma}} d_t \ln T_{tj} + \overset{n}{\underset{i=1}{E}} b_{ii} (\ln X_{ii})^2$$

$$+ \tfrac{1}{2} \overset{n}{\underset{i=1}{E}} \overset{m}{\underset{k-1}{\Gamma}} e_{ik} (\ln X_{ij} \ln P_{kj}) + \tfrac{1}{2} \overset{n}{\underset{i-i}{\Gamma}} \overset{p}{\underset{t-1}{\Gamma}} f_{it} (\ln X_{ij} \ln T_{tj})$$

$$+ \tfrac{1}{2} \overset{m}{\underset{k=1}{\Gamma}} \overset{m}{\underset{k=1}{\Gamma}} h_{kk} (\ln P_{kj} \ln P_{kj}) + \tfrac{1}{2} \overset{n}{\underset{k=j}{\Gamma}} \overset{p}{\underset{t=1}{\Gamma}} r_{kt} (\ln P_{kt}\, l_n T_{tj})$$

$$+ \tfrac{1}{2} \overset{p}{\underset{t=1}{\Gamma}} \overset{p}{\underset{t=1}{\Gamma}} s_{tt} (\ln T_{tj} \ln T_{tj}) + u_j + v_i \qquad (3)$$

where, j = 1,2,3, ……… 169 for crude oil polluted crop farms and

127 for non-polluted crop farms
i = 1,2,3, are physical inputs.

Parameters used include:
Y = Crop output in ton/ha per farmer (crops here refer to different types of crops because of the subsistence nature of production – mixed cropping).
X = Vector of physical inputs used (land area cultivated in hectares, available family and hired labor in man days; fixed and operating capital in dollars).
P = Vector of effect of crude oil pollution index on the farmer;
T=Vector of level of technology index;
vi=Random error due to misspecification of the model;
ui =Ratio of actual value to maximum possible output, that is inefficiency component of error terms.
ln = logarithmic sign
$\forall$ = parameter of intercept,
a_i = parameters of physical inputs,
b_{ijg} = parameters for interaction across ith and gth physical inputs,
c_k = parameters of crude oil pollution variables in indices,
dr = parameters for level of technology variables indices,
b_{ii} = parameters for squared terms of physical inputs.
e_{ik} = parameters for interaction between physical inputs and crude oil pollution variables
f_{it} = parameters for interaction between physical inputs and technology variables.
h_{kk}= parameters for interaction among crude oil pollution variables,
r_{kt} = parameters for interaction between oil pollution and technology variables,
s_{tt}= parameters for interaction across technology variables

It is important to note that Xi and T variables are some conventional physical inputs and technology variables normally considered in transformation process. However, the P variables are the conditioning variables which had been included into the model to capture the negative and detrimental effects of crude oil pollution on crops output or yield.

Translog model specifications

The model specifications for the translog stochastic production function retained in Equation (3) for the crude oil polluted crop farms is as given in Equation (4). As earlier mentioned, the specification for the non-polluted crop farms category excluded the crude oil pollution variables, that is X_5 to X_{12}, and their various interactions.

$$\begin{aligned}\ln Y = {} & \forall_0 + \beta_1 \ln X_1 + \beta_2 \ln X_2 + \beta_3 \ln X_3 + \beta_4 \ln X_4 + \beta_5 \ln X_5 \\ & + \beta_6 \ln X_6 + \beta_7 \ln X_7 + \beta_8 \ln X_8 + \beta_9 \ln X_9 + \beta_{10} \ln X_{10} \\ & + \beta_{11} \ln X_{11} + \beta_{12} \ln X_{12} + \beta_{13} \ln X_{13} + \beta_{14} \ln X_{14} + \beta_{15} (\ln X_1)^2 \\ & + \beta_{16} (\ln X_2)^2 + \beta_{17} (\ln X_3)^2 + \beta_{18} (\ln X_1 \cdot \ln X_2) \\ & + \beta_{19} (\ln X_1 \cdot \ln X_3) + \beta_{20} (\ln X_2 \cdot \ln X_3) + \beta_{21} (\ln X_1 \cdot \ln X_{10}) \\ & + \beta_{22} (\ln X_1 \cdot \ln X_{11}) + + \beta_{23} (\ln X_1 \cdot \ln X_{12}) + \beta_{24} (\ln X_1 \cdot \ln X_{13}) \\ & + \beta_{25} (\ln X_1 \cdot \ln X_{14}) + \beta_{26} (\ln X_2 \cdot \ln X_{13}) + \beta_{27} (\ln X_{10} \cdot \ln X_{11}) \\ & + \beta_{28} (\ln X_{10} \cdot \ln X_{12}) + \beta_{29} (\ln X_{11} \cdot \ln X_{12}) + \beta_{30} (\ln X_{10} \cdot \ln X_{13}) \\ & + \beta_{31} (\ln X_{11} \cdot \ln X_{13}) + \beta_{32} (\ln X_{12} \cdot \ln X_{13}) + \beta_{33} (\ln X_4 \cdot \ln X_{13}) \\ & + \beta_{34} (\ln X_4 \cdot \ln X_{14}) + \beta_{35} (\ln X_{13} \cdot \ln X_{14}) + u + v \end{aligned} \quad (4)$$

It is necessary to explain here that the total number of possible interactions were 75 but these had been drastically reduced to 21 interactions only. In addition to the 14 sets of variables considered in this analysis, the number of parameters estimated increased to 36 (including the intercept). This was done in order to ensure that only economically meaningful and theoretically plausible interactions were retained for the analysis, and also to reduce and ease the computation burden, as well as reduce the risks of multicollinearity.

RESULTS AND DISCUSSION

Table 1 shows the results of maximum likelihood estimation (MLE) for stochastic translog production frontier function in crude oil polluted and non-polluted crop farms in Rivers State, Nigeria. The results will be analyzed following the various types of variables and interactions obtained according to the translog model specification as shown in equation (4).

Stochastic translog production variables interactions

The discussions concentrated on the interactive actions of the translog variables used in the analysis

Squared terms of physical inputs (b_{ii})

Doubling the physical inputs means, using these inputs once again after the initial usage on crude oil spilled farms with the intension of increasing productivity after the application of proper remediation techniques in crude oil polluted crop farms, while in the non–polluted crop farms, it means doubling the usage of these inputs with the sole purpose of increasing production. Therefore, squaring (doubling) the amount of farm land (β_{15}) available for farming in crude oil polluted farms decreased the crop output by 0.093 units, though marginal.

There was an increase of 0.051 units per unit of output experienced in non-polluted crop farms when farmland was doubled (squared), though also marginal inelastic and not significant. These results showed that when amount of farmland available was doubled, there was an increase in output in non-polluted crop farms, and a decrease in farm output in crude oil polluted crop farms category. This could be due to the environmental stress or negative effects of crude oil pollution on crops (Achuba, 2006).

Squaring the labour variable (β_{16}) (labour x labour) in crude oil polluted crop farms reduced output by 0.112 units, and at the same time led to increased cost of production, whereas, if labour was doubled in non-polluted crop farms, crop output increased, though marginally by 0.150 units (significant at 1% level) despite the increased cost of production. Squaring of capital

Table 1. Maximum likelihood estimates of stochastic translog production function in crude oil polluted crop farms in Rivers State, Nigeria.

S/N	Variables	Parameters	Crude oil polluted, Translog MLE		Non-polluted Translog MLE	
			Coefficient value	Standard error	Coefficient value	Standard error
1	Constant (intercept)	∀	7.378***	2.691	9.368***	3.908
			Physical input (ai)			
2	Land (ha)	β_1	-2.584***	0.471	0.403	0.650
	Labour (Mandays)	β_2	0.242	0.525	0.009	0.798
	Capital ($)	β_3	0.145	0.495	0.731	0.635
			Indexes (ai)			
3	Technology index	β_4	0.220	0.191	-0.048	0.314
	Crude oil pollution index	β_5	-7.046***	2.478		
			Crude oil pollution variables (c_k)			
4	**Farmland acquired for:**					
	Flow station	β_6	-0.302***	0.093		
	Borrow pits	β_7	-0.397***	0.128		
	Gas flaring	β_8	-0.189**	0.082		
	Heavy pollution	β_9	-0.766***	0.224		
	Degree of spillage					
	Heavy crude oil spillage	β_{10}	-3.992***	1.983		
	Medium crude oil spillage	β_{11}	5.622**	2.626		
	Light crude oil spillage	β_{12}	5.903***	2.488		
			Technology variables (d_t)			
5	Fertilizers	β_{13}	7.499***	1.119	2.771**	1.167
	Improves seeds	β_{14}	-1.049***	0.276	-2.194***	0.653
			Square terms (b_{ii})			
6	Land x land	β_{15}	-0.093	0.058	0.051	0.074
	Labour x labour	β_{16}	-0.112	0.043	0.150***	0.059
	Capital x capital	β_{17}	-0.040	0.029	0.114***	0.035
			Interaction across inputs (b_{ig})			
7	Land x labour	β_{18}	0.143**	0.067	0.219**	0.091
	Land x capital	β_{19}	0.254***	-0.044	-0.141**	0.070
	Labour x capital	β_{20}	-0.033	0.051	-0.211**	0.058
			Interaction of physical inputs and crude oil pollution variables (e_{lk})			
8	Land x heavy oil spillage	β_{21}	-0.255	0.161		
	Land x medium oil spillage	β_{22}	-1.257***	0.341		
	Land x light oil spillage	β_{23}	-1.027**	0.494		
			Interaction of physical inputs and technology variables (f_{it})			
9	Land x fertilizers	β_{24}	0.363***	0.110	0.059	0.128
	Land x improved seeds	β_{25}	-0.058	0.048	-0.241***	0.058
	Labour x fertilizers	β_{26}	-0.283***	0.100	0.067	0.080
			Interaction among crude oil pollution variables (h_{kk})			

Table 1. Contd.

	Heavy spillage x medium spillage	β_{27}	-1.244*	0.709		
10	Heavy spillage x light spillage	β_{28}	-1.042*	0.647		
	Medium spillage x light spillage	β_{29}	-0.326**	0.153		
			Interaction between crude oil pollution and technology variables (rkt)			
	Heavy spillage x fertilizers	β_{30}	-0.805***	0.224		
11	Medium spillage x fertilizers	β_{31}	-0.586***	0.210		
	Light spillage x fertilizers	β_{32}	-0.792*	0.389		
			Interaction across technology variables (S_{tt})			
	Technology index x fertilizers	β_{33}	-0.312***	0.094	-0.547***	0.109
	Technology index improved seeds	β_{34}	0.656***	0.202	0.060	0.042
	Fertilizers x improved seeds	β_{35}	-0.102*	0.062	0.045	0.062
	Υ		0.863	-	0.9999	-
	λ		3.109***	0.950	190.634	698.59
12	σ		0.621***	0.030	0.883***	0.028
	δ_u^2		0.350	-	0.780	-
	δ_v^2		0.03620	-	0.00002	-
	Log likelihood function		-658.252	-	-733.206	-
	Average technical efficiency		0.586	-	0.664	-
	F-test		26.37***	-	16.21***	-

Source: Field survey, 2003. Asterisks indicate significant level: ***1%; **5%; *10%.

(β_{17}) decreased output in crude oil polluted crop farms by 0.040 units, with a possible increase in cost of production.

However, in non-polluted crop farms output increased when capital was squared by 0.114 (significant at 1%) though marginally regardless of the increase in cost of production expected. These results confirmed the negative effects of crude oil pollution on physical inputs usage in crude oil polluted crop farms (Aade – Ademilua and Mbamalu, 2008).

Interaction among physical inputs (b_{ig})

The interaction between land and labour (β_{18}) showed that for a unit increase in land with corresponding unit increase in labour, crop output level increased by 0.143 units and 0.219 units in crude oil polluted and non-polluted crop farms both significant at 5% level, respectively. This relationship means that in the presence of adequate labour, land productivity could be improved leading to higher output level, especially where the labour is knowledgeable in using remediation techniques in remedying crude oil polluted farmland. The interaction between land and capital (β_{19}) showed that a 10% increase in land area with a corresponding increase in capital resulted in a less than proportionate (inelastic) increase in output by 2.5% (significant at 1%) in crude oil polluted crop farms, whereas in non-polluted crop farms, it lead to a reduction in output by 1.4% (significant at 5%). This reduction in output in non-polluted crop farms could be caused by inadequate supply or even lack of fertilizers and planting materials.

Interaction across physical inputs and crude oil pollution variables (e_{ik})

The relationship between land and heavy crude oil spillage (β_{21}) showed that a unit of heavy crude oil pollution on land, resulted in 0.255 units reduction in crop output (though not statistically significant). The interaction between land and medium crude oil spillage (β_{22}) indicated that a unit of medium crude oil spillage on land, resulted in 1.257 units reduction in crops output, which is more than proportionate (elastic) reduction and was statistically significant at 1%. The interaction between land and light oil spillage indicated that a unit of light crude oil spillage on farmland resulted in a proportionate decrease in crops output of 1.027 units and was statistically significant at 5%. These results obtained in these relationships confirmed that crude oil spillage on farm land had detrimental effects on crops production in Rivers State, Nigeria and affirmed that crude oil spillage reduces land productivity (Ekundayo et al., 2001; Saier, 2006; Okonwu et al; 2010).

Interaction across physical inputs and technology variables (f_{it})

Land and fertilizers interaction (β_{24}) indicated that a unit increase in fertilizers usage resulted in 0.363 units increase in output in crude oil polluted crop farms, significant at 1% level and a very marginal increase of 0.059 units increase in non-polluted crop farms. The interaction between land and improved seeds (β_{25}) showed expected result of negative value (that is, decrease in outputs) in crude oil polluted crop farms by 0.058 units, and also a surprising reduction in output of 0.241 units in non-polluted crop farms for a unit increase in the inputs of production, statistically significant at 1%. The interaction between labour and fertilizers (β_{26}) indicated that a unit increase in the inputs, resulted in a reduction in output by 0.283 units in crude oil polluted crop farms, which was statistically significant at 1%. This could be that on a crude oil spilled crop farm any increase in land, labour, and fertilizers or improved seeds are wasted because of the detrimental effects of crude oil pollution (Uzoho et al., 2004; Dung et al., 2008; Okonwu et al., 2010; Fernandez-Luqueno et al., 2012).

Interaction among crude oil pollution variables (h_{kk})

The interaction among heavy crude oil pollution (spillage) and medium crude oil spillage (β_{27}) showed that a unit increase in medium crude oil spillage on already heavy crude oil spilled land resulted in a more than proportionate (elastic) decrease in output by 1.244 units(significant at 10%). The relationship between heavy spillage and light spillage (β_{28}) showed that a unit increase in light crude oil spillage on the farmland that had been heavily spilled resulted in reduction of output by 1.042 units which was also a proportionate (elastic) decrease in output and also significant at 10%. The relationship between medium crude oil spillage and light crude oil spillage (β_{29}) indicated that a unit increase in light spillage upon farmland that a medium spillage had already occurred resulted in decrease of output by 0.326 units which was statistically significant at 5%. The results of these interactions among crude oil pollution variables stressed the fact that all forms of crude oil spillages on crop farms reduced and/or led to complete lost of crop yield/output on crop farms. These results were similar to the results of Udo and Fayemi (1975); Mubana (1978); Iturbe et al., (2008); Onyenekenwa (2011).

Interactions between crude oil pollution and technology variables r_{kt}

Heavy crude oil spillage and fertilizers (β_{30}) interaction showed that a unit increase in the use of fertilizers on an already heavily crude oil spilled farmland resulted in a 0.805 unit decrease in output of crops significant at 1%. The interaction between medium crude oil spillage and fertilizers usage (β_{31}) indicated that a unit increase in the quantity of fertilizers used on an already medium crude oil spilled farmland, led to a reduction of output by 0.586 units, also (significant at 1%) (Udo and Fayemi 1975; Mubana, 1978). The relationship between light crude oil spillage and fertilizers usage on cropped farmland (β_{32}) indicated that an increase by a unit of fertilizers usage on light crude oil spilled crop farms resulted in a reduction of output by 0.729 units (statistically significant at 10%). These results showed that fertilizers usage had no expected positive effects on crops production on crude oil polluted areas and hence did not lead to expected increase in yields/output (Uzoho et al., 2004; Dung et al., 2008; Okonwu et al., 2010; Fernandez Luqueno et al., 2012).

Interaction across technology variables (S_{tt})

The interaction across fertilizers and improved seeds (β_{35}) showed that a unit increase in the inputs in the crude oil polluted crop farms, resulted in a reduction of output by 0.102 units (significant at 10%), and increase in output by 0.045 units (marginal increase) in non-polluted crop farms. Again, the negative effect of crude oil pollution was felt on the interaction of fertilizers and improved seeds usage in crude oil polluted crop farms. This goes to confirm the negative and detrimental effects of crude oil pollution on crops (Udo and Fayemi, 1975; Mubana, 1978; Achuba, 2006)

The lambda (λ) figure obtained in crude oil polluted farms (3.109) which was statistically significant at 1% and in non-polluted crop farms (190.634) were by far greater than unity. This implies that the one-sided error component, u dominated the symmetric error v as sources of variation. In other words, the discrepancy between actual (observed) output and the maximum (frontier predicted) output in crop farms in Rivers State, Nigeria was primarily due to factors that were within farmers control. The Lambada (λ) result obtained in crude oil polluted crop farms was close to earlier results of Xu and Jeffrey (1998) who had 2.117; Bagi (1984) had $\lambda = 2.438$.

Gamma (Υ) which is the measure of variance of output from the frontier attributed to efficiency was 0.863 in crude oil polluted crop farms and 0.999 in non-polluted crop farms. The random (stochastic) variability accounted for about 13.7% of the variability in crop output in crude oil polluted crop farms and virtually no random variability 0.001 (0.01%) in non-polluted crop farms. From the results obtained in this study, it means that crude oil pollution on crop farms in Rivers State, Nigeria had accounted for about 14% variation in actual output as against the frontier predicted. Therefore, this random variability could have been caused by crude oil pollution on crop farms in Rivers State, Nigeria. This confirmed the fact that crude oil pollution on crop farms is detrimental to crop output, yield and/or production.

Conclusion

The interaction between land and heavy oil spillage resulted in 0.255 units reduction in output, interaction between medium oil spillage and land resulted in 1.257 units reduction in crops output, while interaction between land and light oil spillage resulted in a decrease in crops output by 1.027 units. The interaction between land and improved seeds resulted in decrease in crop output by 0.058 units, while the interaction between labour and fertilizers showed a reduction in output by 0.283 units in crude oil polluted crop farms respectively.

The interactions among crude oil variables showed that heavy oil pollution interacting with medium oil spillage reduced output of crops by 1.244 units, heavy oil spillage and light oil spillage also reduced output by 1.042 units, while medium oil spillage and light oil spillage reduced crops output by 0.326 units. The interaction between heavy oil spillage and fertilizer showed a reduction in output by 0.805 units, medium oil spillage and fertilizers showed a reduction of 0.586 units in crops output, while the interaction between light oil spillage and fertilizers showed crops output reduction by 0.729 units. The interaction between fertilizers and improved seeds showed crops output decrease by 0.102 units in crude oil polluted crop farms. Therefore, crude oil pollution reduced crops output significantly, hence detrimental to crop farms output and production in Rivers State of Nigeria (Achuba, 2006; Aade – Ademilua and Mbamalu, 2008; Dung et al; 2008; Okonwu et al., 2010).

RECOMMENDATIONS

This study therefore, recommends that there is the need to spread widely information on benefits derived from adopting the best and suitable farm practices available in case of crude oil pollution and educating farmers on what functional measures are adoptable incase crude oil spillages or acquisition of farmland for crude oil exploration, exploitation and production occur (which in most cases is inevitable in Rivers State, since Nigeria derives more benefit from the petroleum industry than the agricultural sector). Thus, land will be used for its best and economically viable purposes.

This recommendation could be enforced through extension and rural educational programmes existing in various oil companies' demonstration farms. The Rivers State crop farmers need adequate knowledge and information about farming activities in a crude oil pollution prone environment which may improve their production capacity and thereby productivity (yield/output), farm income and revenue (Okoli, 2006; Adoki and Orugbani, 2007; Orji, et al., 2011).

Conflict of Interests

The author(s) have not declared any conflict of interests.

ACKNOWLEDGEMENTS

We acknowledge the support given to us by our different employers (various university authorities) at the various stages in the collaborative preparation, writing, analysis, discussions and production of this manuscript.

REFERENCES

Aade–Ademilua OE, Mbamalu J (2008). The growth and development of *Phaseoulus vulgaris* L. under petrol and diesel oil polluted waterirrigation. Res. J. Envrion. Toxicol. 2:27–34.

Achuba FI (2006). The effects of sub lethal concentrations of crude oil on the growth and metabolism of cowpea (*Vigna unguiculata*) seedlings. Envron. 26(1):17-20.

Adoki A, Orugbani T (2007). Influence of nitrogenous fertilizer plant efficient on growth of selected farm crops in soils polluted with crude petroleum hydrocarbons. Afr. J. Agric. Res. 2:569– 573.

Ajani OJY, Ugwu PC (2008). Impact of adverse health on agricultural productivity of farmers in Kainji basin North-Central Nigeria using a stochastic production frontier approach. Trends Agric. Econ. 1:1–7.

Ali M (1996). Quantifying the socio-economic determinants of sustainable crop production: An application to wheat cultivation in the Tarui of Nepal. Agric. Econ. 14:45-60.

Atakpo EA, Ayolabi EA (2009). Evaluation of aquifer vulnerability and the protective capacity in some oil producing Western Niger Delta. The Environ. 29:310–317.

Awoyemi TT, Adeoti AI (2006). Gender inequalities and economic efficiency: New evidence from cassava-based farm holdings in rural South-Western Nigeria. Afr. Dev. Rev. 18:428–443.

Bagi FS (1984). Stochastic frontier production function and farm-level technical efficiency of full-time and part-time farms in West Tennessee. N. Cen. J. Agric. Econ. 6:48–55.

Canter LW, Hill LG (1979). Hand book of variables for environmental impact assessment. Ann Arbor, Michigan, USA. Ann Arbor Science Publishers Inc.

Christensen LR, Jorgenson DW, Lau LJ (1973). Transcendental logarithmic production frontiers. Rev. Econ. Stat. 55:282-345.

Dlamini S, Rugambisa J I, Masuku M B, Belete A (2010). Technical efficiency of the small scale sugarcane farmers in Swaziland: A case study of Vuvulane and Big bend farmers. Afr. J. Agric. Res. 5:935-940.

Dung EJ, Bombom LS, Agusomu TD (2008). The effects of gas flaring on crops in the Niger Delta, Nigeria. Geo. J. 73:297–305.

Ekpoh IJ, Obia AE (2010). The role of gas flaring in the rapid corrosion of zinc roofs in the Niger Delta region of Nigeria. The Environ. 30:347-352.

Ekundayo EO, Emede TO, Osayande DI (2001). Effects of crude oil spillage on growth and yield of maize (*Zea mays* L.) in soil of mid western Nigeria. Plant Food Hum. Nutri. (Formerly Quantitas Plantium). 56(4):313-324.

Ekunwe PA, Orewa SI (2007). Technical efficiency and productivity of yam in Kogi State, Nigeria. J. Appl. Sci. 7:1818-1820.

Erhabor PO, Emokaro CO (2007). Relative technical efficiency of cassava farmers in the three agro-ecological zones of Edo State, Nigeria. J. Appl. Sci. 7:2818-2823.

Fernandez – Luqueno F, Corlay-Chee L, Robledo-Santoyo E, Pineda-Pineda J, Vazquez-Alarcon A, Miranda–Romero L A, Cabrera–Lazaro G, Mendez–Bautista J, Lopez–Valdez F, Dendooven L (2012). Growth and development of common velch (*Vicia sativa* L.) in a gasoline–polluted soil amended with organic or inorganic amendments. Afr. J. Agric. Res. 7:1259 -1267.

Harper WM (1971). Statistics MacDonald and Evans Ltd, Estover Plymouth, U.K. pp. 201-202.

Heady D, Alauddin M, Rao SP (2010). Explaining agricultural productivity growth: An international perspective. Agric. Econ. 41:1–4.

Huang Y, Jiang Z, Zeng J, Chen Q, Zhao Y, Liao Y, Shou L, Xu X (2011). The chronic effect of oil pollution on marine phytoplankton in a subtropical bay, China. Environ. Mon. Asses. 176:517–530.

Iturbe R, Castro A, Perez G, Flores C, Torres LG (2008). TPH and PAH concentrations in the subsoil of polyduct segments, oil pipelines pumping stations and right-of-way pipelines from central Mexico. Environ. Geo. 55:1785-1795.

Lachaal L, Karray B, Dhehibi B, Chebil A (2005). Technical efficiency measures and its determinants for olive producing farms in Tunisia: A stochastic frontier analysis. Afr. Dev. Rev. 17:580-591.

Li J, Zhang J, Lu Y, Chen Y, Don S, Shin H (2011). Determination of total petroleum hydrocarbons (TPA) in agricultural soils near a petrochemical complex in Guangzhou, China. Environ. Mon. Asses.

Managi S, Opaluch JJ, Jin D, Grigalunas TA (2006). Stochastic frontier analysis of total factor productivity in the offshore oil and gas industry. Ecol. Econ. 60:204-215.

Mubana EC (1978). An economic analysis of the impact of crude petroleum exploration and production on traditional agriculture in Rivers State. M Phil. Dissertation. University of Ibadan, Nigeria.

Niger Delta Development Commission (2006). Niger Delta Regional Development Master Plan. Niger Delta Development Commission (NDDC), Port Harcourt, Nigeria. 258pp

Nwaichi EO, Uzazobona MA (2011). Estimation of the CO_2 level due to gas flaring in the Niger Delta. Res. J. Environ. Sci. 5:565-572.

Nyagaka DO, Obare GA, Omiti JM, Nguyo W (2010). Technical efficiency in resource use: Evidence from small holder Irish potato farmers in Nyandarua North District, Kenya. Afr. J. Agric. Res. 5:1179-1186.

Okoli CG (2006). Rural households perceptions of the impact of crude oil exploration in Ogba/Egbema/Ndoni Local Government Area of Rivers State, Nigeria. J. Agric. Soc. Res. 6. (Accessed 22/09/11).

Okonwu K, Amakiri JO, Etukudo, MM, Osim SM, Mofunanya AAJ (2010). Performance of maize (*Zea mays*) in crude oil treatment. Glob. J. Pure. Appl. Sci. 16:173-176.

Onyenekenwa CE (2011). Environmental significance of the combustion of liquid petroleum fuels: A review. J. Appl. Sci. 11:2680-2685.

Onyenweaku CE, Okoye BC (2007). Technical efficiency of small-holder cocoyam farmers in Anambra State, Nigeria: A translog stochastic frontier production function approach. Int. J. Agric. Rural Dev. 9:1-6.

Orji JC, Uzoho B, Enwerem JO, Ezurike W (2011). Assesment of the impact of kerosene and diesel on soil nitrogen mineralization. Afr. J. Agric. Res. 6:2486–2493.

Orogun P (2009). Resource control, revenue allocation and petroleum politics in Nigeria: The Niger Delta question. Geo J. 75:459–507.

Otitoloju AA, Are T, Junaid KA (2007). Recovery assessment of a refined oil impacted and fire ravaged mangrove ecosystem. Environ. Mon. Asses. 127 :353–362.

Otitoloju A, Dan-Patrick J (2010). Effects of gas flaring on blood parameters and respiratory system of laboratory mice, Musmusculus. The Environ. 30:340– 346.

Saier M (2006). Pollution. The Environ. 26:205–209.

Udo EJ, Fayemi AAA (1975). The effects of oil pollution of soil on germination, growth and nutrient uptake of corn. J. Envrion. Qual. 4:537-540.

Uzoho BU, Oti NN, Onweremadu EU (2004). Effect of crude oil pollution on maize growth and soil properties in Ihiagwa, Imo State, Nigeria. Int. J. Agri. Rural Dev. 5:91-100.

Xu X, Jeffrey SR (1998). Efficiency and technical progress in traditional and modern agriculture: Evidence from rice production in China. Agric. Econ. 18:157-165.

11

Awareness and adoption of improved cassava varieties and processing technologies in Nigeria

Abdoulaye T., Abass A., Maziya-Dixon B., Tarawali G., Okechukwu R., Rusike J., Alene A., Manyong V. and Ayedun B.

International Institute of Tropical Agriculture (IITA), Ibadan, Nigeria.

Development of high yielding and disease resistant cassava varieties, coupled with the promotion of efficient processing technologies, was the principal intervention aimed at changing the cassava sub-sector in Nigeria. National research and extension programs in Nigeria and IITA have been spearheading efforts to disseminate these varieties alongside improving farmer's access to processing machineries. Several Research-for-Development (R4D) projects were implemented to this effect between early 1980 to date. This paper investigated the effects of improved cassava varieties and processing technologies on adopting households. It also attempts to test and establish the link between adoption of improved cassava varieties and access to processing technologies. The data used in this paper come from a sample household survey of 952 households conducted in four regions of Nigeria. The results showed that in all the study sites farmers grow mixture of improved and local cassava varieties. They process cassava at home using small processing machines and also using services of commercial processors. The most common processed cassava products were found to be garri and fufu. Adopters of improved cassava varieties have higher cassava yield of 16 tons/ha compared to 11 ton/ha for non-adopters. There was also significant yield variation between villages that participated (15 tons/ha) in research for development (R4D) training and those which did not (13 tons/ha). The bivariate probit model estimates showed a strong relationship between adoption of improved cassava varieties and farmers' access to grating machines. Moreover, farmers that were members of either community organizations or cooperative organizations had a higher tendency of using improved varieties than others, suggesting that the introduction of new cassava varieties would be enhanced by farmers' access to processing facilities and services. Moreover, training of farmers and processors through R4D programs has led to increased use of improved technologies.

Key words: Cassava, improved varieties, processing, bivariate probit.

INTRODUCTION

Cassava is an important regional food source for 200 million people – nearly one-third of the population of sub-Saharan Africa. In Nigeria, it is one of the most important food crop. It is the most widely cultivated crop that provides food and income to over 30 million farmers and large numbers of processors and traders. However, in Nigeria, Cassava Mosaic Disease (CMD) poses a serious threat (Alabi et al., 2011). The most vulnerable areas are the South-South and South-East States including the Niger Delta Region (Ogbe et al., 2006; Nweke et al.,

2002). Several initiatives were enacted to address the critical threat of a CMD outbreak in Nigeria and West Africa and to revitalize Nigeria's agricultural economy[1]. Among those efforts was that of IITA and national partners which developed and disseminated high yielding and CMD resistant cassava varieties. Between 2002 and 2010, IITA implemented a research for development (R4D) project called Integrated Cassava Project (ICP) to support the presidential initiative (PI) for cassava launched in 2002 to boost cassava production and processing. Through this project, IITA successfully introduced and promoted cassava varieties via the National Agricultural Research Services (NARs) and Agricultural Development Programs (ADPs). These efforts were complemented with promotion of cassava processing machineries especially for graters. Participants in the project from all major cassava producing regions of Nigeria, were also trained on crop management (density, weed management, fertilizer application etc). In addition, cassava processing centers were established along with introduction of small grating machines. Through these efforts, more than 40 cassava varieties were successfully introduced and promoted to farmers in Nigeria and the establishment of many processing centers and fabricating enterprises was facilitated between 2002 and 2010. It is important to note that local fabricators were trained in producing and maintaining the processing machines.

There is need to understand whether farmers are aware of the improved cassava varieties and processing machines? Also, what is the adoption status of these technologies? Are there any relationship between adoption of improved varieties and processing machines? Similarly, the introduced improved varieties were expected to give higher yields through better varieties with enhanced resistance to biotic stresses. What is the extent of the realization of such yield potentials in farmers' fields?

A number of studies have been carried out on the adoption of improved technologies singly and independently (Shiferaw and Holden, 1998; Zeller et al., 1998; Alene et al., 2000; Oluoch-Kosura et al., 2001; Abdoulaye and Sanders, 2002; Bamire et al., 2002; Akinola et al., 2010). According to von Braun (1988), agricultural growth via technological transformation leads to an expanded food supply which presupposes relationship between production and processing operations in agriculture. Greene (2000) and Maddala (1983) posited that most studies on adoption have reflected farmers-, farm-, institutional and technology-specific factors based on analysis that identified and estimated separately in a single equation model. However, a single equation estimation model could be threatened by bias, inconsistency and inefficiency in estimates. The problem might become worse in decision where simultaneity is detected or observed heterogeneities are correlated. In such situations, possible relationship and synergies in adoption decision are overlooked. Simultaneous estimation makes it possible to establish relationship that can be useful in adoption decisions. Improved cassava varieties and grating machines were often jointly deployed in most areas, but in some cases improved varieties were first demonstrated. Increase in cassava production through better and higher yielding varieties could stimulate more cassava processing and consumption (Braun, 1988). On the other hand, enhanced cassava processing could also lead to increased demand for raw cassava products thereby necessitating greater production. Therefore, a joint estimation method is expected to provide better estimates of the contribution of key variables to either adoption of improved cassava varieties or use of grating machines.

This study was carried out to provide empirical evidence of the state of awareness and adoption of improved cassava varieties and grating machines in Nigeria. Moreover, it attempted to establish likely relationship between the production and processing activities among the Nigeria's farming households.

METHODOLOGY

Study area, sampling method and data collection

The survey was carried out in 4 geopolitical zones in Nigeria known for cassava production. These zones were the South-West (SW), South-South (SS), South-East (SE) and North Central (NC). A total of 952 respondents were selected comprising of 38% who participated in project R4D interventions (participants) and 62% who did not (non-participants). The participants were selected based on their initial participation in the project. These included 160 respondents from the SW, 96 respondents from the SS, 70 respondents from the SE and 35 respondents from the NC. The non-participants were selected randomly from non-participating communities in the regions. They included 262 from SW, 157 from SS, 114 from SE and 58 from NC (Figure 1).

Data analysis

Descriptive statistics and econometric modelling were used in this study. As stated earlier, the joint estimation of adoption of varieties and processing technologies is preferred. The use of joint estimation is expected to reduce the most serious problem in modeling this type of decisions, of variables being endogenous at least for the 2 main variables in the model (use of improved varieties and use of processing machine). Therefore, a Bivariate Probit is used. The model is expressed as follows:

$$Y_{1i}* = X_{1i}\beta_1 + \mu_{1i}$$

$$Y_{1i} = 1 \text{ if } Y_{1i}* > 0$$

$$Y_{1i} = 0 \text{ Otherwise} \qquad (1)$$

[1] For details see Nigerian cassava presidential initiative, Integrated Cassava Project of IITA, RTEP program of the federal government.

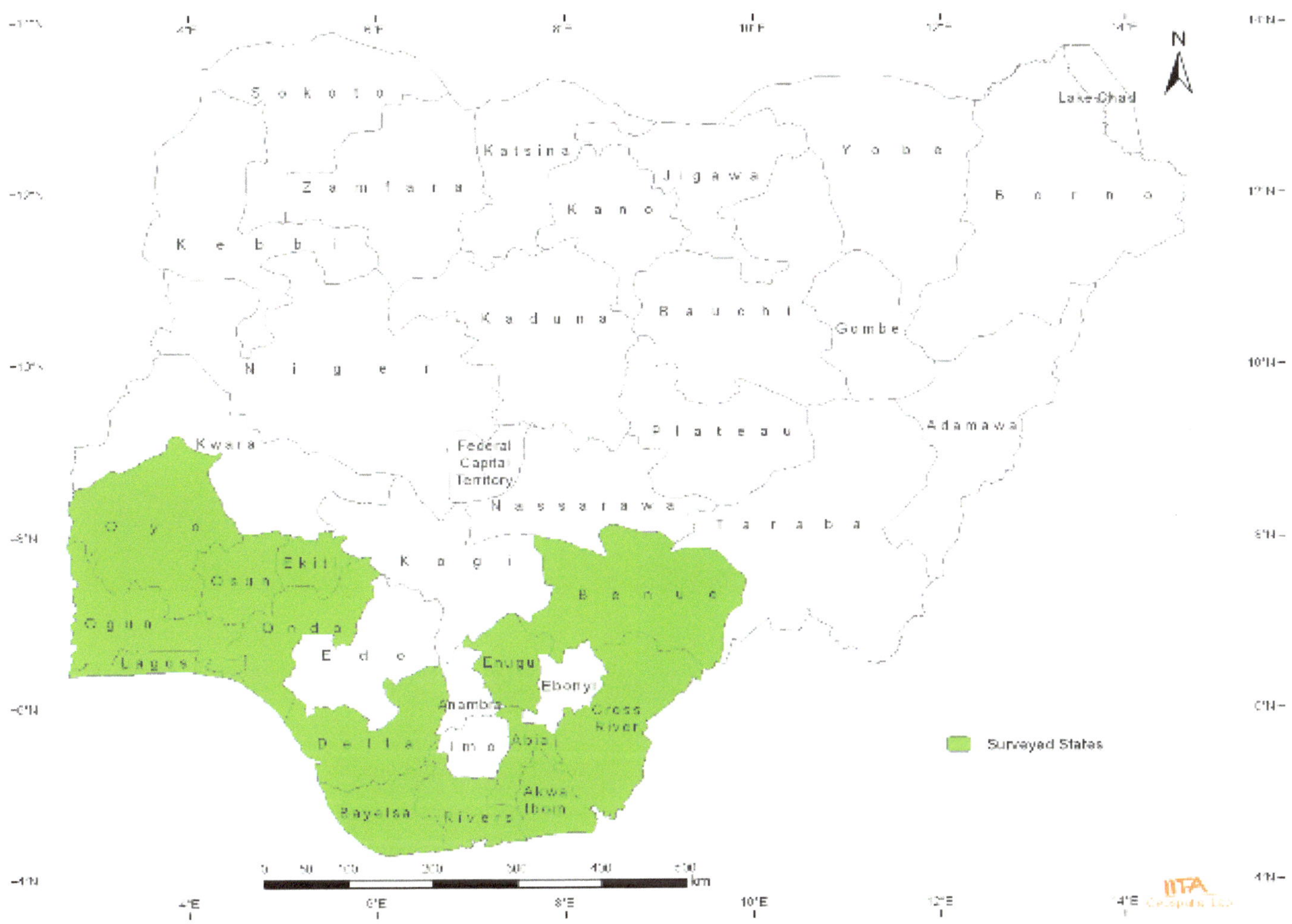

Figure 1. Map of the study area.

$$Y_{2i}^* = X_{2i}\beta_2 + \mu_{2i}$$
$$Y_{2i} = 1 \; Y_{2i}^* > 0$$
$$Y_{2i} = 0 \text{ otherwise} \qquad (2)$$

Where Y_i is the decision to use any of the technology; the two latent variables are decision to use improved cassava varieties and decision to use cassava grating machines. The coefficients β_1 and β_2 are vectors of explanatory variables influencing decision to use improved cassava varieties and grater machines, respectively; and μ_{1i} and μ_{2i} are error terms which are normally distributed but related.

The empirical model of the bi-variate model and each for the single estimation is explicitly stated as follows:

$$Y_1 = \beta_0 + \beta_2 GENDER + \beta_2 AGE + \beta_3 EDUCATION + \beta_4 HHSIZE + \beta_5 CASH + CMSOKAL + COSOKAL + VARAWARE + TRAINING + GRATER + \mu_1$$

And

$$Y_2 = \beta_0 + \beta_2 AGE + \beta_3 CASH + \beta_4 CMSOKAL + \beta_5 COSOKAL + \beta_6 ARLAND + EXTENSION + TRAINING + VARIETY + \mu_2$$

The independent variables included farmer, farm and institutional factors postulated to influence technology adoption. These variables were sex (gender) of the household head, age (age) of the household head in years, the household size (HHsize), measure of social interaction resulting from membership in farmers' organization and cooperative societies (CMSOCKAP and COSOCKAP), cash available at hand measured in dummy, education of household head (education) measured by farmers' ability to read and write, effective extension contacts (extension) measured in dummies by the regularity of visits by extension agents, size of arable land (ARland). Other variables included were the percentage of land planted improved cassava varieties (variety) and proportion of cassava grated by grating machine (grater) as well as farmers' awareness of existence and benefits of improved cassava varieties.

The rationale for inclusion of these factors was based on a priori expectation of agricultural technology adoption literature. The effect of age on technological adoption decisions may be negative or positive. Younger farmers have been found to be more knowledgeable about new practices and may be more willing to bear risk and adopt new technology because of their longer planning horizons. The older the farmers, the less likely they are to adopt new practices as they place confidence in their old ways and methods. On the other hand, older farmers may have more experience, resources, or authority that may give them more possibilities for trying a new technology. Thus, for this study, there

Table 1. Description of variables.

Variable	Description	Unit
Gender	Gender of the respondent (Male = 1, Female = 0)	
Age	Age of respondent in years	Years
Education	A measure of ability to read and write. Ability to read and write = 1, 0 otherwise	
HHsize	Number of people living under the same roof and taking joint decision about their welfare	Number
Cash	Cash saving: 1 = if having saving in cash form, 0 = Otherwise	
CMSOCKAP	Membership of community association, 1 = member, 0 = non-member	
COSOCKAP	Membership of cooperative society, (1 = member, 0 = non-member)	
Varaware	Respondent's awareness of improved cassava varieties, 1 if aware, 0 = non aware.	
Training	Respondent's participation in either training on improved cassava varieties or use of grating machines	
Grater	% of tuber grated by grater machine	%
Arland	Household arable land	ha
Extension	Contact with extension services on the use of improved varieties and grading machines, 1 = access, 0 = non access	
Variety	% of land cultivated to improved cassava	%

is no agreement on the sign of this variable as the direction of the effect is location-or technology-specific (Feder et al., 1985; Nkonya et al., 1997; Oluoch-Kosura et al., 2001; Bekele and Drake, 2003). Education was hypothesized to influence the adoption of decisions positively since, as farmers acquire more, their ability to obtain, process, and use new information improves and they are likely to adopt. Education increases the ability of farmers to use their resources efficiently and that will enhance their ability to obtain, analyse and interpret information. Several studies reviewed by Feder et al. (1985) indicate positive relationship between education and technology adoption (Alene et al., 2000; Nkonya et al., 1997; Oluoch-Kosura et al., 2001).

Institutional factors of social capitals and farmers' awareness of the benefits derivable from improved cassava varieties, participation in R4D programs (training*)* and extension contact were hypothesized to influence the adoption positively as these support services facilitate the uptake of new technologies. Membership in associations (CMSOCKAP), such as cooperative societies (COSOCKAP), has been found to enhance the interaction and cross-fertilization of ideas among farmers (Bamire et al., 2002). Farmers who are not members of associations are expected to have lower probabilities of adoption and a lower level of use of either improved cassava varieties or grating machine. The extension contact variable incorporates the information that the farmers obtain on their production activities on the importance and application of innovations through counselling and demonstrations by extension agents on a regular basis. It is hypothesized that the respondents who are not frequently visited by extension agents have lower possibilities of adoption than those frequently visited (Adesina and Zinnah, 1993; Shiferaw and Holden, 1998; Oluoch-Kosura et al., 2001; Bamire et al., 2002). The variable was measured as dichotomous with respondents 'contact during the period scoring one, and zero for no extension contact on the use of the technologies (Table 1).

Measures of wealth such as off-farm income and income from other sources apart from processing were also hypothesized to influence adoption positively. They are generally considered to be capital that could be used either in the production process or be exchanged for cash or other productive assets. They are expected to influence the adoption of the technologies positively (Shiferaw and Holden, 1998; Zeller et al., 1998; Negatu and Parikh, 1999). To the extent that liquidity is a constraint to adoption, off-farm income and income from other sources will have a positive effect on adoption. The level of off-farm income, however, may not be exogenous but be affected by the profitability of the farming operation that in turn depends on technology adoption decisions. Thus, the adoption of the technologies and the level of off-farm income may be determined simultaneously. This arises due to the labor allocation decisions of the households about farm and non-farm activities. However, the off-farm income of the household surveyed is mostly derived from the remittances of family members in non-farm business activities and from employment in non-farm sector. As the skill requirements for these jobs are likely to be different from those of farming, the farm and non-farm employment may be considered as non-competitive activities. In this situation, the level of non-farm income would be largely exogenous to the adoption decision (Lapar and Pandey, 1999).

Household size, which includes all people living under the same roof and who eats from the same pot as the household head, has been identified to have either a positive or a negative influence on adoption (Manyong and Houndekon, 1997; Zeller et al., 1998; Oluoch-Kosura et al., 2001; Bamire et al., 2002; Bekele and Drake, 2003). Larger family size is generally associated with greater labor force availability for the timely operation of farm activities. The negative relationship of the variable with adoption has been linked to the increased consumption pressure associable with a large family. It is therefore difficult to predict 'a priori' the sign for this variable in this study. In addition, percentage of improved cassava varieties was expected to be positively related to the percentage of cassava grated and vice versa. The size of arable land is also expected to be positively related to technology adoption.

RESULTS AND DISCUSSION

Socioeconomic characteristics of the sample households

Demographic and socioeconomic characteristics of our sampled households are summarized in Table 2. These characteristics play important role in understanding the

Table 2. Socioeconomic characteristics of the sample households.

Variables	Values
N	**952**
Region (%)	
South-South(SS)	27
South-East (SE)	19
South-West (SW)	44
North-Central (NC)	10
Gender (%)	
Male	89
Female	11
Marital status (%)	
Single	3
Married	89
Divorced	1
Separated	1
Widowed	7
Age of household heads	
<20	1
21-40	26
41-60	61
61-80	12
>80	1
Age of household heads (average)	49
Years of farming experience (%)	
1 – 10	18
11- 20	32
21-30	24
31-40	17
>40	9
Farming experience (average)	24
Farming experience in cassava production (average)	22
Cassava processing experience (average)	19
Years of processing experience (%)	
1-20	66
21-40	29
41-60	4
>60	1
Household size (average)	8
Dependency ratio (average)	1.5
Main decision maker (%)	
HH head	69
Spouse	2
Children	0.1
HH head and spouse	24
HH head and kids	3
Spouse and kids	1
All members	2
Level of education (%)	
Educated	77
Year of education	
1-5	6
6-10	35
11-15	51
16-20	8
>20	0.1
Average number of years of education	10
Association	
Cooperative association (%)	27.4
Processing association (%)	9.9
Growers associations (%)	20.4
Marketers association (%)	4.9
Transporter association (%)	1.1
Total association group (%)	63.7
Cooperative (average years)	22
Processing association (average years)	8
growers associations (average years)	6
marketers association (average years)	6
Transporter association (average years)	7
Intervention village	38
Counterfactual village	62

Source: Data analysis (2012).

differences among households and hence explaining their behaviour regarding technological change. The major characteristics of households covered in the survey included are those related to the relative frequency distribution of heads of the households by gender, age, years of formal education, marital status. Also included were household asset ownership structures, distribution of household farms, land tenure types, sources of farm credit, and household consumption patterns. The sampled household heads were 81% men and 89% of them were married having family responsibility. Family responsibility presupposes their willingness to get involved in productive activities to meet family demands. The average family size of 8 suggested availability of family labour on the farm. In addition, the dependency ratio of 1.5 which showed that there were more number of dependants (children below 15 years old and adult above 64 years old) compared to working population (> = 15 years and < = 64 years old) in all the zones. Education level of the respondents was high with an average of 10 years of formal education and with 77% of respondents responding to be having formal education.

Average farming experience was 22 years indicating that study sample was composed of experienced farmers (Table 2). Most of the respondents belong to grower and cooperative groups among others. These groups normally encourage their members sometimes with moral

Table 3. Percentage of household making different products from cassava.

Products	Pooled	Percentage of total	Intervention	Non-intervention	Participation	Non-participation
N	**952**		**358**	**594**	**145**	**807**
Garri	82	52	75	85	90	80
Fufu making	48	30	39	53	55	45
Flour making	10	6	10	10	15	9
Starch	2	1	3	2	10	1
Abacha	10	6	16	6	0	0
Lafun	6	4	15	1	0	0

Source: Data analysis (2012).

and sometimes financial support (credit) for adopting technologies.

Agricultural production in the study area

The main land tenure system was by inheritance (53%) followed by one being rented (29%) among others. Majority of the respondents cultivated farm size of 2 ha or less (80%). This is an indication that they were mainly small scale farmers. The respondents were engaging in cultivation of many crops including roots and tuber, cereals, legumes among others. When arable crops were ranked according to most important crop grown, 70% of the respondents indicated that cassava was their most preferred crop, followed by yam, maize and plantain among others. Percentages of area of land cultivated for different crops also indicated that cassava had the biggest area and occupied the largest percentage of land used for cropping by the farmers irrespective of village types considered.

Household cassava processing

Cassava tubers are processed by households into different cassava products. Almost all the products were previously known to farmers, the work done by change agents was just to improve their processing activities, thus adding value to it and increasing market value. Looking at Table 5, the products increased over the years. However, by disaggregating by village groups, it can be seen that the control villages is better than intervention ones. There were higher percentages of participants processing all these products than non-participants. This same trend is observed when considering alternative ways of utilizing cassava at home. Garri and fufu (foufou) were the most common products made by households constituting 52 and 30%, respectively, while the remaining percentage was shared by other products like cassava flour and starch (Table 3).

Technological awareness and use of technologies

Awareness and use of cassava production technology

Awareness and knowledge of a technology is a prerequisite for its use. Information on level of awareness and use of production technologies is presented in Table 4. The level of use for improved cassava was relatively high (68%) than other production technologies like fertilizer. The results suggest that awareness and use of improved variety of cassava was skewed towards intervention villages which have higher percentages for both variables compared to non-intervention villages.

Increased in awareness and use of improved technologies as shown in the table increased with how closer the respondents were to the change agents with participating farmers having highest awareness and use levels, followed by non-participating farmers from intervention villages and then farmers from non-intervention villages.

Awareness and use of cassava processing technology

The study found that among the promoted innovation, awareness of grating and pressing was the highest. Farmers with first-hand information from research and extension agent (participating respondents) have higher awareness and use in all introduced technologies. Table 5 shows that the spread of information about the technologies was a collective effort by many stakeholders. Results indicate that farmers to farmers' technological diffusion played the greatest role in dissemination of the technologies. Also, it is expected that with better use of production and processing technologies by farming households from intervention villages and participating respondents, these farmers would have positive impact on their farm output and productivity.

Table 4. Awareness and use of inputs used in cassava production.

Variable input	Pooled		Intervention		Non-intervention		Participation		Non-articipation	
	Aware (%)	Use (%)	Aware (%)	Use (%)	Aware (%)	Use (%)	Aware (%)	Use (%)	Aware (%)	Use (%)
Improved planting materials	75	68	88	74	67	65	100	94	70	64
Basal (NPK)	45	25	45	30	44	23	61	50	42	21
Topdress_Urea	27	7	25	8	29	6	39	17	25	5
Herbicides	37	17	33	14	40	19	46	26	35	16
Insecticides	30	8	29	10	31	7	36	16	29	7
Manure	33	13	30	14	35	12	36	15	33	12

Source: Survey data (2012).

Table 5. Sources of information about different technologies (% of respondents).

Technology/source	N	IITA (%)	NGO (%)	Farmer (%)	Media (%)	Ext. agent (NARS) (%)	Agro-dealer (%)	Others (%)
Improved planting materials	629	16	1	28	3	46	3	4
Peeling	153	8	3	44	7	28	9	1
Washing	121	7	3	46	3	33	9	1
Grating	360	10	1	59	2	22	5	2
Chipping	42	12	2	17	12	43	12	2
Extracting	289	8	1	61	5	19	4	2.3
Pressing	71	13	3	56	3	20	4	1
Sifting	70	6	1	43	9	23	13	6
Drying	36	8	-	31	11	33	14	3
Boiling	8	25	-	25	-	38	-	13
Distilling	45	7	-	31	9	42	9	2
Fermenting	164	15	1	51	2	23	7	2
Frying	10	20	-	10	-	50	10	10
Pelletizing	123	7	2	53	3	24	6	5
Grinding	78	10	3	37	15	23	5	6
Milling	39	5	3	44	13	28	3	6

Source: Data analysis (2012).

Table 6. Average reported cassava yields among farmers.

Variable	Adopter	Non-adopter	Difference	Participating villages	Non-participating villages	Difference
Yield (ton/ha)	16.1±4	11±5	4.9** (114)	15.0±4	13.0±8	2** (5.2)

Figures in the bracket are t-values; ***, ** means significant at 1 and 5%, respectively. Source: Survey data (2012).

Cassava productivity

It is expected that investment in inputs such as improved cassava cuttings along with complementary agronomic practices would lead to higher yields for adopting farmers. Survey results indicate that the cassava tuber was higher for adopting households compared to non-adopting ones. The difference between the two groups was also found to be statistically significant (Table 6). Also, as expected, yields were higher in villages that participated in R4D programs compared to on-participating ones. This might be related to the higher use of improved cassava varieties and the trainings received by farmers in those villages.

Determinants of adoption of improved cassava varieties and grating machines

Both single equation and joint estimation results are presented in Table 7. Results from single probit

Table 7. Probit and Bivariate Probit model estimates of the determinants of adoption of improved cassava varieties and grating machines.

Determinants	Single equation estimation				Joint estimation			
	Probit model: Adoption of improved cassava varieties		Probit model: Use of cassava grating machines		Y1 = Decision to use improved cassava variety		Y2 = Decision to use cassava grating machine	
	Estimates	P>|z|	Estimates	P>|z|	Estimates	P>|z|	Estimates	P>|z|
Gender	0.236	0.351			0.139	0.179		
Age	0.011	0.500	0.004	0.575	0.002	0.738	-0.003	0.479
Education	0.151***	0.000			-0.076	0.431		
HHsize	0.071	0.015			0.017	0.100		
Cash	0.408	0.046	0.322*	0.040	0.127	0.220	0.091	0.328
CMSOCKAP	0.286	0.193	-1.311***	0.000	0.275*	0.015	-0.521***	0.000
CPSOCKAP	0.834***	0.000	0.811***	0.000	0.206*	0.059	0.067	0.513
Varaware	6.330***	0.000			1.462***	0.000		
Training	0.115	0.672	0.761***	0.001	0.050	0.716	0.096	0.475
Grater	0.452**	0.048			1.411***	0.000		
Arland			0.014	0.199			-0.002*	0.057
Extension			0.410***	0.001			0.009*	0.059
Variety			0.205	0.195			1.196***	0.000
Constant	-8.614	0.001	0.511	0.121	-2.787	0.000	-0.019	0.925
Athrho					12.694			
Rho					1.000			
Chi^2	368.00		105		206.338			
$Prob>chi^2$	0.000		0.000		0.000			
Pseudo R^2	0.384		0.091					
Log likelihood	295.27		-532.44					

***, **, * Significant at 1, 5 and 10%, respectively. Source: Data analysis (2012).

estimations for both varieties and machines are shown to help understand the joint relationship between adoption of improved varieties and access to cassava grating machines. The joint estimation results are emphasized and preferred for interpretation because both use of improved cassava varieties and use of grating machines have shown positive effects on each other in the single equation estimation. Results from joint estimation using bivariate probit regression showed that different variables affected the probability of adoption of improved cassava varieties and probability of use of grating machines. For the first equation on use of improved cassava varieties, significant variables that affected probability of such use included membership in community organizations, membership in cooperatives organizations, awareness of the benefits associated with the adoption of the improved varieties, and the proportion of cassava grated. Farmers that were members of either community organizations or cooperative organizations had a higher tendency of using improved varieties than others. Membership in community organization showed a positive effect in increasing the probability of adopting improved cassava varieties by 0.28. Also, membership in cooperative societies increased the probability of adoption of improved cassava varieties by about 0.21. In addition, greater increase in the probability of adoption of improved cassava varieties is indicated by model results for awareness of the importance and benefits associated with the use as well as the proportion of cassava grated using grating machines in the household. As expected, those farmers who were made aware of the potential of improved cassava varieties were more likely to adopt them. Awareness about the benefits of improved cassava varieties had a positive coefficient of 1.46, while the coefficient on the proportion of cassava grated using machines was slightly lower at 1.41 (Table 7). This underscores again the importance of giving farmers opportunity to experience and learn about new technologies in the adoption process. Also, since the majority of households (about 82%) use mainly small grating machines at home, these results indicate that promotion of such small scale processing would have great impact in increasing adoption of improved cassava varieties.

For the second equation, model estimation results indicate that the most significant variable influencing the use of grating machine was the proportion of land planted to improved cassava varieties. A 10% increase in the proportion of land planted to cassava varieties increased the probability of using grating machine by 12%. Frequency of access to extension services showed a positive effect on the probability of adopting grating machine by about 1 percentage point. Farmers with smaller farm showed higher tendency of using grating machines. This was expected as farmers with smaller farms would likely have more time for processing activities than others. Membership in community organization was negative but significant in influencing the use of grating machine. This result might be related to the availability of small individual grating machines that most farmers are using instead of relying on big community level processing centres.

Conclusion

Awareness and adoption of improved cassava varieties was relatively high. Adopting farmers have high cassava yield of 16 tons/ha compared to non-adopters (11 ton/ha). However, yields are still low when compared to potential yield of 30 to 40 tons/ha from research trial plots. Introduction of smaller grating machines has helped increase awareness and use of cassava grating and pressing machines by households. However, there is a need to mechanize peeling of cassava roots in order to address the increasing labor constraints in rural areas. Mechanizing peeling is the next big leap that is needed for cassava industry to continue to grow in Nigeria. The results of bivariate probit regression showed that adoption of improved cassava varieties had an effect on farmers' access to grating machine and vice versa. That is, the most significant variable influencing the use of grating machine was the adoption of improved cassava varieties.

The results confirm the strong complementarity between improved cassava varieties and processing machines. Since the use of grating machines is having greater effect on adoption of improved cassava varieties, introduction of processing machine should precede that of improved cassava varieties. Also, the introduction of new technologies should be backed up by training and provision of complementary services. Finally, promotion of processing should also include small processing machines as their availability in the study areas has been an important factor explaining the observed differences in adoption.

REFERENCES

Abdoulaye T, Sanders JH (2002). Stages and determinants of fertilizer use in semiarid African agriculture: the Niger experience. Agric. Econ. 32:167-179.

Alabi OJ, Kumar PL, Naidu RA (2011). Cassava mosaic disease: A curse to food security in Sub-Saharan Africa. Online. APS*net* Features. doi:10.1094/APSnetFeature-2011-0701.

Alene AD, Poonyth D, Hassan RM (2000). Determinants of adoption and intensity of use of improved maize varieties in the central highlands of Ethiopia: A Tobit analysis". Agrekon (South Afr. J. Agric. Econ. 39(4):633-643.

Akinola AA, Alene AD, Adeyemo R, Sanogo D, Olanrewaju AS, Nwoke C, Nziguheba G, Diels J (2010). "Determinants of Adoption and Intensity of use of Balanced Nutrient Management Systems Technology in the Northern Guinea Savanna of Nigeria". Q. J. Int. Agric. 49(1):25-45.

Adesina AA, Zinnah, MM (1993). "Technology characteristics, farmers' perceptions and adoption decisions: A Tobit model application in Sierra Leone". Agric. Econ. 9:297-311.

Bamire AS, Fabiyi YL, Manyong VM (2002). Adoption pattern of fertilizer technology among farmers in the ecological zones of South-Western Nigeria: A Tobit analysis. Austr. J. Agric. Res. 53:901-910.

Bekele W, Drake L (2003). Soil and water conservation decision behavior of subsistence farmers in the Eastern Highlands of Ethiopia: A case study of the Hunde-Lafto area. Elsevier Science Ltd. Ecol. Econ. 46(3):437-451.

Feder G, Just RE, Zilberman D (1985). Adoption of agricultural innovations in developing countries: A survey. The University of Chicago, USA. pp. 255-299.

Greene WH (2000). Simulated Likelihood Estimation of the Normal-Gamma Stochastic Frontier Function," Working Papers 00-05, New York University, USA.

Lapar MLA, Pandey S (1999). Adoption of soil conservation: the case of the Philippine uplands. Agric. Econ. 21:241-256.

Maddala GS (1983). Limited-Dependent and Qualitative Variables in Economics, New York: Cambridge University Press. pp. 257-291.

Manyong VM, Houndekon AV (1997). Land tenurial systems and the adoption of mucuna planted fallows in the derived savannas of West Africa". Paper presented at the workshop on property rights, collective action and technology adoption. ICARDA, November 22-25. Aleppo, Syria.

Negatu W, Parikh A (1999). The impact of perception and other factors on the adoption of agricultural technology in the Moret and Jiru Woreda (district). Ethiopia Agric. Econ. 21:219-229.

Nkonya E, Schroeder T, Norman D (1997). Factors affecting adoption of improved maize seed and fertilizer in Northern Tanzania." J. Agric. Econ. 4:1-12.

Nweke FI, Spencer DSC, Lynam JK (2002). The cassava transformation: Africa's best-kept secret. Michigan State University Press, East Lansing.

Ogbe FO, Dixon AGO, d'A Hughes J, Alabi OJ, Okechukwu R (2006). Status of cassava begomoviruses and their new natural hosts in Nigeria. Plant. Dis. 90:548-553.

Oluoch-Kosura WA, Marenya PP, Nzuma MJ (2001). Soil fertility management in maize-based production systems in Kenya: Current options and future strategies. Paper submitted at the Seventh Eastern and Southern Africa regional maize conference, Nairobi. pp. 350-355.

Shiferaw B, Holden ST (1998). Resource degradation and adoption of land conservation technologies in the Ethiopian Highlands: A case study in Andit. Tid, North Shewa. Agric. Econ. 21:241-256.

Zeller M, Diagne A, Mataya C (1998). Market access by smallholder farmers in Malawi: Implications for technology adoption, agricultural productivity and crop income. Agric. Econ. 19:219-229.

12

Financial profitability and technical efficiency of horticultural crops in the Nakanbé River Watershed in Burkina Faso

Daniel P. KABORE

Centre d'Analyse des Politiques Economiques et Sociales, 01 BP 1919 Ouagadougou 01, Burkina Faso.

In the watershed of the Nakanbé River in Burkina Faso, horticultural crops are financially profitable with high gross incomes, cost/revenue ratios and attractive hourly returns to labor. Based on a 6-h work day, profitability indicators of horticultural crops show that farmers are better-off compared to the 'less than one dollar' situation of people in Burkina Faso; horticultural crop production could thereby be an important part of a poverty reduction strategy in the studied sites and other rural areas. The proportion of efficient producers ranged from 35% for tomato in Ouonon to 100% for onion in Zekeze. Considering efficiency for individual crops, onion production in Zekeze was the most efficient, with a maximum score of (100%), followed by the production of cucumber in Ouonon and green beans in Kongoussi with respective scores of 98%. Inputs (seeds, fertilizers and pesticides) are used in excessive quantities which tend to reduce technical efficiency and could have negative environmental impacts. Improving horticultural crop producers' capacities to use appropriate rates of chemicals and organic inputs or biological treatments is recommended since the survival as well as the development of this kind of farming increases rural households' incomes and supplies fresh food to an increasing urban population at a reasonable price.

Key words: Horticulture, cost/revenue ratio, hourly return, efficiency, input excess, environment, Burkina Faso.

INTRODUCTION

Burkina Faso is one of the poorest countries in the world; the incidence of poverty is 43.9% at the national level (INSD, 2010). Poverty is greater in rural than urban areas. In order to ensure their economic and nutritional survival, rural, urban as well as peri-urban populations produce horticultural crops under irrigation, particularly during the dry season. In Burkina Faso, only 12% of the total irrigable land is under irrigation (Government of Burkina Faso, 2005), and 30% of the irrigated area produces horticultural crops. Burkina Faso has been an exporter of fruit and vegetables such as fresh green beans since 1970, but in recent years the competitiveness of this sector has sharply declined. This decline in competitiveness of horticultural products and the sub-sequent reduction of exports has been 'the result of an intrinsic decline in the quality of the product" (Government of Burkina Faso, 2005). Any loss of competitiveness leads to a loss of market and therefore, to a reduction of exports earnings. Since the daily income per capita is less than one U.S. dollar in Burkina Faso, any decline in export earnings reduces per capita income and then worsens poverty.

Reduced quality of horticultural products not only has a negative influence on the competitiveness of the sector, but also leads producers to use excessive amounts of inputs (fertilizers and pesticides) during production, which poses environmental problems.

According to the definition of Koopmans (1951), 'a producer is efficient if he maximizes output given the input he uses; that is, transforming inputs to outputs he needs a certain kind of technology. Since this type of efficiency deals solely with technology this type of efficiency is called technical efficiency', while allocative efficiency measures the ability of the farmer to use inputs in optimal proportions, given input prices.

Technical efficient is rather the ability of the farmer to produce maximum output from a given level of inputs, whatever these inputs are.

It is therefore important to identify bottlenecks and weaknesses in horticultural production to understand the production decline in recent years in order to improve production in the future. This objective of this study is to analyze the profitability and technical efficiency[1] of horticultural crop production in the Nakanbé River watershed in Burkina Faso.

METHODOLOGY

Data collection

The first step in data collection was the choice of study sites in the watershed with farmers' cooperatives and associations. The selection criteria were (i) the volume of horticultural crop production during the dry season of candidate sites, (ii) ease of access to the production area, and (iii) the level of cooperation of the producers in the sites.

Using the above criteria, four (4) sites were chosen (Figure 1): Kongoussi, in the Bam province; Titao, in the Loroum province; Ouonon, in the Passore province and Zekeze in the Boulgou province.

The survey was conducted during the 2004 dry season to a total of 131 randomly selected farmers: 31 in Kongoussi, 24 in Titao, 36 in Zekeze and 40 in Ouonon. The questionnaire used for data collection included the (i) farmers' socio-demographic characteristics (ii) land parcel identification, crops and cropped areas, and land tenure, (iii) cropping calendar describing the succession of farming operations (iv) quantity and cost of inputs used, (v) family labor indicated by the number of persons and amount time spent working by gender, age, crop produced, and cropping operation, (vi) amount of hired labor in hours and payments in cash and in-kind (vii) inventory of equipment used (purchased or borrowed), (viii) cost of equipment repair and maintenance, (ix) quantities harvested and sales by crop, and (x) marketing problems.

Method of analysis

The method of analysis included profitability, technical efficiency and measures of excess inputs use.

[1] This technical efficiency differs from the agronomic efficiency (fertilizer or irrigation).

Indicators of financial profitability

Production budgets were used to assess profitability, especially cost/revenue ratio and hourly payment, during the dry season. For a given crop, the household average production was computed and valued at average sale prices to get the gross income (GI).

The corresponding cost/revenue ratio (C/RR) is calculated as: C/RR= GI/VC where VC is total variable costs to produce the crop. The related hourly payment (HP) is obtained, dividing GI by the number of family hours (FL), that is HP=GI/FL.

The higher the gross income and hourly payment, the more profitable is the crop[2].

Technical efficiency

The Data Envelopment Analysis (DEA) method was used to analyze the technical efficiency of horticultural crops producers which is a non-parametric method, a deterministic measurement of technical efficiency. This method is based on models of optimization and linear programming techniques. The analysis was performed using the DEAP (Data Envelopment Analysis Program) (Coelli, 1996); in addition to the level of efficiency it provides, this software also gives information on the variables which are the sources of the inefficiency.

The method is based on the measurement of the ratio output/input in the form $u'y_i/v'x_i$ where u is a Mx1 vector of relative weights of outputs and v a Kx1 vector of relative weights of inputs. Optimal weights are obtained by solving a problem of mathematical programming, specified as follows:

$$\max_{u,v} \; (u'y_i/v'y_i),$$

$$\text{st } u'y_j/v'y_j \le 1,\; j= 1.2, \ldots, N$$
$$u,v \ge 0.$$

The objective is to find values for u and v such that the measure of the efficiency of the unit of decision i (a farmer in our case) is maximized under the constraint that all measures of efficiency are less than or equal to 1. This formulation of ratio has the disadvantage of having an infinite number of solutions since if (u^*, v^*) is a solution of the program, then so is (αu^*, αv^*).

To overcome such a difficulty, an additional constraint is imposed and the problem is reformulated after transformation; the DEAP model developed by Coelli (1996) uses the dual program as reformulated (Coelli, 1996).

Some authors such as Boles (1966) and Afriat (1972) proposed methods of linear programming to estimate the isoquant frontier; such methods did not receive attention until Charnes et al. (1978) introduce the term "Data *Envelopment Analysis*". The model developed by Charnes et al. (1978) focused on the inputs[3] and based on the assumption of the constancy of returns to scale. Later on, alternative hypotheses were introduced; Banker et al. (1984) proposed a model rather based on the assumption of variable returns to scale.

Inputs excess

Referring to Figure 2 illustrating this idea of input excess, it can be seen that on the same linear segment of the isoquant convex frontier SS', you can have an optimal point with the possibility of

[2] Gross income was preferred to net income due to difficulties to get required data, that is, opportunity cost of labor in the absence of a rural labor market, costs related to crop production.

[3] Some models are centered on the output but we use here the option input-oriented DEA software.

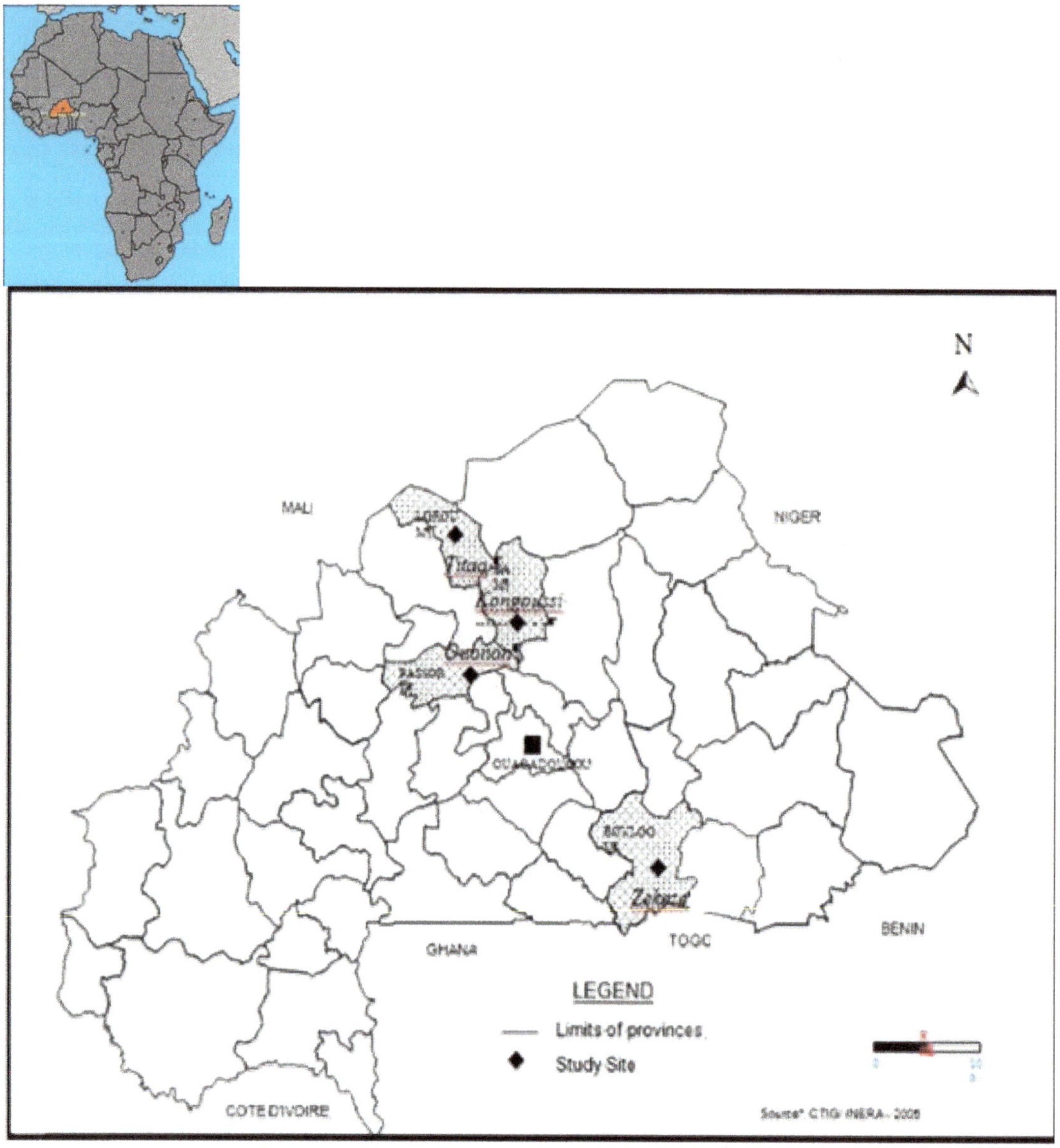

Figure 1. Location of study sites.

reducing the quantity of inputs (x_1 and x_2) used without reducing the level of production (y). We may then wonder if such a point is optimal; it is an excess of inputs used, that could have been saved, since this does not increase the level of production. The technical efficiency for decision units A and B are respectively of OA'/OA and OB'/OB; the question that can be raised is whether point A' is efficient since the quantity of input x_2 can be reduced by CA' without reducing the level of production; CA' is therefore the input excess of x_2. The Data Envelopment Analysis software provides options for the calculation of the excess inputs.

RESULTS AND DISCUSSION

Main crops production

The main crops in Titao are onion and potato; green beans predominate in Kongoussi while at Zekeze, onion, tomato and banana are the most important vegetable crops. In Ouonon, the main crops are tomato, eggplant and cucumber (Table 1).

Profitability of horticultural crops[4]

Potato and onion in Titao have the highest gross incomes per hectare, followed respectively by onion in Zekeze, green beans in Kongoussi, banana in Zekeze, onion and tomato in Ouonon, tomato in Zekeze, eggplant and cucumber in Ouonon (Table 2). However, the cost/revenue ratio of 8.77 reveals the greatest profitability for onion in Ouonon. Potato and green beans, despite good gross incomes, recorded relatively poor cost/revenue

[4] US dollar 1.00= FCFA 500.

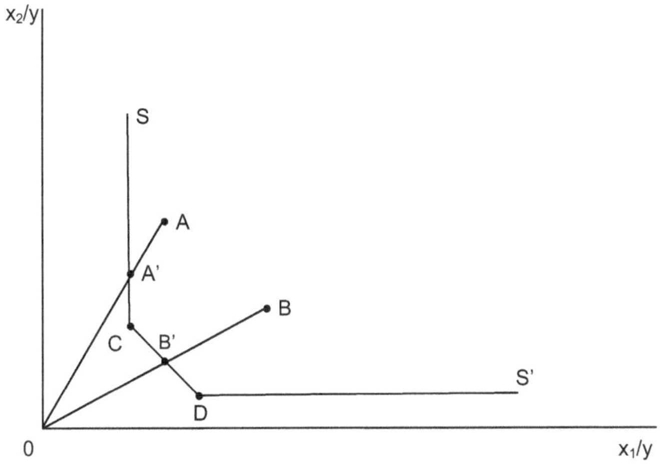

Figure 2. Measure of efficiency and excess inputs. x_1/y : quantity of input x_1 to produce a unit of production y; x_2/y: quantity of input x_2 to produce a unit of production y. Source: Coelli (1996).

Table 1. Main crops areas by study site.

Sites	% of households	Area (m²)
Kongoussi-Bam		
Green Beans	87	2,978
Titao-Lorum		
Onion	58	433
Potato	39	511
Ouonon-Passoré		
Tomato	80	8,113
Onion	75	7, 993
Eggplant	53	3,640
Cucumber	40	3,700
Zekeze- Boulgou		
Banana	56	11,000
Tomato	47	6,147
Onion	47	4,755

Source: Author's computations.

ratios due to high input costs, especially for seeds and water. Green beans producers had to pay a water fee of 400,000 FCFA/year and the seed cost was estimated at 225,650 FCFA per hectare, while potato required 2,126,600 FCFA of seed per hectare for its production.

The cost/revenue ratio was significantly higher for tomato in Ouonon than in Zekeze. Eggplant and cucumber had the lowest gross revenues, but still financially rewarding with revenue/cost ratios of 1.72 and 1.77, respectively.

Potato had the highest gross margin per hour (1733 FCFA). The hourly remuneration of green beans was intermediate in Kongoussi (525 FCFA), and lowest for cucumber in Ouonon (206 FCFA) and banana in Zekeze (260 FCFA) which was the most labor-demanding.

Onion had a higher hourly payment in Ouonon than in Titao and Zekeze. Tomato recorded a higher hourly remuneration in Ouonon than in Zekeze.

Based on a 6-h work day, profitability indicators show that farmers are better-off compared to the 'less than one dollar' situation of people in Burkina Faso. Pro-poor rural economic policy could therefore be promoted through the production of horticultural crops.

Technical efficiency

Scores of efficiency

Scarce studies are available on horticulture efficiency in Africa and in Burkina Faso. However, Jema (2008) conducted an interesting one in Ethiopia and found that mean technical efficiency for vegetable production was 96% in two districts of eastern Ethiopia.

In our study, it is found that technical efficiency scores (Table 3) showed that the proportion of efficient producers ranged from 35% for tomato in Ouonon to 100% for onion in Zekeze.

Cucumber production in Ouonon, green beans in Kongoussi, tomato in Zekeze, potato in Titao and eggplant in Ouonon also recorded a technical efficiency score greater than 70%.

Considering efficiency for individual crops, onion production in Zekeze was the most efficient, with a maximum score of 100%, followed by the production of cucumber in Ouonon and green beans in Kongoussi with respective scores of 98%. Onion had a lower efficiency score in the northern site of Titao, and in Ouonon (88% and 84% respectively); the lowest technical efficiency score was recorded in Ouonon for tomato with a score of 68.4%. Green beans and potato producers were less efficient than onion producers in the southern site of Zekeze and cucumber producers in Ouonon. The reasons for this inefficiency should be investigated in order to improve the profitability and competitiveness of horticultural crops.

Excess inputs

The use of excessive levels of inputs is one source of technical inefficiency. In Kongoussi, the inputs that most farmers used in excessive quantities, included seeds (probably because of poor quality) and fertilizers (NPK and urea) (Table 4).

The use of fertilizers in excess quantities reflects the decline in soil fertility that producers tried to compensate yield decrease by increasing application rates which might result in environmental pollution. Less important excess uses were observed on family labor, energy and

Table 2. Per hectare production budgets (FCFA).

Item	Kongoussi	Titao		Ouonon				Zekeze		
	Green beans	Potato	Onion	Onion	Eggplant.	Tomato	Cucumber	Banana	Tomato	Onion
Gross Income (FCFA)	2 091 486	8 369 262	3 534 948	1 463 926	447 564	1 201 404	388 080	2 074 317	708 985	2 148 133
Variable costs (FCFA)	856 747	3 080 201	1 141 320	166 857	260 427	181 767	219 372	673 327	354 995	598 704
Seeds	225 650	2 126 600	291 288	15 844	13 968	15 996	15 189		33 000	175 882
NPK fertilizer	77 750	480 066	428 260	32 743	45 120	39 432	31 654	107 803	40 500	82 536
Urea	35 424	-	65 869	11 508	24 000	12 393	20 003	177 680	26 059	44 469
Organic manure	28 703	196 929	121 580	18 333	25 038	22 386	24 720	-	8 437	56 666
Insecticide	23 350	124 645	100 595	12 450	25 717	13 743	14 167		17 312	17 500
Water fee	400 000	-	-	-	-	-	-	-	-	-
Labor Cost	20 000									
Diesel/gasoline	-	131 361	131 361	15 180	38 909	18 935	26 500	224 607	132 326	138 300
Repairs and maintenance	4 100	20 600	2 367	9 522	2 511	9 960	2 524	93 467	20 173	23 133
Hired labor	41 770	-	-	51 277	85 164	48 922	84 615	69 770	77 188	60 218
Cost/revenue ratio	2.44	2.72	3.10	8.77	1.72	6.61	1.77	3.08	2.00	3.59
Family labor (hours)	3 980	4 830	4 025	1 654	2000	1549	1884	7967	2979	3878
Hourly payment based on gross margin (FCFA)	525	1 733	878	885	224	776	206	260	238	554

Source: Author's computations.

Table 3. Technical efficiency by site and crop in the Nakanbé River watershed.

Item	Kongoussi	Titao		Ouonon				Zekeze		
	Green beans	Potato	Onion	Onion	Egg-plant	Tomato	Cucumber	Banana	Tomato	Onion
Technical efficiency (%)	97.5	95.8	88.2	84.3	91	68.4	98.2	88.5	91.1	100
Proportion of efficient producers (%)	74	82	61	60	79	35	94	68	82	100

Source: Author's computations.

depreciation for 43% inefficient farmers.

In Titao, excess use of inputs was observed in the production of potato and onion: excess use of seeds and NPK fertilizer for 513,059 FCFA/ha and 150,111 FCFA/ha respectively for potato. For onion production, excess levels were of 293,697 FCFA/ha for seed, 203,695 FCFA for NPK fertilizers, 61,919 FCFA for urea, 201,739 FCFA for the organic manure and 106,795FCFA for pesticides. On the production site of Ouonon, not only inputs were used in excess quantities; hired labor also recorded a peak of excess level of 39,778 FCFA/ha and 50,117 FCFA/ha for onion and tomato, respectively.

In Zekeze, banana inefficient producers recorded on average an excess inputs use of 112,076 FCFA/ha for the NPK fertilizer, 138,237 FCFA for urea and 125,064 FCFA for energy. Unlike banana, the recorded major excess input use was in tomato production: 169,889 FCFA/ha (100% of farmers) for energy and 116,780 FCFA

Table 4. Excess production factors (in FCFA/ha).

Input/factor	Kongoussi	Titao		Ouonon				Zekeze		
	Green Beans	Potato	Onion	Onion	Eggplant	Tomato	Cucumber	Banana	Tomato	Onion
Seeds	18016 (57)	513059 (100)	293697 (71)	9924 (25)	2492 (100)	8489 (55)	0	0	7667 (67)	0
NPK fertiliser	21374 (71)	150111 (100)	203695 (71)	14069 (67)	27352 (40)	21461 (80)	110 (100)	112076 (100)	6891 (100)	0
Urea	7957 (71)	0	61919 (86)	5265 (75)	24000 (20)	8937 (65)	555 (100)	138237 (67)	6482 (100)	0
Organic manure	7700 (85)	22211 (50)	201739 (100)	10274 (58)	20984 (100)	12985 (90)	13 (100)	0	2221 (67)	0
Insecticide	6291 (71)	60173 (50)	106795 (57)	4648 (75)	4927 (40)	9138 (55)	625 (100)	0	5040 (33)	0
Energy	3347 (43)	0	46864 (43)	5995 (92)	1343 (20)	15072 (100)	0	125054 (67)	169889 (100)	0
Hired labor	0	0	0	39778 (33)	0	50117 (60)	0	50345 (33)	116780 (33)	0
Depreciation	700 (43)	18120 (50)	4704 (71)	4490 (75)	2181 (60)	8772 (75)	0	6067 (33)	0	0
Family labor	1021 (43)	0	14501 (71)	584 (67)	2245 (40)	738 (70)	110 (100)	6695 (33)	2330 (67)	0

Source: Author's computations. Figures in parentheses are the proportion of inefficient producers in the use of input or factor of production. Energy=diesel/gasoline repairs and maintenance.

for hired labor (33% of farmers).

In general, all sites recorded excess inputs including fertilizers (NPK, urea) and insecticides. This means that producers are losing money by applying such high quantities of fertilizers and insecticides. In addition to the reduced profitability, excessive use of fertilizer and insecticides can reduce the sustainability of their production system due to negative long-term effects on the environment. Furthermore, this would contribute to reduce market demand as consumers are more and more demanding for use of ecological and healthy production practices.

CONCLUSION AND RECOMMENDATIONS

The results of the study show that production of horticultural crops is financially profitable in the Nakanbé River watershed in Burkina Faso. Satisfactory cost/revenue ratios were found, especially in Ouonon for tomato and onion. Horticulture has relatively high hourlypayment: 525 FCFA/hour for green beans in Kongoussi, 1,733 FCFA for potato in Titao, 885 FCFA and 776 FCFA in Ouonon, respectively. Promotion of horticultural crop production could be part of a sound pro-poor strategy in rural areas. Despite good financial performance, horticultural crop producers are not totally efficient and improvements are possible. Farmers exhibited technical inefficiencies, which indicates that they have not mastered existing technologies and that improvements are needed. These inefficiencies should be addressed before new costly investments are made and future research should investigate the causes of the technical inefficiency. Agronomic studies are necessary to assess the impact of excess input use (particularly of fertilizer and insecticides) on the environment. Producers need training on input rates, substitution of organic for chemical inputs, and use of biological techniques, and sensitization on the environmental risks associated with excessive chemical rates.

The survival as well as development of horticultural crop production is important to increase rural household income while supplying fresh food to an increasing urban population at a reasonable price.

Conflicts of Interest

The authors have not declared any conflict of interest.

REFERENCES

Afriat SN (1972). Efficiency Estimation of Production Functions. Int. Econ. Rev. 13:568-598.

Banker RD, Charnes, A, Cooper WW (1984). Some Models for Estimating Technical and Scale Inefficiencies in Data Envelopment Analysis. Manage Sci. 30:1078-1092.

Boles JN (1966). Efficiency Squared-efficient Computation of Efficiency. Proceedings of the 39th Annual Meeting of the Western Forum Economic Association, pp.137-142.

Charnes AWW, Cooper, Rhodes E (1978). Measuring the Efficiency. Eur. J. Oper. Res. 2:429-444.

Coelli TJ (1996). A guide to DEAP version 2.1: A Data Envelopment Analysis (computer) Program. CEPA,

Department of Econometrics, University of New England, Armidale, NSW 2351, Australia
Jema H (2008). Economic efficiency and marketing performance of vegetable production in the eastern and central parts of Ethiopia. Diss. (sammanfattning/summary) Uppsala: Sveriges lantbruksuniv., Acta Universitatis agriculturae Sueciae, 1652-6880;2008:17
Government of Burkina Faso (2005). Competition, investment and competitiveness: Constraints and Recommendations. ISAF, the International Finance Corporation and the
World Bank.
INSD (2010). Analyse de quelques résultats des données de la phase principale de l'enquête intégrale sur les conditions de vie des ménages, EICVM 2009.
Koopmans TC (1951). An analysis of production as an efficient combination of activities. In Koopmans, T. C. editor, Activity Analysis of Production and Allocation, John Wiley and Sons, Inc.

13

Impact of macroeconomic policies on poverty alleviation in Sub-Saharan African countries

Mogos Teweldemedhin

Polytechnic of Namibia, School of Natural Resources and Spatial Sciences, Department of Agriculture and Natural Resources Sciences, Namibia.

The objective of this paper is to examine the relationship between macroeconomic variables and poverty alleviation in Sub-Saharan Africa, by applying descriptive illustration and weighted least square (WLS) regression econometric analysis using the multidimensional poverty index (MPI) taken from the oxford poverty and human development initiative (OPHI) as dependent variable. Furthermore, principal component analysis (PCA) was performed to avoid multicollinearity problems and to improve the estimation power of the regression. Long-term annual gross domestic product (GDP) growth trends were analysed by dividing countries into four groupings, namely upper income, lower middle income, lower income and conflict countries. The results show that post-conflict nations experience good progress in economic growth. With the exception of the ratio of government expenditure to GDP (GEXPGDP), foreign direct investment, net inflows (% of GDP) (INFGDP), agriculture, value added (% of GDP) (AGR.GDP) and the Gini coefficient (GINICOEF) (not significant and not reported), all other variables were found to be statistically significant at the specified significance level. Furthermore, population growth (annual %) (POPGRWTH) holds greater positive magnitude, and shows that economic growth is moving at a slower pace compared to population growth, which complicates the economic development agenda on this continent. The major factors limiting growth are restrictive fiscal policy, contractionary monetary policy in most countries, and balance of payments constraints. Furthermore, it is important to improve local capabilities and inter-firm linkages, thus achieving well-managed privatisation, while it is equally important to have subsidies reaching the poor.

Key words: Sub-Saharan Africa, multidimensional poverty index (MPI), macroeconomic variables, poverty alleviation.

INTRODUCTION

Since poverty is not an easy concept to define, there is a wide range of definitions influenced by different disciplinary approaches and ideologies. The dominant Western definition since World War II has defined poverty in monetary terms, using levels of income or consumption as measurements (Grusky and Kanbur, 2006) and defining the poor by a headcount of those who fall below a given income/consumption level or 'poverty line'

(Handley et al., 2009). Subramanian (1997) subsequently devised a multidimensional definition comprising the basic needs approach (Streeton et al., 1981; Handley et al., 2009), the capabilities approach (Sen, 1999; Handley et al., 2009) and the human development approach (UNDP, 2006). The acceptance thereof is reflected in the widespread use of the human development index (HDI) of the United Nations development programme (UNDP), which is a composite measure of three dimensions of human development, namely: (i) life expectancy, (ii) educational attainment and (iii) standard of living, measured by income in terms of its purchasing power parity (UNDP, 2006). The conceptualisation of HDI is used in this paper to define poverty.

Sub-Saharan Africa (SSA) is afflicted by many forms of poverty, with HDI scores in most SSA countries having stagnated or declined since 1990, making this region the poorest in the world. Indeed, 28 of the 31 low human development countries are in SSA (UNDP, 2006). An analysis of income poverty is similarly disappointing in that since 1990, income poverty has fallen in all regions of the world except SSA, where there has been an increase in both the incidence and the absolute number of people living in income poverty. This sees some 300 million people in SSA almost half the region's population living on less than US$1 per day (UNDP, 2006).

While the rest of the world has made significant progress towards poverty alleviation, Africa in particular Sub-Saharan Africa continues to lag behind. This trend is projected to increase unless preventative measures are taken. Many factors have contributed to this trend, including the high prevalence of HIV/AIDS, civil war and the associated strife and poor governance, frequent drought and famine, and agricultural dependency on the climate and environment. Food security on the continent has worsened since 1970, and the proportion of the population this is malnourished in SSA has remained within the 33 to 35% range (Velde et al., 2004). The prevalence of malnutrition within the continent varies by region, being lowest in Northern Africa (4%) and highest in Central Africa (40%) (Mwaniki, 2009). Over 70% of the food insecure population in Africa lives in the rural areas. Ironically, smallholder farmers, the producers of over 90% of the continent's food supply, make up the majority (50%) of the food insecure population, while the remainder consists of the landless poor in rural areas (30%) and the urban poor. Food security has three aspects, namely food availability, food access, and food adequacy (Mwaniki, 2009)

To date, regardless of the intensity of development programmes, very little has changed in SSA economies in the past 10 years Although food aid, technical arrangements, and financial and other humanitarian assistance continue to flow in from the developed countries to SSA, problems of drought, famine, inflation, international debt and unemployment continue to escalate. The existing social and infrastructure facilities such as health, education, transportation and many other institutional structures are relatively weak and inadequate (Velde et al., 2004).

According to the 2010 ranking of the top 47 poorest countries worldwide, 32 of those are in SSA. Although the global headcount is between $1.25 and $2.00/day, the multidimensional poverty index (MPI) is considered below $1.50/day poverty line. Niger, Ethiopia and Mali, at 93, 90 and 88% respectively, are the top three countries on the list (Alkire and Santos, 2010).

Therefore, the objective of this paper is to test the impact of macroeconomic policies on the variables of poverty alleviation in SSA. The paper is divided into two main parts: First, the growth performance of SSA is illustrated and described, while the poverty situation is quantified with reference to the MPI, and some of the main reasons for Africa's stagnation in economic development are pointed out. Thereafter, the relationship between the MPI and different macroeconomic variables was tested.

This paper thus gives insight into the areas where research and attention by policy makers and donors are likely to prove more valuable at this point in time. This result is therefore partial rather than comprehensive; it is an agenda focusing closely on the non-macroeconomic factors constraining Africa's poor.

Problem statement and motivation of the study

Economic growth does not come risk-free. Although material progress can be measured by the growth in national output, income and spending, the rapid economic growth of developed countries is accompanied by several short-term and long-term problems (Riley and Eton, 2006), including inflation risks, inequalities and regional disparities. The idea of economic degrowth amongst rich nations is emerging as a response to the triple crisis (environmental, social and economic); it did not appear out of the blue. Sustainable degrowth may be defined as an equitable downscaling of production and consumption that increases human wellbeing and enhances ecological conditions at the local and global level, over both the short and long term (Rull, 2010).

However, according to Weeks (2009), the major causes of growth instability in SSA are: (1) Fluctuations in the terms of trade, which impact directly on aggregate demand via export and import prices, thus affecting the fiscal balance through trade taxes, and tightening or loosening the balance of payments constraint; (2) variations in weather that largely determine the performance of rain-fed agriculture in a region where irrigation is limited; and (3) Low investment confidence towards SSA (mainly due to perceptions of poor quality of governance, legal protection of private property, and institutional limits on leaders). Therefore, the study is directive from a policy perspective, as macroeconomic

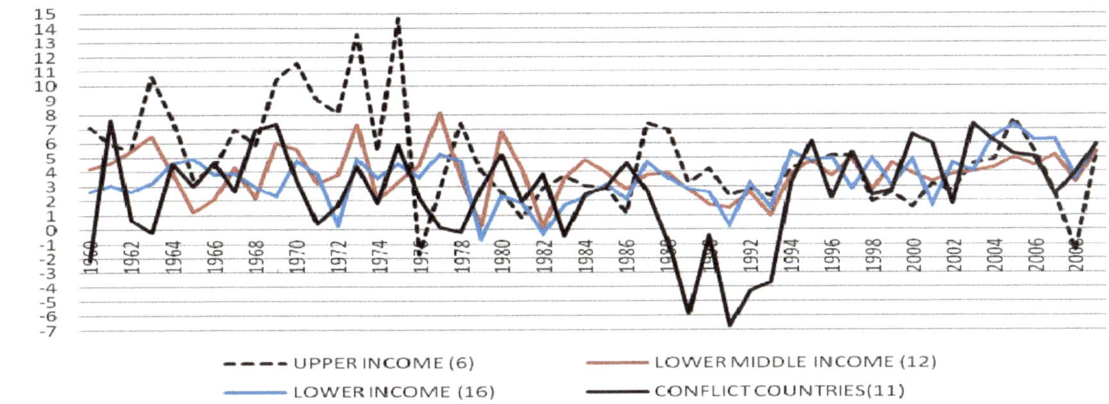

Figure 1. GDP annual percentage growth for SSA, 1960 to 2010. Source: Own representation based on World Bank dataset (2012) dataset.

policy constitutes an important element in the government's efforts to boost the underlying supply capacity of the economy. From a research perspective, the empirical results of this study would be timeous, as SSA affords the opportunity for an in-depth case study on account of significant variations in trade policy orientation and productivity performance across economies. In addition:

1. This is the time when most African nations are showing progress in governance and democracy exercises;
2. This is the time that Africa is encountering a number of challenges, such as climate change, high rates of transmitted diseases (such as HIV/AIDs), and requiring different strategies;
3. This is also the time when there is a need for sustainability to address the triple crisis (economic, social and environmental) using effective and efficient policy instruments; and
4. As mentioned above, if developed nations are willing to downscale their production and consumption to increase human wellbeing and enhance ecological conditions, this will have certain implications for Africa.

METHODOLOGY AND DATA

A descriptive illustration and weighted least square (WLS) regression econometric analysis was applied to build the influence of variables in the modelling. Descriptive statistics were used to assess differences in the basic characteristics of the macroeconomic variables that can influence or affect the MPI. Furthermore, principal component analysis (PCA) was performed to avoid multicollinearity problems and to improve the estimation power of the regression. Out of 21 variables, PCA extracted 12 variables relevant to the analysis. The econometric equation is constructed as follows:

$$MPI = \alpha_1 EXDEBTG + \alpha_2 GEXP.GDP + \alpha_3 INFL.GDP + \alpha_4 GDPGRWTH + \alpha_5 SAVGRWTH + \alpha_6 CREDP.GDP + \alpha_7 MILL.GDP + \alpha_8 TRAD.GDP + \alpha_9 AGR.GDP + \alpha_{10} HELTH.GDP + \alpha_{11} POP.GRWTH + \alpha_{12} GINICOEF + \varepsilon$$

Where: MPI is as calculated for 104 developing countries by Alkire and Santos (2010) in the oxford poverty and human development initiative (OPHI). This is identified the first multidimensional poverty estimation using micro datasets (household surveys) for such a large number of countries, covering about 78% of the world's population. The MPI has the mathematical structure of one of the Alkire and Santos poverty multidimensional measures and is composed of 10 indicators, corresponding to the same three dimensions as the HDI: Education, Health, and Standard of Living. The MPI captures a set of direct deprivations that afflict a person at the same time. This tool could be used to target the poorest, to track the millennium development goals, and to design policies that directly address the interlocking deprivations experienced by the poor.

The MPI reveals the combination of deprivations that afflict a household at the same time. A household is identified as multi-dimensionally poor if, and only if, it is deprived in some combination of indicators with a weighted sum of 30% or more of the dimensions. The dimensions, indicators and deprivation criteria are contained in the Appendix. The MPI is the product of two numbers: Headcount (H), or percentage of people who are poor, and average intensity of deprivation (A), reflecting the proportion of dimensions in which households are deprived. Alkire and Santos show that this measure is very easy to calculate and interpret, is intuitive yet robust, and satisfies many desirable properties.

The independent variables sourced from the World Bank (2012) database were those relevant variables extracted by PCA, namely: External debt stocks (% of GNI) (EXDEBTG), Government expenditure ratio to GDP (GEXPGDP), Foreign direct investment, net inflows (% of GDP) (INFGDP), GDP growth (annual %) GDPGRWOTH, Gross domestic savings (% of GDP) (SAVGRWTH), Domestic credit to private sector (% of GDP) (CREDP.GDP), Military expenditure (% of GDP) (MILL.GDP), Trade (% of GDP) (TRAD.GDP), Agriculture, value added (% of GDP) (AGR.GDP), Health expenditure, total (% of GDP) (HELTH.GDP), Population growth (annual %) (POPGRWTH), Gini coefficient (GINICOEF), and Error term.

RESULTS AND DISCUSSION

Figure 1 illustrates the 1960 to 2010 growth performance for four groupings of countries in the region: conflict-affected countries (11), non-conflict lower-middle-income countries (12), upper-income countries (6), and non-conflict

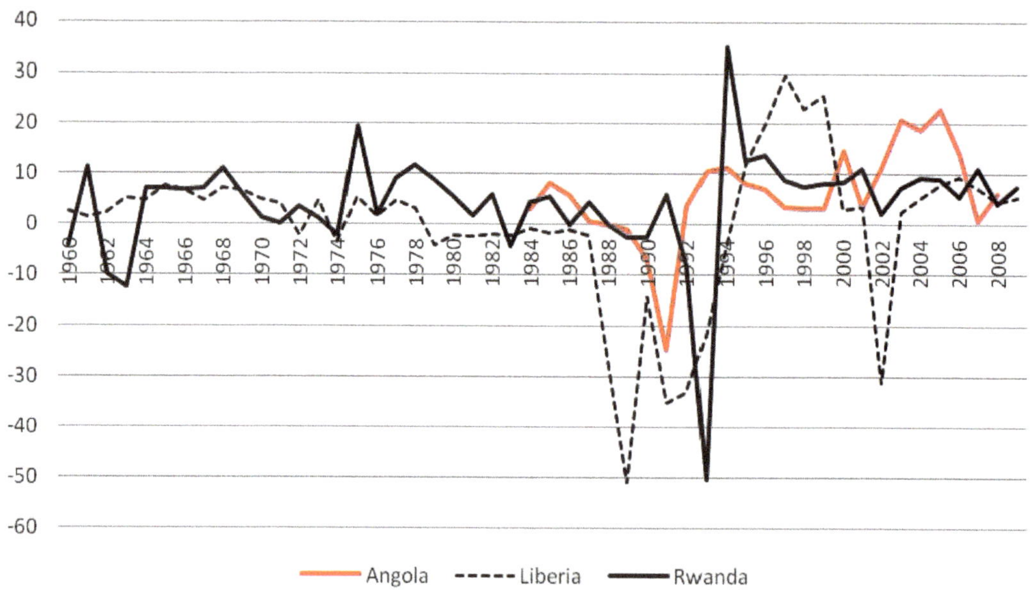

Figure 2. GDP annual percentage growth among conflict-history countries (Angola, Liberia and Rwanda), 1960 to 2010. Source: Own representation based on World Bank (2012) dataset.

low-income countries (16).

In the upper-income categories, Botswana's growth gave a misleading impression of significant improvement; however, Botswana showed exceptional growth performance of 21 to 26 % from 1970 to 1972 and again in 1987. According to a study by Maipose (2008), this was mainly due to the following reasons:

1. From 1975 to 1989, known as the second period in the history of Botswana, and characterised by the end of colonial rule, Botswana experienced the introduction of a multiparty democratic system of government under the inherited market-based economy, as well as the integration of traditional institutional structures into modern institutions, underlined by a policy stance that sought to maximise the flow of foreign capital, aid and private investment, resulting in the moderate growth of the time; and
2. The third period, covering the whole of the 1990 and the new millennium (though overlapping to some degree with the end of the 1980s), saw the start of a new policy environment, signalling the end of the state-led development strategy and a new reorientation towards private-sector-led development with the emphasis on economic diversification, export competitiveness, and privatisation options (Maipose, 2008).

The records for the low-income group of 16 countries fluctuate less than 5%, with the exception of the period 2006 to 2007, with growth of around 6% (Figure 2). This result was not a true reflection of the whole low-income grouping of SSA; rather it was the result of good economic growth in Guinea in 2006 and 2007 (22 and 18% respectively). Growth in per capita income averaged a mere 0.2% during the 1990s, rising to only 1.2% from 2000 to 2005. This lacklustre performance underscores the need for more expansionary and investment-focused macroeconomic policies. The recovery of gross domestic product (GDP) growth in conflict-affected countries, specifically after 1992, shows the annual GDP growth trend in most of those countries catching up with the rest of Africa's groupings, mainly due to certain policy changes. On the other hand, middle-income countries have been showing a drop in performance since 1994, while only modest improvements have been observed in low-income countries.

Figures 3 show the global debt as a percentage of GDP. As indicated in Figure 3, the majority of the countries' debt-to-GDP ratio is very high. The rationale behind acquiring high debt in most SSA countries is that it is believed to: (i) promote private consumption; (ii) promote public investment; (iii) increase total factor productivity (TFP), and (iv) raise sovereign long-term nominal and real interest rates. As a result, the government policy perspective is geared towards financial injections to support longer-term economic growth, but this policy becomes unsustainable and dangerous and may lead to a collapse in the economic prospective of nations and the wellbeing of societies.

The SSA international debt situation is growing ever more serious, while military expenditure is increasing simultaneously. According to WFP (2009) records, SSA remains the top food-aid receiver worldwide. Erratic climatic conditions along with other problems have placed SSA economies in the vicious cycle of economic underdevelopment (Figure 4). Worst of all, existing

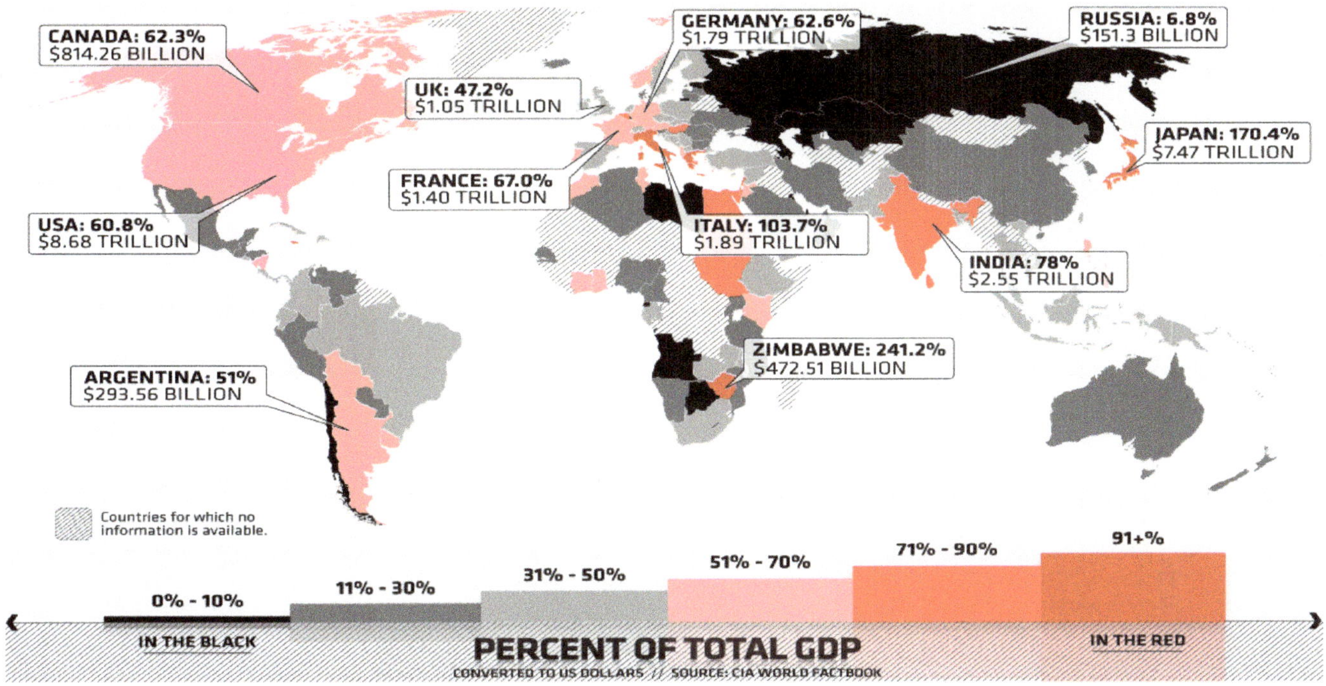

Figure 3. National debt by country, 2009. Source: CIA (2012).

ambitions of war and a lack of effective and efficient policies for development planning make SSA the front runner for a triple crisis (environmental, social and economic). The crucial questions are therefore: Why has economic development policy and cultural change been so difficult to attain in SSA? What might have gone wrong? Have the problems associated with SSA economic theory been diagnosed incorrectly, or have economic theories of development been applied incorrectly? Or are there other factors that have not yet been explored or identified that need to be addressed through other appropriate policies? And where does Africa go from here?

Some of the main reasons for Africa's stagnation in economic development can be summarised as follows:

1. Good leadership, as a prerequisite for economic growth, includes the need to create a secure environment for property, political stability, social harmony, and a respected legal code that protects the rights of owners. Additionally, SSA nations need to offer infrastructure such as roads, ports, airports, railways, electricity, water, telecommunications, and a well-educated and skilled labour force (Friedman, 2006; Sandbrook, 1985). However, the majority of African states have failed to supply the aforementioned basic services, and little has been done thus far in this regard (Mills, 2010);
2. Failure to diversify out of primary product exports led to serious policy failures in poor countries following independence. In 1950, SSA accounted for 3% of world exports, but five decades later that share had dropped to 1.5% (Mills, 2010). For example, although oil revenue in Nigeria increased by 885% over 35 years, the number of people living on less than $1 per day increased by 535%, which means an annual increment of 25% and 15%, respectively. In addition, SSA is a net importer of food, whereas Asia and Latin America have doubled or tripled their agricultural production over 30 years. Another example of policy failure is Zambia, which saw a number of sectors collapse due to poor government policy choices, such as: i) government investing in certain sectors rather than acting as a regulatory body; and ii) government later privatising those sectors while the industries were still in their infancy (Mills, 2010);
3. Notable structural factors include political and ethnic bias, excessive control of political power, economic policies that are discriminatory to the poor (Bates, 2003), and ineffective conflict resolution mechanisms with regard to disputes;
4. According to Ayittey (1998), many indebted African countries have a debt service ratio of about 40%, meaning that for every US dollar earned on exports of domestically produced goods, 40% go towards servicing the debt, while the rest goes towards covering imports, military expenditure, production improvements, education, healthcare and other expenses; and
5. African governments were pushed/forced by international monetary fund (IMF) policies to take certain actions that were not domestically feasible.

To support the above arguments from the empirical evidence, the study applied econometric equations to test

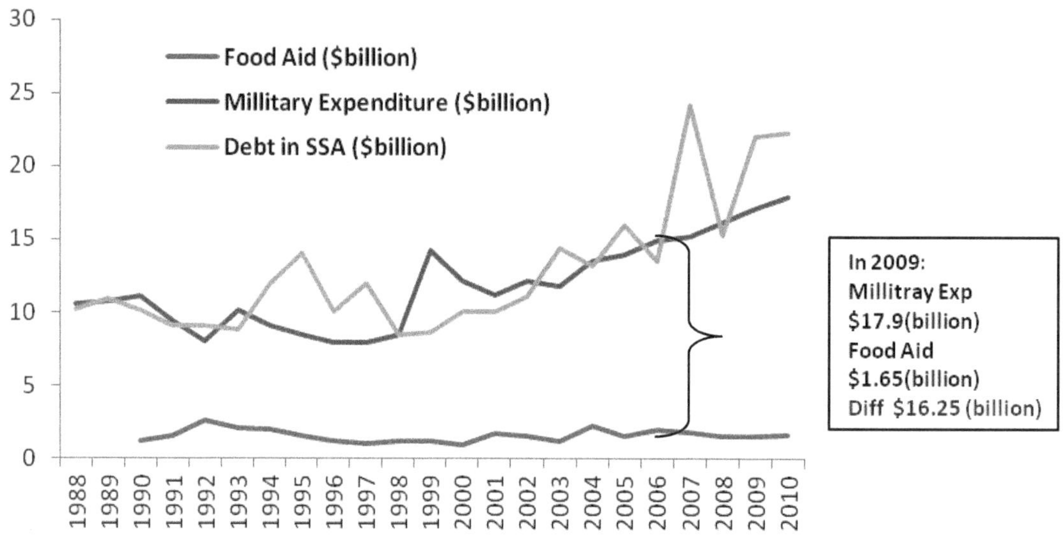

Figure 4. Comparison of debt, food aid and military expenditure in SSA, 1988 to 2010. Source: Own representation based on IMF and UNDP data.

Table 1. Macroeconomic factors influencing levels of poverty in Sub Saharan Africa: Weighted Least Square (2009).

Coefficients	B	Std Error	Beta	t	Sig "t"
EXDEBTG	0.001	0.001	0.096	0.795***	0.095
GEXPGDP	-0.001	0.002	-0.071	-0.478	0.638
INFLGDP	0.001	0.003	0.033	0.240	0.812
GDPGROWTH	0.014	0.007	0.249	1.964***	0.062
SAVGRWTH	-0.003	0.001	-0.634	-2.244**	0.035
CREDP.GDP	-0.004	0.002	-0.262	-2.175**	0.041
MILL.GDP	-0.057	0.020	0.440	2.908*	0.008
TRAD.GDP	-0.003	0.001	-0.727	-0.339*	0.003
AGR.GDP	0.002	0.002	0.211	1.110	0.279
HELTH.GDP	-0.014	0.006	-0.288	-2.250	0.0358
POPGRWTH	0.102	0.025	0.445	4.080	0.001
GINICOEF	0.138	0.169	0.097	0.816	0.423
(Error Term)	0.512	0.174		2.936	0.008
R -Square					0.800
Adjusted R Square					0.692
ANOVA					0.00
F-test					7.36
No. of Observations					34

*, ** and *** significance level at 1, 5 and 10% respectively.

the hypothesis to 34 SSA countries. After conducting the necessary statistical tests, including the relationship among the macroeconomic variables and key determinants/attributes that can contribute to MPI in the SSA region, the ordinary least square (OLS) econometrical model was applied to the cross-sectional 2009 dataset of the World Bank related to MPI. Furthermore, due to the presence of heteroscedasticity and multicollinearity, WLS was the selected estimator.

As shown in Table 1, the overall adjusted explanatory power for export determinants was estimated at 69%. The ANOVA result shows that it is significant at the 1% level, suggesting that there is a linear relationship among the variables (Table 1). With the exception of FEXPGDP, INFLGDP, AGR.GDP and GINICOEF (not significant and not reported in Table 1), all other variables were found to be statistically significant at the specified level of significance. Furthermore, all variables were found to hold the expected sign. However, POPGRWTH was determined to hold a greater positive magnitude of its

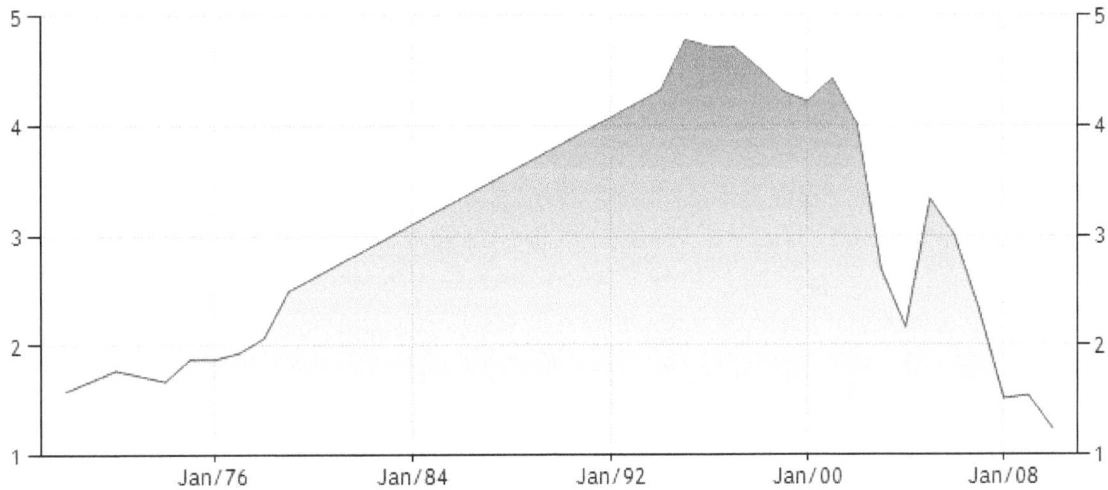

Figure 5. Total debt service (% of GNI) in SSA. Source: World Bank (2012).

estimated coefficient at 0.102, showing that population growth is the greatest challenge to economic growth in SSA.

External debt ratio to GDP (EXDGDP) was found to be significant at 10% and positive, thus implying that external debt contributed significantly to the MPI in SSA (Figure 4). The long-term debt crisis, along with a multitude of other problems, has crippled economic growth in SSA. The World Bank (2012) reported some good news, with debt servicing showing a significant improvement, amounting to 4.6% of GNI in 1996 to 1997. This is due to substantial external debt relief, which has liberated fiscal space in SSA (Figure 5). The IMF (2011) reported that deficits have been increased beyond sustainable medium-term paths; these should be revisited so that policy buffers can be restored. Whereas output remains well below potential, there is a strong case for fiscal policy to help sustain demand in the near term, subject to financing availability.

The four core components of macroeconomic policy that can drive higher GDP growth include government spending, investment spending, savings, and trade balance. This model shows that the first two variables were found not to be significant, meaning that these variables do not play a role in driving higher economic growth, whereas the second two (savings and trade balance) were found to be negatively related and significant at 5 and 1%, respectively. However, the annualised GDP growth is at a very slow pace compared to population growth, implying that SSA economic growth is not growing at the same pace as population growth to support poverty reduction strategies in SSA. According to a recent IMF (2011) report, the limited integration of many countries in the region into the global economy may have helped, but only marginally. Previous (milder) global economic slowdowns had a much more damaging impact. This time, the global downturn was much sharper, but the dislocation was far less. The main factor distinguishing this slowdown from previous cycles has been the stronger macroeconomic position of most countries in the region.

The credit to private sectors ratio to GDP (CREDP.GDP) is negatively related to MPI and is significant at 5%, implying that there is improvement in credit access to private sectors, which can stimulate trade and investment and thereby have a positive impact on good economic growth performance. Military expenditure showed a negative and significant influence at 1% to the MPI, thus implying that if SSA can reduce military expenditure, resources could be redirected to other basic services, such as health, education and R&D, which would best support economic growth. As reported by the World Bank (2012), military expenditure as a percentage of GDP in SSA accounts for 1.57%, whereas there is zero allocated budget for R&D. Additionally, Figure 4 shows a comparison between the military expenditure (estimated at $25 billion) and food aid (estimated at $1.65 billion) in SSA in 2009. Reducing the military budget would have allowed SSA to fill the food deficit from its own resources.

Conclusions

The objective of this paper is to test the relationship between macroeconomic variables and poverty alleviation in SSA. The study applied descriptive analysis and a WLS econometric model to test the relationship between the MPI and basic macroeconomic variables, and further applied PCA to extract the relevant variables that explain the dependent variable (MPI) (Annexure Figure A1 to A5). The paper analysed some of the key trends of macroeconomic variables, and further tested the relationship between macroeconomic indicators and the MPI.

With the exception of GEXPGDP, INFLGDP, AGR.GDP and GINICOEF (not significant and not reported) all other variables was found to be statistically significant at the specified significance level. Furthermore, all variables were found to hold the expected sign. However, POPGRWTH was found to hold greater positive magnitude with an estimated coefficient of 0.102, showing that economic growth is moving at a slower pace than population growth, thus complicating the economic development agenda on the continent.

The major causes of instability in economic growth for SSA can be summarised as: (i) fluctuations in the terms of trade, which impact directly on aggregate demand via export and import prices, affecting the fiscal balance through trade taxes, and tightening or loosening the balance of payments constraint; and (ii) variations in weather, which largely determine the performance of rain-fed agriculture in a region where irrigation is limited. Both these factors are beyond the direct management of SSA governments in the short and medium term, although the effect of the latter could be reduced in the long run by structural changes in the agricultural sector.

The simplest element to specify is how to raise the economic growth rate. Setting aside exogenous factors such as weather effects, the major limitations to more rapid growth are: (1) restrictive fiscal policy; (2) contractionary monetary policy in most countries; and (3) a balance of payments constraint. Furthermore, it is important to improve local capabilities and inter-firm linkages in view of achieving well-managed privatisation, while subsidies to reach the poor are equally important.

It is highly recommended that SSA countries have effective human capital policy, improved infrastructure and good governance, thus:

1. Human capital policy, investing in education and health, focusing on quality and outreach for the poorest, e.g. by providing public goods and institutions;
2. Infrastructure, enabling the poorest to take part in growth opportunities as a result of trade liberalisation, e.g. by providing infrastructure;
3. Addressing asset and income inequality directly through redistribution via transfer and safety nets; and
4. Good governance, focusing on institutions and other factors that drive pro-poor policies and outcomes.

Lastly, it is important for SSA countries to manage debt crises properly, to reduce military expenditure, to improve conflict resolution mechanisms, to invest more in basic services and to emphasise R&D as a means to focus on export orientation capacity.

Conflict of Interests

The author(s) have not declared any conflict of interests.

REFERENCES

Alkire S, Santos ME (2010). Acute multidimensional poverty: A new index for developing countries. OPHI Working p. 38. Oxford: OPHI.

Ayittey G (1998). Africa in chaos. New York, NY: St. Martin's Press.

Bates R (2003). Rental havens and protective shelters. In: T. Young (Ed.). Readings in African Politics. London: International African Institute, pp. 16-23. PMid:14615955

CIA (Central Intelligence Agency) (2012). World Fact Book. Available online at: http://www.creditloan.com/infographics/national-debt-vs-gdp/ (Accessed on 25 April 2012).

Friedman T (2006). The world is flat: A brief history of the twenty-first century. New York, NY: Macmillan.

Grusky D, Kanbur R (2006). Introduction: The conceptual foundations of poverty and inequality measurement. In: D. Grusky & R. Kanbur (Eds.). Poverty and inequality. Stanford: Stanford University Press.

Handley G, Higgins K, Sharma B, Bird K, Cammack D (2009). Poverty and poverty reduction in Sub-Saharan Africa: An overview of the issues. Working London: Overseas Development Institute, P. 299.

International Monetary Fund (IMF) (2011). World Economic Outlook: September 2011. Washington, DC: IMF.

Maipose GS (2008). Policy and institutional dynamics of sustained growth in Botswana. Available online at: www.growthcommission.org/storage/cgdev/.../gcwp035web.pdf (Accessed on 16 May 2012).

Mills G (2010). Why Africa is poor and what Africans can do about it. Johannesburg: Penguin.

Mwaniki A (2009). Achieving food security in Africa: Challenges and issues. Available online at: www.un.org/africa/.../Achieving %20Food %20Security %20in %20Afr (Accessed on 7 May 2012).

Riley G, Eton C (2006). Costs and benefits of economic growth. Available online at: http://tutor2u.net/economics/revision-notes/a2-macro-economic-growth-costs-benefits.html (Accessed in May 2012).

Rull V (2010). The candid approach. EMBO Rep. 11:14-17.

PMCid:PMC2816632.

Sandbrook R (1985). The politics of Africa's economic stagnation. Cambridge: Cambridge University Press.

Subramanian S (1997). Measurement of inequality and poverty. Oxford: Oxford University Press. PMCid:PMC158335

United Nations Development Programme (UNDP) (2006). Human Development Report 2006: Beyond scarcity – Power, poverty and the global water crisis. New York, NY: UNDP.

Velde DW, Warner M, Page S (2004). Growth and investment in Sub-Saharan Africa. London: Commission for Africa Overseas Development Institute.

Weeks J (2009). Forty years of ODA and conditionality in Africa. London: School of Oriental and African Studies.

World Food Programme (WFP) (2009). Food and aid flows. Available online at: www.wfp.org/fais/. (Accessed on 30 April 2012).

World Bank (2012). http://www.tradingeconomics.com/sub-saharan-africa/total-debt-service-percent-of-gni-wb-data.html. (Accessed on 16 May 2012.

ANNEXURE

Variables used in calculating MPI

1. Health (each indicator weighted equally at 1/6):

- **Child Mortality:** If any child has died in the family;
- **Nutrition:** If any adult or child in the family is malnourished.

2. Education (each indicator weighted equally at 1/6):

- **Years of Schooling:** If no household member has completed 5 years of schooling;
- **Child School Attendance:** If any school-aged child is out of school in years 1 to 8.

3. Standard of Living: (each of the six indicators weighted equally at 1/18):

- **Electricity:** If household does not have electricity;
- **Drinking Water:** If it does not meet MDG definitions, or is more than 30 min walk;
- **Sanitation:** If it does not meet MDG definitions, or the toilet is shared;
- **Flooring:** If the floor is dirt, sand or dung;
- **Cooking Fuel:** If the household cooks with wood, charcoal or dung;
- **Assets:** If the household does not own more than one radio, television, telephone, bicycle, motorbike or refrigerator, and does not own a car or truck.

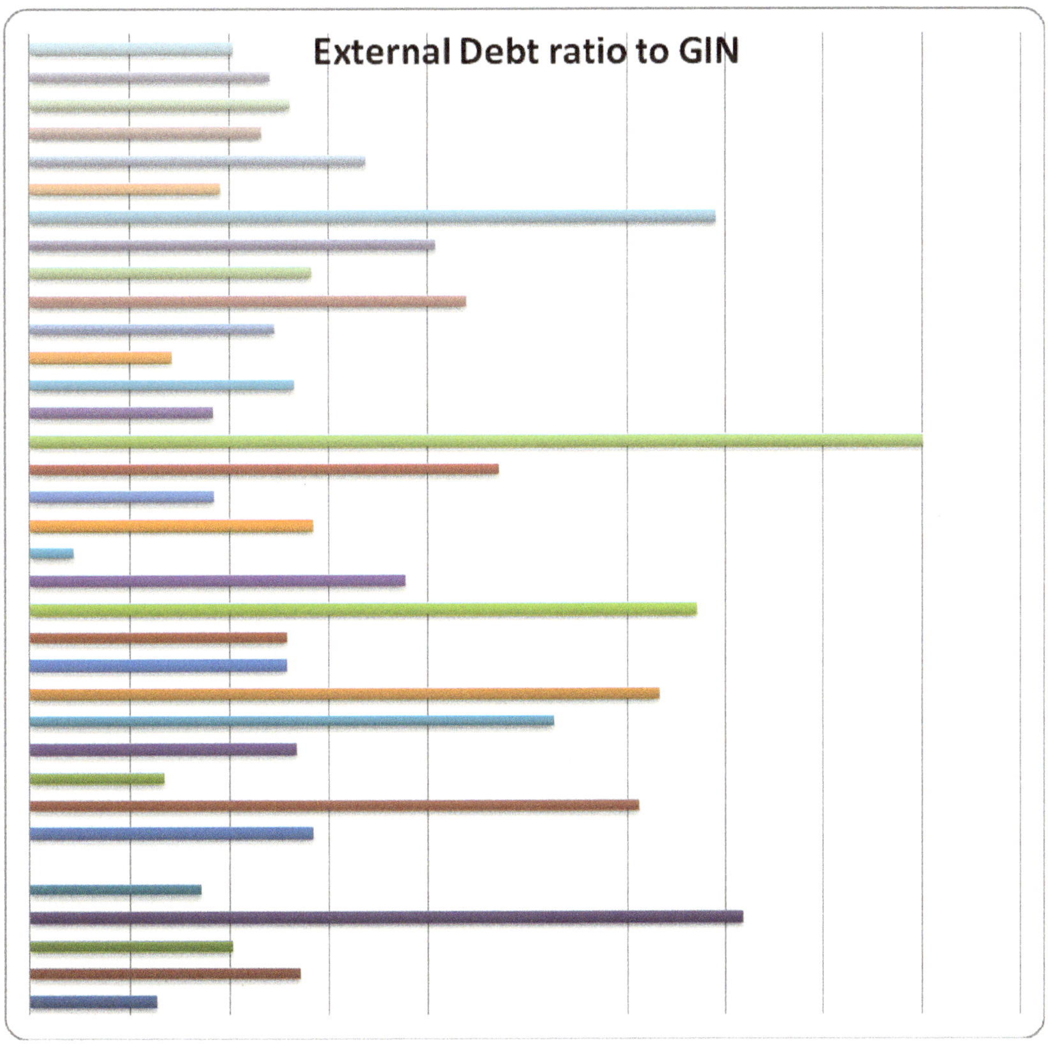

Figure A1. External debt ratio to GNI, 2009.

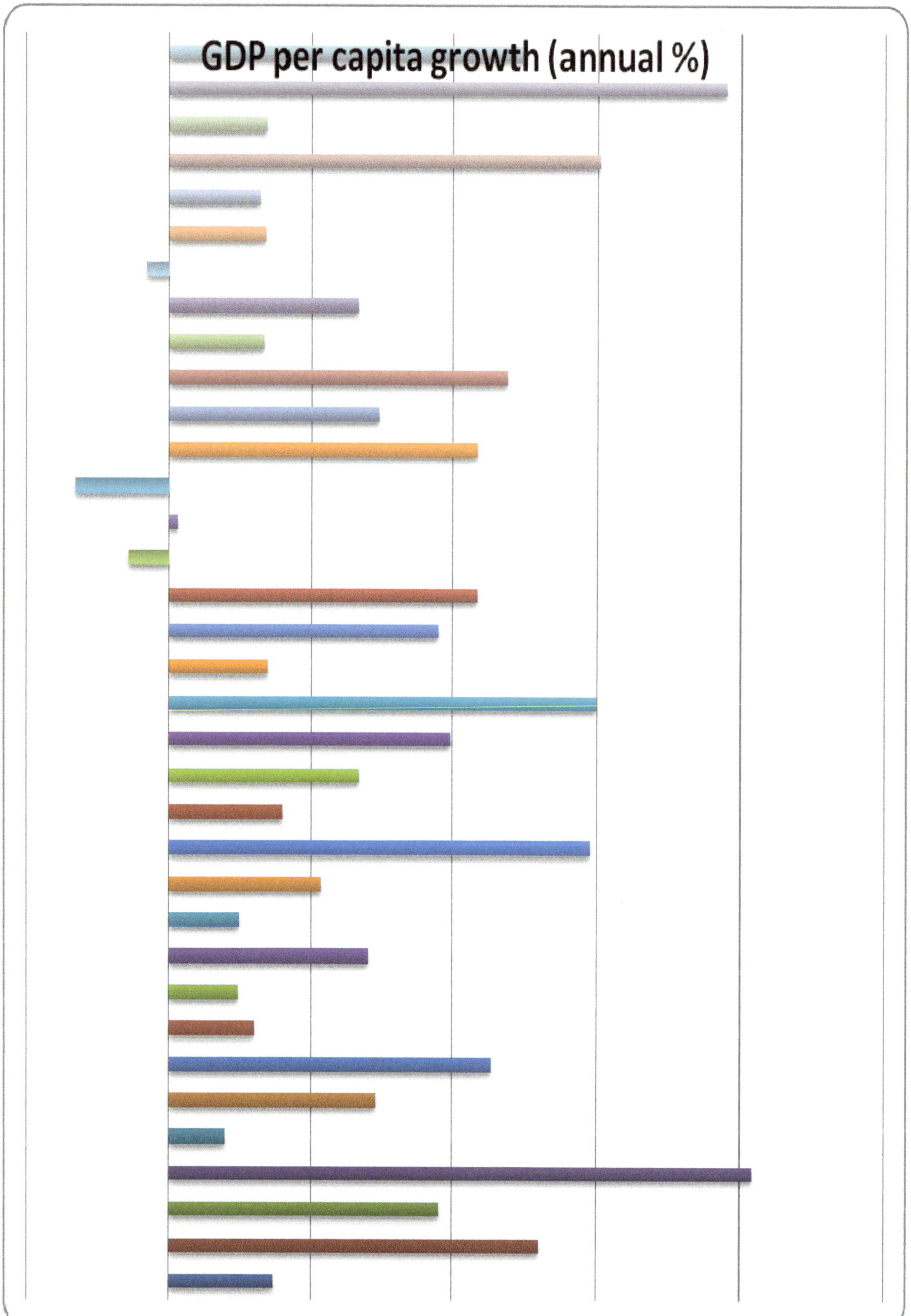

Figure A2. GDP per capita growth (annual %), 2009.

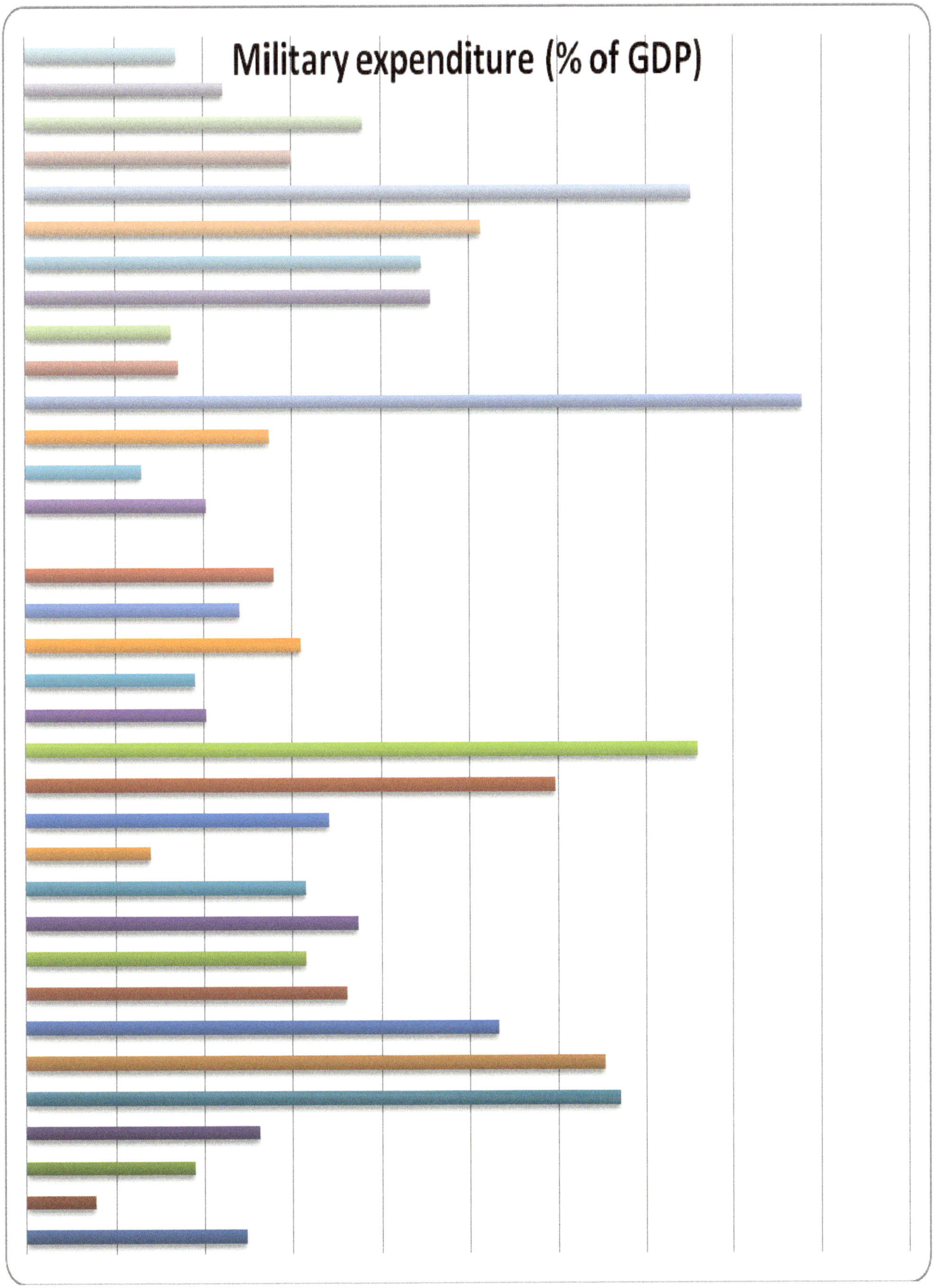

Figure A3. Military expenditure (% of GDP), 2009.

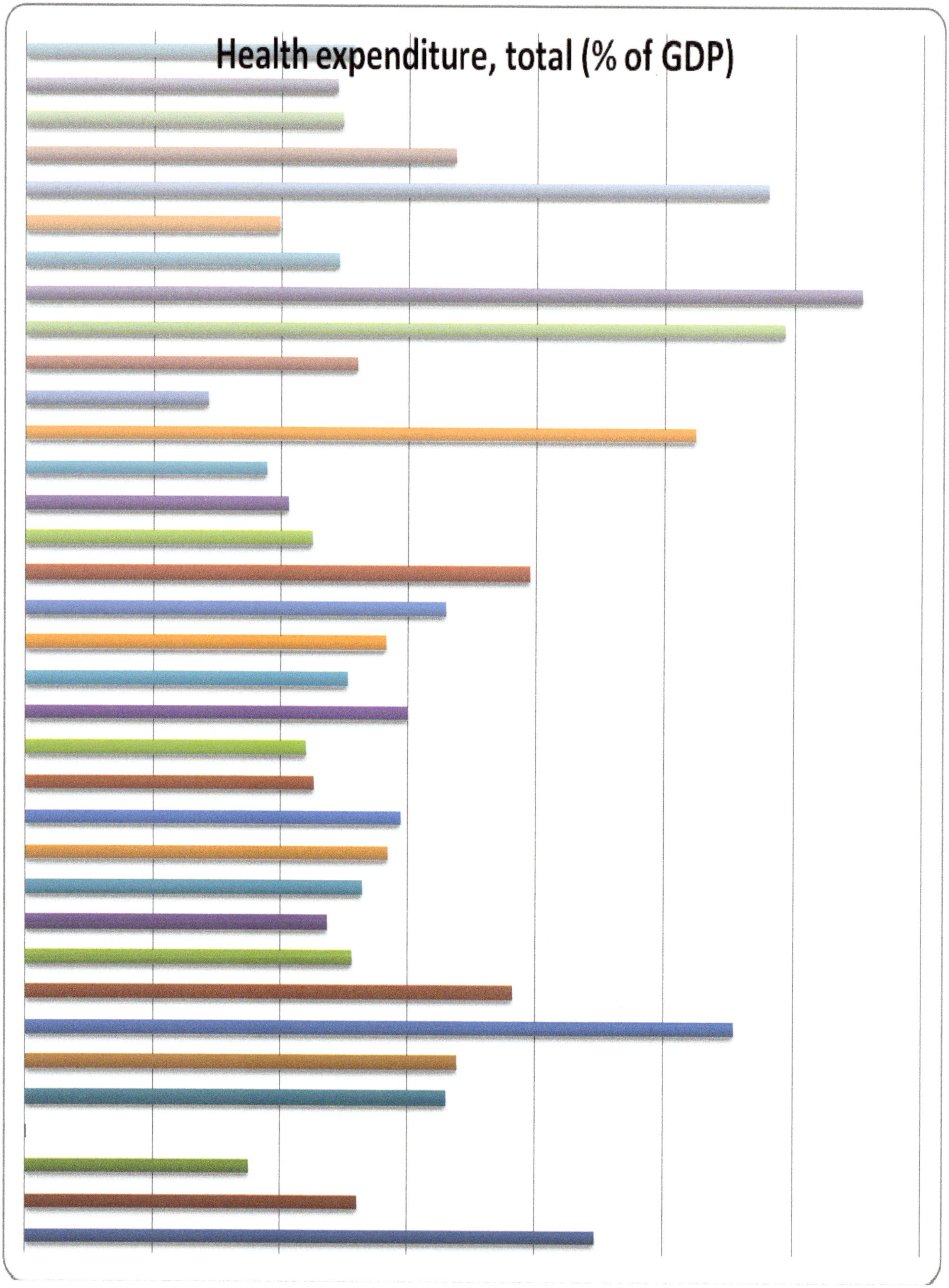

Figure A4. Health expenditure (% of GDP), 2009.

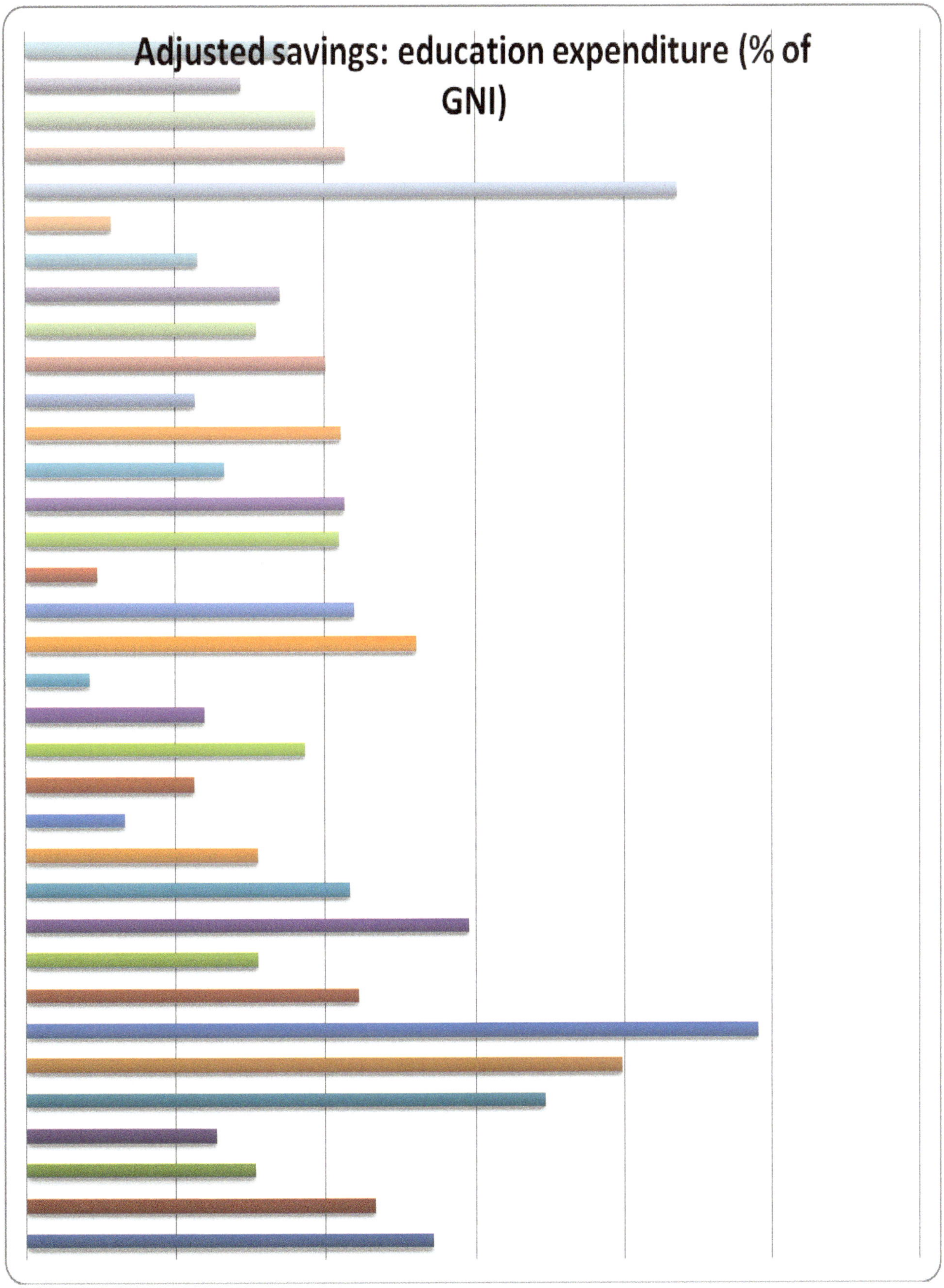

Figure A5. Education expenditure (% of GNI), 2009.

Willingness to pay for irrigation water and its determinants among rice farmers at Doho Rice Irrigation Scheme (DRIS) in Uganda

Namyenya Angella[1], Sserunkuuma Dick[2] and Bagamba Fred[3]

Department of Agribusiness and Natural Resource Economics, School of Agricultural Sciences, Makerere University, P. O. Box 7062, Kampala Uganda.

The government of Uganda is currently rehabilitating its irrigation schemes. The largest of these is Doho Rice Irrigation Scheme (DRIS), where farmers will after rehabilitation bear the costs of its maintenance through payment of user fees. This study analyzes farmer's willingness to pay (WTP) user fees and its determinants, using data gathered from 200 rice farmers at DRIS in 2012. The contingent valuation (CV) bidding game approach and Ordinary Least Squares (OLS) methods were used to elicit WTP and analyze the determinants of WTP, respectively. The study findings show that while farmers are willing to pay Ush 20,000 (USD 8)/acre/season on average, Ush 15,000 (USD 6) acre/season is actually needed to cover maintenance costs as per the 2013/2014 work plan for DRIS. The study recommends charging Ush 15,000/acre/season, however, which not only generates sufficient revenue to cover the maintenance costs, but also lies below the average WTP, which several farmers should be willing to pay without coercion. However, because not all farmers are willing to pay Ush 15,000, it is necessary to incentivize voluntary payment and strong enforcement of penalties against non-payment among those with low WTP. The OLS regression results suggest need for additional intervention that enhances private benefits to farmers, such as improved access to credit, markets and training in soil/water management and rice growing.

Key words: Irrigation water, user fees, willingness to pay, rice, Uganda.

INTRODUCTION

Uganda has of late witnessed erratic rainfall seasons (MWE, 2007) and increasing occurrence of drought conditions which has frustrated rain-fed agriculture and rendered irrigation investment critical for increased agricultural production in Uganda (MAAIF, 2012). Consequently, the government of Uganda (GOU) has prioritized rehabilitation of the existing irrigation schemes whose infrastructure broke down over a long period of misuse and poor maintenance (MWE, 2012a; MWE, 2009). Currently, the schemes under rehabilitation include

Doho, Mubuku and Agoro irrigation schemes in Butaleja, Kasese and Lamwo districts, respectively (MAAIF, 2012). Construction of the major irrigation schemes in Uganda started in the 1960's. Doho Rice Irrigation Scheme (DRIS) in particular was constructed between 1976 and 1985 by the GOU to promote rice production in eastern Uganda through the provision of irrigation water,improved rice seeds, farm tools, marketing and milling services. Following its completion, the GOU partitioned DRS into ten blocks covering a total area of 1,012 ha; and each block was partitioned into smaller plots (0.10 to 0.40 ha) that were leased to individual farmers on a first come, first served basis. GOU retained the role of maintaining the irrigation structures through the Ministry of Agriculture, Animal Industry and Fisheries (MAAIF) up until the early 1990s (MWE, 2012b). During this period, the irrigation and drainage channels were regularly de-silted by the GOU which enabled sustainable flow of irrigation water to the rice fields.

However, driven by budgetary constraints around 1994 and examples of successful collective action in irrigation water management in other parts of the world (Meinzen-Dick et al., 2000), the GOU withdrew its support[1] and devolved management of the irrigation scheme to Doho Rice Scheme Farmers' Association. The association adopted an earlier resolution made by farmers, district officials and local leaders, which required all farmers to pay an irrigation user fee of Ush 5,000 (USD 5.1)[2]/acre per season towards the cost of mechanized de-silting of the irrigation and drainage channels. In addition, farmers were required to contribute labor towards the collective cleaning and weeding of the channels. A committee composed of an elected Chairperson and 10 block-level executive members and counselors was set up to collect user fees; mobilize farmers for the collective cleaning of the channels; and monitor collective action on each block. A bylaw was enacted stating that those who did not comply with user-fee payment or participate in collective channel maintenance in any cropping season would have their plots of land withdrawn from them the following two seasons and rented out to willing farmers, and the money realized would go toward the cost of maintaining the scheme.

However, following the devolution of management from the GOU to the farmers' association, a collective action problem arose and hindered achievement of the desired outcome of adequate supply of irrigation water to rice plots through collective effort. This was attributed to shortage of funds to de-silt the channels caused by the failure of farmers to comply with the by-law requiring each farmer to pay the irrigation user-fee and participate in collective channel maintenance, coupled with poor enforcement of this bylaw[3]. Literature shows that only two thirds (66%) of the farmers at DRIS fully complied with the bylaw on user-fee payment in 2001 (Sserunkuuma et al., 2009) and only about 40% of the irrigation fees are collected on average. In addition, active participation by farmers in collective channel maintenance is limited and the penalty of barring obstinate farmers from growing rice on their plots for the following two seasons is rarely enforced (Nakano and Otsuka, 2011).

The main factors emasculating compliance were found to be poor awareness of the bylaw and the associated benefits; poor enforcement of the bylaw[4]; and the negative perception by farmers of the private benefits they derived from compliance. One fifth to one quarter of the farmers surveyed in 2001 perceived the private benefits derived from the scheme not to be worth the cost incurred; and the study found a significant negative relationship between compliance with the bylaw and the perception that benefits of compliance are lower than the costs.

This negative perception was caused by the extensive silting of the channels, which significantly reduced water conveyance to some rice fields. The lack of sufficient incentives (in form of water supply) for payment of user fees partly explains why one-third of the farmers did not comply with the user-fee bylaw (Sserunkuuma et al., 2009). Failure to adequately de-silt the channels had set up a cycle of failure in which an insufficient number of farmers paid user fees in a given season, which translated into inadequate de-silting of the channels, which in turn lowered the amount of irrigation water supplied to the rice plots, limiting rice yields and farmers' ability and willingness to pay (WTP) the user fees in the following season.

To break this cycle, rehabilitation of the entire irrigation system was recommended to increase water supply to farmers and improve rice yields on their fields as well as their willingness and ability to pay the user fees. In accord with this recommendation, GOU has since October, 2011 embarked on the rehabilitation and revitalization the irrigation scheme at Doho as well as those at Mubuku and Agoro. After completion of the rehabilitation process, the responsibility of maintaining the scheme at Doho will again revert to the farmers (MWE, 2012b); and it is envisaged that a user fee will be charged per acre per season to raise funds for operating and maintaining the irrigation scheme. Poor awareness and enforcement of the user-fee and collective action bylaw at Doho and the associated poor compliance

[1] except for payment of salaries of a few staff like irrigation engineers and agricultural extension agents

[2] The official average mid-rate for 1994 is Ush 979 to 1USD.

[3] To manage irrigation facilities effectively and allocate water resources efficiently, it is critically important to enforce the rules of water allocation and maintenance of irrigation channels and drainages (Ostrom, 1990).

[4] The administration at DRIS attributes poor bylaw enforcement to the physical characteristics of the irrigation system, with no means of blocking water supply to individual defaulters as a way of incentivizing them to pay the user fees. For similar reasons, it is not possible to levy user fees based on the volume of water received.

cited above can be attributed to the manner in which the bylaw was enacted, with limited involvement, sensitization and consultation of farmers, which led to low farmer buy-in.

Nkonya et al. (2001) observe that it is difficult to effectively enforce and educe compliance with bylaws that are not clearly understood or ratified by farmers. With the impending transfer of management responsibility to farmers after rehabilitation of DRIS and the accompanying need for farmers to contribute towards the maintenance costs, it is imperative to determine how much farmers are willing to contribute; and to use this information to guide the setting of appropriate user fees.

This study was undertaken with the objective of determining farmers' WTP user fees; and how this varies across rice farmers at DRIS. The study shows that while the provision of private incentives to farmers is important for improved management of devolved irrigation schemes, it is not a panacea but must go hand in hand with strong enforcement of penalties against free riders, and investment in provision of supporting services that enhance the private benefits to farmers. While this study shows that education attainment enhances farmers' willingness to contribute money towards maintenance of the irrigation scheme at DRIS, available literature shows that education emasculates willingness to participate when farmers are required to contribute labor.

WTP is a commonly used Contingent Valuation Method (CVM) approach for valuing goods and services that are not traded in the markets, including natural resources and resource services (Lipton et al., 1995) such as water for household use and irrigation; amenities such as national parks; and private non-market commodities such as reductions in the risk of death or days of illness avoided. It is the economic value of a good to an individual (Yang et al., 2007) or the maximum sum of money an individual is willing to part with in exchange for an increase in the quantity or quality of a natural resource good or service (Agudelo, 2001).

Akter (2007) estimated the value of irrigation water in a small scale irrigation project in the Homna sub-district in Bangladesh. He used CVM to elicit farmers' WTP for the irrigation water, using irrigation charges per decimal land area per cropping season as the payment vehicle. He found the mean WTP to be 1670 Taka (US$ 27.83) per kani (30 decimals of land) per cropping season; and a significant impact of age, education, family size, number of income sources and ownership of farmland on WTP. Basarir et al. (2009) used the Torbit and Heckman sample selection models to study the WTP of vegetable producers for high quality irrigation water in the Turhal and Suluova regions of Turkey and found a significant relationship between WTP and gender and water quality.

Whittington et al. (1990) used the bidding game format to estimate WTP for water services in Laurent, a rural community in Haiti and found it to be 5.7 gourdes (US$ 1.14) per month; and observed that developing countries are likely to produce better quality CVM surveys compared to industrialized countries.

Casey et al. (2005) studied WTP for improved water services in Manaus, Amazonas, Brazil, using both open-ended and bidding game approaches. They found the mean WTP to be R$11 (US$ 5.61) per month; and also observed that the respondents were willing to pay more for drinking water than the current charges. Other studies have found significant relationships between WTP for water or other natural resources and education of the household head, household size, farming experience, farm size, proximity to the resource, access to markets, extension services, credit and training, peoples' attitudes and perceptions on payment (Adepoju and Omonona, 2009; Mezgebo et al., 2013; Ogunniyi et al., 2011; Wendimu and Bekele, 2011; Addis, 2010; Moffat et al., 2012; Calkins et al., 2002; Rodriguez and Southgate, 2003; Kassahun, 2009; Latinopoulos, 2001; Ulimwengu and Sanyal, 2011; Farolfi et al., 2007; Alhassan, 2012; Illukpitiya and Gopalakrishnan, 2004; Calatrava and Sayadi, 2005). These studies guided the choice of variables used in the model explaining variation in WTP user fees across the sampled households at DRIS.

METHODOLOGY

Study area, sampling procedure and data

This study was conducted at Doho Rice Irrigation Scheme (DRIS) located 34° 02'E and 0° 50'N on the right bank of river Manafa in Mazimasa and Kachonga sub-counties of Butaleja district in Eastern Uganda (Figure 1).

DRIS occupies an area of 2,500 acres (1,012 ha), sub-divided into 10 blocks of unequal size, namely; 1A, 1B, 2A, 2B, 3, 4A, 4B, 5A, 5B and 6 (Figure 2). The 10 blocks are connected by three layers of channels, namely; main, sub and tertiary channels. The main channel provides irrigation water from River Manafwa to the scheme and branches out into the sub-channels, which provide irrigation water to each of the 10 blocks. Basically, each block has one sub-channel and consists of 5 to 15 smaller zones called strips, each surrounded by a tertiary channel that provides irrigation water to plots belonging to 20 to 30 farmers by a tertiary drainage channel. The tertiary drainage channel for one strip serves as the tertiary irrigation channel for the strip next to it. After flowing through paddy fields, water is collected in the main drainage channel through the tertiary and sub-drainage channels and drained back into River Manafwa (Nakano and Otsuka, 2011).

This study involved a survey of 200 households randomly drawn from among the rice farmers at DRIS in September 2012. A stratified random sampling procedure was employed, using the 10 blocks that make up DRIS as the strata to ensure that farmers on all blocks are represented in the study sample. Using the list of households for each block, a proportionate number of households was randomly drawn based on the household population of that block relative to the total number of households at DRIS[5]. Data was gathered from the sampled farmers using a structured questionnaire administered through in-person interviews with the household head. To elicit farmers' responses on WTP for irrigation

[5] 14 households were drawn from the smallest block and 33 households from the largest

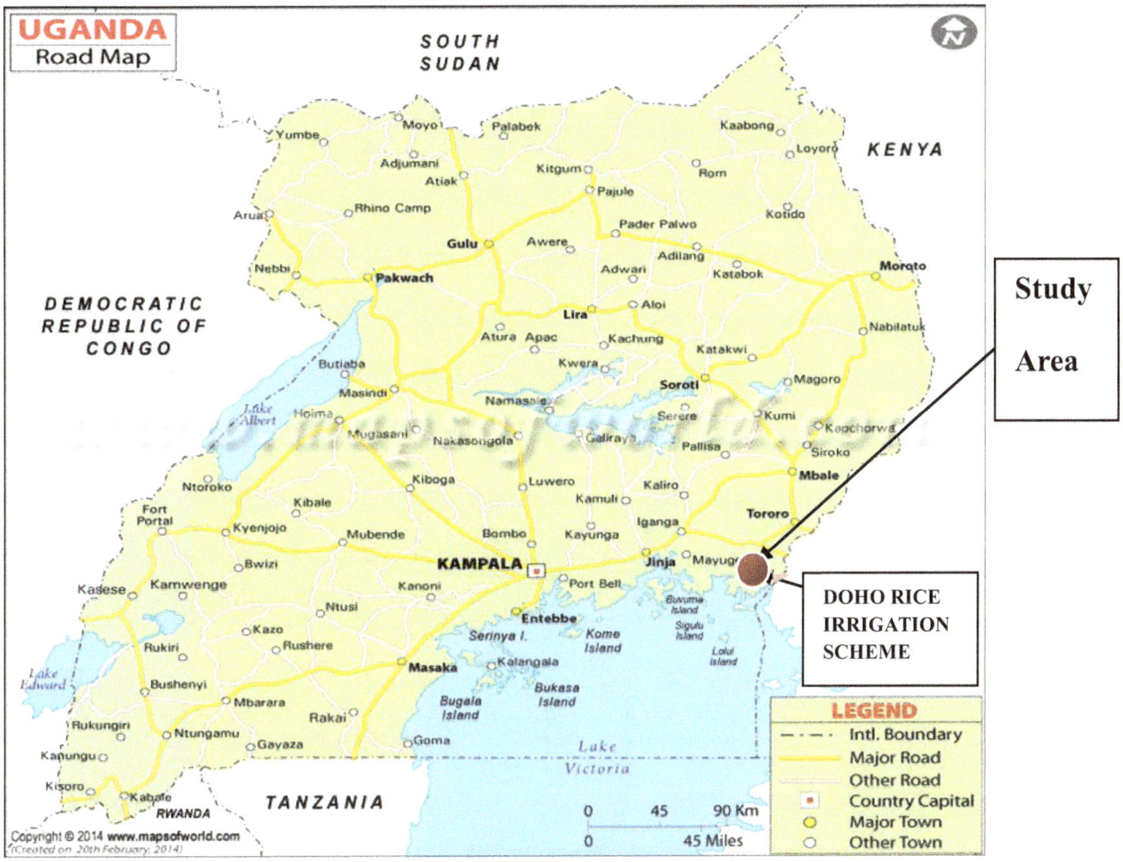

Figure 1. Map of Uganda showing the study area; Source: www. mapsofworld.com.

water, the study used a contingent valuation (CV) approach involving the iterative bidding game (Randall et al., 1974).

The game starts by querying individuals at some initial monetary value and keeps raising (or lowering) the value until the respondent declines (accepts) to pay. The final amount of money is interpreted as the respondent's WTP. Despite criticism of the bidding game approach as being prone to starting point bias, which makes the final WTP amount at the end of the bidding game systematically related to the initial bid value, Whittington et al. (1990) argue that the bidding game produces better quality WTP data in developing countries than in industrialized countries. This is because it is well understood and accepted by respondents in developing countries, who are used and prepared to negotiating over the price of just about any item they purchase on a regular market, unlike their cohorts in the industrialized countries.

In this study, the starting bid price was set at Ush 5000/acre per season, which the farmers at DRIS were required to pay according to the existing bylaw enacted in 1994. Since the commodity to be valued (irrigation water) was familiar to the respondents, the bidding game was not framed in a probabilistic sense, but rather the respondent was asked if they were willing to pay the starting bid price of Ush 5000/acre per season to experience adequate supply of irrigation water following the de-silting of irrigation and drainage channels. If the respondent answered "yes", the bid was increased until the respondent answered no. The highest yes response value was recorded as the maximum WTP. If the respondent answered "no", the bid was reduced until the respondent answered yes, and the highest yes response value was recorded as the maximum WTP. Farmers were not actually required to pay the bid amount they stated, which could have rendered this measure of WTP biased and subjective. This is a key limitation of this study. However, the fact that the study involved valuation of a familiar commodity for which they were already paying helped to purge some of the bias. Additional data was collected on household-level characteristics (age, gender, education, household size, years of irrigation farming), farm size, rice production and marketing in the first cropping season of 2012, access to training and extension related to rice production and irrigation water management, access to credit, and farmers' perceptions and attitudes about who should be responsible for paying the cost of maintaining the supply of irrigation water.

The theoretical model

The economic value of a non-market good to an individual can be measured by the magnitude of their WTP for the good. Formally, WTP is defined as the amount that must be taken away from an individual's income (to meet the costs of providing the non-market good) while keeping their utility constant as shown in the equation below:

$$V(y - WTP, p, q_1; Z) = V(y, p, q_0; Z)$$

Where *V* denotes the indirect utility function, *y* is the income of the

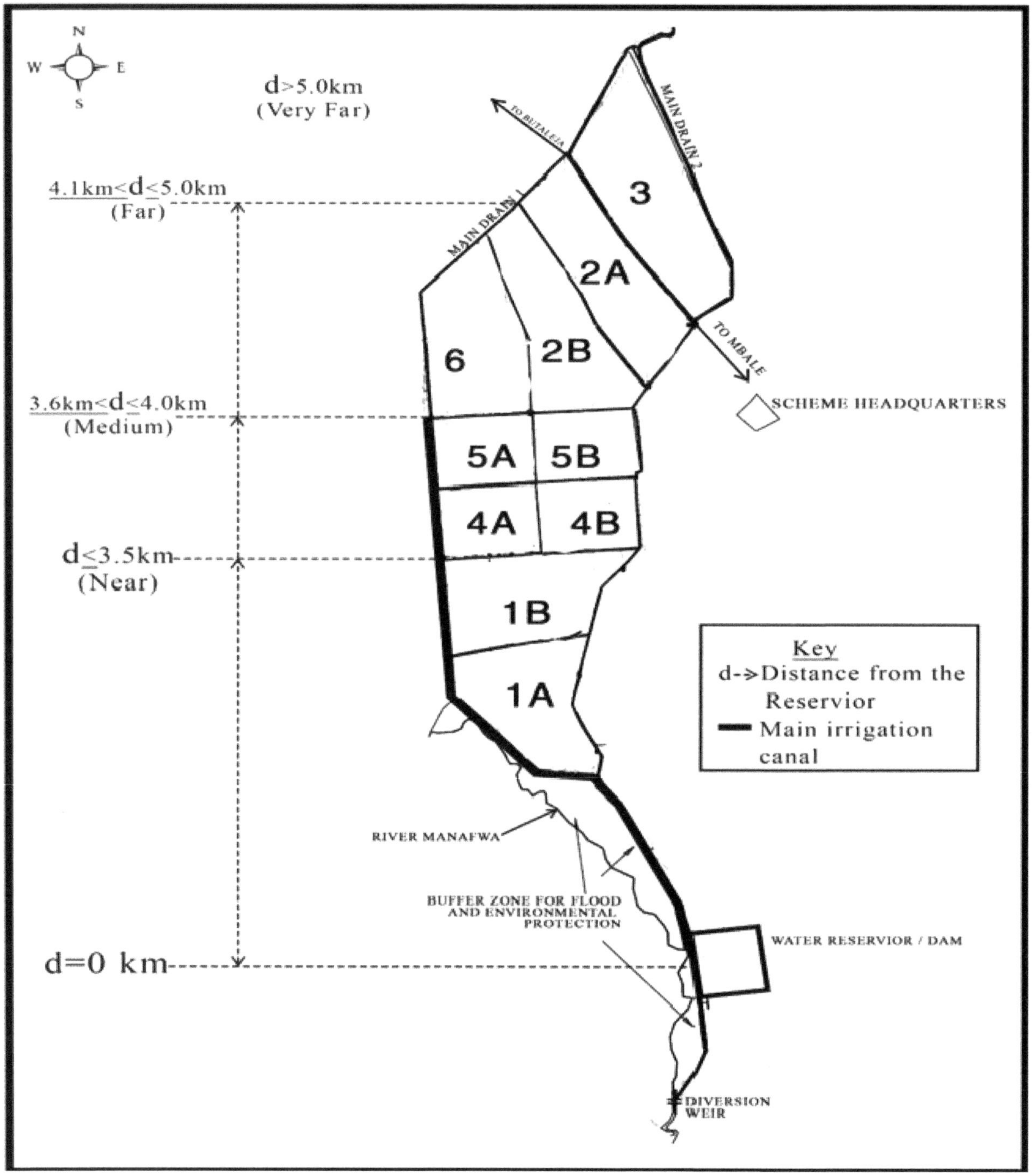

Figure 2. Map showing location of the 10 blocks with respect to water source/ reservoir.

individual, p is a vector of prices faced by the individual, q_0 and q_1 are the alternative levels of the non-market good under baseline and improved conditions, respectively (with $q_1 > q_0$ indicating an improvement from q_0 to q_1), and Z is a vector of individual characteristics affecting the trade-off that the individual is prepared to make between income and the non-market good. This equation implies that WTP depends on (i) the initial and final level of the good in question (q_0 and q_1); (ii) respondent income; (iii) prices faced by the respondent; and (iv)other respondent socio-economic characteristics.

The empirical model

Determination of the factors influencing farmers' WTP for irrigation water at DRIS was achieved through estimation of a double-log

Table 1. Budget estimates for maintenance and operation of DRIS as per the 2013/2014 work plan.

Expenditure item	Estimated cost per season (Ush.)
Excavator maintenance (servicing)	3,000,000
Maintenance of canal gates (main, medium and small)	2,000,000
Maintenance of farm roads	7,500,000
Maintenance of irrigation canals	10,000,000
Maintenance of drainage canals	5,000,000
Maintenance of broken pedestrian or foot bridges	2,000,000
Servicing of other machines	5,250,000
Meetings	2,570,000
Total	37,320,000

Source: DRIS annual work plan for 2013-2014 (1 USD equals Ush 2,500).

Ordinary Least Square (OLS) regression model. The general form of the model is specified as:

$$lnY_i = \beta_0 + \beta_i lnX_i + \mu_i$$

Where; *ln* is natural logarithm, Y_i is the dependent variable, X_i is a vector of explanatory variables, β_0 and β_i are the parameters to be estimated, and μ_i is the random error term. Thus, the estimated OLS model explaining variation in WTP across sampled rice farmers at DRIS is specified as:

$$\ln WTP = \beta_0 + \beta_1 \ln EDU + \beta_2 \ln EXP + \beta_3 \ln HHS + \beta_4 \ln FSIZE + \beta_5 \ln DMKT + \beta_6 TRA + \beta_7 EXT + \beta_8 CRE + \beta_9 OFFA + \beta_{10} ATT + \beta_{11} PSOURCE + u_i$$

Where; WTP= Farmers willingness to pay for irrigation water, EDU= Education of household head measured in years of schooling, EXP= Practical experience in rice farming under irrigation measured in years, HHS= Household size, FSIZE= Farm size or total area of land owned at DRIS measured in acres, DMKT= Distance in kms from the household to the nearest market where rice is sold. TRA=Participation in training related to soil and water conservation, rice growing or irrigation water management (1= Trained, 0= Otherwise), EXT=Access to extension services (1= Accessed extension, 0= Otherwise), CRE=Access to credit in the past two years (1= Accessed credit, 0= Otherwise), OFFA=Involvement in an off-farm activities by at least one household member (1=Involved in off-farm activities, 0= Otherwise), ATT=Attitude towards payment for irrigation water (1= Positive attitude, 0= Otherwise), PSOURCE = Proximity to irrigation water source [(1= Very far (>5 km); 2= Far (4.1-5 km); 3=Medium (3.6-4.0 km); 4=Near (≤ 3.5 km)], ln= Natural logarithm, β_i= regression parameters, u_i = random error term.

RESULTS AND DISCUSSION

Findings from the contingent valuation survey show that rice farmers at DRIS are willing to pay an average of Ush 20,000 (USD 8) per acre or USD 20 per hectare as user fees per season. Based on the total acreage of DRIS of 2,500 acres (1,012 ha), this implies that charging Ush 20,000 per acre per season would generate Ush 50 million (USD 20,000) in total revenue per season.

However, the budget estimates for maintenance and operation costs of DRIS as per the 2013/2014 work plan (Table 1) show that only three quarters (75%) of this (Ush 37.32 million) is needed to cover the costs. This implies that charging Ush 15,000/acre per season would generate enough revenue for maintaining and operating DRIS. Before explaining the variation in WTP user fees across sampled rice farmers, their socio-economic characteristics which are hypothesized to influence WTP are hereby briefly examined. The sampled farmers are grouped into two categories based on whether or not the money they are WTP as user fees is adequate to cover the maintenance and operation costs of DRIS (Ush 15,000/acre per season). Analysis of farmers' WTP shows that 58% of the sampled farmers (N=200) are WTP at least Ush 15,000/acre per season as user fees; and these constitute the first category of farmers defined as "adequate WTP" (ADWTP).

The second category is composed of the rest of the farmers (42% of the sample) whose WTP is inadequate to cover the costs; and this is referred to as the "inadequate WTP" (INADWTP) category. Table 2 shows that a typical rice-growing household at DRIS is male-headed (94% of the sample) and of medium size (7.3 people).

However, a significantly higher proportion of households in the ADWTP category are male-headed (98.3%) than their cohorts in the INADWTP category (88.1%); and the average household size in the ADWTP category (8.1 people) is higher than in the INADWTP category (6.3 people). The average age of the head of a typical rice-growing household at DRIS is estimated at 42.2 years and does not differ between the two categories; but the education of the household head (estimated at 7.3 years of schooling for the entire sample sample) is significantly higher among households

Table 2. Selected socioeconomic characteristics of sampled rice-growing households at DRIS.

Variables	Entire sample (N=200)	ADWTP (N=116)	INADWTP (N=84)	Chi-square/ t-value
Percentage of male headed households	94	98.3 ^{a}	88.1 ^{b}	8.953
Age(years) of household head	42.2	43.3 ^{a}(1.211)	40.4 ^{a}(1.680)	1.420
Number of years of formal education of household head	7.3	8.5 ^{a}(0.283)	5.6 ^{b}(0.377)	6.270
Household size	7.3	8.1 ^{a}(0.309)	6.3 ^{b}(0.377)	3.587
Total area of land owned at DRIS	2.7	3.3 ^{a}(0.242)	2.0 ^{b}(0.162)	4.195
Practical experience (years) in rice farming under irrigation	13	15.8 ^{a}(0.855)	9.3 ^{b}(0.853)	5.163
Percentage of households trained in soil/water conservation/rice growing	58	76.7 ^{a}	32.1 ^{b}	39.75
Percentage of households accessing extension services on rice	53.5	61.2 ^{a}	42.9 ^{b}	6.594
Percentage of households who had access to credit in the past two years	29.5	36.2 ^{a}	20.2 ^{b}	5.974
Percentage of households with at least one member involved in off-farm activities	29.5	27.6 ^{a}	32.1 ^{a}	0.486
Distance from the household to the nearest rice market	1.5	1.5 ^{a}(0.425)	1.6 ^{a}(0.389)	-1.446
Percentage of households with a positive attitude towards payment for irrigation water	85.5	90.5 ^{a}	78.6 ^{b}	5.608
Rice output (Kgs/household) per season	683.9	792.63 ^{a}	533.75 ^{b}	2.721
Rice yield (Kg/Acre) per season	740.9	748.37 ^{a}	730.58 ^{a}	0.330
Rice Net-Income (Ushs/household) per season	1,071,800	1,254,300 ^{a}	819,830 ^{b}	2.704
Rice Net-Income (Ushs/Acre) per season	1,184,400	1,225,600 ^{a}	1,127,500 ^{a}	0.843

Numbers in parentheses are standard errors. Different superscripts (a, b) reflect statistically significant differences in variable between the farmers categories; while same superscripts signify no difference. The 3rd column labeled ADWTP presents summary statistics for households with adequate WTP (WTP $\geq$ Ush 15,000/acre per season); while the 4th column labeled INADWTP presents summary statistics for households with inadequate WTP (WTP $<$ Ush 15,000/acre per season).

in the ADWTP category (3.5 years) than their cohorts in the INADWTP category (5.6 years).

Households in the ADWTP category are also endowed with bigger farmland at DRIS (3.3 acres) and longer practical experience in irrigated rice farming (15.8 years) than those in the INADWTP category (estimated at 2 acres and 9.3 years, respectively). Furthermore, significantly higher percentages of households in the ADWTP category had prior to the contingent valuation survey accessed training in rice growing and soil and water conservation (76.7%), rice-related extension services (61.2%) and credit (36.2%) than their cohorts in the INADWTP category, for which the corresponding percentages were 32.1, 42.9 and 20.2%, respectively.

During the contingent valuation survey, respondents were asked for their opinion about compelling farmers to pay for the maintenance of the irrigation scheme at DRIS. Majority (85.5%) had a positive attitude towards payment of user fees, but the proportion of such households was significantly higher in the ADWTP (90.5%) than in the INADWTP (78.6%) category. 30% of the sampled households had at least one household member engaged in off-farm activity as their main occupation but the proportion of such households did not differ significantly between the two categories. The distance from the home of the sampled households to the nearest market where they sell rice was estimated at 1.5 km but this also did not differ between the ADWTP (1.5 km) and INADWTP (1.6 km) categories

Recent studies conducted at DRIS show that the availability of irrigation water has a positive and significant impact on rice yield and income which in turn significantly affects household contribution of labor to the cleaning of irrigation and drainage channels (Nakano and Otsuka, 2011;

Table 3. Determinants of farmers' WTP user fees at DRIS.

Variable	Coefficient	Robust standard errors	T-Value
Constant	7.762	0.271	28.65
ln Education of household head	0.397***	0.059	6.76
ln Household size	0.083	0.092	0.90
ln Total area of land owned at DRIS	0.250***	0.059	4.22
ln Practical experience (years) in rice farming under irrigation	0.156**	0.071	2.22
ln Distance from the household to the nearest rice market	-0.444*	0.251	-1.77
Training in soil/water conservation/rice growing	0.361***	0.115	3.15
Access to extension services on rice	0.140	0.114	1.23
Access to Credit	0.214**	0.086	2.49
Involvement in Off-farm Income Activity	0.002	0.093	0.02
Positive Attitude towards payment of user fees	0.126	0.121	1.05
Proximity to Irrigation Water Source	0.076	0.059	1.27
Number of observations		200	
Adjusted R-Squared		0.51	
Prob>F		0.0000	
F(11, 188)		26.37	
Breusch-Pagan test for heteroskedasticity Prob> chi2		0.0487	
Mean VIF		1.62	

*,**,***,Significance at the 10, 5 and 1% levels, respectively.

Nakano et al., 2013). This study also gathered data on rice production and marketing by the sampled households in the first cropping season of 2012. The results show an average rice yield of 741kg/acre per season (1.83 mt/ha per season) for the entire study sample (N=200), which does not significantly differ between the ADWTP (748.4 kg/acre) and INADWTP (730.6 kg/acre) categories. This seems to suggest that the availability of irrigation water did not differ significantly across the sampled households in the first season of 2012, likely because as noted by Nakano and Otsuka (2011), rainfall and water supply at the scheme are abundant in the first cropping season and farmers occasionally suffer from flooding, although water supply is scarce in the second season. The average net income from an acre of rice (measured by gross margin) was estimated at Ushs 1,184,400 (USD 474) per season; and this as well did not significantly differ between the ADWTP (Ush 1,225,600/acre) and INADWTP (Ush 1,127,500/acre) categories.

However, the average net income per household in the first season of 2012 (estimated at Ush 1,071,800 for the entire sample) was significantly higher among households in the ADWTP category (Ush 1,254,300) than their cohorts in the INADWTP category (Ush 819,830). This is attributed to the fact that households in the ADWTP category are endowed with bigger land at DRIS (3.3 acres) than their cohorts in the INADWTP category (2 acres), which enables the former to earn higher rice income. Therefore, larger farmers earn more income from rice and as a result have better willingness and ability to pay user fees than smaller farmers. This corroborates the findings of Nakano and Otsuka (2011); and supports the notion that private benefits conferred by plot size are the prime motivation for participation in collective irrigation water management (White and Runge, 1994), because farmers with larger plots enjoy greater income benefits from abundant water supply, hence the greater incentive to pay user fees.

The descriptive results discussed above suggest a significant relationship between socio-economic characteristics of rice-growing households at DRIS and their WTP user fees. These relationships are examined further using regression analysis; and the results are summarized in Table 3. The adjusted coefficient of determination (Adjusted R-Squared) value of 0.51 means that 51% of the variation in farmers' WTP user fees is explained by the variables included in the regression model. The regression results show statistically significant relationships between farmers' WTP user fees and formal education of the household head; farm size; practical experience in irrigated rice farming; participation in training related rice growing in general or soil and water conservation and irrigation water management in particular; and access to credit and markets. In this study, access to credit and participation in training are treated as exogenous despite being choice variables to the respondents. This is because they are pre-determined in the sense that they both happened in the past, well before the respondents were asked to respond to the bids on user fees; in the same way that education is treated as pre-determined and exogenous yet the respondent made the choice of the level at which to end schooling.

The regression results imply that a one percent increment in education of the household head, land endowment and practical experience in irrigated rice farming increases farmers' WTP user fees by 0.4, 0.25 and 0.16% respectively; while a 1% reduction in distance to the rice market increases WTP by 0.44%. A switch from having "no access" to "access" to credit and training (in rice growing, soil and water conservation and irrigation management) is associated with a 43.5% ($e^{0.361}$) and 23.8% ($e^{0.214}$) increase, respectively in the geometric mean of farmers' WTP user fees.

These findings imply that more educated farmers have higher WTP irrigation fees, likely because higher education is associated with better understanding of the benefits of adequate supply of irrigation water in agricultural production. Education is also believed to increase farmers' ability to obtain, analyze and assimilate information that helps them to make prudent decisions related to the management of their farming enterprises.

Also, education is a good proxy for off-farm income because it enables agricultural households to pursuit alternative income opportunities outside agriculture (e.g.,salary or business), which increases their ability and WTP irrigation fees. These results are consistent with the findings Adepoju and Omonona (2009); Mezgebo et al. (2013); Ogunniyi et al. (2011); Wendimu and Bekele (2011) who found a positive relationship between formal education and WTP.

However, Nakano and Otsuka (2011) found education attainment to be negatively correlated with household contribution of labor towards the collective maintenance of DRIS, because of the higher opportunity cost of labor associated with non-farm income among more educated households. This renders more educated households less keen to contribute labor to collective action than their less educated cohorts. The implication of these findings is that the impact of education attainment on participation in irrigation water management plays out differently depending on whether the users are required to contribute labor or money to ensure the supply of irrigation water.

The positive relationship between farm size and WTP is likely because farmers with larger land endowment also cultivate larger rice plots at DRIS and earn higher income from rice when the supply of irrigation water is adequate. These findings are consistent with those of Mezgebo et al. (2013), Ulimwengu and Sanyal (2011); and Nakano and Otsuka (2011); and illustrate the prime importance of private benefits conferred by farm size in collective irrigation water management (White and Runge, 1994). Also, given that rice production is the most important income source for over 80% of the households at DRIS (Sserunkuuma et al., 2009), farm size is a good proxy for household income, which enhances ability and WTP user fees. The positive correlation between practical experience in rice farming under irrigation and WTP is likely because farmers with longer experience are more familiar with the benefits of adequate supply of irrigation water enjoyed when DRIS was properly maintained and have also observed the decline in rice output through the years as the scheme deteriorated. This enables them to better appreciate the importance of their contribution towards improved water supply, hence the higher WTP. This result is consistent with Addis (2010); Kassahun (2009); and Latinopoulos (2001).

Access to credit and training related to rice growing in general or soil and water conservation and irrigation water management in particular are associated with higher WTP of user fees, likely because training tends to increase farmers' awareness of the dangers of unabated siltation of the irrigation channels and appreciation of their role in abating these dangers through payment of user fees, as well as appreciation of the ensuing benefits.

This finding is consistent with Calatrava and Sayadi (2005) who found that farmers who attended agricultural training courses had significantly higher WTP for water in tropical fruit production in South Eastern Spain. The positive relationship between access to credit and WTP is likely because credit enables cash constrained farmers to earn more income from agribusiness and other micro-enterprises, (Zeller, 2000), which enhances their ability and WTP user fees. The need to earn money to pay back the acquired credit also likely contributed to the higher WTP bids among farmers who accessed credit, with the hope that this will lead to increased rice output and income to enable them to pay back the credit.

This result corroborates the findings of Addis (2010) and Illukpitiya and Gopalakrishnan (2004). Distance to the rice market and WTP are negatively correlated because farmers closer to the markets incur less transaction costs and earn more from their rice compared to those further away and are, thus, willing and able to pay more to ensure adequate supply of irrigation water.

This finding is consistent with Ulimwengu and Sanyal (2011) who found a negative impact of travel distance on the WTP for agricultural services. The rest of the explanatory variables (household size, access to extension services, involvement in off-farm activities, having a positive attitude towards payment of user fees and proximity to the irrigation water source) have positive but statistically insignificant relationships with WTP user fees. This is likely because of the way these variables were captured in the contingent valuation survey, which doesn't reflect their true impact on WTP.

CONCLUSIONS AND RECOMMENDATIONS

This study analyzed farmers' WTP user fees and the determinants of WTP at DRIS, which is currently undergoing rehabilitation by the government of Uganda (GOU) and is due to revert to farmers who are expected to manage and maintain it through payment of user fees and contribution of labor to collective action. The study was motivated by the need to determine how much farmers are willing to contribute towards the maintenance costs; and to use this information to guide the setting of appropriate user fees.

The study found that while farmers are on average WTP Ush 20,000/acre per season as user fees, Ush 15,000/acre per season is actually needed to cover maintenance and operation costs as per DRIS' work plan for 2013/2014. The higher WTP notwithstanding, the study recommends charging Ush 15,000/acre per season, which not only generates sufficient revenue to cover the costs, but also lies below the average WTP, implying that several farmers would willingly pay this amount without coercion.

However, because the WTP for some farmers (42% of the study sample) is below Ush 15,000 (for some it is as low as Ush 1,000), there is need for continued sensitization of farmers on the importance of their contribution towards the cost of supplying water to ensure farmer buy-in. This has to go hand in hand with provision of incentives for voluntary payment as well as strong enforcement of penalties against non-payment of the user fees among those with low WTP. Available literature shows a history of poor enforcement of penalties against uncooperative farmers at DRIS, leading to incessant shortfalls in the collection of user fees (only 40% of irrigation fees being collected on average) and inadequate contribution of labor towards collective maintenance of the irrigation and drainage channels.

Fujiie et al. (2005) recommends incentivizing community leaders who mobilize farmers for collective action and collect the user fees as a way of reversing the shortfalls. An incentive such as a monetary reward for chairpersons and counselors based on the number of farmers from their respective blocks who pay user fees promptly and participate in the cleaning of the channels could help to improve the performance of those leaders in collection of user fees and mobilization farmers for collective action. Another strategy for addressing the shortfalls is to switch from the current self-enforcement mechanisms to private third-party agencies to enforce compliance with the existing bylaw. The regression results show statistically significant relationships between farmers' WTP user fees and formal education of the household head; farm size; practical experience in irrigated rice farming; participation in training related rice growing in general or soil and water conservation and irrigation water management in particular; and access to credit and markets. These results imply that in addition to provision of incentives for voluntary payment and enforcement of penalties against non-payment of user fees, appropriate interventions related to these factors influencing farmers' WTP are necessary. For example, the positive relationship between WTP and participation in training in soil and water conservation, rice growing or irrigation water management implies that intensifying training in these areas is important to increase farmer awareness of the dangers of unabated siltation of the irrigation channels and appreciation of the importance of their contribution towards the cost of de-silting to ensure adequate supply of irrigation water. Interventions that promote farmers' access to affordable credit are also recommended, based on the positive and significant relationship between having acquired credit and WTP. These may include establishment of an agricultural bank or risk-sharing guarantee schemes to motivate financial institutions with a rural branch network to provide credit to farmers at more affordable rates. In light of the findings of a positive correlation between farmers' WTP user fees and market access, there is need to invest in increasing farmers' access to rice markets to reduce transaction costs and enable farmers to receive better returns to rice production; which will in turn enhance their ability and WTP user fees.

The positive relationship between farm size and WTP implies that development of a land rental or lease market at DRIS would enable interested farmers to expand the sizes of their rice farms; which will in turn increase their income and ability to pay as well as WTP user fees.

While this study shows that education attainment enhances farmers' willingness to contribute money towards maintenance of the irrigation scheme, available literature shows that education emasculates willingness to participate in irrigation water management when farmers are required to contribute labor.

Therefore, switching from the current practice of compelling all farmers to contribute both money and labor to the alternative involving giving them the option of contributing larger amounts of either labor or money (in lieu of the other) depending on the opportunity cost of their labor in other income generating activities may help to reduce the shortfalls in user fee collection and labor contribution towards collective maintenance of the irrigation system at DRIS.

Conflicts of Interest

The authors have not declared any conflict of interest.

REFERENCES

Addis T (2010). Economic valuation of irrigation water: The case of Erere Woldia irrigation project in Harari Regional State, Ethiopia. MSc. Thesis, Addis Ababa University, Ethiopia.

Adepoju AA, Omonona BT (2009). Determinants of willingness to pay for improved water supply in Osogbo Metropolis, Osun State, Nigeria. Department of Agricultural Economics, Ladoke Akintola University of Technology.

Agudelo JI (2001). The economic valuation of water: Principles and methods. Value of water research report series No.5. IHE Delft, Netherlands.

Akter S (2007). Farmers' willingness to pay for irrigation water under government managed small scale irrigation projects in Bangladesh. J. Bangladesh Stud. 9:21-31
Alhassan M (2012). Estimating farmers' willingness to pay for improved irrigation: An Ecomomic study of the Bontanga irrigation scheme in Northern Ghana. MSc. Thesis, Colorado State University.
Basarir A, Sayili M, Muhammed S (2009). Analyzing producers' willingness to pay for high quality irrigation water. Bulgarian J. Agric. Sci. 15(6):566-573.
Calatrava JL, Sayadi S (2005). Economic valuation of water and willingness to pay analysis with respect to tropical fruit production in south eastern Spain. Spanish J. Agric. Res. 3(1):25-33.
Calkins P, Larue B, Vézina M (2002). Willingness to pay for drinking water in the Sahara: The case of Douentza in Mali. *Cahiers d' Econ. Sociol. Rurales* 64:37-56.
Casey JF, Kahn JR, Rivas A (2005). Willingness to pay for improved water services in Manaus, Amazonas, Brazil. Ecological Economics.
Farolfi S, Mabugu R, Ntshingila S (2007). Domestic water uses and values in Swaziland: A Contingent Valuation Analysis. *Agrekon* 46(1):157-170.
Fujiie M, Hayami Y, Kikuchi M (2005). The conditions of collective action for local commons management. Agric. Econ. 33(2):179-189.
Illukpitiya P, Gopalakrishnan C (2004). Decision-making in soil conservation: Application of a behavioural model to potato farmers in Sri Lanka. Land Use Policy. 21(4):321-331.
Kassahun HT (2009). Payment for environmental service to enhance resource use efficiency and labour force participation in managing and maintaining irrigation infrastructure, the case of upper blue Nile basin. PhD. Thesis, Cornell University.
Latinopoulos PZM (2001). Willingness to pay for irrigation water: A case study in Chalkidiki, Greece. Aristotle University of Thessaloniki, Greece.
Lipton DW, Wellman K, Sheifer IC, Weiher RF (1995). Economic valuation of natural resources: A handbook for coastal resource policy makers. NOAA Coastal Ocean Program Decision Analysis series No.5. NOAA Coastal Ocean office, Silver Spring.
MAAIF (2012). Policy statement for the financial year 2012/2013. Ministry of Agriculture, Animal Industry and Fisheries (MAAIF), Entebbe, Uganda.
Meinzen-Dick RS, Raju KV, Gulati A (2000). What affects organization and collective action for managing resources? Evidence from canal irrigation systems in India. Environment and Production Technology Division Discussion Paper 61. Washington, D.C. International Food Policy Research Institute.
Mezgebo A, Tessema W, Asfaw Z (2013). Economic Values of Irrigation Water in Wondo Genet District, Ethiopia: An Application of Contingent Valuation method. J. Econ. Sustain. Develop. 4(2):23-36.
Moffat B, Motlaleng GR, Thukuza (2012). Households' willingness to pay for improved water quality and reliability of supply in Chobe Ward, Maun. Botswana J. Econ. 8(12):45-61.
MWE (2012a). An irrigation master plan for Uganda. Ministry of Water and Environment (MWE), Kampala, Uganda.
MWE (2012b). Proposed institutional arrangements and regulatory framework for sustainable management of irrigation schemes implemented under FIEFOC project. Final report. Ministry of Water and Environment (MWE), Kampala, Uganda.
MWE (2009). Strategic sector investment plan for the water and sanitation sector in Uganda. Ministry of Water and Environment (MWE), Kampala, Uganda.
MWE (2007). Climate change: National Adaptation Programmes of Action, Ministry of Water and Environment (MWE), Kampala, Uganda.
Nakano Y, Bamba I, Diagne A, Otsuka K, Kajisa K (2013). The Possibility of a Rice Green Revolution in Large-scale Irrigation Schemes in Sub-Saharan Africa, in Otsuka K and Larson D, eds. An African Green Revolution-Finding Ways to Boost Productivity on Small Farms.
Nakano Y, Otsuka K (2011). Determinants of household contribution to collective irrigation management: Case of the Doho Rice Scheme in Uganda. Environ. Develop. Econ. 16(5):527-551.
Nkonya E, Babigumira R, Walusimbi R (2001). Soil conservation by-laws: Perceptions and enforcement among communities in Uganda. Mimeo. Washington, D.C. Environment and Production Technology Division, International Food Policy Research Institute.
Ogunniyi LT, Sanusi WA, Ezekiel AA (2011). Determinants of rural household willingness to pay for safe water in Kwara State, Nigeria. Aquaculture, Aquarium, Conservation and Legislation. Int. J. Bioflux 4(5):660-669.
Ostrom E (1990). Governing the Commons - The Evolution of Institutions for Collective Action. Cambridge University Press.
Randall A, Ives BC, Eastman C (1974). Bidding games for valuation of aesthetic and environmental improvements. J. Environ. Econ. Manage. 1:132-149.
Rodriguez F, Southgate D (2003). Water resources management and willingness to pay: The case of Cotacachi, Ecuador. Sustainable Agriculture and Natural Resource Management Collaborative Research Support Program. Res. Brief. P.15.
Sserunkuuma D, Ochom N, Ainembabazi JH (2009). Collective Action in the Management of Canal Irrigation Systems: The Doho Rice Scheme in Uganda. In Institutional Economics Perspectives on African Agricultural Development, eds. Kirsten JF, Dorward AR, Poulton C, Vink N (Chapter 17, pp. 375-387). Washington, D.C. International Food Policy Research Institute.
Ulimwengu J, Sanyal P (2011). Joint estimation of farmers' stated willingness to pay for agricultural services. International Food Policy Research Institute Discussion Paper. P. 1070.
Wendimu S, Bekele W (2011). Determinants of individual willingness to pay for quality water supply: The case of Wonji Shoa sugar estate, Ethiopia. J. Ecol. Nat. Environ. 3(15):474-480.
White TA, Runge CF (1994). Common property and collective action: lessons from cooperative watershed management in Haiti. Econ. Develop. Cult. Change 43(1):1-41.
Whittington D, Briscoe J, Mu X, Barron W (1990). Estimating the willingness to pay for water services in developing countries: A case study of the use of contingent valuation surveys in southern Haiti. Econ. Develop. Cult. Change 38(2):293-311.
Yang JC, Pattanayak S, Choe KA (2007). Good practices for estimating reliable willingness-to-pay values in the water supply and sanitation sector No.23. Philippines: Asian Development Bank.
Zeller M (2000). Product innovation for the poor: The role of Microfinance, Policy Brief No. 3: Rural Financial Policies for Food Security of the Poor. Int. Food Policy Res. Institute. Washington. PMCid:PMC92312

Assessing the extent and determinants of adoption of improved cassava varieties in south-western Nigeria

B. A. Awotide[1], T. Abdoulaye[2], A. Alene[2] and V. M. Manyong[3]

[1]Department of Agricultural Economics, University of Ibadan, Nigeria.
[2]International Institute of Tropical Agriculture (IITA), Malawi.
[3]International Institute of Tropical Agriculture (IITA), Tanzania.

This paper investigates the determinants of adoption of improved cassava varieties in south- western Nigeria. The data come from a farm household survey of 841 households selected using a three-stage stratified random sampling procedure. The data collection was conducted in 2011 by the International Institute of Tropical Agriculture (IITA), Ibadan, Nigeria. Empirical estimates of a Double-Hurdle model revealed that adoption increases with the age of the household head and is influenced by the gender of the household head, hired labour, cultivated land, and access to credit. The results further showed that the intensity of adoption is influenced by hired labour and farm size; access to information about the improved cassava varieties is determined by the age, gender, and level of education of the household head, and by off-farm income.

Key words: Adoption, improved cassava, double-hurdle, Nigeria.

INTRODUCTION

Cassava plays key roles in African development as a famine-reserve crop, rural food staple, cash crop for urban consumption, and raw material for livestock and industry (Nweke et al., 2002). Cassava is a staple food for over 200 million people in sub-Saharan Africa and an important food and cash crop in several tropical African countries, especially Nigeria where it plays a principal role in the food economy (Agwu and Anyaeche, 2007). Approximately hundred million Nigerians eat cassava-based foods at least once a day and the per capita consumption exceeds 200 kg/year in the north central, southwest, southeast, and south-south parts of the country (Africa Agriculture News, 2013). Cassava is the most important source of carbohydrates for human consumption in the tropics after maize. The high level of carbohydrates is an advantage in Africa because it makes cassava the cheapest source of food calories (Nweke et al., 2002).In most countries, cassava is becoming an important cash crop that has a high potential for use as an industrial raw material in the manufacture of starch, flour, and many other important products.

For decades, Nigerian farmers relied solely on the traditional varieties and this reliance generated concern.The limitations of these varieties included a low yield,long maturity period, and high susceptibility to diseases such as Cassava Mosaic Disease(CMD) and brown streak Disease (CBSD). Achieving a substantial increase in cassava productivity which has been one of the major goals of successive Nigerian Governments over several decade require ability to overcome the above limitations.

To accomplish this objective, the government initiated modern research into cassava in 1954. This research led to the development of some improved cassava varieties. Subsequently, the severe attack of the Cassava Bacterial Blight (CBB) the years that followed necessitated a collaboration between the International Institute of Tropical Agriculture (IITA) and its partners that let to the development of resistant improved cassava varities to the CBB (Akoroda et al.,1985). IITA releases the first two IITA clones in 1976, namely TMS 30211 and TMS 30395, which were rapidly followed by TMS 30572, TMS 30001, TMS 300017, TMS 30110, TMS 30337, TMS 30555, TMS 4(2)1425 and others (IITA 1984). Since then, efforts to improve cassava have continually increased such that IITA working with national partners has developed more than forty ICVs in the last forty five years (Eke-Okoro and Njoku, 2012).

In a recent study conducted by Abdoulaye et al (2014) ICVs adopters were observed to have a higher yield of about 16 tons/ha compared with 10 tons/ha for non-adopters. The implication is that the desired increase in productivity due to the ICVs and the subsequent impact on poverty reduction will not be achieved unless the ICVs are widely adopted by the Nigerian farmers. However, evidence from the literature shows that the adoption of ICVs is not yet universal in Nigeria. This implies that some farmers cultivate the improved cassava varieties (adopters) and some do not (non-adopters). Also, the level of adoption among the adopters also varies. This implies that there are some farmers among the adopters that utilized all their available farmland for ICVs whereas others only plant ICVs on a share of their farmland. This raises two pertinent questions: first, why are some cassava farmers adopting ICVs and others are not. Second, why does the intensity or the size of the area of farmland devoted to the cultivation of ICVs vary among the adopters?, third, what is the role of access to information on ICVs adoption in Southwestern Nigeria.

Studies that have attempted to provide answers to the above questions which are needed for future agricultural planning is still very scanty and in particular no recent information on ICVs adoption in the southwest is available. The most recent studies on ICVs adoption was conducted in the 1980s (Ay et al., 1983; Ikpi et al., 1986; Keyser, 1984). Thus, leaving a gap in the literature that this study intends to fill. Therefore, the broad objective of this study is to examine the determinants and intensity of adoption of ICVs in Southwestern Nigeria. Specifically, the study assess the influence of the farmers' socio-economic/demographic characteristics on the decision to adopt the ICVs and also examine the effect of access to information on the adoption of ICVs in Southwestern Nigeria .

Most importantly, in contrast to most adoption studies in Nigeria that adopted either logit, probit or tobit models (Igodan et al., 1988; Saka et al., 2005; Eze et al., 2006; Saka and Lawal, 2009; Junge et al., 2009; Okoedo-Okojie and Onomolease, 2009; Odoemenen and Obinne, 2010; Kudi et al., 2011), we employ a Double-Hurdle model to deal with the two-stage decision process involved in improved agricultural technology adoption and assess the effect of access to information using the Heckman Probit selection model. In order to achieve the stated objectives of this study, we therefore tested the following hypotheses: The extent and determinants of ICVs do not depend on the farmers' socio-economic characteristics and access to information has no significant effect on the adoption of improved cassava varieties in South western Nigeria.

CONCEPTUAL, ANALYTICAL FRAMEWORK AND ESTIMATION TECHNIQUES

Modeling the intensity and determinants of improved cassava varieties adoption

Rogers and Shoemaker (1971) defined adoption as the decision to apply an innovation and to continue using it. According to Wale and Yallew (2007), farmers' decisions about adoption are either discrete (whether or not to take up the technology) or continuous (the intensity of use of the technology). The theory of utility maximization is generally used to explain farmers' responses to new technology (Adesina and Seidi, 1995; Adesina and Baldu-forson, 1995).

According to this theory, a farmer will adopt a given technology such as ICVs if the utility obtained from it exceeds that of the traditional varieties. For instance, if U_{i0} is the utility derived from the use of the traditional cassava variety while U_{i1} is the expected utility from the adoption of ICVs; although not observed directly, the utility that farmer i will derive from adopting a given measure of the ICVs (j) can be expressed as:

$$U_{ij} = X_i\beta_j + \tau_{ij} \qquad j = 1, 0; \quad i = 1, \ldots\ldots, n \tag{1}$$

Where X_i is a farm–specific function, β_j is a parameter to be estimated, τ_{ij} is a disturbance term with mean zero and constant variance. In addition, adoption of any agricultural technology may also be measured by both the timing and extent of utilization by individuals (Sunding and Zilberman, 2001). In this study, a farmer is defined as an adopter if he or she is found to be growing at least one ICV. This implies that an adopter could still be growing the traditional cassava varieties alongside the improved varieties. We defined the adoption variable as a dummy with 1 indicating adoption and 0 otherwise. A farmer would adopt an ICV, that is., j=1 if $U_{i1} > U_{i0}$.

The intensity of adoption is measured by the proportion of farmland devoted to the production of ICVs.The literature suggests several theoretical or conceptual models on farmers' decisions to adopt new technology (Feder and Slade, 1984; Abadi and Panned, 1999; Negatu and Parikh, 1999; Isham, 2002). Many of the numerous studies that assessed the determinants of adoption of improved agricultural technology have utilized the Logit, Probit, or Linear probability models.

The objective of this study goes beyond the determinants of adoption to analyze the intensity of ICV adoption in Nigeria. The Tobit model has been employed by many authors to assess the intensity of agricultural technology adoption (Adesina and Baldu-Forson, 1995; Roos et al., 2000; Alene et al., 2000; Abadi-Ghadim et al., 2005; Jensen et al., 2007). One of the major drawbacks of the Tobit model is the fact that the decisions on whether or not to adopt ICVs and how much to adopt are assumed to be made jointly and hence the factors affecting the two decisions are assumed to be the same. However, it is believed that the adoption process is in two stages; the first stage involves the decision to adopt and the second stage involves the decision on the proportion of the area to be devoted to ICVs. Hence, the explanatory variables in the two stages may differ. Against this backdrop, the use of a single model may be erroneous, since the factors influencing the two-stage decisions will be difficult to analyze using just one model. In this study, we believe that it is likely the decisions on adoption and intensity of adoption of ICVs in southwestern Nigeria may not be made jointly, and the factors affecting each decision may not be the same. Thus we used the double-hurdle model proposed by Cragg (1971) in which the event of a farmer being a potential adopter and the intensity of adoption are treated separately. Furthermore, empirical results by both Moffatt (2003) and Martínez-Espiñeira (2006) reveal that the double-hurdle model gives results superior to those obtained from Tobit and P-Tobit models.

According to Cameron and Trivedi (2009), a double-hurdle model has the interpretation that it reflects a two-stage decision-making process, each part being a model of one decision. The two parts are functionally independent. The double-hurdle model is a parametric generalization of the Tobit model, in which two separate stochastic processes determine the decision to adopt and the level of adoption of the technology (Green, 2000; Martínez-Espiñeira, 2006). In addition, the double–hurdle model allows for the possibility of zero observations in both outcomes (Wooldridge, 2001; Cameron and Trivedi, 2005). The model has an adoption (D) equation presented below:

$$D_i = 1 \text{ if } D_{i^*} > 0 \text{, and } D_i = 0 \text{ otherwise} \tag{2}$$
$$D_{i^*} = \lambda Z_i + \pi_i$$

Where *D** is a latent variable that takes the value 1 if the farmer adopts ICVs and 0 otherwise, Z is a vector of household characteristics and λ is a vector of parameters. The level of adoption (Y) has an equation of the following:

$$Y_i = Y_{i^*} = \gamma' X_i + \tau_i \text{ if } Y_{i^*} > 0 \text{ and } D_{i^*} > 0 \text{, 05 otherwise} \tag{3}$$

Where: Y_i is the observed answer to the proportion of area planted with improved cassava varieties. X is a vector of the individual's characteristics and γ is a vector of parameters. The error terms, π_i and τ_i are distributed as follows:

$$\left.\begin{aligned} \pi_i &\sim N(0,1) \\ \tau_i &\sim N(0,1) \end{aligned}\right\} \tag{4}$$

The log-likelihood function for the double-hurdle model is:

$$LogL = \sum_0 \ln\left[1 - \Phi(\lambda_i Z_i^1)\left(\frac{\gamma X_i^{'}}{\theta}\right)\right] + \sum \ln\left[\Phi(\lambda Z_i^{'})\frac{1}{\theta}\phi\left(\frac{Y_i - \gamma X_i^{'}}{\theta}\right)\right] \tag{5}$$

The independent double hurdle model assumes that the two error terms from the two hurdles are normally distributed and uncorrelated. This suggests that the two stage ICVs adoption decision and the intensity of use/adoption are done independently by the farmers. Under the assumption of independency between the error terms π_i and τ_i the model as originally proposed by Cragg (1997) is equivalent to a combination of a truncated regression model and a univariate probit model.

The double-hurdle and the closely related two-part model have been used extensively to assess agricultural technologies adoption by Cooper and Keim (1996), Uri (1998), Teklewold et al. (2006), Shiferaw et al. (2008), Langyintuo and Mungoma (2008), Legese et al. (2009), Kassie et al. (2009), Gebregziabher and Holden (2011), Smith et al. (2011) and Alamerie et al. (2013) among many others. Empirically, the double model contains logit and Tobit model estimnated as a single equation in STATA and the estimated equations are presented implicitly below:

D = f(age, age2, educ, hlab, ownland, extconta, gender, moccup, fasize, error term)
Y = f(age, age2, educ, hlab, ownland, extconta, gender, moccup, fasize, error term)

Where D and Y are the adoption status and the proportion of area devoted to ICVs production, respectively.

Access to information and adoption

Awareness or exposure to improvred agricultural technologies through information either from the extension agents, mass media or mobile phone has been identified as one of the vital eterminants of technology adoption (Diagne and Demont, 2007; Dontsop-Nguezet et al., 2011). In addition, information source have been reported as important stimulus to individuals in the adoption process (Rogers, 1995). Certainly, the adoption of ICVs is not likely to be possible if the farmers are not aware of or exposed to ICVs through access to information. Hence, the adoption of ICVs can be described as a two-stage process (Cragg, 1971).

The first involves obtaining all the available relevant information about ICVs and the second involves taking a critical decision whether to adopt ICVs or not. This leads to a sample selectivity problem, since only those who obtain information about the varieties are in a better position to adopt it, whereas it is mandatory to make an inference about ICV adoption among the rural population as a whole. Thus, we adopt Heckman Sample Selectivity model (Maddison, 2006). The Probit model for sample selection assumes that an underlying relationship exists between the independent (socio-economic and demographic characteristics of the farmers) and the dependent variables (access to information) (Deressa et al., 2008), the latent equation being given by:

$$y_j^* = x_j \alpha + \tau_{1j} \tag{6}$$

Such that we observed only the binary outcome given by the probit

Table 1. Description of the variables included in the analysis.

Variables	Description	Expected sign
Dependent variable		
Adoption status(binary)	1 if farmer adopt at least one ICV, 0 otherwise	
Proparea	The share of ICVs area to total farmland (%)	
Independent variables		
Age	Age of household head in years	-/+
Age2	Square of age of household head	-/+
Gender	Dummy (1=male)	-/+
Offinc	Off-farm income. Dummy (1=yes)	+
Educ	Years of formal education of household head (years)	+
Extconta	Contact with extension agents. Dummy (1=yes)	+
amtcredit	Total amount of credit obtained in Naira	+
Fasize	Total farmland cultivated (ha)	+
Reland	Rented land. Dummy (1=yes)	+
Patecheva	Participation in technology evaluation. Dummy (1=yes)	+
Moccup	Main occuipation. Dummy (1= farming, 0 otherwise)	+
Ownland	Ownership of farmland . Dummy (1=yes)	+
Hlab	Hired labour. Dummy (1 if cost of hired labor is greater than mean of the group and 0 otherwise)	+/-

model as:

$$y_j^{probit} = (y_j^* > 0) \quad (7)$$

The dependent variable is observed only if j is observed in the selection equation

$$y_j^{select} = (w_j \varphi + \tau_{2j} > 0) \quad (8)$$

$$\tau_1 \sim N(0,1)$$
$$\tau_2 \sim N(0,1)$$
$$Corr(\tau_1, \tau_2) = \rho$$

The selection equation is (6), while (8) is the outcome equation. Where: x is a k-vector regressor, w is a m vector of repressors. τ_1 and τ_2 are the error terms. In cases where $\rho \neq 0$, standard probit techniques applied to Equation (6) will generate biased estimates. However, the Heckman Probit (heckprob) provides consistent, asymptotically efficient estimates for all parameters in such models (Statacorp, 2003).

Therefore, in thus study, the Heckman Probit selection model is employed to analyze the effect of access to information on the adoption of ICVs in southwestern Nigeria. The first part of model is the probit model, estimating the determinants of access to information. In the second part, we estimate the determinants of adoption of ICVs with access to information as one of the explanatory variables. The description and definition of the variables used in the models are presented in Table 1.

Data and descriptive statistics

The study area is Southwestern Nigeria. The data for this study originated from a survey conducted by IITA. Five (Ekiti, Osun, Ogun, Ondo and Oyo) out of the six States that comprise the Southwestern geopolitical zone were selected for the study. A three-stage stratified random sampling procedure was employed, whereby States were used as strata to improve sampling efficiency and account for possible major differences in the adoption of ICVs across States. Rural Local Government Areas (LGAs) were used as primary sampling units (PSUs). Enumeration areas (Eas), defined as a cluster of housing units, were used as secondary sampling units (SSUs) and households were the final sampling units.

LGAs were selected from each State based on probability proportional to size, where size is measured in terms of the number of Eas. The Eas that formed the sampling frame were obtained from the Nigerian Bureau of Statistics which uses the 2003/2004 master sampling frame of the National Integrated Survey of Households. The advantage of using Eas as sampling units is that each EA is approximately the same size. This ensured that all farmers had an equal probability of being selected. Within each LGA, four Eas were selected at random from a sampling frame classified as rural or semi-urban, giving a total of 80 Eas.

Finally, a list of households was developed for selected Eas and a sample of at least 10 farming households was selected randomly in each of the sampled Eas, giving a total of at least 841 households (Table 2). The survey was carried out over three months from August to October 2011. Community and household questionnaires were administered by trained enumerators with a senior agricultural economist in the field and the general supervision of IITA's economist. Data collection involved Focus Group Discussion (FGD), farmers' interviews, field observation of varieties, and plot area measurements.

Socio-economic characteristics of the respondents

The percentage distribution of adopters and non-adopters of ICVs by State (Table 3) shows that Ogun has the highest number of adopters (94%) followed by Osun (87%), Ondo (86%), Ekiti (81%)

Table 2. Distribution of the sampling households across the selected states.

Characteristics	State					
	Ekiti	Ogun	Ondo	Osun	Oyo	All
All enumeration areas (EAs)	11561	12754	19213	25910	31137	100575
All local government areas (LGAs)	16	20	18	30	33	117
Sample LGAs	2	3	4	5	6	20
Sample EAs or communities	8	12	16	20	24	80
Sample households	88	125	175	209	244	841

Source: IITA/DIIVA Adoption and Impact Survey (2011).

Table 3. Percentage distribution of adopters and non-adopters of ICVs by State.

State	Adopters (N=670)	Non-adopters (N=155)
	Percentage	Percentage
Ogun	94.35	5.65
Osun	87.44	12.56
Ondo	86.39	13.61
Ekiti	81.40	18.60
Oyo	65.27	34.73

Source: IITA/DIIVA Adoption and Impact Survey (2011).

and Oyo (65%). Table 4 shows the main socio-economic characteristics of the farmers by adoption status. As revealed by the t-test there is no significant difference between the adopters and non- adopters in age, total area of farmland cultivated, share of cassava in the farmland cultivated, and amount of credit obtained for cassava planting material and fertilizer. More importantly, there is no significant difference in the cost of planting material.

This shows that the average cost is the same for ICVs and traditional varieties and has a negative implication for the seed sector. However, the adopters and non-adopters of ICVs are statistically significantly different in the number of years of education, number of mobile phones, and cost of hired labor, herbicide, and fertilizer.

RESULTS AND DISCUSSION

Determinants and intensity of improved cassava varieties adoption

The result of the double-hurdle model is presented in Table 5. A positive significant coefficient in the first Hurdle-Logit model signifies that the corresponding regressor increases the probability of a positive observation in the adoption process. Similarly, in the second part, a positive coefficient means that, conditional on a positive count, the corresponding variable increases the value of the count (Cameron and Trivedi, 2009). The results of the first part of the model show that the log-likelihood of -77.83 and the LR chi^2 (10) 481.25 (significant at 1% level), imply that the model is fitted and the explanatory variables used in the model are collectively able to explain the extent and determinants of ICV adoption in southwestern Nigeria.

The results of the first part of the double-hurdle model are basically the Logit model of determinants of ICVs adoption and show that the coefficient of the gender of the household head is negative and statistically significant. This implies that adoption of ICVs is higher among female-headed than male-headed households. Labor is one of the main inputs in cassava production. Improved practices are labor intensive, hence availability (both hired and farm labor) is necessary for improved technology adoption.The coefficient of hired labor is positive and statistically significant. This shows that those farmers that have access to labor are more likely than not to adopt ICVs.

This finding is consistent with that of Hailu (2008) for the adoption of improved technologies for teff and wheat production in Ethiopia, Land is an important variable in agricultural production. The size of the land available for farming is usually a major factor in explaining technology adoption (Just and Zilberman, 1983). If farmers are land constrained, the probability of adoption would be very low. Owned farmland is more important than rented farmland in crop production. Hence, farmers producing crops on their own farmland are expected to have a higher probability of adopting ICVs. The result shows that

Table 4. Socio-economic/demographic characteristics of adopters and non-adopters.

Characteristics	Adopter (A) (N=670)	Non-adopter (NA) (N=155)	Mean difference (A-NA)	T-test statistics	P- value
Age (years)	50.00	49.00	0.82	0.56	0.57
Years of formal education	6.00	5.00	0.96	2.08**	0.04
Farmland (ha)					
Total farmland cultivated	3.16	2.85	0.31	0.72	0.47
Own land cultivated	2.43	2.34	0.08	0.19	0.85
Rented land cultivated	1.85	1.76	0.09	0.19	0.85
Sharecropped land cultivated	0.16	0.21	0.05	0.50	0.62
Cassava share of farm land cultivated (%)					
Cassava share of total farm land cultivated	64.55	63.61	0.94	0.38	0.70
Cassava share of owned land cultivated	46.45	47.98	1.53	0.39	0.69
Cassava share of rented land cultivated	38.29	39.59	1.29	0.28	0.78
Cassava share of sharecropped farm land cultivated	8.10	12.56	4.47	1.18	0.24
Household asset endowment					
Number of radios	2.00	2.00	0.04	0.39	0.09
Number of television sets	1.00	1.00	0.09	1.05	0.01
Number of mobile phones	2.00	1.00	0.23	3.59***	0.00
Access to credit					
Amount of credit borrowed for planting material	2113.32	2018.07	95.25	0.11	0.55
Amount of credit borrowed for fertilizer	1316.92	387.09	929.82	1.45	0.35
Estimated cost of cassava production (₦)					
Hired labor for land preparation	17918.06	22425.47	4507.41	1.66*	0.06
Hired labor for planting	6645.68	8202.02	1556.33	1.01	0.85
Hired labor for weeding	13346.46	15951.26	2604.79	1.48	0.82
Hired labor for harvesting	6526.78	6806.92	280.14	0.22	0.98
Cost of cassava planting material	6355.51	2473.08	3882.45	1.43	0.35
Cost of herbicide/pesticide	3328.91	2557.69	771.22	2.87***	0.002

*,**,*** implies significant at 10%, 5% and 1%, respectively, Source: IITA/DIIVA Adoption and Impact Survey (2011).

the coefficient of owned farmland is positive and statistically significant. This reveals that farmer-owners are more likely to adopt than those that practice farming on rented farmland.

The age of the household head, regarded as a primary variable in technology adoption, is negative and statistically significant thus indicating that younger farmers are more likely than older farmers to adopt ICVs (Rämö et al., 2009; Jensen et al., 2007). This is in line with the general literature on technology adoption and has been explained by the fact that older farmers are usually more reluctant to change. In addition, the young farmers are less risk-averse (Rogers, 1983; Alavalapati et al., 1995). This finding is similar to those of Jensen et al. (2007) and Rämö et al. (2009), but in contrast to the findings of Teklewold et al. (2006) and Hailu (2008).

However, the positive coefficient of age-square reveals that age shows a quadratic pattern in the adoption of ICVs. This implies that the adoption of ICVs among the younger farmers would increase to a certain level and then start to decrease as age increases in line with the life cycle hypothesis. Access to credit and specifically the amount of credit obtained are very important in agricultural production as credit allows farmers to invest in new technology or acquire other productivity enhancing inputs such as agro-chemicals and fertilizer. Thus, the amount of credit obtained is expected to increase the probability of adoption. The result shows that the amount of credit obtained by the farmers significantly increases the adoption of ICVs in the study area.

Table 5. Determinants and intensity of adoption of improved cassava varieties: Double-hurdle.

Variables	First-Hurdle		Second-Hurdle	
	Coefficient	Std. Error	Coefficient	Std. Error
Amount of credit	0.000*	0.000	9.35E-07	1.36E-06
Age2	0.002**	0.001	-0.0001	8.44E-05
Age	-0.257**	0.116	0.005	0.009
Education	0.023	0.055	-0.005	0.006
Hired labor	1.009*	0.521	0.136*	0.055
Own farmland	8.943***	1.309	-0.309	0.376
Extension	0.023	0.47	-0.005	0.053
Gender	-3.187**	1.393	-0.024	0.079
Main occupation	0.652	0.691	-0.045	0.094
Total farmland	0.001	0.096	-0.069***	0.011
Constant	2.446	3.008	4.539	0.461
LR Chi2(10)	481.25***		46.96***	
Pseudo R2	0.76		0.012	
Log likelihood	-77.83		2009.5594	
lnalpha			-1.33	0.076
Alpha			0.264	0.019

***, **, and *, implies significant at 1, 5, and 10% respectively, Source: IITA/DIIVA Adoption and Impact Survey (2011).

This implies that adoption would increase as farmers gained more access to credits related to agricultural production. A significant positive effect of access to credit on the adoption of improved maize varieties was reported by Feleke and Zegeye (2006) and Paudel and Matsuoka (2008). Similar effects on the adoption of fish enterprises were observed by Matiya et al. (2005). In the same vein Beshir et al. (2012) also obtain a positive effect of credit on determinants of chemical fertilizer technology adoption in North eastern highlands of Ethiopia. Beke (2011) found that the coefficient for predicted probability of being credit constrained has a negative and significant effect on the adoption and use intensity of improved rice varieties in Ivory Coast. This suggests that credit constraints tend to reduce the adoption of improved agricultural technologies. The implication is that farmers should also be granted access to adequate credit to achieve increased adoption of ICVs.

The result of the second part of the double-hurdle shows that the coefficient of hired labor has a positive and significant effect on the probability of increasing the proportion of total farmland devoted to cassava production in the study area. However, a negative and significant coefficient was observed for total farmland. This could be due to the fact that as the area of available farmland increases, there is a tendency for the farmers to go into multiple cropping, thereby reducing the land for cassava production. This is in agreement with the findings of Roos et al. (2000), Villami et al. (2008), Breen et al. (2009) and Rämö et al. (2009) on perennial energy crop adoption, but contrary to the findings of Doss and Morris (2001) for the adoption of inorganic fertilizer.

The effect of information on adoption of ICVs

Information is an essential component of agricultural technologies adoption. A farmer that is not aware of the existence of ICVs will not be likely to adopt. In this paper we empirically examine the effect of access to information on the adoption of ICVs in Nigeria using the Heckman Probit Selection model. To justify the use of this model we evaluate its appropriateness over the standard probit model by examining the presence of any sample selection. This is done by checking if there is any correlation between the error terms of the outcome (regression)and selection (Probit) models. The result shows that the rho is significantly different from zero (Wald χ^2 =24.66, with ρ =0.003), thus justifying the use of this model to assess the effect of information on ICV adoption in the study area.

The results of the Heckman Probit selection model are presented in Table 6. The first stage is referred to as the selection model and takes into account whether or not the farmer has access to information about ICVs. The second stage, known as the outcome model, examines whether the farmer adopted any ICV, conditional on whether any information was obtained about the ICV. The dependent variable of the first stage model (access to information) is specified as binary, which is equal to one if the farmer has access to information about the ICVs and

Table 6. The results of the heckman probit selection model.

Variables	Selection model		Outcome equation	
	Coefficient	Std. Err.	Coefficient	Std. Err.
Age	-0.016***	0.004	0.013**	0.006
Gender	0.333**	0.161	-0.257	0.243
Years of formal education	0.067***	0.015	0.025	0.018
Rented farmland	-0.287*	0.147	0.531*	0.317
Own farmland	-0.034	0.158	0.045	0.157
Off-farm income	0.577**	0.233	-0.308**	0.156
Participate in technology evaluation	0.219	0.231	0.241	0.172
Hired labor	0.600***	0.124	0.353	0.236
Contact with extension agents	0.072	0.180	-0.036	0.175
Constant	0.617	0.405	-0.094	0.555
Wald(X^2)	24.66***	ρ =0.003		

***, **, and *, implies significant at 1, 5, and 10% respectively, Source: International Institute of Tropical Agriculture (IITA)/DIIVA Adoption and Impact Survey (2011).

0 otherwise. The dependent variable of the adoption model is also binary, equal to one and if the farmer planted at least one ICV and 0 otherwise. The results of the first stage show that the factors that tend to significantly affect farmers' access to information are age, gender, years of formal education, rented farmland, and off-farm income. The coefficients of gender, education, off-farm income, and hired-labor variables are positive and statistically significant. Since, gender is one if the household head is male and 0 otherwise.

Therefore, the positive significance of gender implies that the male-headed households have a higher probability of having access to information than female-headed households. The significance of education at 1% suggests that education is a very important determinant of access to information. The implication is that the educated household head, the primary decision-maker, is more capable of obtaining and assimilating information about the advantages of the adoption of an ICV and the negative effects that could result from not adopting it. Participation in off-farm activities could further predispose the farmers to getting access to information.

The coefficients of age and rented farmland are both negative and statistically significant in determining farmers' access to information. This implies that younger farmers are more likely than older farmers to have access to information. Those farmers that operate on rented farmland are likely to experience limited access to information. The result of the outcome model reveals that the adoption of ICVs based on access to information is positively and significantly determined by the age of the household head and use of rented farmland; off-farm income has a negative and statistically significant effect on the decision of farmers to adopt ICVs in the study area.

SUMMARY, CONCLUSION AND POLICY RECOMMENDATIONS

The collaboration between IITA and its partners has resulted into the development of ICVs and their subsequent dissemination to farmers in Nigeria.This study provides empirical information concerning the factors that determine the adoption of ICVs and the intensity of adoption in southwestern Nigeria, using sub-nationally representative data collected by IITA from about eight hundred and fourty one households for the study.

The results of the Double-Hurdle model reveal that the adoption of ICVs is higher among female-headed than male-headed households. Those farmers that have access to labor are more likely to adopt ICVs than those who are labor-constrained. In addition, farmers that own their farmland are more likely to adopt than those that practice farming on rented farmland. Younger farmers are more likely than older farmers to adopt ICVs. Access to credit increases ICV adoption tremendously and access to abundant hired labor is important in the study area. As the area of available farmland increases, there is a tendency for the farmers to go into multiple cropping, thereby reducing the area for cassava production.This suggests that an increase in area could have the tendency to encourage multiple cropping and thus reduce the intensity of ICV adoption.

The Heckman Probit selection model is employed to analyze the two-stage process of access to information and adoption of ICVs-having access to information, which creates awareness about the ICVs, in the first stage and then in the second stage adopting the IVCs, based on the information about the attributes and benefits inherent in adoption. The results further indicate that age of the household head, gender, education, and off-farm income are the variables that are positive and statistically significant in determining access to information on ICV adoption. Factors determining adoption are age, rented farmland, and off-farm income.

Finallay, this study has been able to empirically establish that cassava farmers in southwestern Nigeria are capable of intensively adopting ICVs if they have access to credit and hired labor, and own their farmland. Therefore, we recommend that access to credit should be improved and the present land tenure system in the rural areas should be re-examined to ensure that farmers have adequate access to land for agricultural production.

Conflict of Interest

The authors have not declared any conflict of interest.

REFERENCES

Abadi AK, Pannell DJ (1999). A conceptual framework of adoption of an agricultural innovation. Agric. Econ. 21(2):145-154.

Abadi Ghadim AK, Pannell DJ, Burton MP (2005). Risk, uncertainty, and learning in adoption of a crop innovation. Agric. Econ. 33(1):1-9.

Adesina AA, Baldu-Forson J (1995). Farmers' perception and adoption of new agricultural technology: Evidence from analysis in Burkina Faso and Guinea, West Africa. Agric. Econ. 13 (1):1-9.

Adesina AA , Seidi S (1995). Farmers' perception and adoption of new agricultural technology: Analysis of modern mangrove rice varieties in Guinea-Bissau. Q. J. Int. Agric. 34(4):358-371.

Abdoulaye T , Abass A, Maziya-Dixon B, Tarawali G, kechukwu R, Rusike J, Alene A, Manyong V, Ayedun B (2014). Awareness and adoption of improved cassava varieties and processing technologies in Nigeria. J. Dev. Agric. Econ. 6(2):67-75.

Agwu AE, Anyaeche CL (2007). Adoption of improved cassava varieties in six rural communities of Anambra State, Nigeria. Afr. J. Biotechnol. 6 (2):89-98.

Akoroda MO, Gebremeshel I, Oyinlola AE (1985). Impact of IITA Cassava varieties in Oyo State of Nigeria. IITA, Ibadan, Nigeria. P. 105.

Alamerie K, Ketema M, Gelaw F (2013). Risk management strategies and pesticides use in vegetable production: The case of smallholder farmers in Kombolcha Woreda, East Hararge Zone, Oromia national regional state, Ethiopia. J. Econ. Sust. Dev. 4(7).

Alavalapati JRR, Luckert MK, Gill DS (1995). Adoption of agroforestry practices: a case study from Andhra Pradesh, India. Agroforest. Sys. 32:1-14.

Alene AD, Poonyth D , Hassan RM (2000). Determinants of adoption and intensity of use of improved maize varieties in the Central Highlands of Ethiopia: A Tobit analysis. Agrekon 39(4):633-643.

Ay P, Oyediran OR, Ogunsakin DA (1983). IITA cassava now part of local farming system: variety 30572 supplements local varieties and opens new production opportunities. Farming Systems Program Report, IITA, Ibadan, Nigeria.

Beke TE (2011). Institutional constraints and adoption of improved rice varieties: Econometric evidence from Ivory Coast. Rev. Agric. Environ. Stud. 92(2):117-141.

Beshir H, Emana B, Kassa B and Haji J (2012). Determinants of chemical fertilizer technology adoption in North eastern highlands of Ethiopia: the double hurdle approach. J. Res. Econ. Int. Fin. (JREIF). 1(2):39-49.

Breen J, Clancy D, Moran B, Thorne F (2009). Modelling the potential supply of energy crops in Ireland: Results from a probit model examining the factors affecting willingness to adopt. The Rural Economy Research Centre Working Paper Series.

Cameron AC, Trivedi PK (2005) . Micro-econometrics methods and applications: Cambridge University Press, Unite Kingdom.

Cameron AC, Trivedi PK (2009). Micro-econometrics using STATA : Revised edition. Cambridge University Press, Unite Kingdom.

Cooper JC, Keim RW (1996). Incentive payments to encourage farmer adoption of water quality protection practices. Am. J. Agric. Econ. 78(1):54-64.

Cragg JG (1997). Using higher moments to estimate the simpler errors-in –variables model. Rand J. Econ. 28(3):571-591.

Cragg JG (1971). Some Statistical models for limited dependent variables with application to the demand for durable goods. Econometrica 39(5):829-844.

Deressa T, Hassan RM, Alemu T, Yesuf M, Ringler C (2008). Analysing the determinants of farmers' choice of adaptation methods and perceptions of climate change in the Nile Basin of Ethiopia. International Food Policy Research Institute (IFPRI) Discussion paper 00798, Washington D.C., USA.

Diagne A, Demont M (2007). Taking a new look at empirical models of adoption: average treatment effect estimation of adoption rate and its determinants. Agric. Econ. 37(2-3):201-210.

Dontsop-Nguezet PM, Diagne A, Okoruwa VO, Ojehomon VET (2011). Impact of improved rice technology adoption (NERICA varieties) on income and poverty among rice farming households in Nigeria: a local average treatment effect (LATE) approach. Q. J. Int. Agric. 50(3):267-291.

Doss CC, Morris ML (2001). How does gender affect the adoption of agricultural innovations?: The case of improved maize technology in Ghana. Agric. Econ. 25(1):27-39.

Eke-Okoro ON, Njoku DN (2012). A Review of cassava development in Nigeria 1940-2010. ARPN J. Agric. Bio. Sci. 7(1):59-65.

Eze CC, Ibekwe UC, Onoh PO, Nwajiuba CU (2006). Determinants of adoption of improved cassava production technologies among farmers in Enugu State of Nigeria. G. Appro. Ext. Pract. 2(1):37–44.

Feder G, Slade R (1984). The acquisition of information and adoption of new technology. Ame. J. Agric. Econ. 66(3):312-320.

Feleke S, Zegeye T (2006). Adoption of improved maize varieties in Southern Ethiopia: factors and strategy options. J. Food Policy 31(5):442-457.

Gebregziabher G, Holden S (2011). Does irrigation enhance and food deficits discourage fertilizer adoption in a risky environment? Evidence from Tigray, Ethiopia. J. Dev. Agric. Econ. 3(10):514-528.

Green WH (2000). *Econometric analysis* (4th edition). Englewood Cliffs, NJ: Prentice Hall.

Hailu B (2008). Adoption of improved teff and wheat production technologies in crop and livestock mixed systems in Northern and Western Shewa zones of Ethiopia, PhD Dissertation, University of Pretoria, South Africa.

Igodan CO, Oheji PE, Ekpere JA (1988). Factors associated with the adoption of recommended practices for maize production in Kainji Lake Basin in Nigeria. J. Agric. Admin. Ext. 29(2):149-156.

Ikpi A, Gebremeskel T, Hahn ND, Ezumah H, Ekpere JA (1986). Cassava a crop for household food security. IITA-UNICEF Collaborative Program Report, IITA. Ibadan, Nigeria.

Isham I (2002). The effect of social capital on fertiliser adoption: evidence from rural Tanzania. J. Afr. Econ. 11(1):39-60.

IITA (International Institute of Tropical Agriculture) (1984). Annual report 1984. Ibadan, Nigeria.

Jensen K, Clark CD, Ellis P, English B, Menard J, Walsh M, de la

Torre Ugarte D (2007). Farmer willingness to grow switch grass for energy production. Bioma. Bioener. 31(11-12):773-781.

Junge B, Deji O, Abaidoo R, Chikoye D, Stahr K (2009). Farmers' adoption of soil conservation technologies: A case study from Osun state, Nigeria. J. Agric. Educ. Ext. 15(3):257-274.

Just RE, Zilberman D (1983). Stochastic structure, farm size and technology adoption in developing agriculture. Ox. Econ. Papers-New Series. 35(2):307-328.

Kassie M, Mahmud Y, Köhlin G (2009). The role of production risk in sustainable land-management technology adoption in the Ethiopian Highlands. Working papers in Economics (University of Gothenburg). Number 407.

Keyser J (1984). Yoruba names mirror Cassava traits. IITA Research Briefs 5(4):3, 6, 8.

Kudi TM, Bolaji M, Akinola MO, Nasa IDH. (2011). Analysis of adoption of improved maize varieties among farmers in Kwara State, Nigeria. Int. J. Peace Develop. Stud. 1(3):8-12.

Legese G, Langyintuo AS, Mwangi W, Jaleta M, La Rovere R (2009). Household Resource Endowment and Determinants of Adoption of Drought Tolerant Maize Varieties: A Double-hurdle Approach. Contributed Paper prepared for presentation at the International Association of Agricultural Economists Conference, Beijing, China, August 16-22.

Langyintuo AS, Mungoma C (2008). The effect of household wealth on the adoption of improved maize varieties in Zambia. Food Policy 33(6):550-559.

Matiya G, Wakabayashi Y, Ng'ong'ola D, Takenouchi N (2005). A logit analysis of socio-economic factors influencing people to become fishermen around Lake Malombe in Malawi. J. Appl. Sci. Res. 1(1):18-32.

Martínez-Espiñeira R (2006). A Box-Cox Double-Hurdle Model of Wildlife Valuation: The Citizen's Perspective. Ecol. Econ. *58*(1):192-208.

Moffatt PG (2003). Hurdle models of loan default. School of Economic and Social Studies University of East Anglia. Available on-line at http://www.crc.ems.ed.ac.uk/conference/presentations/moffat.pdf

Negatu W, Partkh A (1999). The impact of perception and other factors on the adoption of agricultural technology in the Moret and Jiru *woreda* (district) of Ethiopia. Agric. Econ. 21(2):205-216.

Nweke FI, Spencer DSC, Lynam JK (2002). The Cassava transformation: Africa's best kept secret. Michigan State University Press, East Lansing, Mich. USA.

Odoemenen IU, Obinne CPO (2010). Assessing the factors influencing the utilization of improved cereal crop production technologies by small-scale farmers in Nigeria. Ind. J. Sci. Tech. 3(1):180-183.

Okoedo-Okojie DU, Onemolease EA (2009). Factors affecting the adoption of yam storage technologies in the Northern Ecological zone of Edo State, Nigeria. J. Hum. Ecol. 27(2):155-160.

Paudel P, Matsuoka A (2008). Factors influencing adoption of improved maize varieties in Nepal: A Case Study of Chitwan District. Aust. J. Bas. Appl. Sci. 2(4):823-834.

Rämö AK, Järvinen E, Latvala,T, Toivonen R, Silvennoinen H (2009). Interest in energy wood and energy crop production among Finnish non-industrial private forest owners. Biom. Bioener. 33(9):1251-1257.

Rogers EM (1983). Diffusion and innovations (3rd. ed.) New York: Free press.

Rogers EM (1995). Diffusion of Innovation (4th edition) The free Press New York, pp. 21-45

Rogers EM, Shoemaker FF (1971). Communication of innovation: A cross-cultural approach". New York: the free press.

Roos A, Rosenqvist H, Ling E, Hektor B (2000). Farm-related factors influencing the adoption of short-rotation willow coppice production among Swedish farmers. Acta Agric. Scand. Sect. B- Plt. Soil Sci. 50(1):28-34.

Saka JO, Okoruwa VO, Lawal BO , Ajijola S (2005). Adoption of improved rice varieties among smallholder farmers in southwestern Nigeria. W. J. Agric. Sci. 1(1):42-49.

Saka JO , Lawal BO (2009). Determinants of adoption and productivity of improved rice varieties in southwestern Nigeria. Afr. J. Biotechnol. 8(19):4923-4932.

Shiferaw BA, Kebede TA, You L (2008). Technology adoption under seed access constraints and the economic impacts of improved pigeon-pea varieties in Tanzania. Agric.Econ. 39(3):309-323.

Sunding D, Zilberman D (2001). The agricultural innovation process: Research and technology adoption in a changing agricultural sector. Pages 207–261 *in:*Handbook of Agricultural Economics edited by B.L. Gardner and G.C. Rausser. Elsevier, Oxford.

Smith DJ, Schulman C, Current D, Easter KW (2011). Willingness of agricultural landowners to supply perennial energy crops. Agricultural and Applied Economics Association & NAREA Joint Annual Meeting, Pittsburgh, Pennsylvania, July 24-26.

Teklewold H, Dadi L, Yami A, Dana N (2006). Determinants of adoption of poultry technology: A Double-Hurdle approach. Livestk Res. Rural Develop. *18*(3):40.

Uri ND (1998). Conservation tillage and the use of energy and other inputs in us agriculture. Ener. Econs. 20 (4):389-410.

Villami MB, Silvis AH , Bollero GA (2008) . Potential miscanthus' adoption in Illinois: Information needs and preferred information channels. Biom. Bioener. 32(12):1338-1348.

Wale E , Yallew A (2007). Farmers' variety attribute preferences: implications for breeding priority setting and agricultural extension policy in Ethiopia. Afr. Dev.Rev. 19 (2):379-396.

Wooldridge JM (2001). Econometric Analysis of Cross Section and Panel Data. 1 ed: The MIT Press.

Adoption of New Rice for Africa (NERICA) technology in Ogun State, Nigeria

Adedeji T. O.[1], Nosiru M. O.[1], Akinsulu A. A.[1], Ewebiyi I. O.[1], Abiona B. G.[2] and Jimoh T. S.[1]

[1]Department of Agricultural Production and Management Science, Tai Solarin University of Education, Ijebu-Ode, Ogun State, Nigeria.
[2]Department of Agricultural Extension and Rural Sociology, University of Agriculture, Abeokuta, Ogun State, Nigeria.

This study was carried out with the aim to find out the level and socio-economic determinants of the adoption of New Rice for Africa (NERICA) using Ewekoro Local Government Area of Ogun State in Nigeria as a case study. One hundred and sixteen rice farmers were randomly selected in this area, twenty eight of which adopted NERICA rice varieties within 2009/2010 cropping season. Information was gathered through the use of well structured questionnaire and personal interview of the farmers. Descriptive statistics, ordinary least-squares (OLS) regression analysis and T-test statistics were used for the analysis of the data collected. It was discovered through the findings that the rate of adoption was 33.36%, while the socio-economic determinants of the adoption of the technologies include farming experience, age of the farmer, frequency of contact with the officials of Agricultural Development Programme (ADP) in the state and farm size. Moreover, the findings show that the non-adopters of NERICA technological package had higher average output and average yield than the adopters of the technologies. Lastly, the study gave valuable recommendations which could be helpful toward enhancing improved rate of adoption of NERICA rice technologies by the farmers in the study area.

Key words: New Rice for Africa (NERICA), adoption, rice, socio-economic determinants, technologies.

INTRODUCTION

Since the introduction of New Rice for Africa (NERICA) in the mid 1990's, the NERICA has carved a special niche for itself among upland rice farmers in sub-Saharan Africa (SSA). Today, it is a symbol of hope for food security in SSA, the most impoverished region in the world where a staggering one-third of the people are under-nourished and half of the population struggle to survive on US $1 a day or less (Think Quest Team, 2006). The NERICA is a group of rice varieties resulting from the inter-specific crossing between the Asian rice (*Oryza sativa*) and the African rice (*Oryza glaberrima*). They are the output of the hybridization breeding programme started in 1991 by the Africa Rice Centre (WARDA). The NERICA varieties promises to raise significantly the productivity, income and food security of rain fed upland rice farmers in SSA. The NERICA varieties were introduced to rice farmers starting from 1996 in West Africa through participatory varietal selection (PVS) trials (Tiamiyu, 2008). Till date, up to eighteen (18) NERICA varieties have been disseminated in numerous countries across SSA such as Nigeria, Sierra-Leone, Gambia, Ghana, Guinea, Mali, Cote d'Ivoire and Togo through informal channels by farmers and by development projects. These rice varieties which are suited to dry lands were distributed and sown on more than 200,000 ha during the last five years in several African countries notably Guinea, Nigeria, Cote d'Ivoire and Uganda, according to African Rice Centre (2008). Though this represents a major advancement, it is still projected to fall short of meeting the growing demands for rice as a food staple.

In order to improve and encourage the increased production of rice in Nigeria, some varieties bred by WARDA for upland ecologies were introduced to farmers through the participatory varietals selection (PVS) trials in 1999 and 2001. The PVS was conducted in nine States which included Ogun State and Federal Capital Territory. The apparent success of these trials led to the release of NERICA varieties. The Federal Government launched the Multinational NERICA Rice Dissemination Project (MNRDP) in 2003. This was aimed at promoting the use of NERICA seed varieties and complementary production technologies packaged with it among farmers (Tiamiyu, 2008). In 2003, MNRDP started the implementation activities for the dissemination of NERICA in six States including Ogun State where the area of this study was located. According to the MNRDP project, the NERICA rice varieties embody improved seed technology and management practices of agrochemicals in terms of biological and chemical technologies. The NERICA varieties cultivated in the study area are NERICA 1 to 8. The production parameters of the adopters of these technologies can be used to solve the problem of low yield, low productivity and rice self-insufficiency in Nigeria.

Statement of problem

The world food problem is a serious issue especially in Africa. Population increase is escalating, and increase in crop production is becoming an uphill task. Recently, in order to resolve the problem, NERICA was developed in West Africa. However, certain problems bother the mind of the researchers. The growth in consumption which has become most substantial in Africa's rapidly growing cities, where rice is increasingly becoming the staple diet of the poor urban households. Rice has therefore become a staple of considerable strategic importance. At present, rice imports are still significant because the region is yet to be self sufficient in rice production. As a result of the increment in the consumption of rice in the country, the imported rice is being used to bridge the gap existing between production of the local rice and the domestic demand of the consumers. The demand-supply gap of rice is becoming a major problem which requires attention in Nigeria's agricultural economy. Moreover, high exchange rate involved in importation of rice is a serious issue that requires quick attention if an economy would achieve any significant development. Hence, the adoption of NERICA rice by the farmers is believed to have potential to lead to an increase in productivity of rice in the country, result in reducing demand-supply gap, increase rice self sufficiency and enhance improved foreign exchange earnings.

Objectives of study

The objective of this study is to determine the socio-economic factors influencing the adoption of the NERICA technology and to examine the level of adoption of NERICA rice by farmers in the study area.

Statement of hypotheses

i) There is no statistically significant difference in the output level of NERICA adopters and non-adopters.
ii) There is no statistically significant difference in the level of yield of NERICA adopters and non-adopters.

LITERATURE REVIEW

Much evidence has been provided attesting the productive performance of the agricultural sector in Africa and factors influencing it, but there is still little evidence on crop-specific and sub-regional productive performance. An assessment of crop-specific efficiency, productivity and adoption analysis should be of more interest to policy makers implementing liberalization policy than overall aggregates. The adoption of the new agricultural technology that led to the green revolution in Asia can also lead to significant increases in agricultural productivity in Africa, and stimulate the transition from low productivity subsistence agriculture to a high productivity agro-industrial economy (World Bank, 2008). Mendola (2006) observes that the adoption of the high yielding variety has positive effect on household well-being in Bangladesh. More recently, Kijima et al. (2008) conducted a study on the impact of NERICA in Uganda and found that NERICA adoption reduces poverty without deteriorating the income distribution. A study carried out by De Janvry and Sadoulet (1992) shows a positive impact of adoption of agricultural technologies. In contrast, a study in Bangladesh by Hossain et al. (2003) shows that the adoption of high yielding varieties of rice has a positive effect on the richer households but had negative effects on the poor. In Zimbabwe, Bourdillon et al. (2002) observe that the adoption of high yielding varieties of maize increases the crop incomes of adopters only modestly. These conflicting findings justify the need for further research on this issue. There is suggestion that the promotion of NERICA cultivation can contribute to improving expenditure/income of farmers and consequently to poverty reduction. This is consistent with the study of Irz et al. (2001), who show that a close relationship exists between farm productivity and household poverty.

In his study, Diagne (2006a) determines the rates of adoption and rates of diffusion of NERICA rice varieties in some West African countries which include Cote d'Ivoire, Guinea and Benin. He also employs econometric analysis to show the socio-economic determinants of adoption of NERICA in Cote d'Ivoire. He shows that the main factors which positively influence the adoption of NERICA include growing rice partially for sale, household size, growing upland rice, past participation in PVS trials and living in a PVS-hosting village. On the other hand, he discovers that age of the farmers and having a secondary occupation had a negative impact on adoption (Diagne, 2006b). In Guinea, Diagne et al. (2007) show that the main socio-economic determinants of adopting NERICA with positive effects were participation in training programme and living in a village where NGO SG2000 has had activities. In Benin, land availability and living in PVS-hosting village were found to have positive effects on adoption. It was also found that varietal attributes such as swelling capacity and short growing cycle were important determinants of adoption of NERICA (Adegbola et al., 2005).

MATERIALS AND METHODS

Study area and sampling technique

This study was conducted in Ewekoro Local Government Area of Ogun State, Nigeria. This is a major rice-producing area in the state, comprising of Wasimi and Arigbajo which serve as major villages in the area. Major crops cultivated include maize, rice, cassava and melon while sugarcane is the major cash crop of the inhabitants of the Local Government Area. Majority of the people living in this locality were involved in farming activities either on a part time or full time basis. One hundred and sixteen small-scale rice farmers were sampled in the study area by using simple random sampling technique.

The data were collected during the 2008/2009 farming year through the use of well structured questionnaire which was administered by personal interview conducted for the selected farmers. Descriptive statistics and multiple regression analysis were used in meeting the objectives of this study. The descriptive statistics entails the frequency tables which show the distribution of the socio-economic characteristics of the farmers such as sex, age, household size, farming experience, educational qualification and proximity to ADP official village of residence. The regression analysis shows the socio-economic factors that determine the level of adoption of NERICA technology. The model of the regression analysis is as follows:

$$Q = a_o + a_1X_1 + a_2X_2 + a_3X_3 + a_4X_4 + a_5X_5 + a_6X_6 + a_7X_7 + a_8X_8$$

Where Q = level of NERICA technology adoption of farmers; the level of adoption was measured using the proportion of utilisation of each component that constitutes the NERICA technological package; these include cultivation of NERICA varieties, appropriate fertilizer application, mechanization and pesticide/herbicide application. The summation of the proportion of utilization of each of these components determines the level of NERICA technology adoption of the farmers. X_1 = sex (dummy: male = 1, female = 0); X_2 = age (years); X_3 = household size; X_4 = farm size (acres); X_5 = proximity to ADP official village of residence (dummy: farmer residing in ADP official resident village = 1; otherwise = 0); X_6 = education (years); X_7 = farming experience (years); X_8 = frequency of ADP contact; a_o = constant; $a_1 - a_8$ = parameters to be estimated.

RESULTS AND DISCUSSION

The result of the descriptive statistics of the sampled rice farmers

From Table 1, it could be observed that over 55% of the farmers were male while less than 45% were female. The average age of the farmers is observed to be 41.42 years, as majority of them (about 69%) falls within ages 30 to 59 years. There is a relatively low level of education among the farmers: the average year of formal education among the sampled paddy farmers was 2.86, as almost 70% of the farmers had no formal education. The average household size is 6.14, and most of the paddy farming households had sizes ranging between 1 and 10. The average year of farming experience was found to be 19.71 years, while the average farm size was 4.41 acres. Most of the farmers (about 80%) cultivate on 5 acres of land or lesser. About two-third of the farmers had no contact with Agricultural Development Project (ADP) officials, while less than 10% had contact with the ADP officials at least once a month. Consequently, this could be the reason for the low adoption level of NERICA by less than 25% of the rice farmers.

The items that are involved in NERICA technology adoption level include cultivation of NERICA seed, fertilizer application, mechanization and herbicide/pesticide application. Moreover, the stages of adoption are itemized to range from 'not aware', 'aware', 'thinking about it', 'interested', 'ready to adopt', to 'adopted'. These are coded from 0 to 5. A farmer that adopts all the four components of the technology mentioned earlier is scored 20 (4 × 5) scores, while a farmer that is not aware of all the four technologies is scored 0 (4 × 0). The adoption level (*x*) of a farmer could be determined using the following formula:

$$x = \frac{a}{20} \times 100$$

a represents the adoption score given to an individual farmer. For this study, any farmer that does not cultivate NERICA seed variety was not considered as an adopter of NERICA technology. The value *x* serves as the regress and for the exogenous variables. To determine the mean adoption level, this study adopts the formula:

Average adoption level = $\sum fx/N$

Where *f* = frequency of each value observed; *N* = number of observations of the variable *x*.

Table 1. Socio-economic characteristics of the selected rice farmers.

Socio-economic characteristics	Frequency	Percentage	Mean
Sex			
Male	64	55.17	
Female	52	44.83	
Total	116	100.00	
Age (years)			
20-29	16	13.79	
30-39	24	20.69	
40-49	30	25.86	
50-59	26	22.41	41.42
60-69	14	12.07	
70 & above	6	5.17	
Total	116	100.00	
Educational qualification			
Non-formal	80	68.97	
Primary	30	25.86	2.86
Secondary	6	5.17	
Total	116	100.00	
Household size			
1-5	48	41.38	
6-10	54	46.55	6.14
11-15	14	12.07	
Total	116	100.00	
Farm size (acres)			
0.01-2.50	62	53.45	
2.51-5.00	30	25.86	
5.01-7.50	4	3.45	4.41
7.51-10.00	6	5.17	
10.01-12.50	10	8.62	
12.51-15.00	4	3.45	
Total	116	100.00	
Farming experience (years)			
1-10	44	37.93	
11-20	28	24.14	
21-30	30	25.86	19.71
31-40	4	3.45	
41-50	6	5.17	
51-60	4	3.45	
Total	116	100.00	
Frequency of ADP contact			
Weekly	4	3.45	
Monthly	6	5.17	
Quarterly	14	12.07	
Yearly	16	13.79	
Not at all	76	65.52	
Total	116	100.00	

Table 1. Contd.

Adopted rice variety		
Local	88	75.86
NERICA	28	24.14
Total	116	100.00

Source: Field survey (2010).

Table 2. Comparison of the output (in bags) and average yield (bags/acre) of paddy farmers based on NERICA rice adoption type.

Farmer category	Average output (bags)/average yield (bags/acre)	Standard deviation	Standard error
Adopter	3.5043	3.3830	0.3606
	8.2045	2.3643	0.2520
Non-adopter	9.6741	9.5578	1.0189
	18.3409	11.7678	1.2545

Source: Field survey (2010).

The average level of adoption in the area of study was observed to be 33.36%.

The result of the regression analysis showing the socio-economic factors influencing the level of adoption of NERICA technology in the study area

$$Q = 71.125^{**} + 2.435X_1 - 1.807X_2^{*} - 0.124X_3 + 1.310X_4^{**} + 2.987X_5 - 0.068X_6 - 0.459X_7^{**} + 5.954X_8^{**}$$

$$(9.249) \quad (2.447) \quad (1.280) \quad (0.401) \quad (0.364) \quad (3.075) \quad (0.376) \quad (0.117) \quad (1.235)$$

Where, ** - 1% significance level; * - 5% significance level.

From the regression analysis computed using SPSS 15.0, it was observed that age of the farmers, farm size, farming experience and frequency of ADP contact were the most significant socio-economic variables influencing the level of adoption of the farmers. The younger farmers tend to have higher adoption level than the older farmers; this is in conformity with literatures. Also, farmers with larger farm holdings seem to have higher adoption level than those cultivating on smaller pieces of land. Moreover, rice farmers with fewer years of experience tend to adopt NERICA technologies more than the more experienced farmers in the study area. More importantly, increased contact with the ADP officials has a tendency to improve the level of adoption of the farmers in the area of study.

Comparison of the performance of the adopters and non-adopters of NERICA technologies in the study area

From Table 2, it could be observed that the non-adopters had higher output and yield than the adopters of NERICA technologies. This may not be as a result of lower potential of the technology but rather due to the incidence of pests such as rodents and birds, according to the information obtained through personal interviews with the farmers. It was observed that the pests have preference for NERICA varieties paddy output over the local rice varieties paddy output; hence, the lower output and yield were recorded by the rice farmers in the study area. Moreover, through personal interviews, some of the adopters of the technologies were observed to show possibility of discontinuation of the technologies in the preceding seasons. The mean values of the output and yield of the adopters and non-adopters of NERICA varieties were used to conduct the T-test statistics presented in Table 3. From the result of the T-test conducted, it could be deduced that there is statistically significant difference in the output and yield levels of NERICA adopters and non-adopters in the study area.

CONCLUSION AND RECOMMENDATIONS

The study shows that younger farmers have tendency to adopt improved technologies than the older farmers, and it was discovered that about two-thirds of them were 40 years and above. This is not a good omen for instituting an improved technology if the discoveries in the literatures are anything to go by. Farmers that operate on relatively small scale level are discovered to have lower adoption level, and it is a known fact that the agricultural production in developing economies mainly depends on small production scale. Almost four-fifth of the sampled farmers in the study area operates on less than 5 acres of land, while the average farm size was 4.41 acres. It

Table 3. Test of hypotheses using T-test.

Hypothesis (H_0)	Degree of freedom	Critical value	T-value	Decision
There is no statistically significant difference in the output level of NERICA adopters and non-adopters.	0.05, 43	2.02	5.412	Reject H_0
There is no statistically significant difference in the level of yield of NERICA adopters and non-adopters.	0.05, 43	2.02	7.658	Reject H_0

Source: Data analysis (2010).

was also revealed that those paddy farmers with fewer years of experience tend to adopt NERICA technologies. A little less than 40% of the rice farmers in the study area had farming experience of ten years or less. Contact with the ADP officials serves as an important factor that could enhance higher level of adoption of improved technologies by the farmers in the area of study. However, only about 8% have contact with the ADP officials at least once in a month, and about two-thirds have no contact at all with any ADP or extension agents. Consequently, there appears to be a relatively low level of adoption of NERICA technology among the paddy farmers.

In conclusion, NERICA technological package would only be of great benefit to our agricultural economy, if and only if the technologies are directed towards the right direction, and there is increase in farmers-extension agent ratio. Also, it is not enough to introduce a new technological package but to sustain such package is very essential for a successful agricultural production system. Therefore, this study offers the following recommendations toward helping the agricultural policy makers:

i) Strategies should be developed toward reaching the older farmers on the field, since they form a larger proportion of the farmers remaining in agricultural production. This also applies to the relatively more experienced farmers; extension agents should endeavor to introduce every approach possible with the aim of convincing them to adopt improved technological packages.

ii) Improved agricultural technologies should be made available and affordable to the small scale farmers since they produce the bulk of the agricultural produce in the developing world economies.

iii) Any serious-minded government should consider a significant increase in the number of agricultural extension agents, if there would be a successful and timely dissemination of newly introduced agricultural technological packages. It is only through such attitude would there be an initiation of a move toward economic development. Also, NGOs with agricultural-related functions should consider the enhancement of information dissemination and extension services as part of their obligations, as this would help to bridge the gaps between already existing farmers and government-employed extension workers.

iv) The agricultural researchers, especially those involved in the development of NERICA rice varieties, should look into developing varieties that would pass the test of pest attacks especially rodents which form an obstacle against increased productivity of the varieties in the study area.

ACKNOWLEDGEMENT

Special thanks goes to the members of staff and students of the Department of Agricultural Production and Management Science, Tai Solarin University of Education, Ijagun, Ijebu-Ode, Ogun State, Nigeria.

REFERENCES

Adegbola P, Arouna A, Diagne A, Adekanmbi SA (2005). Determinants socio-economic et taux d'adoption et d' adoption des nouvelles varieties de riz NERICA au Centre du Benin. *Papier a soumettre pour publication.*

Africa Rice Centre (WARDA)/FAO/SAA (2008). NERICA: the New Rice for Africa – a Compendium. Somado EA, Guei RG, Keya SO (eds.). Cotonou, Benin: Africa Rice Center (WARDA); Rome, Italy: FAO; Tokyo, Japan: Sasakawa Africa Association.

Bourdillon M, Hebinck P, Hoddinott J, Kinsey B, Marondo J, Mudege N, Owens T (2002). Assessing the Impact of HYV Maize in Resettlement Areas of Zimbabwe. Discussion paper 161, Washington, D.C: Int. Food Policy Res. Inst.

De Janvry A, Sadoulet E (1992). "Structural Adjustment under Transaction Costs". Paper presented at the 29th Meeting of the European Association of Agricultural Economists on Food and Agriculture Under Structural Adjustment, University of Hohenheim, Stuttgart, 1992 (Frankfurt am Main, Germany: P. Lang, 1994).

Diagne A (2006a). The diffusion and adoption of NERICA rice varieties in Côte d'Ivoire'. Dev. Econ. 44(2):208-231.

Diagne A (2006b). Le Nerica, myth et réalité. Grain de Sél, Numéro Spéciale, 34-35 Mars- Aout, 2006.

Diagne A, Sogbossi MJ, Sékou D, Diallo AS, Bacar BA (2007). Evaluation de la diffusion et de l'adoption des variétés de riz NERICA en Guinée. Miméo ADRAO, 2007. Paper presented at the AAAE 2nd Conference, 19-22 August 2007, Accra.

Hossain M, Lewis D, Bose ML, Chowdhury A (2003). Rice research technological progress, and impacts on the poor: the Bangladesh case. Discussion paper 110, Int. Food Policy Res. Inst. (IFPRI).

Irz X, Lin L, Thirtle C, Wiggings S (2001). Agricultural Productivity Growth and Poverty Alleviation Development. Policy Rev. 19(4):449.

Kijima Y, Otsuka K, Sserunkuuma D (2008). Assessing the impact of NERICA on income and poverty in central and western Uganda. Agric.

Econ. 38:327-337.
Mendola M (2006). Agricultural technology adoption and poverty reduction: A propensity-score matching analysis for rural Bangladesh. Food Policy. 32:372-393.
Think Quest Team (2006). Finding solutions to poverty. library.thinkquest.org/A Dollar a Day/Poverty Overview.
Tiamiyu SA (2008). Efficiency and technology use among growers of NERICA rice varieties in the Savanna zone of Nigeria, Unpublished Ph.D Thesis, University of Ibadan.
World Bank (2008). Agriculture for Development. World Development Report.

17

Adoption of drought-tolerant rice in Thailand: Participatory varietal selection and implications for breeding programs

Orachos Napasintuwong[1] and Carl Pray[2]

[1]Department of Agricultural and Resource Economics, Faculty of Economics, Kasetsart University, Bangkok, Thailand.
[2]Department of Agricultural, Food, and Resource Economics, Rutgers: the State University of New Jersey, New Brunswick USA.

Rice production in Northeast Thailand has been suffering because of drought and limited irrigation. In this area, glutinous rice is generally produced for household consumption. The new drought-tolerant glutinous rice, RD12, developed to target drought-prone area in Northeast Thailand, was approved by the Rice Department in 2007. The breeding of this new variety has incorporated farmers' participation to identify preferred traits. This study aims to determine key factors contributing to the adoption of RD12 by focusing on farmers' preference for traits subjectively selected from the farmer participatory varietal selection (PVS) program. Key findings are that farmers who are exposed to more late-season drought because their fields are at higher elevations are more likely to adopt drought-tolerant varieties despite their preference for other better tasting varieties. Also, those who prefer the cooking characteristics of RD12 are more likely to adopt it, instead of the existing drought-tolerant variety. These two characteristics are significant for the adoption of RD12 and were identified as a result of PVS. This suggests that PVS is essential in the breeding program to promote new and appropriate technology to farmers.

Key words: Rice, drought tolerant, Thailand, adoption, participatory varietal selection, participatory breeding.

INTRODUCTION

Thailand is the sixth largest rice producer after China, India, Indonesia, Bangladesh, and Vietnam, and since the 1980s has been the largest exporter in the world market - until 2012 when it was surpassed by Vietnam and India. In spite of this, rice productivity in Thailand is the second lowest in Asia after Myanmar (International Rice Research Institute, 2010).Insufficient water is one important reason for Thailand's low productivity, particularly in the Northeast which produces more than 40% of total rice production in the country and where more than 60% of the total area cultivated is in rice (Office of Agricultural Economics of Thailand, 2011).

Northeast Thailand has the largest share of agricultural land and the largest number of farms. Due to salinity, low soil fertility and inadequate water infrastructure, this area is one of the least developed in the country's economy. Water storage and irrigation is a key production constraint in the Northeast Thailand. Despite investments in small, medium and large-scale irrigation systems (that is, dams, pumping stations, and irrigation) in the past sixty years, irrigation is concentrated on supplementary irrigation in the wet season and Northeast Thailand remains a predominantly rainfed area (Floch and Molle, 2007). Although rice can be produced two to three times a year, the majority of rice production in the Northeast is limited to the wet season (planting during May to October; harvesting during August to April) because of water accessibility during the dry season (planting during November to April; harvesting during February to October). With limited irrigation, farmers have to leave their land idle during the dry season. In 2010, 6.05 million hectares of rice were cultivated during the wet season in the Northeast compared to only about 0.48 million hectares during the dry season (Office of Agricultural Economics of Thailand, 2011).

Drought can adversely affect rice productivity at different times during the production cycle. In Northeast Thailand, it can occur both early, late and intermittently during the cropping season. Drought early in the cropping season causes a delay in rice transplanting and subsequently results in yield loss. It also increases the probability of late season yield loss from a delay in flowering. Yield loss during late-season drought, which develops at the end of the wet season before crop maturation is more severe than early-season drought. The estimates of yield reduction from late-season drought were 45 to 50% and 15 to 20% for the upper and middle top sequence, respectively (Jongdee, 2003). A twenty year simulation of yield loss in the Northeast showed that drought was more likely to develop in the latter stages of crop development; thus, high rainfall could still result in low yield if a drought occurs at a critical stage (Jongdee et al., 1997).

Prapertchob et al. (2007) found that during the late wet season, the upper Northeast has the lowest rainfall but the highest coefficient of variation. Although this zone was identified as low-risk of drought based on rainfall, hydrology and physical aspects of landscape according to Mongkolsawat et al. (2001), it is exposed to higher probability of yield and economic losses from late-season drought. Drought caused by climate change is expected to continue in the future and will have a significant impact on agriculture.The current impact of rainfall on agriculture based on the Generalized Monsoon Index shows that Northeast Thailand is experiencing severe drought, particularly in August and September (Thai Meteorological Department, 2012) when wet season crop planting occurs. Rainfall in Northeast Thailand has fluctuated in the past few decades and is expected to be lower by 2090 (Thai Meteorological Department, 2010). Adaptation strategies such as changing planting and harvesting time, switching to different cultivars and developing drought-tolerant crops were suggested as options (Asian Development Bank, 2009).

The Rice Department of Thailand successfully developed RD12, glutinous drought-tolerant rice suitable for the Northeast. It was approved by the Rice Department in March, 2007. Nongkhai province is the first area where RD12 has been disseminated. This province is bordered by Laos PDR to the North and by the Mekong River to the East. The mountains to the West cause intermittent and late droughts during the wet season. The climate in this area is dry during the planting time of the dry season (November to April) with the average rainfall of less than 100 mm and humid during the planting time of the wet season (May to October) with the average rainfall almost 300 mm; however, rainfall drops dramatically in October to about 108 mm on average (Nongkhai Rice Research Center, 2013). Even though low fertility and coarse-textured and loamy sandy soils constrain production, weather especially variable rainfall distribution plays an important role in crop productivity and stability. Despite several large rivers, including the Mekong River, and small creeks and irrigation projects, less than 7% of agricultural land benefits from water management projects (Nongkhai Rice Research Center, 2013). Most farmers in this target drought-prone area are small farmers whose glutinous rice production is mainly for household consumption. Thus, preferences toward cooking characteristics are just as important as other traits.

Furthermore, as Prapertchob et al. (2007) found, farmers in this zone allocate their land to rice cultivation more than in other zones, but their yield and net returns from rice cultivation are the lowest. The development of drought-tolerant rice varieties for this target area will not only alleviate poverty, but also will ensure food security in the households. The breeding program of RD12 took into account farmer preferences by integrating participatory varietal selection (PVS) in the breeding process as a key determinant towards variety approval. Despite several released varieties, not all of them have been well-accepted by farmers mainly due to the superiority of some traits of the existing varieties. RD12 is one of the first released varieties recommended for drought-prone areas. PVS is relatively new to the rice breeding program in Thailand. To date the impact of varieties developed using PVS has not been assessed. This paper aims to reveal whether the preferred traits of glutinous rice discovered by the PVS for rice breeding in drought-prone areas influence the adoption of the new variety. Specifically it seeks to identify factors affecting the adoption of RD12 by focusing on farmer preferences, particularly for early maturity and eating quality.

Participatory varietal selection for drought-tolerant rice

Drought-tolerance traits can be classified into primary traits (such as, root depth), secondary traits (such as, leaf rolling), integrative traits, phenology (such as, flowering time) and plant-type traits (such as, plant height) (Kamoshita et al., 2008). Though simple, early flower genotype is often the most effective way of increasing yield under late-season drought (Kamoshita et al., 2008). Early flowering and early maturity varieties can escape from a late-season drought (Jongdee et al., 2006) which is a common problem for rice cultivation in Northeast Thailand. In non-irrigated areas, higher elevated paddies are likely to lose standing water earlier than those in lower positions. Thus earlier-flowering varieties would also reduce the risk of late-season drought and increase potential yield in drought-prone areas of the Northeast. Because glutinous rice is grown mainly for household consumption in upper Northeast Thailand, eating quality is likely as important as agronomic traits for farmers.

Despite the availability of recommended rice varieties for rainfed areas, the adoption of new varieties has had only partial success, to some extent due to the lack of traits important to farmers. Recently International Rice Research Institute (IRRI) recommended PVS be included as a standard part of all rainfed rice breeding programs (International Rice Research Institute, 2006). PVS trials are conducted on farms under the complete management of farmers; thus, it helps breeders and agronomists learn about the performance of new varieties under the real conditions faced by farmers and which varieties are preferred by them. In recognizing farmers' preferred traits, PVS has been increasingly implemented in several breeding programs (Bellon and Reeves, 2002; Ceccarelli and Grando, 2007; Doward et al., 2007; Efisue et al., 2008; Manzanilla et al., 2011; Morris and Bellon, 2004; Rasabandit et al., 2006; Witcombe et al., 1999; Wurzinger et al., 2011). Evidence shows that PVS improves the exposure to and adoption of new varieties (Diagne, 2006; Tshewang and Ghimiray, 2010; Witcombe et al., 1999); however, it has not been evident in the case of drought-tolerant glutinous rice in Thailand.

The "mother-baby" model was used in the breeding program of drought-tolerant rice in Thailand (Pantuwan et al., 2006; Jongdee et al., 2006) adopted from Snapp's technique (2002). The "mother" trial is replicated within-site to test a range of varieties and research hypotheses under a breeder's management either located on a research station or on-farm. The "baby" trial comprises a number of satellite trials of large plots under farmers' management and farm resources. The objective of "mother" trials is to assess the agronomic characteristics of different lines designed and managed by breeders while the "baby" trials are designed by breeders and managed by farmers to elicit farmer perceptions. In this rainfed lowland rice breeding program, four hundred and seventy one field experiments were conducted in three provinces in the North and twelve provinces in the Northeast by selecting existing local varieties, existing modern varieties and promising lines from the breeding program for each location. Both upper toposequence position (upper paddy) and lower toposequence position (lower paddy) were selected for trials. In the mother trials, agronomic characteristics including yield, number of panicles, height, flowering days were analyzed, and the farmers vote for the variety they most preferred. In the baby trials, the experiment varieties were compared to farmers' own varieties based on agronomic characteristics. Grain characteristics (example, size, color) of both paddy rice and milled rice, and eating quality (example, softness, sweetness, and aromatic) of cooked rice were evaluated by farmers in both mother and baby trials.

Farmers' preference for agronomic characteristics revealed appropriate maturity matching with water conditions in paddies. Varieties with appropriate maturity (that is, early maturity for upper paddy), resistance to disease and insect pests, having good tillering with erect-strong stems, tall stature, similar level of height of panicles, droopy leaves and small number of leaves, big and long panicle, even panicle size, many grains per panicle, dense grains within panicle and long-slender grains are preferred. Furthermore, long slender and white grains are preferred while fragrance was farmers' preference, but not a strong requirement. PVS also showed that cooking and eating quality are the most important characteristics. Farmers could reject varieties preferred as paddy and milled rice if they dislike the cooking and eating qualities (Pantuwan et al., 2006).

RD12 is a cross between RD6 and Hahng Yi71. RD6, photoperiod-sensitive glutinous rice, has been well-preferred in this area due to its supreme cooking quality and higher selling price. However, RD6 is susceptible to blast disease, especially when growing in the upper terrain. Hahng Yi71, photoperiod-sensitive early-maturity and blast-resistant glutinous rice has been commonly grown in upper terrain to avoid the risk of late-season drought (Saleeto et al., 2009).

Nevertheless, Hahng Yi71 has poor cooking quality and is not favored for consumption. The perceived advantageous traits of RD12 include early maturity, high yield and good milling; whereas, cooking quality is poorer than the popular RD6 (Jongdee et al., 2006). RD12 is resistant to blast and has earlier maturation than RD6; however, it is less resistant to drought than Hanhng Yi71. From the PVS in rainfed lowland rice breeding program discussed above, RD12 is recommended for rainfed areas in the Northeast, particularly in short rainy season or upper terrain area (Rice Department, 2009).

If the PVS were to improve the adoption of the new variety, Given the joint density of random vector

$\varepsilon_n = (\varepsilon_{n1}, \ldots, \varepsilon_{nJ})$, preferred traits (particularly early-maturity) would be one major factor influencing the adoption of RD12 in the intermittent and late season drought environment. The following area discusses economic concept and model of technology adoption.

ANALYTICAL FRAMEWORK

Several of adoption studies have focused on farm and farmer characteristics with little attention to the characteristics of the varieties (Feder et al., 1985). Doss (2006) emphasized the advantage of micro-level cross-sectional analysis to understand farmer preferences, growing conditions in specific areas and what varietal characteristics are important to farmers. Recent studies accentuated farmer perceptions of varietal traits and their influences on the adoption behavior (Adesina and Zinnah, 1993; Adesina and Seidi, 1995; Adesina and Baidu-Forson, 1995; Edmeades et al., 2008; Hintze et al., 2003; Joshi and Pandey, 2006; Wale and Yalew, 2007; Ramasamy et al., 1999), with particular interest in hedonic pricing (Dalton, 2004; Pingali et al., 2001) and revealed preferences of trait valuation (Useche et al., 2009). In this study, farmer preferences for drought-tolerant glutinous rice are of particular interest; thus, traits subjectively selected from PVS breeding program are hypothesized to influence the adoption of RD12.

By taking into account production characteristics (example, yield, duration, disease resistance) and consumption characteristics (example, taste), the adoption of glutinous rice variety in the drought-prone areas assumes a utility maximization behavior. An individual farmer (or household), n, would obtain a certain level of utility (and/or profit) from each variety alternative, j, and will choose one that provides the greatest utility. The true utility that farmer n obtains from variety j is U_{nj}, $j = 1, 2, \ldots J$. He will choose variety i if and only if $U_{ni} > U_{nj}$ $\forall j \neq i$. The true utility of farmers is unknown, but varietal traits and perception on traits, labeled x_{nj} $\forall j$, and farm and household characteristics, labeled s_n, can be observed. Following technology adoption based on technology attributes (Rahm and Huffman, 1984; Adesina and Zinnah, 1993), the representative utility, denoted $V_{nj} = V(x_{nj}, s_n)$ $\forall j$, depends on these observed variables. Since $U_{nj} \neq V_{nj}$, true utility is decomposed as $U_{nj} = V_{nj} + \varepsilon_{nj}$, where ε_{nj} is assumed to be random. The probability that farmer n chooses variety i (Train, 2009) can be written as:

$$P_{ni} = Prob(U_{ni} > U_{nj} \ \forall j \neq i)$$
$$= Prob(V_{ni} + \varepsilon_{nj} > V_{nj} + \varepsilon_{nj} \ \forall j \neq i)$$
$$= Prob(\varepsilon_{nj} - \varepsilon_{ni} < V_{ni} - V_{nj} \ \forall j \neq i). \quad (1)$$

The cumulative probably in Equation (1) can be written as:

$$P_{ni} = \int_{\varepsilon} I(\varepsilon_{nj} - \varepsilon_{ni} < V_{ni} - V_{nj} \ \forall j \neq i) f(\varepsilon_n)\, d\varepsilon_n . \quad (2)$$

where $I(.)$ equal 1 when the expression in parentheses is true and 0 otherwise. For this study, we assume that ε_{nj} is independently, identically distributed extreme value, and the cumulative distribution of $\varepsilon_{nj} - \varepsilon_{ni}$ *follows* the logistic distribution

$$F(\varepsilon_{nj} - \varepsilon_{ni}) = \frac{e^{\varepsilon_{nj} - \varepsilon_{ni}}}{1 + e^{\varepsilon_{nj} - \varepsilon_{ni}}} \quad (3).$$

Given that the logit probability of V_{ni} is sigmoid, the logit model provides appropriate implications for this study. It implies that a small increase in V_{ni} (presumably from improved traits) has little effect on the choice probability when V_{ni} of a variety is either very low or very high, compared to other varieties. The greatest effect of an increase in V_{ni} on the probability of it's being chosen is when the probability is close to 0.5. For example, in the drought-prone area if the utility from growing a drought-tolerant variety i is very low (or very high) compared to other varieties, a small improvement in the drought-tolerant trait will have little effect on the probability that a farmer will adopt variety i. The change of probability that variety i is adopted from an improved drought-tolerant characteristic will be greatest when there is 50-50 chance that it is being chosen. Omitting the proof of algebraic manipulation (Train, 2009), the logit choice probabilities of Equation (3) are given as:

$$P_{ni} = \frac{e^{V_{ni}}}{\sum_j e^{V_{nj}}} \quad (4)$$

The representative utility is specified to be linear in parameters: $V_{nj} = \alpha' x_{nj} + \beta' s_{nj}$. Thus, a logit choice probability in Equation (4) is defined as:

$$P_{ni} = \frac{e^{\alpha' x_{ni} + \beta' s_{ni}}}{\sum_j e^{\alpha' x_{nj} + \beta s_{nj}}} \quad (5)$$

Parameter estimates from Equation (5) are interpreted as a pairwise comparison between the effects of changes in independent variable on alternative i and the base alternative. The change in probability that farmer n chooses variety i given a change in an observed variable x_{ni} (or s_{ni}) is:

$$\frac{\partial P_{ni}}{\partial x_{ni}} = \frac{\partial \left(\frac{e^{\alpha' x_{ni} + \beta s_{ni}}}{\sum_j e^{\alpha' x_{nj} + \beta s_{nj}}} \right)}{\partial x_{ni}} \quad (6)$$

This marginal effect evaluated at the sample mean is given as:

Table 1. Sampling design and sample size.

Stage I: Rice production potential zone*	Stage II: Districts by RD12 promotion intensity	Total rice farming households	Population proportion	Expected sample size	Actual sample size	Actual sample proportion
A (High)	High: Muang Nongkhai	11,209	0.18	44	36	0.14
B (Medium)	High: Phon Phisai, Ratanawapi	19,516	0.32	77	103	0.40
	Low: Pak Khat	3,743	0.06	15	24	0.09
C (Low)	High: So Phisai	9,704	0.16	38	38	0.15
	Low: Fao Rai	6,749	0.10	27	21	0.08
D (Unsuitable)	Low: Bung Kan	10,973	0.18	43	36	0.14
	Total	61,894	1.00	244	258	1.00

*, Based on rice production potential zoning defined by Nongkhai Rice Research Center (2007).

$$\frac{\partial P_{ni}}{\partial x_n} = P_{ni}(\alpha_n - (\bar{\alpha}_n + \bar{\beta}_n)) \quad (7)$$

Where

$$\bar{\alpha}_n + \bar{\beta}_n = \sum_l P_{nl}\alpha_l + \sum_k P_{nk}\beta_k \quad (8)$$

Data

The target area for RD12 is the upper Northeast of Thailand where intermittent and late-season drought has been a major constraint on rice productivity. Since the adoption of the new variety is assumed to reveal preferences, the scope of the study area is limited to Nongkhai province in the upper Northeast of Thailand where farmers have access to RD12. Two stage stratified sampling technique was adopted. In the first stage, districts are grouped based on rice production potential zoning defined by Nongkhai Rice Research Center (2007).

The rice production potential zones were classified soil and water conditions combined with climate, top sequence, and farm management. In the second stage, each rice production potential zone is stratified by the intensity of RD12 accessibility. The number of farmers who received RD12 is used as a proxy to capture the extent where RD12 has been promoted. The intensity of RD12 promotion is classified low (less than twenty five farmers who received RD12 seeds during 2003 to 2008) to high (less than 25 farmers who received RD12 seeds during 2003 to 2008). The data are collected from households in seven districts of Nongkhai province in upper Northeast Thailand.

This study assumes a 5% statistical significant level and 20% distribution of RD12 in Nongkhai during the pre-survey. Adopting Tyfos (1996), the sample size is identified as 244 based on the total of 61,894 rice farming households in seven selected districts (Nongkhai Statistical Office, 2004). The two-stage stratified proportional sampling technique (to total number of rice farm households) was used for the 2009/2010 cropping season. Table 1 shows the actual number of samples in each district. Farmers were interviewed to obtain rice farming information on every plot owned (or rented) by a household, including both during the wet and dry seasons.

EMPIRICAL MODEL

Based on collected data the four major glutinous rice varieties grown in the sampling area are Hanhg Yi71, RD12, RD10, and RD6. The summary of traits comparison among these varieties is shown in Table 2. A multinomial logit model derived from Equation (5) is estimated for choices of glutinous rice variety *j* based upon the assumption of a linear utility function. Thus,

$$U_{nj} = \sum_{l=1}^{L} \alpha_{nl}\, x_{jl} + \sum_{k=1}^{K} \beta_{nk}\, s_{jk} + \varepsilon_{nj} \quad (9)$$

In the multinomial logit model, traits and farmers perception on traits (x) includes the yield and taste preference of the new variety (RD12), and the popular high cooking quality rice (RD6), relative to the existing drought-tolerant variety (Hanhg Yi71). The extension of this logit model is to include taste variation associated with observed variables (Train, 2009). We assume that the utility a farmer receives from yield varies depending on the proportion household consumption of glutinous rice to total rice production (CONSUMP). As a result differences in taste could be reflected in the parameter of yield; thus $\beta_{n,yield} = \rho$*(CONSUMP).

Farm and household characteristics (s) includes the production condition in the drought prone area such as toposequence, access to water, cropping season, recent experience of blast disease, and whether rice production is the main source of income. Access to water is limited in the rainfed areas, particularly during the dry season so these conditions are hypothesized for the adoption of the drought tolerant variety. Aside from drought, blast disease is also the major production problem in sampling areas so the experience of blast disease may also influence the adoption decision. Focusing on the occurrence of RD12 adoption, a binomial logit model is

Table 2. Traits comparison of popular glutinous rice varieties.

Variety	Photo-period sensitive	Dry season	Harvest time	Appropriate for upper terrain	Drought-tolerant	Potential yield (ton/ha)	Cooking quality	Blast tolerant
Hanhg Yi71	Yes	No	Nov. 4	Yes	Drought escape	3.16	Soft/chewy	Yes
RD12	Yes	No	Nov. 7-17	Somewhat	Drought escape	3.26	Good, soft/chewy	Somewhat
RD10	No	Yes	130 days	No	No	4.12	Soft	Somewhat
RD6	Yes	No	Nov. 21	No	Yes	4.16	Very good, soft/aromatic	Susceptible

Source: Rice knowledge bank, Rice Department, 2009.

Table 3. Description of variables in choice models.

Dependent variable	
VARIETY	Glutinous rice variety: Hanhg Yi71, RD6, RD10, RD12
ADOPT	RD12 adoption: 1 if adopt, 0 otherwise.
Independent variables	
Traits and Perception on Traits, X_l	
YIELD	Rice yield (kg/rai)
CONSUMP*YIELD	Proportion household consumption of glutinous rice to total rice production * YIELD
PRICE	Output price (baht/kg)
RD12TASTE	Farmer perception of RD12 taste: 1 if farmer thought RD12 has better cooking quality than Hahngyi71, 0 otherwise
RD6TASTE	Farmer perception of RD6 taste: 1 if farmer thought RD6 has better cooking quality than Hahngyi71, 0 otherwise
RD12HARVEST	Farmer perception of the ease to harvest RD12: 1 if farmer thought RD12 is easier to harvest than Hahngyi71, 0 otherwise
RD12PLANTING	Farmer perception of the ease to transplant RD12: 1 if farmer thought RD12 is easier for crop establishment than Hahngyi71, 0 otherwise
Farm and household characteristics, S_k	
CONSUMP	Proportion household consumption of glutinous rice to total rice production
TOPOSEQUENCE	Toposequence of rice production plot: 1 if upper sloping terrain, 0 otherwise.
WATER	Access to water source: 1 if irrigated or availability of other water resources, 0 otherwise.
SEASON	Cropping season: 1 if dry season, 0 otherwise
BLASTEXP	Experience of blast disease: 1 if farmer experience blast disease in the past five years, 0 otherwise
MAININC	Main income source: 1 if rice is major source of income, 0 otherwise.

1 ha=6.25 rai.

estimated (1 if adopted, 0 if not adopted). Since RD12 is photoperiod-sensitive, the binomial logit model of RD12 adoption includes only the wet season. Farmers' acceptance of RD12 is hypothesized be influenced mainly by trait perceptions of RD12 relative to Hanhg Yi71—the existing drought tolerant variety. Table 3 lists all variables of the choice models.

RESULTS AND DISCUSSION

Tables 4 to 6 show the summary of glutinous rice varieties grown in the study areas. Based on four hundred and forty four plots, RD12 and RD6 are the most preferred varieties (about 35% individually), followed by Hanhg Yi71 and RD10, respectively. Both RD12 and RD6 are photoperiod-sensitive and have good cooking quality, compared to the others. San Pah Tawng and RD4 were other glutinous rice varieties found in this area, but they are not popular.

The two varieties together accounted for less than 2%. This probably is because San Pah Tawng and RD4 have

Table 4. Glutinous rice varieties grown in Nongkhai province, 2009/2010.

Variety	No. of plots	Percentage
Hanhg Yi71	97	21.85
RD12	155	34.91
RD10	26	5.86
RD6	159	35.81
RD4	6	1.35
San Pah Tawng	1	0.23
Total	444	100.00

Table 5. Glutinous rice varieties in Nongkhai province by toposequence, 2009/2010*.

Variety	Lower terrain		Upper terrain	
	No. of plots	Percentage	No. of plots	Percentage
Hanhg Yi71	7	3.52	90	37.82
RD12	34	17.09	121	50.84
RD10	14	7.04	12	5.04
RD6	144	72.36	15	6.30
Total	199	100.00	238	100.00

*Excluding RD4 and San Pah Tawng.

Table 6. Glutinous rice varieties grown in Nongkhai province by cropping season, 2009/2010*.

Variety	Wet		Dry	
	No. of plots	Percentage	No. of plots	Percentage
Hanhg Yi71	96	22.97	1	5.26
RD12	154	36.84	1	5.26
RD10	11	2.63	15	78.95
RD6	157	37.56	2	10.53
Total	418	100.00	19	100.00

* Excluding RD4 and San Pah Tawng.

poorer cooking quality, are susceptible to blast and are not drought resistant. These findings are similar to the studies by Isvilanonda and Hossain (2000), Gypmantasiri et al. (2003) and Vejpas et al. (2005). Prior to the release of RD12, RD 6 was the most common glutinous rice variety in the rainfed area of Northeast Thailand. Early maturing varieties, including RD15 which has shorter duration than KDML105 (non-glutinous Jasmine rice varieties) were becoming more popular for rainfed lowland area during the wet season (Vejpas et al., 2005). In the upper terrain, RD12 is the most popular variety and accounted for about half of all plots. Hanhg Yi71 which was the recommended variety for upper terrain before the development RD12 is the second most popular (Table 5). In the lower terrain, however, RD6 which has the highest cooking quality is the most popular variety and accounted for over 70% of all lower terrain production. Rice production in the study area is very limited during the dry season as is evident from Table 6. Again, RD12 and RD6 are the most popular varieties in the wet season. Unsurprisingly compared to RD10 which is non-photoperiod sensitive; the farmers are less likely to choose Hanhg Yi71, RD12 and RD6 in the dry season.

The parameter estimates of the multinomial logit model are presented in Table 7. More useful results are the estimates of marginal effects presented in Table 8. One of the most interesting findings is that the probability of choosing Hahng Yi71 or RD12 significantly increases by35% for upper terrain while there is only about 4% probably of choosing RD10. The probability of choosing RD6, however, decreases by about 7% for the upper terrain. This implies that to minimize the risks from late-

Table 7. Parameter estimates of multinomial logit model (RD10=base).

Variable	Hanhg Yi71		RD12		RD6	
	Parameter (β)	Std. Err.	Parameter (β)	Std. Err.	Parameter (β)	Std. Err.
CONST	0.229	1.319	2.618**	1.226	3.949**	1.241
CONSUMP*YIELD	0.004	0.003	0.003	0.003	0.001	0.003
RD12 TASTE	-0.103	1.000	-0.159	0.981	-1.781	1.130
RD6 TASTE	3.204**	1.523	1.443	1.537	3.355**	1.565
TOPOSEQUENCE	0.218	1.130	-1.062	1.067	-4.914**	1.097
WATER	-0.255	0.763	0.031	0.721	0.069	0.769
SEASON	-4.490**	1.573	-6.040**	1.477	-6.997**	1.348
BLAST	0.812	0.674	0.459	0.629	1.119	0.704
MAININC	0.072	0.876	0.218	0.830	-0.147	0.891
Log likelihood	-342.039					
Number of plots	434					

Table 8. Marginal effects of multinomial logit model.

Variable	Hanhg Yi71		RD12		RD6		RD10	
	Coeff	Std. Err.	Coeff	Std. Err.	Coeff	Std. Err.	Coeff	Std. Err.
CONSUMP* YIELD	0.0003*	0.000	0.0002	0.000	-0.00004**	0.000	-0.00006	0.000
RD12 TASTE	0.077	0.073	0.150*	0.089	-0.240**	0.061	0.013	0.032
RD6 TASTE	0.147*	0.081	-0.381**	0.068	0.268	0.111	-0.034**	0.017
TOPOSEQUENCE	0.353**	0.040	0.352**	0.047	-0.743**	-0.035	0.0384**	0.017
WATER	-0.046	0.056	0.022	0.079	0.024	0.082	0.0003	0.017
SEASON	-0.100	0.118	-0.440**	0.072	-0.286**	0.039	0.826**	0.139
BLAST	0.026	0.055	-0.114	0.070	0.109	0.067	-0.021	0.024
MAININC	-0.003	0.068	0.066	0.084	-0.061	0.081	-0.002	0.019

** = significance at 5% level, * = significance at 10% level.

season drought, the early maturity trait of Hanhg Yi71 and RD12 is the key reason for adopting drought tolerant varieties in the upper sloping terrain.

Another interesting result is the perception that RD12 has better cooking quality than Hanhg Yi71. This increases the probability of adopting RD12 by about 15% and lowers the probability of adopting RD6 by 2.4%. It suggests that if farmers have a higher preference for the cooking quality of RD12 than for Hanhg Yi71, they will be more likely to adopt RD12 if they believe that RD12 is appropriate for protection against drought. In contrast, the perception that RD6 has a preferred cooking quality than Hahng Yi71 decreases the probability of RD12 by 38%. This is because RD6 is superior for consumption in the existing market. The adoption of new variety RD12 is diminished when farmers have strong cooking preferences for existing varieties.

It is hypothesized that yield could be an important determinant of variety adoption. By allowing variability of yield preference by the proportion of household consumption, it was found that yield as valued by household consumption increases the probability of cultivating Hahng Yi71, but lowers the probability of cultivating RD6. This could be because RD6 is premium glutinous rice that also has high market price. As demand of household consumption increases despite increasing yield, the probability of adopting RD6 decreases. In other words, as household rice demand increases, farmers are more likely to grow Hanhg Yi71 and less likely to grow RD6. Farming in the dry season is possible only for the non-photosensitive variety. Predictably, it increases the probability of growing RD10 (non-photosensitive variety) and decreases the probability of growing RD12 and RD6 (photosensitive varieties).

The results of RD12 adoption excluding dry season from the binomial logit model are shown in Table 9. It is found that for the upper terrain, the probability of adopting RD12 increases by 41%, compared to the lower terrain. Similar implication drawn from this binomial model is that for upper sloping terrain, increasing risk of late-season drought results in a higher likelihood of new early maturity variety. The studies by Joshi and Bauer (2006),

Table 9. Parameter estimates and marginal effect of binomial logit model of RD12 adoption in wet season.

Variable	Parameter (β)	Std. Err.	Marginal effect	Std. Err.
CONST	-5.328**	1.722		
YIELD	-0.0004	0.001	-0.00009	0.000
PRICE	0.381**	0.186	0.078**	0.038
RD12 TASTE	0.685	0.982	0.156	0.239
RD12 HARVEST	-0.799	0.864	-0.139	0.122
RD12 TRANSPLANT	-2.041	1.526	-0.248**	0.087
TOPOSEQUENCE	2.238**	0.518	0.418**	0.079
CONSUMP	1.434**	0.671	0.295**	0.137
Log likelihood	-79.409			
Number of plots	147			

** = significance at 5% level, * = significance at 10% level.

Nanfumba et al. (2013), and Ward et al. (2013) also suggest that farmer's preferences towards early-maturity or short duration significantly affects rice variety choice in the rainfed area. Because glutinous rice production in this region is important for household consumption, taste preference can be important to the adoption decision. The probability of adopting RD12 increases as the percentage of household consumption increases, perhaps owing to the superior taste and higher potential yield than the existing drought tolerant variety. Similarly, increasing the selling price for rice with the improved cooking quality makes farmers more likely to adopt RD12.

The physical appearance of rice paddy RD12 is more similar Hahng Yi71 (long, skinner, darker yellowish brown) than RD6.Since RD12 is new to the market and not easy to distinguish by appearance, the market price of RD12 is generally the same as Hahng Yi71 although it's cooking quality is similar to RD6. This may suggest that if RD12 were to become more recognized in the market with a higher price, farmers would be more likely to adopt it.

Between two competing drought tolerant varieties: RD12 and Hahng Yi71, RD12 has stronger stalks than Hahng Yi71 so it's believed to be easier to harvest, especially in environment where wind and storms can cause fallen stalks. However, the perception that RD12 is easier to harvest than Hahng Yi71 insignificantly affects the adoption decision of RD12. This is not surprising because from the PVS, tillering is not one of the key preferred traits and our result implies that it may not be important to farmers. Surprisingly, perception of superior cooking quality of RD12, compared to Hahng Yi71, does not affect the probability of RD12 adoption. This could be because in the binomial model, the dry season is excluded and the cooking quality of the new drought tolerant variety is not as important as the drought escape characteristic thus, even if RD12 is superior to Hahng Yi71 for consumption, the taste alone does not influence the adoption of RD12. The result is similar to Joshi and Bauer (2006) that taste is not significant toward variety choice in the rainfed area of Nepal. It is surprising that the perception that RD12 is easier to transplant than

Hahng Yi71 negatively affects RD12 adoption. It is possible that the ease of transplanting is mistakenly perceived as weaker rooting and germinating and negatively has impacted the adoption.

CONCLUSION AND RECOMMENDATION

The breeding program of the Rice Department released the new glutinous rice variety RD12, to target the drought problem in Northeast Thailand in 2007. In the RD12 breeding program, farmers' participatory selection was the key in the selection process to ensure the acceptance of the new variety. The major improved characteristics of RD12 which were identified from PVS included early maturity and good cooking quality. RD12 has been promoted in Nongkhai province in the upper Northeast of Thailand where late season drought is a major constraint, particularly for upper terrain. Hahng Yi71 had been the most appropriate variety to target the drought problem prior to the release of RD12 and it has been widely adopted despite its poor cooking quality. Our results show that major traits of RD12 identified by the farmer PVS significantly influences the adoption of the new variety. Farmers in the upper terrain production are more likely to adopt RD12 to minimize the risk from late-season drought. The taste preference towards RD12 compared to existing drought tolerant variety, Hahng Yi71, also significantly increases the probability of RD12 adoption and lowers the probability of adopting RD6, the superior cooking quality variety. However, since RD12 is photoperiod-sensitive, it is less preferred for cultivation in the dry season. The current market still does not recognize the quality of the RD12 and often suppresses its price because of its similar physical appearance to the lower quality variety. The result shows that a higher price increases the adoption of RD12; this suggests that when RD12 becomes more recognized in the market and when its price reflects its superior quality, it will be more accepted by farmers. The adoption RD12 has been found to be closely related to farmer perception of its traits. The preferred traits identified by farmer PVS program in rice breeding of RD12 effectively enhance its adoption.

Conflict of Interests

The author(s) have not declared any conflict of interests.

ACKNOWLEDGEMENTS

This study is part of the "Measuring the Impact of Rockefeller Foundation's Funding for the Development of Drought-tolerant Rice in Thailand" project supported by the Rockefeller Foundation. The authors greatly appreciate the insights on drought-tolerant rice development in Thailand from the Rice Department breeders: Dr. Boonrat Jongdee, Dr. Grienggrai Pantuwan, and Dr. Suwat Jearakongman. They also thank Mr. Somjai Saleeto from Nongkhai Rice Research Center for information on RD12 extension program; and Ms. Pitchaya Chaiyapoom and Mr. Grianggrai Tubtong for data collection.

REFERENCES

Adesina AA, Zinnah MM (1993). Technology characteristics, farmers' perceptions and adoption decisions: A Tobit model application in Sierra Leone. Agric. Econ. 9:297-311.

Adesina AA, Baidu-Forson J (1995). Farmers' perceptions and adoption of new agricultural technology: evidence from analysis in Burkina Faso and Guinea, West Africa. Agric. Econ. 13:1-9.

Adesina AA, Seidi S (1995). Farmers' perceptions and adoption of new agricultural technology: analysis of modern mangrove rice varieties in Guinea Bissau. Q. J. Int. Agric. 34:358-371.

Asian Development Bank (2009). The Economics of Climate Change in Southeast Asia: A Regional Review. [online] Available from: http://www.adb.org/sites/default/files/economics-climate-change-se-asia.pdf. [Accessed 10 June 2013].

Bellon MR, Reeves J (Eds.) (2002). Quantitative Analysis of Data from Participatory Methods in Plant Breeding. Mexico, DF: CIMMYT.

Ceccarelli S, Grando S (2007). Decentralized-participatory plant breeding: an example of demand driven research. Euphytica 155:349-360.

Dalton TJ (2004). A household hedonic model of rice traits: economic values from farmers in West Africa. Agric. Econ. 31:149-159. http://dx.doi.org/10.1016/j.agecon.2004.09.003

Diagn A (2006). Diffusion and adoption of Nerica rice varieties in Côte d'Ivoire. Dev. Econ. 44:208-231.

Doss C (2006). Analyzing technology adoption using microstudies: limitations, challenges, and opportunities for improvement. Agric. Econ.34:207-219.

Doward P, Craufurd P, Marfo K, Dogbe W, Bam R (2007). Improving participatory varietal selection processes: participatory varietal selection and the role of informal seed diffusion mechanisms for upland rice in Ghana. Euphytica 155:315-327.

Edmeades S, Phaneuf DJ, Smale M, Renkow M (2008). Modelling the crop variety demand of semi-subsistence households: bananas in Uganda. J. Agric. Econ. 52:329-349.

Efisue A, Tongoona P, Derera J, Langyintuo A, Laing M, Ubi B (2008). Farmers' perceptions on rice varieties in Sikasso region of Mali and their implications for rice breeding. J. Agron. Crop. Sci. 194:393-400.

Feder G, Just RE, Zilberman D (1985). Adoption of agricultural innovations in developing countries: a survey. Econ. Dev. Cult. Change. 33:255-297. [online] Available from: http://www.jstor.org/stable/1153228. [Accessed 10 June 2013].

Floch P, Molle F (2007). Marshalling water resources: a chronology of irrigation development in the Chi-Mun river basin, Northeast Thailand. Mekong program on water, environment and resilience, IRD/IWMI. Working Paper.

Gypmantasiri P, Limnirankul B, Muangsuk C (Eds.) (2003). Integration of farmer participatory plant breeding for rainfed lowland rice improvement in North and Northeast Thailand: I Bio-physical and socio-economic characterization of rainfed lowland rice production systems of the North and the Northeast of Thailand. Ubon (Thailand): Ubon Ratchathani Rice Research Center. [online] Available from: http://www.mcc.cmu.ac.th/agsust/publication_SA/Biophysical-socioeconomic%20report_full%20paper.pdf. [Accessed 10 June 2014].

Hintze LH, Renkow M, Sain G (2003). Variety characteristics and maize adoption in Honduras. Agric. Econ. 29:307-317.

International Rice Research Institute (2006). Rice Breeding Course. International Rice Research Institute. [online] Available from: http://www.knowledgebank.irri.org/ricebreedingcourse/bodydefault.htm#Participatory_Variety_Trials_For_Rainfed_Rice_Cultivar_Evaluation.htm. [Accessed 20 October 2013].

International Rice Research Institute (2010). Rough rice production and yield. In World Rice Statistics. International Rice Research Institute. [online] Available from: http://ricestat.irri.org:8080/wrs/. [Accessed 20 June 2013].

Isvilanonda S, Hossain M (2000). Thailand's Rice Economy and Constraints to Increasing Production. Paper presented at the IRRI-ICAR International Workshop on Constraints to Increasing Rice Production in Asia: Insights from a Study on Farmers' Perceptions, Hyderabad, India, 7-9 June, 2000.

Jongdee B (2003). Designing a national breeding program for developing drought tolerant rainfed lowland varieties: the Thailand experience. In KS Fischer, R Lafitte, S Fukai, G Atlin and B Hardy (Eds.), Breeding Rice for Drought- prone Environments. Los Baños: Int. Rice Res. Institute, pp. 64-69.

Jongdee B, Pantuwan G, Fukai S, Fischer K (2006). Improving drought tolerance in rainfed lowland rice: an example from Thailand. Agric. Water. Manage. 80:225-240.

Jongdee S, Mitchell, JH, Fukai S (1997). Modeling approach for estimation of rice yield reduction due to drought in Thailand. In S Fukai, M Cooper and J Salisbury (Eds.), Breeding Strategies for Rainfed Lowland Rice in Drought-prone Environments. Canberra: Austrian Centre for International Agricultural Research, pp. 65-73.

Joshi GR, Bauer S (2006). Farmers' choice of the modern rice varieties in the rainfed ecosystem of Nepal. J Agric. Rural. Dev. Trop. 107:129-138. [online] Available from: http://www.jarts.info/index.php/jarts/article/view/127/110. [Accessed 10 June 2014].

Joshi GR, Pandey S (2006). Farmer's perceptions and adoption of modern rice varieties in Nepal. Q. J. Int. Agric. 45:171-186.

Kamoshita A, Babu RC, Boopathi NM, Fukai S (2008). Phenotypic and genotypic analysis of drought-resistance traits for development of rice

cultivars adapted to rainfed environments. Field. Crop. Res. 109:1-23.
Manzanilla DO, Paris TR, Vergara GV, Ismail AM, Pandey S, Labios RV, Tatlonghari GT, Acda RD, Chi TTN, Duoangsila K, Siliphouthone I, Manikmas MOA, Mackill DJ (2011). Submergence risks and farmers' preferences: implications for breeding Sub1 rice in Southeast Asia. Agric. Syst. 104:335-347.
Mongkolsawat C, Thirangoon P, Suwanweramtorn R, Karladee N, Paiboonsank S, Champathet P (2001). An evaluation of drought risk area in Northeast Thailand using remotely sensed data and GIS. Asian. J. Geoi. 1:33-43.
Morris ML, Bellon MR (2004). Participatory plant breeding research: opportunities and challenges for the international crop improvement system. Euphytica 136:21-34.
Nanfumba D, Turyahabwe N, Ssebuliba J, Kakuru W, Kaugule J, Omio S, Samuka M (2013). Participatory identification of farmer acceptable improved rice varieties for rain-fed lowland ecologies in Uganda. Afr. Crop. Sci. J. 21:683-691. [online] Available from: http://www.ajol.info/index.php/acsj/article/view/98439/87716. [Accessed 20 June 2014]
Nongkhai Rice Research Center (2007). Rice Production Potential Zoning in Nongkhai. Khonkaen: Khonkaen Printing.
Nongkhai Rice Research Center (2013). Nongkhai Province: Knowledge Management Corner for Rice Production. Nongkhai Rice Research Center. [online] Available from: http://nki.brrd.in.th/km/index.php?option=com_content&view=article&id=45. [Accessed 20 June 2014].
Nongkhai Statistical Office (2004). Rice Production Area, Villages, and Households by District.
Office of Agricultural Economics of Thailand (2011). Area, yield and total output of rice in Rice Production Statistics, 2010/11. [online] Available from: http://www.oae.go.th/ewt_news.php?nid=9704 [Accessed 20 June 2014]
Pingali PL, Rozelle SD, Gerpacio R (2001). The farmer's voice in priority setting: a cross-country experiment in eliciting technological preferences. Econ. Dev. Cult. Change 49:591-609.
Pantuwan G, Jongdee B, Jearakongman S, others (2006). Integration of farmer participatory plant breeding for rainfed lowland rice improvement in North and Northeast. Unpublished report, Ubonratchathani: Rice Department. a copy of this report is on file with the authors. (in Thai)
Prapertchob P, Bhandari H, Pandey S (2007). Economic costs of drought and rice farmers' drought-coping mechanisms in Northeast Thailand. In S Pandey, H Bhandari, B Hardy, (Eds.), Economic costs of drought and rice farmers' coping mechanisms: a cross-country comparative analysis. Los Baños: International Rice Research Institute, pp. 113-148.
Rahm MR, Huffman WE (1984). The adoption of reduced tillage: the role of human capital and other variables. Am. J. Agric. Econ. 66:405-413.
Ramasamy C, Bantilan CS, Elangovan S, Asokan M (1999). Perceptions and adoption decisions of farmers in cultivation of improved pearl millet cultivars - a study in Tamil Nadu. Indian. J. Agric. Econ. 54:139-154.
Rasabandit S, Jaisil P, Atlin G, Vera Cruz C, Jongdee B, Banterng P (2006). Participatory variety selection (PVS) to assess farmer preferences of traditional glutinous rice varieties in the Lao PDR. Khon Kaen Agric. 34:128-141.
Rice Department (2009). Registered Rice Varieties of Thailand Database. Bureau of Rice Research and Development. [online] Available from: http://www.brrd.in.th/rvdb/ (in Thai) [Accessed 20 October 2013].
Saleeto S, Jumpaget R, Gasemrat C, others (2009). RD12 (Nong Khai 80) rice variety. Thai. Rice. Res. J. 3:5-20.
Snapp S (2002). Quantifying farmer evaluation of technologies: the mother and baby trial design. In MR Bellon, J Reeves (Eds.), Quantitative Analysis of Data from Participatory Methods in Plant Breeding. Mexico, DF: CIMMYT.
Thai Meteorological Department (2012). The assessment of rainfall impact on crops in Thailand. August- September, 2012. 10. Meteorological Department. (in Thai) [online] Available from: http://www.tmd.go.th/climate/climate.php?FileID=10. [Accessed 20 October 2013].
Thai Meteorological Department. (2010). Climate variability and climate change projection. Technical Document No. 551-524-01-2010. Climatological Center, Meteorological Department. (in Thai) [online] Available from: http://www.tmd.go.th/info/climate_future.pd. [Accessed 20 October 2013].
Train KE (2009). Discrete Choice Methods with Simulation. Cambridge: Cambridge University Press.
Tshewang S, Ghimiray M (2010). Participatory variety selection: increasing rice varietal diversity. Bhu. J. RNR. 6:1-10. [online] Available from: http://www.corrb.gov.bt/images/RNR%20Journals/RNR%20Journal%20of%20Bhutan%202010.pdf . [Accessed 20 June 2014].
Tyfos P (1996). Sampling Methods for Applied Research: Text and Cases. New York: Wiley.
Useche P, Barham BL, Foltz JD (2009). Integrating technology traits and producer heterogeneity: a mixed-multinomial model of genetically modified corn adoption. Am. J. Agric. Econ. 91:444-461.
Vejpas C, Bousquet F, Naivinit W, Trébuil G, Srisombat N (2005). Companion modeling and multi-agent systems for integrated natural resource management in Asia. In B François, T Guy, H Bill (Eds.), Metro Manila: IRRI, 141-163.
Wale E, Yalew A (2007). Farmers' variety attribute preferences: implications for breeding priority setting and agricultural extension policy in Ethiopia. Afr. Dev. Rev. 19:379-396.
Ward PS, Ortega DL, Spielman DJ, Singh V (2013). Farmer Preferences for Drought Tolerance in Hybrid versus Inbred Rice: Evidence from Bihar, India. IFPRI Discussion Paper 01307. December 2013. International Food Policy Research Institute. CGIAR Research Program on Policies, Institutions and Markets. [online] Available from: http://www.ifpri.org/sites/default/files/publications/ifpridp01307.pdf. [Accessed 15 June 2014]
Witcombe JR, Petre R, Jones S, Joshi A (1999). Farmer participatory crop improvement. IV. The spread and impact of a rice variety identified by participatory varietal selection. Exp. Agric. 35:471-487.
Wurzinger M, Solkner J, Iniguez L (2011). Important aspects and limitations in considering community-based breeding programs for low-input smallholder livestock systems. Small Ruminant Res. 98:170-175.

Profit efficiency of dairy farmers in Kenya: An application to smallholder farmers in Rift Valley and Central Province

Leone Iga Mawa[1], Mutuku Muendo Kavoi[2], Isabelle Baltenweck[1] and Jane Poole[3]

[1]Department of Markets, Gender and Livelihoods, International Livestock Research Institute (ILRI), P. O. Box 30709-00100, Nairobi, Kenya.
[2]Faculty of Agriculture, Jomo Kenyatta University of Agriculture and Technology, P. O. Box 62000-00200, Nairobi, Kenya.
[3]Research Methods Group, International Livestock Research Institute, P. O. Box 30709-00100, Nairobi, Kenya.

The dairy industry in Kenya is an important source of livelihood among the smallholder farmers who supply over 70% of the total milk. However, there is a growing concern on rising costs of milk production among farmers. The study assessed profit efficiency of smallholder dairying in the Rift Valley and Central Provinces of Kenya using stochastic frontier analysis for estimating farm level profit efficiency and identifying the specific determinants of efficiency. The results showed that the farmers are fairly profit efficient with an average of about 68%. Cost of fodder produced on farm significantly improved profit efficiency among farmers. However dairy profit efficiency can be enhanced if fodder production is embraced, as well as other supplemental feed technologies that are commensurate with local conditions. Institutional policy reforms on smallholder dairying will help protect the industry and its sustainability for smallholders.

Key words: Dairy, profit efficiency, stochastic frontier, smallholder farmer.

INTRODUCTION

Kenya is the second largest dairy producer and consumer in sub-Sahara Africa and is relatively self-reliant (USAID, 2010). About 60% of the total milk production in Kenya is produced by farmers in the Rift Valley and Central Province who own about 80% of the exotic and cross-breed cattle (Omore et al., 1999). Previous studies have identified that smallholder farmers supply over 70% of the total milk, mainly from cattle, but some little quantities from camels and goats (USAID, 2008; Muriuki, 2003). Smallholder dairying is a potential contributor of income and employment generation both on-farm and off-farm, as well as improved nutrition of households (Staal et al., 2008; USAID, 2010). At the macro level, the dairy industry contributes an estimated 14% of agricultural gross domestic product (GDP) and approximately 4% of overall Kenya's national GDP (USAID, 2010). It is anticipated that the demand for milk is likely to double due to the growing world population

(FAO, 2011, 2003). Despite the demand prospects, farmers and stakeholders have increasingly expressed concern over the growing costs of milk production which ultimately result in diminishing milk profits in Kenya. FAO (2003) reported that there is competition for land, water and other environmental resources due to the increasing population. It is apparently plausible that these constraints are shifting farmers' practices from traditional pasture grazed systems to intensive production technologies which are often more expensive than extensive systems. This study proposed to assess profit efficiency of smallholder dairy production and identify its determinants among farmers in the Rift Valley and Central Provinces using Stochastic Frontier approach (SFA). Profit efficiency is a wider concept than cost efficiency since it takes into account the effects of the choice of a certain vector of production both on costs and on revenues, thus offering complementary information useful for the analysis of dairy farm efficiency.

METHODOLOGY

Theoretical framework

Stochastic production frontier models were introduced by Aigner et al. (1977) and Meensen and van den Broeck (1977). Battese and Coelli (1995) and Coelli (1996), extended the stochastic production frontier model, suggesting that the inefficiency effects can be expressed as a linear function of explanatory variables, reflecting farm-specific characteristics. Farm level inefficiency measurements are common among researchers (Kumbhakar et al., 1989; Wang et al., 1996; Abdulai and Huffman, 2000; Rahman, 2003). The advantage of the stochastic frontier model is that it allows estimation of farm specific efficiency scores and the factors explaining efficiency differentials among farms in a single stage estimation procedure. Three common efficiency measures include Technical efficiency, Allocative efficiency and Economic efficiency. However, the profit efficiency measure combines the concepts of Technical and Allocative efficiencies into the profit relationship and as such any errors in the production decision are assumed to be translated into lower profits or revenue for the producer (Ali et al., 1994). Kumbhakar and Lovell (2000) provided a detailed account of stochastic frontier models. Accordingly, the stochastic frontier function is defined by Equation 1 as follows:

$$\Pi = f(P_{ij}, Z_{ik})\ exp(\varepsilon) \qquad (1)$$

The error term, $\varepsilon_i = v_i\_u_i$, is assumed to behave in a manner consistent with the stochastic frontier concept (Ali and Flinn, 1989)
Where:
Π = normalized profit of the ith firm;
P_{ij} = price of jth variable input faced by the ith farm divided by output price;
Z_{ik} = level of the kth fixed factor on the ith firm.
v_i is assumed to be identically and normally distributed with mean zero and constant variance as:
$N(0, \sigma_v^2)$.
u_i is the one-sided disturbance form representing profit inefficiency and it is independent of v_i;
and i = 1,2 ….., n, representing the individual firms.

Profit efficiency of the ith firm can be presented as:

$$EFF = E[exp(-u_i)/\varepsilon_i] = E[exp(-\delta_0 - \sum_{d=1}^{D}\delta_d W_{di})/\ \varepsilon_i] \qquad (2)$$

Where:
E = expectation operator, which is achieved by obtaining the expressions for the conditional expectation u_i upon the observed value of ε_i;
W_{di} = d^{th} explanatory variable associated with inefficiencies on firm i.
δ_0 and δ_d = unknown parameters jointly estimated using the maximum likelihood method with the stochastic frontier and the inefficiency effects functions simultaneously.
The likelihood function is expressed in terms of the variance parameters: sigma squared;

$(\sigma^2) = \sigma_v^2 - \sigma_u^2$ and; gamma (γ) = $\frac{\sigma_u^2}{\sigma^2}$ (Battese and Coelli 1995).
The parameter γ represents the share of inefficiency in the overall residual variance with values in the interval of 0 and 1. A value of 1 suggests the existence of a deterministic frontier, whereas a value of 0 can be seen as evidence in the favor of ordinary least square (OLS) estimation.

Specification of the empirical model

Profit efficiency is defined as the gain from operating on the profit frontier, while taking into account farm-specific prices faced and factor endowments. Assuming a farm that maximizes profit is operating in a perfectly competitive input and output markets and uses a singular output technology, the actual normalized profit function is derived as:

$$GM(\Pi) = TR - TVC = PQ - WX_i \qquad (3)$$

Where: GM = Gross margin; TR = total revenue; TVC = total variable cost - (as opposed to fixed costs as they remain fixed whether or not production has taken place, and to what scale production has been). Normalizing the profit function is achieved by dividing both sides of Equation 3 by the output market price, that is:

$$\frac{\Pi(p,z)}{p} = \sum\frac{(PQ-WX_i)}{P} = \frac{PQ-WX_i}{P} = f(X_i, Z)\sum P_i X_i \qquad (4)$$

Where: - $f(X_i, Z)$ is the production function.
The profit function in implicit form which specifies efficiency is expressed as:

$$\Pi = f(P_{ij}, Z_{ik})\ exp\ (v_i - u_i) \qquad (5)$$

I = 1,2,3………….n represent the individual firms.
The profit efficiency is expressed as the ratio of predicted profit to the predicted maximum profit for a best firm and is expressed as:

$$\text{Profit efficiency } (E\Pi) = \frac{\Pi}{\Pi_{max}} = \frac{exp[\Pi(p,z)]exp(\ln v)exp(\ln u)0}{exp[\Pi(p,z)]exp(\ln v)} \qquad (6)$$

Firm specific profit efficiency is again the mean of the conditional distribution of u_i given by $E\Pi$ and is defined as:

$$E\Pi = E[\frac{exp(\ln u_i)}{E_i}] \qquad (7)$$

$E\Pi$ takes the value between 0 and 1.
If u_i is = 0, that is, on the frontier, the firm is obtaining potential

maximum profit given the price it faces and the level of fixed factors. If $u_i > 0$, the firm is inefficient and losses profit as a result of inefficiency. The inefficiency effect model can only be estimated if the efficiency effects are present. Given that u_i is present in the model it implies that it is justifiable to employ the SFA as similarly argued by Aneani et al. (2011). In this study, Battesse and Coelli (1995) and Coelli (1996) models were used to specify the stochastic frontier function with behavior inefficiency components and used to estimate all parameters together in one step maximum likelihood estimation.

Different functional forms have been used by scholars for measuring firm level efficiency, namely the Cobb-Douglas function, normalized quadratic, normalized translog and generalized Leontif, but the commonly used forms are the Cobb-Douglas and Translog forms. A detailed literature is presented by Abdulai and Huffman (2000) on the weaknesses of the two functional forms. It is argued that the Cobb-Douglas form is restrictive compared to the more flexible functional forms such as the translog and quadratic forms. Upton (1979) also added that the Cobb-Douglas function cannot show both increasing and diminishing marginal productivity in a single response curve. As a result it does not give a technical optimum and may lead to the over estimation of the economic optimum. An ideal option would be the translog. However, the drawbacks of the translog model are that it has potential problems of insufficient degrees of freedom due to the presence of interaction terms; even though such interaction terms do have important economic implications and meaning (Abdulai and Huffman, 2000). Scholars such as Ọlayide and Heady (1982) used the quadratic function to measure the direct effects of inputs on output. However, the transcendental function and the quadratic functional models seem unpopular among researchers due to limited application.

Despite the restrictive nature of the Cobb-Douglas function, many scholars and researchers have found it relevant, especially when there are many variables in the model (Taru et al., 2011; Ojo et al., 2006; Rahman, 2003; Ekpebu, 2002; Abdulai and Huffman, 2000; Saleem, 1988; Kalirajan and Obwona, 1994; Dawson and Lingard, 1991; Yilma, 1996; Nsanzugwanko et al., 1996; Battese and Hassan, 1999). Ekpebu (2002) on the other hand argued that the Cobb-Douglas functional form is useful in analysis of surveys where many variable inputs are involved and it is necessary to measure returns to scale, intensity of factors of production and overall efficiency of production. It is also argued that it provides a means of obtaining coefficients for testing hypotheses (Cobb and Douglas, 1928; Erhabor, 1982). Akighir and Shabu (2011) cited Ellebu, Koku and Ogidi (2004) that the evidence of the superiority of Cobb-Douglas functional form is supported by its satisfaction of the economic, statistical and econometric criteria required unlike the other functional forms.

In view of the above arguments, the Cobb-Douglas functional form was applied for estimating dairy profit efficiency of the smallholder farmers in the study area, whose empirical model is specified below. However, for comparison purposes, the translog, quadratic and transcendental forms were equally applied (though their empirical models have not been specified here).

The explicit Cobb-Douglas functional model for this study is specified as follows:

$$\ln \pi_i = \alpha_0 + \alpha_1 \ln Z_{1i} + \beta_k \sum_{k=1}^{12} \ln P_{ki} + (v_i - u_i) \quad (8)$$

Where π is the normalized profit computed as total milk revenue per litre less variable cost per litre divided by farm specific milk price per litre; α and *β* represent the individual variable coefficients in the stochastic frontier model; *Z* is the total number of cows on the farm; *P*s represent the cost of production inputs per unit of each respective input (that is, P_{1i} = artificial insemination price per cow, P_{2i} = price of veterinary services per administration, P_{3i}= price of extension services per visit, P_{4i} = monthly labor wage, P_{5i} = purchase fodder price per kilogram (kg), P_{6i} = imputed price of produced fodder per kg, P_{7i} = price of dairy concentrates per kg, P_{8i}= price of conserved feeds per kg, P_{9i} = grazing price per cow, P_{10i} = price of water per litre, P_{11i} = price of milk transport (per litre), P_{12i} = price of milk (per litre) for feeding calves. $(v_i - u_i)$ is the composed error term.

The inefficiency model (u_i) is defined by the equation as follows:

$$u_i = \delta_0 + \sum_{z=1}^{12} \delta_z T_z \quad (9)$$

Where *δ* represent the respective regressor coefficients; *T*s represent farm, household and institutional socioeconomic characteristics (that is, T_1 = age of farm owner, T_2 = size of fodder land, T_3 = size of grazing land, T_4 = hourly wage rate, T_5 = production system(dummy), T_6 = production scale(dummy), T_7 = gender of farm owner (dummy), T_8 = hired labor(dummy), T_9 = extension service access (dummy), T_{10} = paid extension service (dummy), T_{11} = paid water (dummy), T_{12} = rented land (dummy).

Data and variables measurement

The study was conducted in Kenya's Rift Valley and Central Provinces where the East Africa Dairy Development (EADD) project has earmarked interventions, excluding the pastorally dominant cattle keeping communities. Data was collected from smallholder dairy farmers using structured household questionnaires. Details of production costs and revenues generated from milk in the past 3 months from the date of interview (August, 2012) were identified. This was to ensure accurate recall of production situations by the farmer, which is highly unlikely for an entire annual period due to lack of record keeping as argued by Staal et al. (2008). Detailed data was collected on feeds and their sources, other inputs used in production, the costs involved in acquiring these inputs and related services accessed by the farmer. Additionally, the amount of milk obtained from lactating cows and the revenues generated were obtained. Milk that was consumed by the farm families, milk fed to calves and milk given to family friends/neighbors were valued at the cost price, while that sold was valued at the going market price of the respective channels. The gross profit of milk was used as the main outcome variable. The cost determinants and inefficiency factors in the frontier model were then examined against the outcome variable as earlier specified. Multistage sampling procedure was used to select a representative number of farmers into the study. Farmers were stratified according to the main production system that is, mainly extensive or intensive system. A further stratification was based on the production scale that is, small-scale and medium-scale, assigned according to the number of cows managed under the respective production systems. These were arrived at based on the EADD precept for categorizing production scales of the smallholder farmers based on the baseline data for Rwanda, Uganda and Kenya (Table 1).

The study sample size was computed using the formula (Equation 10) for obtaining sample sizes of each group for comparison in the study (if the outcome variable is a continuous univariate data). An assumption of this study was that unit profits among the farmers is a continuous variation and follows a normal distribution pattern, varied unit costs, herd sizes, production systems, parity differences of lactating cows, and management efficiency differences across farms, to mention but a few.

$$n = 2 \times \frac{\left(Z_{\frac{\alpha}{2}} + Z_{\beta}\right)^2}{d^2} \sigma^2 \quad (10)$$

Where: n = approximate sample size; d = margin of error (mean unit profit difference between the 2 groups, mean ± d = confidence interval) (assumed at 0.1); σ^2 = assumed std deviation of 0.2 for unit profit; $Z_{\alpha/2}$ = 1.96 signifying a 2-sided sample size at 95% confidence level; Z_β = power of the test in identifying a significant difference (that is, 'chance' of this happening - 80%)

A total of 122 farmers (half for intensive and half for extensive production systems) were therefore approximated to be an adequate representative sample for the study. However, there were relatively fewer numbers of intensive farms identified during the field data collection process. Hence, only a total of 85 farmers were therefore ultimately surveyed.

Table 1. Production scales per production system.

Production scale	Production system	
	Intensive	Extensive
Small-scale	≤ 3 Cows	≤ 15 Cows
Medium-scale	≥ 4 Cows	≥ 16 Cows

Source: EADD Field Survey (2009).

EMPIRICAL RESULTS AND DISCUSSION

Descriptive summaries

Table 2 shows the descriptive summaries of the variables that were structured in the study questionnaire. The study found that the average number of cows owned per farm is 3 with a standard deviation of 2.0. The average land size under fodder production was found to be 2.3 acres with a standard deviation of 2.2, while that of grazing land was 6.0 acres with a high standard deviation of 52.4.

The proportion of farmers practicing extensive system was only 27% of the total farmers. This is mainly attributed to the competitive use of land resources influenced by the rising population (FAO, 2003). It was found that the intensive farms incurred higher production costs than the extensive farms on average. This difference was mainly attributed to feed costs and labour expenses incurred in intensive systems than extensive systems. The average total cost of milk production incurred per household in the past 3 months was Kshs 41,070, while total revenue amounted to Kshs 42,210. The average household milk produced in 3 months amounted to 1617 L, while the average total milk sold was 1100 L. Households also consumed about 520 L at home on average. Feed costs constituted the greater proportion of farmers' cost of milk production. Among the cost components, fodder produced on farm constituted the greater proportion of variable costs with an average of Kshs 23,000. The average amount of milk produced by households was 1,617 L though with a high standard deviation. This variation was mainly attributed to the number of cows in lactation, varying lactation lengths and parity effects, among others.

In terms of marketing of milk outputs, majority of farmers sold a greater portion of milk to the Chilling Plants (CPs) (local fresh milk buying centers) in the past 3 months amounting to 930 L on average. The least quantity was sold to private traders (35 L). Although the study did not prioritize producer choice for market channels, the preference for CPs by farmers would be largely attributed to a couple of factors, despite the lower prices offered by this channel. These could range from: a) the capability and the reliability of CPs to buy and pay for every quantity supplied compared to small scale traders and individual consumers; b) the relatively stable prices offered by CPs; c) the input incentives and extension services by some CPs to farmers on contractual terms; d) the belonging of some of the farmers' to the cooperatives owning these CPs. Overall, the average household revenue earned in a quarter of a year from milk amounted to Kshs 42,000. The revenue from direct milk sales approximated to Kshs 29,000 and that of unsold milk amounted to KShs13,000 (Table 2).

Table 3 presents a summary statistics of the normalized unit variable costs and gross margin per liter of milk in the stochastic frontier model. Generally, the cost of feeds constituted the greater proportion of cost of milk production. The average cost of fodder production amounted to Kshs 0.74. This was followed by cost of labour and conserved feeds (Kshs 0.20, respectively). Among the feed costs, the cost of grazing was the cheapest (Kshs 0.10). Overall, the average cost of milk transport was the least of all the variable costs (Kshs 0.03). The normalized average gross margin per litre of milk amounted to Kshs 0.62.

Smallholder dairy profit efficiency in Kenya

Table 4 shows the stochastic model estimation results for the four functional forms: Cobb-Douglas, translog, quadratic and transcendental forms. The results showed that the average profit efficiency estimated by the Cobb-Douglas functional model was 68%. Compared to the other functional forms, the quadratic function estimated a similar mean efficiency (68%). The translog form estimated an average efficiency of 67%, while the transcendental form estimated an average of 71% (Table 4).

The likelihood ratio test was used to compare between the functional forms. The translog form was taken as the unrestricted log likelihood function (ULLF) and the rest as the restricted log likelihood functions (RLLF) since the coefficient estimates of some variables were hypothesized to be 0. The test statistic used to determine whether there was any difference between the translog function and any one of the other forms was:

Table 2. Summary statistics of variables used in the stochastic frontier model.

Variable Name	Minimum	1st Quartile	Median	Mean	3rd Quartile	Maximum	Standard deviation
Physical inputs							
No. of cows on farm	1.0	2.0	3.0	3.3	4.0	11.0	2.0
No. of hired labour	0.0	0.0	0.0	0.5	1.0	2.0	0.6
No. of household labour	0.0	2.0	3.0	2.6	3.0	7.0	1.5
Acres of fodder	0.0	0.6	1.5	2.3	3.4	10.0	2.2
Acres of grazing land	0.0	0.0	0.0	6.0	0.0	500.0	52.4
Variable costs							
Breeding	0.0	0.0	500.0	950.6	1300.0	10500.0	1608.4
Health	0.0	935.0	1990.0	2394.0	3190.0	9560.0	2015.0
Extension	0.0	0.0	0.0	102.3	0.0	1500.0	318.4
Hired labour	0.0	0.0	0.0	3551.0	6000.0	15000.0	4553.2
Purchased fodder	0.0	0.0	0.0	1004 .0	0.0	44000.0	4910.7
Concentrates	0.0	1673.0	2600.0	3745.0	5530.0	17120.0	3619.6
Farm produced fodder	0.0	5500.0	15000.0	23000.0	34380.0	100000.0	22444.6
Conserved feeds/forage	0.0	0.0	0.0	1248.0	0.0	22950.0	3782.9
Grazing	0.0	0.0	0.0	682.4	0.0	9000.0	1756.3
Water	0.0	0.0	0.0	216.9	0.0	13500.0	1468.6
Transport	0.0	0.0	180.0	596.6	967.5	4500.0	882.7
Milk to calves	0.0	0.0	0.0	2298.0	2866.0	22240.0	4488.3
Total variable costs	**1600.0**	**16100.0**	**32020.0**	**38610.0**	**51490.0**	**174400.0**	**30084.7**
Fixed costs							
Depreciation	14.7	182.8	544.0	1458.0	1799.0	9729.0	2008.7
Household labour	0.0	668.9	1698.0	2209.0	3307.0	15790.0	2262.4
Total fixed costs	14.7	282.6	1618.0	2465.0	4379.0	14050.0	2771.4
Total cost	**2078.0**	**16620.0**	**33920.0**	**41070.0**	**57770.0**	**177900.0**	**31299.0**
Milk sales							
Qty sold to Chilling plant	0.0	450.0	697.5	930.3	1080.0	4230.0	831.7
Qty sold to Individuals	0.0	0.0	0.0	128.7	180.0	1080.0	236.1
Qty sold to Private traders	0.0	0.0	0.0	34.5	0.0	990.0	148.8
Total quantity sold	**0.0**	**540.0**	**855.0**	**1094.0**	**1508.0**	**4230.0**	**877.7**
Price							
Unit price - Chilling plant	22.0	24.0	26.0	26.1	27.0	32.0	*
Unit price - individuals	20.0	25.9	30.0	28.9	30.0	60.0	*
Unit price - private traders	10.0	24.8	25.5	26.8	30.5	39.0	*
Milk price	**21.0**	**25.0**	**27.0**	**26.7**	**28.6**	**32.0**	**2.4**
Quantity							
Quantity fed to calves	0.0	0.0	0.0	101.0	168.8	1350.0	211.0
Quantity consumed at home	0.0	180.0	270.0	317.6	360.0	1080.0	211.0
Quantity offered to workers	0.0	0.0	0.0	104.7	0.0	2700.0	426.4
Total quantity unsold	**0.0**	**202.5**	**360.0**	**523.3**	**607.5**	**3510.0**	**578.1**
Milk revenue							
Total milk output	**180.0**	**821.2**	**1260.0**	**1617.0**	**2205.0**	**6210.0**	**1240.2**
Total revenue (sold milk)	0.0	14580.0	22070.0	29170.0	39760.0	127400.0	24171.6
Total revenue (unsold milk)	0.0	4840.0	9511.0	13040.0	18380.0	50600.0	11774.0
Total milk revenue	5890.0	24130.0	35550.0	42210.0	53690.0	157500.0	27779. 7

Source: EADD field survey (2012).

Table 3. Descriptive statistics of variable costs (Kshs) in the stochastic frontier model.

Variable	Frequency	Minimum	1st Quartile	Median	Mean	3rd Quartile	Maximum
Breeding	45	0.01	0.03	0.04	0.05	0.06	0.19
Health	78	0.00	0.04	0.06	0.08	0.10	0.37
Extension	8	0.01	0.02	0.03	0.04	0.07	0.11
Labour	41	0.03	0.08	0.13	0.20	0.28	0.83
Purchased fodder	12	0.01	0.03	0.08	0.12	0.15	0.56
Produced fodder	69	0.04	0.26	0.60	0.74	1.15	3.70
Concentrate	74	0.00	0.05	0.11	0.13	0.16	0.60
Conserved feeds	20	0.00	0.03	0.07	0.20	0.31	0.90
Grazing	17	0.00	0.05	0.07	0.10	0.12	0.28
Water	7	0.01	0.01	0.02	0.06	0.04	0.31
Transport	48	0.00	0.01	0.02	0.03	0.04	0.07
Calf milk	25	0.05	0.10	0.12	0.13	0.15	0.33
Gross margin	85	-0.10	0.45	0.62	0.62	0.75	1.86

Source: EADD field survey, August (2012).

$$\lambda = 2(ULLF - RLLF) \quad (11)$$

The test statistic λ follows the Chi-square (χ^2) distribution with degrees of freedom (df) equal to the number of restrictions imposed by the null hypothesis on the RLLF. Equation 11 was used to compute λ from the log likelihood values of the estimated functional forms in Table 4. If the *apriori* restrictions are valid, the restricted and the unrestricted (log) LF should not be different, in which case λ will be 0. But if that is not the case, the two LFs will diverge. The hypothesis tested was that the translog function was not different from the Cobb-Douglas function (where λ = 72.6598 and df = 32), the quadratic function (where λ = 18.3322 and df = 8), the transcendental function (where λ = 16.0474 and df = 21). The statistics calculator was then used to compute the p-values for the Chi-square test of the given Chi-square values and the df. The respective computed p-values were 0.0000, 0.0189 and 0.7156. The results showed that the translog function was statistically significant from the Cobb-Douglas function (p-value = 0.000) and the quadratic function (p-value = 0.0189); leading to the conclusion that the restrictions should not have been imposed. However, there was no difference between the translog and the transcendental functions (p-value = 0.7156).

As alluded to the earlier discussion, each of the functional forms had strengths and weaknesses. Nevertheless, scholars and researchers have found the Cobb-Douglas functional form useful in analysis of surveys where many variable inputs are involved like in this study. In spite of restrictions, the superiority of the Cobb-Douglas functional form in the results is supported by its satisfaction of the economic, statistical and econometric criteria required unlike the other functional forms. A panoramic view over the results of the models gives the impression that the Cobb-Douglas functional form resonates with and underscores the significance of the socioeconomic and institutional factors better than the other functional forms. Hence, the Cobb-Douglas functional model output was adopted for further discussion of the study findings.

The study found that the specified stochastic frontier model was adequate in estimating profit efficiency, in that the variance parameter, sigma-squared (0.86), was statistically significant (p-value < 0.001). This implied that the composed error term ($\varepsilon = v_i - u_i$) strongly dominated the measurement error. The gamma value of 0.91 was also found to be significant. A value of 1 would suggest a deterministic approach for the efficiency estimates since there is no random noise, while a value of 0 would mean OLS model is a best estimator because there is no inefficiency. A variance ratio (γ^*)[1] of 0.79 was computed, implying that 79% of the differences in actual and the observed frontier efficiency is attributable to the farmers inefficient practices. Additionally the null hypothesis (H_0: $\gamma = 0$); specifying that the inefficiency effects in the stochastic frontier are not stochastic was rejected because the value of gamma is significantly different from 0 (p-value < 0.001). Further evidence of "goodness of fit" of the stochastic frontier model was proven by the likelihood ratio test (Table 4) with a significant p-value (<0.001), signifying that the stochastic frontier model was a better estimator of profit efficiency in this study than the traditional OLS model.

The estimated coefficients of the Cobb-Douglas function can best be explained as the elasticity of the respective variables. As argued by Abdulai and Eberlin (2001) the first-order coefficients of the functional forms are of less significance in interpreting the outputs of the model because they are not very informative, but rather

[1] $\gamma^* = \gamma / [\gamma + (1-\gamma)\pi/(\pi-2)]$ (Coelli et al., 1998)

Table 4. Profit efficiency estimates of the respective functional forms.

Variable name	Parameter	Cobb-Douglas	Translog	Quadratic	Transcendental
Stochastic frontier model					
Intercept	β_0	-1.6511***	-2.9966***	-0.6766	0.6755
In Cows (size on farm)	β_1	0.0021	0.5325*	-0.2048	-0.2780
In Breeding	β_2	-0.0175	0.0164	0.0033	-0.0068
In Health	β_3	-0.0005	-0.1733	-0.0417	-0.0233
In Extension	β_4	-0.0043	-0.2497**	-0.0715.	0.0009
In labour	β_5	-0.0057	-0.0291	0.0084	-0.0106
In Fodder purchased	β_6	-0.0521***	0.0712	-0.0474*	0.0134
In Produced fodder	β_7	0.0902***	1.3983***	0.1035***	0.0743***
In Concentrates	β_8	-0.0056	-0.1756*	0.0410	0.0171
In Conserved feeds	β_9	-0.0390***	0.0694	0.0454	0.0299
In Grazing	β_{10}	-0.0144	-0.0249	-0.0021	-0.0283
In Water	β_{11}	-0.0385*	0.1380	0.0875	0.0204
In Transport	β_{12}	-0.0280***	-0.0232	-0.0798*	-0.0494*
In Calf milk	β_{13}	-0.0121	-0.1612***	-0.0973*	-0.0070
In Cows (size on farm) squared	β_{14}		-0.5474**	-0.0244	
In Breeding squared	β_{15}		0.1056*	-0.0191	
In Health squared	β_{16}		-0.0339	-0.0302	
In Extension squared	β_{17}		-0.4671**	-0.1885	
In Labour squared	β_{18}		-0.0444	0.0292	
In Fodder purchased squared	β_{19}		0.3257***	0.0194	
In Produced fodder squared	β_{20}		0.0979*	0.0352*	
In Concentrates squared	β_{21}		0.0507***	0.0308 *	
In Conserved feeds squared	β_{22}		0.1529*	0.0882	
In Grazing squared	β_{23}		0.0041	0.0081	
In Water squared	β_{24}		0.5899***	0.2915	
In Transport squared	β_{25}		0.0828	-0.1473*	
In Calf milk squared	β_{26}		-0.2425*	-0.1230.	
In Breeding * In Extension	β_{27}		-0.0069***		
In Breeding * In Health	β_{28}		0.0091		
In Fodder purchase * In fodder Produced	β_{29}		-0.1255***		
In Fodder Produced *In Concentrates	β_{30}		-0.0141.		
In Cows* In Labour	β_{31}		0.0215*		
In Health * In Concentrates	β_{32}		0.0080		
In Cows * In Health	β_{33}		0.0504*		
In Cows* In fodder produced	β_{34}		0.1230		
Cows (total number on farm)	β_{35}				0.0030
Breeding	β_{36}				-0.0570
Health	β_{37}				0.1705
Extension	β_{38}				-0.0582
Labour	β_{39}				0.0774
Fodder purchased	β_{40}				-5.5331***
Fodder produced	β_{41}				0.0215
Concentrates	β_{42}				-0.2885
Conserved feeds	β_{43}				-0.2920
Grazing	β_{44}				-0.0025
Water	β_{45}				-1.9768*
Transport	β_{46}				0.8178
Calf milk	β_{47}				-0.1616
Breeding * extension	β_{48}				0.1281
Breeding * health	β_{49}				-0.0157

Table 4. Contd.

Fodder purchased* produced fodder	β_{50}				0.7646***
Produced fodder * concentrates	β_{51}				10.1260
Cows * Labour	β_{52}				-0.0006
Health * concentrates	β_{53}				0.0001
Cows * health	β_{54}				-0.0044
Cows * produced fodder	β_{55}				0.0570
Inefficiency frontier model					
Intercept	δ_0	-4.94633**	-0.9067	-1.7096*	-0.4652
Age of household head	δ_1	0.00802*	-0.0576***	-0.0348.	-0.0291
Acres of fodder land	δ_2	-0.44717***	-0.4073***	-0.3249**	-0.2713.
Acres of grazing land	δ_3	-0.02519***	-0.0510***	-0.0572***	-0.0516*
Hourly wage rate	δ_4	0.20042	0.2463	0.3882	0.0530**
System (dummy:1 = Int., 0 = Ext.)	δ_5	0.99896*	-0.2764	-2.0451**	-1.5180*
Scale (dummy:1 = Sml., 0 = Med.)	δ_6	0.54996	0.1873	1.3588.	0.8707
Gender (dummy:1 = Male, 0 = Fem.)	δ_7	0.04999	0.8587	1.0500	0.9295
Hired labor (dummy:1 = Yes,0 = No)	δ_8	1.38063**	-1.5717***	0.1484	-0.2401
Access extn.(dummy:1 = Yes, 0 = No)	δ_9	1.41616***	0.6302	-0.1252	-0.4361
Paid extn.(dummy:1 = Yes, 0=No)	δ_{10}	0.48777	-0.8412	-0.7587	-0.4195
Paid water (dummy:1 = Yes, 0 = No)	δ_{11}	-0.06407	0.0700	0.7039	-0.4242
Hired land (dummy:1 = Yes, 0 = No)	δ_{12}	1.03041***	0.3016	0.8440.	0.7041
Age * Size fodder land	δ_{13}		-0.0394***	-0.0162.	-0.0104
Age * Size grazing land	δ_{14}		-0.0026***	-0.0025***	-0.0022***
Age * wage rates	δ_{15}		0.0472*	-0.0086	0.0089
System * Scale	δ_{16}		0.1928	0.5424	0.6655
Gender * hired labour	δ_{17}		0.0400	-0.1351	0.1376
Extension access * paid extension	δ_{18}		0.9995	-0.7587**	-0.4195
System * access to extension	δ_{19}		0.9634	2.2072	1.6131*
Scale * Gender	δ_{20}		-0.2463	-1.0247	-1.0294
Gender * access to extension	δ_{21}		-0.0536	-1.1522	-0.6770
System * gender	δ_{22}		-1.1091	1.0000**	1.0000
Hired labour * paid extension	δ_{23}		4.2317***	2.8884***	2.6968**
Model diagnostics					
Sigma-squared ($\sigma^2 = \sigma_v^2 + \sigma_u^2$)	σ^2	0. 8593***	0.2095***	0.3232***	0.1690***
Gamma ($\gamma = \sigma_u^2/(\sigma_v^2 + \sigma_u^2)$)	γ	0. 9099***	0.9455***	0.9610***	0.9382***
Total number of observations		85	85	85	85
Log likelihood value		-24.4045	11.9254	2.7593	3.9017
Mean efficiency		0.6768	0.6727	0.6796	0.7122

Source: EADD field survey (August 2012). Extn.= Extension

the output elasticity for each of the inputs calculated at the variable means. The principle underlying elasticity is the measure of responsiveness of an output to unit increase in input. From the Cobb-Douglas model the elasticity with respect to normalized unit cost of purchased fodder, conserved feeds and milk transport were significant (p-value <0.001). The elasticity with respect to water was significant at 5% (p-value <0.05). The negative coefficient signs of these variables imply that they are significant at reducing profits among smallholder dairy farmers. The absolute values of these coefficients suggest that a unit increase in the price of purchased fodder, conserved feeds, milk transport and purchased water have the ability to decrease efficiency by 5, 4, 3 and 4%, respectively. Despite the fact that the parameter coefficients for cost of breeding, health, extension, labour, concentrates, grazing and milk to calves were negative, implying that a unit increase in the price of these inputs decreases profit efficiency, their effects were found not to be significant (p-value >0.05).

The elasticity with respect to cost of farm produced fodder was significant (p-value <0.001) and with a positive coefficient. A unit increase in the cost of on-farm fodder led to 9% increase in profit efficiency. This indicates that investments in on-farm fodder production are a better feed choice in current dairy production in Kenya since it enhances profitability of the smallholder farmers. It was also observed that a unit increase in the number of cows of a household positively increased frontier function for milk gross margin, though not significant (p-value >0.05).

Determinants of profit efficiency among smallholder dairy farmers

This study considered several socio-economic characteristics that have the potential of influencing profit efficiency of the farmers (Table 4). The classical interpretation of inefficiency variables in the stochastic frontier function is based on the signs of the parameter estimates which are interpreted in the opposite manner to those of the general stochastic frontier model. A negative sign implies that there is an increase in profit efficiency associated with the inefficiency factor, while a positive sign means a decrease in efficiency. The socio-economic variables that significantly increased profit efficiency among the smallholder farmers were size of fodder land and size of grazing land. Efficiency increments associated with size of land for fodder and for grazing are in line with theoretical expectations. This is attributed to the lower unit feed costs in milk production of farmers with fair land sizes under fodder production and for grazing.

The study found that profit efficiency significantly decreased with age of the farm owner (p-value <0.05). This was mainly attributed to the high production costs associated with the elderly farmers especially due to hired labor services used on farm as a result of reduced mobility in farming, limited application of new technologies that are more efficient than the traditionally inefficient technologies, among others. The positive and the significant (p-value <0.05) impact of the production system variable indicated that farmers engaged in intensive system of dairy keeping tend to exhibit higher levels of inefficiency. This difference could arise due to the higher costs incurred in intensive systems, as well as inefficiencies in resource allocation that renders them less profitable as argued by Kumbhakar and Lovell (2000). Farmers' access to extension services was significant (p-value <0.0001) and decreased profit efficiency. In theory, the access to extension services paid for by the farmer can result into increased cost of milk production hence a decrease in milk gross margins. Hired labor and hired land were significant (p-value <0.01 and p-value <0.001, respectively) inefficiency factors at reducing profit efficiency among smallholder farmers. These findings tally with theoretical expectations, in that a farmer who relied entirely on household labor incurred no costs on labor and therefore their unit gross margins remained high and hence were more profit efficient. It is also true that paying for land for dairy activities increased the cost of milk production among farmers and hence caused such farmers to be less profit efficient. The study identified that factors such as gender, scale of production, wage rates, paid water and paid extension services were insignificant at influencing dairy profit efficiency among the smallholder farmers.

Efficiency distribution among dairy farmers

The mean profit efficiency of the smallholder dairy farmers approximated to 68% as earlier presented. It was also found that 54% of the farmers were distributed to a profit efficiency of greater than 70% (Table 5). The least farmer was 6.5% profit efficient, while the best farmer was 99% efficient. A general distribution of efficiency among other socio-economic characteristics is summarized in Table 6. Most of the observed differences are attributed to the costs incurred in production by the respective groups, and are in line with theoretical expectations.

CONCLUSION AND RECOMMENDATIONS

There has steadily been an increasing concern on feed resources for smallholder dairying in Kenya especially in the past decade. Stakeholders often cite high costs in milk production as farm technology is changing so fast to intensive systems. The study explored profit efficiency of smallholder dairy farmers in the Rift Valley and Central Provinces in Kenya. It was found that the farmers were 68% profit efficient. Despite this mean efficiency, there was a wide variation among the farmers, with 54% of them achieving a profit efficiency of more than 70%. The best farmer attained a maximum profit efficiency of 99%, while the least had 6.5%. Fodder produced on farm significantly improved profit efficiency among the dairy farmers. On the other hand, costs such as fodder purchased, conserved feeds, milk transport and water for cattle significantly reduced profit efficiency among the farmers.

The socio-economic variables that significantly influenced profit efficiency among the dairy farmers were size of fodder land, size of grazing land, age of farm owner, production system, extension service, hired labour and hired land. With an average profit efficiency of only 68%, the smallholders can further improve their profit efficiency if farmers' inefficient practices are improved. With the increasing competition for land resources, an average profit efficiency of only 68% among the farmers is an indication that the farmers are not very efficient in dairy production currently. Hence, efficient feed

Table 5. Profit efficiency indices of dairy farmers in Kenya.

Efficiency Class Index	Frequency	Percentage (%)	Cumulative Percentage
0.00 - 0.10	2	2.4	2.4
0.11 - 0.20	3	3.5	5.9
0.21 - 0.30	3	3.5	9.4
0.31 - 0.40	6	7.1	16.5
0.41 - 0.50	6	7.1	23.5
0.51 - 0.60	8	9.4	32.9
0.61 - 0.70	11	12.9	45.9
0.71 - 0.80	14	16.5	62.4
0.81 - 0.90	14	16.5	78.8
0.91 - 1.00	18	21.2	100.0
Total	85	100.0	

Source: EADD field survey (August 2012).

Table 6. Efficiency distribution among socio-economic groups.

Socio-economic characteristic	Dummy	Frequency	Efficiency
Gender	Female	29	0.67
	Male	56	0.68
Age below 51 years	No	26	0.64
	Yes	59	0.69
Extension services	No	24	0.71
	Yes	61	0.66
Paid extension services	No	75	0.67
	Yes	10	0.72
Employed labor	No	45	0.72
	Yes	40	0.63
Paid water services	No	77	0.67
	Yes	8	0.71
Paid land services	No	64	0.68
	Yes	21	0.66

Source: EADD field survey (August 2012).

technologies commensurate with farm local conditions are required. Institutional policy reforms targeting competitive dairy sector performance are needed so as to expand the productivity and profitability of the smallholder dairy farmers.

Conflict of Interest

The author(s) have not declared any conflict of interests

REFERENCES

Abdulai A, Eberlin R (2001). Technical efficiency during economic reform in Nicaragua: Evidence from farm household survey data. Econ. Syst. 25(2):113-125.

Abdulai A, Huffman W (2000). Structural adjustment and economic efficiency of rice farmers in Northern Ghana. Econ. Dev. Cult. Change 48(3):503-520.

Aigner DJ, Lovell CAK, Schmidt P (1977). Formulation and estimation of stochastic frontier production function models. J. Econ. 6:21-37.

Akighir DT, Shabu T (2011). Efficiency of resource use in rice farming enterprise in Kwande Local Government Area of Benue State,

Nigeria. Int. J. Hum. Soc. Sci. 1(3):215-220.

Ali F, Parikh A, Shah M (1994). Measurement of profit efficiency using behavioural and stochastic frontier approaches. Appl. Econ. 26(2):181-188.

Ali M, Flinn JC (1989). profit efficiency among basmati rice producers in Pakistan Punjab. Am. J. Agric. Econ. 71(2):303.

Aneani F, Anchirinah VM, Asamoah M, Owusu-Ansah F (2011). Analysis of economic efficiency in cocoa production in Ghana. Afri. J. Food Agric. Nutri. Dev. 11(1):4507-4526.

Battese GE, Coelli TJ (1995). A model for technical inefficiency effects in a stochastic frontier production function for panel data. Emp. Econ. 20(2):325-332.

Battese GE, Hassan S (1999). Technical efficiency of cotton farmers in the Vehari District of Punjab, Pakistan. Pakistan J. Appl. Econ. 15(1&2):41–53.

Cobb C, Douglas P (1928). A theory of production. Am. Rev. 18(1):139-165.

Coelli TJ (1996). A guide to Frontier 4.1: A Computer program for stochastic frontier production and cost Function Estimation, CEPA Working Paper 96/07, University of New England, Armidale, NSW, 2351, Australia. P. 33. Available at: https://files.itslearning.com/data/ku/103018/publications/coelli96.pdf?

Dawson PJ, Lingard J (1991). Approaches to measuring technical efficiency on Philippine rice farms. J. Int. Dev. 3(3):211-228. Available at:

Ekpebu I (2002). "Resources Productivity in Cassava Yam based Farm Enterprises in Benue Nigeria" Ph.D Thesis. Department of Agriculture Economics, University of Nigeria, Nsukka.

Erhabor P (1982). Efficiency of resource use under small-scale irrigation technology in Nigeria. Purdue University, Water Resources Research Center. "June 1982," Available at: https://openlibrary.org/books/OL3002834M/Efficiency_of_resource_use_under_smallscale_irrigation_technology_in_Nigeria

FAO (2003). World agriculture: towards 2015/2030. FAO, Rome. Available at: http://www.fao.org/docrep/005/y4252e/y4252e00.htm

FAO (2011). Dairy development in Kenya, by H.G. Muriuki. Rome, Available at: http://www.fao.org/docrep/013/al745e/al745e00.pdf

Kalirajan KP, Obwona MB (1994). A measurement of firm- and input-specific technical and allocative efficiencies. Appl. Econ. 26(4):393-398. Available at:
[Accessed January 13, 2014]

Kumbhakar SC, Biswas B, Bailey DV (1989). "A study of economic efficiency of Utah Dairy Farmers: A system approach". Rev. Econ. Stat. 71:595-604.

Kumbhakar SC, Lovell CAK (2000). Stochastic frontier analysis. Cambridge: Cambridge University Press.

Meensen W, Van den Broeck J (1977). "Efficiency estimation from Cobb-Douglas production functions with composed error". Int. Econ. Rev. 9:43-64.

Muriuki H (2003). Milk and dairy products, post-harvest losses and food safety in sub-Saharan Africa and the Near East: A review of the small-scale dairy sector - Kenya. Rome, Italy: Food Agric. Organ. Prevention of Food Losses Programme. Rome, Italy, p.60. Available at: http://www.fao.org/fileadmin/templates/ags/docs/dairy/P1assessment kenya.pdf [Accessed January 13, 2014].

Nsanzugwanko MD, Battese GE, Fleming EM (1996). The technical efficiency of small agricultural producers in Central Ethiopia. Working Paper No. 5/96, Centre for efficiency and productivity analysis, University of New England, Australia.

Ojo SO, Afolabi JA, Fagbenro OA (2006). Profitability and technical efficiency of Artisanal fisheries production in Nigeria. Mar. Res. Econ. 19:23-30.

Ọlayide SỌ, Heady EO (1982). Introduction to agricultural production economics, Ibadan University Press, University of Ibadan. Available at: http://books.google.co.ke/books/about/Introduction_to_Agricultural_Production.html?id=RKVHAAAAYAAJ&pgis=1 [Accessed January 9, 2014]

Omore A, Muriuki H, Kenyanjui M, Owango M, Staal SJ (1999). The Kenya dairy sub-sector; A rapid appraisal. Smallholder Dairy (Research and Development) Project Report. Nairobi, Kenya, Available at: http://ilri.org/infoserv/Webpub/fulldocs/InvestingInDairy/DOX/Omore et al-1999-Kenya dairy sector rapid appraisal.pdf

Rahman S (2003). Profit efficiency among Bangladeshi rice farmers. Food Policy 28(5):487-503.

Saleem ST (1988). Relative efficiency of cotton farms in Sudanese irrigated agriculture. World Dev. 16(8):975-984.

Staal SJ, Nin Pratt A, Jabbar M (2008). Dairy development for the resource poor. Part 3: Pakistan and India dairy development case studies. Available at: http://cgspace.cgiar.org/handle/10568/1582 [Accessed January 9, 2014]

Taru VB, Lawal H, Tizhe I (2011). Technical efficiency of sole cowpea production in Adamawa State, Nigeria: A Cobb-Douglas stochastic frontier function. J. Econ. Int. Fin. 3(8):504-507.

Upton M (1979). Farm Management in Africa: The principles of production and planning, English Language Book Society.

USAID (2008). Kenya Dairy Sector Competitive Program: Annual report June 2008-September 2008, Available at: http://pdf.usaid.gov/pdf_docs/Pdacm614.pdf

USAID (2010). Consumer milk quality perception/preferences and an assessment of willingness to pay for quality. USAID dairy sector competitiveness program, Available at: http://pdf.usaid.gov/pdf_docs/PDACS668.pdf

Wang JC, Wailes EJ, Cramer GL (1996). A shadow-price frontier measurement of profit efficiency in Chinese agriculture. Am. J. Agric. Econ. 78:146-156.

Yilma M (1996). Measuring Smallholder efficiency: Uganda coffee and food crop production. Unpublished dissertation, Goteborgs. Universitet: Ekonomiska Studier Utgivnaav Nationalekonomiska Institutionen Handelsh, Kompendiet-Goteborg. Sweden.

Economic optimal allocation of irrigation water in Egypt

Dawoud, Seham D. Z.

Department of Agricultural Economics, Faculty of Agriculture, Damietta University, Egypt.

The water scarcity problem is globally getting worse especially in the light of increase in water demand among its competing uses. Thus, it is an important to optimize the water allocation to crops. In this paper, a linear programming model has been formulated to ensure the efficient allocation of scarce water resources among the competing crops. This model was constrained by land, water, labour, production costs, and organization constraints, determining the optimal plan for two possible future scenarios. The mathematical analysis was based on statistical data for the years (2009-2011) from the official statistical institutions in Egypt. The results of the two scenarios are as follows: Under the maximization of the net return per unit of land, there is an increase in total net returns by 3.56% more than the actual net returns. The optimized cropping pattern has been coupled with about 3.24% water saving and about 3.13% reduction in production costs compared to actual cropping pattern. However, under the minimization of irrigation water requirements, the total net returns decreased by 10.20% indicating losses below the actual situation. It has resulted in about 11.05% water saving and 11.24% reduction in the costs of production compared to the existing situation. These results can be used as a reference for indicative cropping pattern and irrigation water management in Egypt.

Key words: Linear programming, efficient water allocation, optimal cropping pattern, water management.

INTRODUCTION

Water is one of the most important natural resources for the world's economic development. In many areas around the world, conflicts have risen due to increase in water demand among its competing uses (World Bank, 2002; Young, 2005). Particularly, agriculture is becoming the sector to which policy makers are pointing out as the core of the water problem (Koundouri et al., 2006). This is clearly the case in Egypt, where water resources are limited to the *Nile* water. It is the major source of fresh water, supplying 96% of renewable fresh water resources. Egypt relies on the availability of its annual share of *Nile* water, which is stored in Lake Nasser. This is approximately 55.5 billion cubic meters annually following agreement between Egypt and Sudan in 1959. Water demand is increasing as a result of the rapid population growth, agricultural expansion, as well as industrial development, and higher standards of living. This increase in demand for the limited water resources puts pressure on the decision-makers to formulate policies to improve the allocation of the scarce water resources. Because agriculture is the major water consumer in Egypt, it will be important to ensure efficient allocation of irrigation water resources.

Linear Programming LP is a widely used mathematical modelling technique to determine the optimum allocation of scarce resources among competing demands. Some examples are presented. Kheper and Chaturvedi (1982) applied a linear programming model to make decisions about options of groundwater management in conjunction with optimal cropping pattern and production functions of water. Panda et al. (1983) applied linear programming models for conjunctive use of surface and groundwater to canal command area of Punjab by adopting an optimal cropping. Further, to resolve the complex problem of irrigation management within a large heterogeneous basin, Paudyal and Gupta (1990) applied a multilevel optimization technique. They determined the optimal cropping patterns in various sub-areas of the basin, the optimal design capacities of irrigation facilities, including surface and groundwater resources, and the optimal allocation policies of water for conjunctive use. Mainuddin et al. (1997) used an LP model to determine the cropping pattern to ensure optimal use of available land and water resources in a groundwater irrigation project.

Various reports (Amir and Fisher, 1999; Al-Weshah, 2000; Salman et al., 2001; Singh et al. 2001; Samei Tabieh, 2007) address optimal cropping pattern and optimal allocation of water by using LP model. They observe considerable improvement in the economic return as well as in the utilization of land and water resources by adopting an optimal cropping pattern. Abdelaziz et al (2010) obtained the optimal cropping pattern in North Darfur state, Sudan using the Linear Programming (LP) technique. The optimal plan was different from the farmers' plan. The LP model resulted in a profitable objective function while the farmers' plan gained a loss. Igwe et al. (2011), argue that linear programming technique is relevant in optimization of resource allocation and achieving efficiency in production planning particularly in achieving increased agricultural productivity. They applied LP technique to determine the optimum enterprise combination. The actual land use and the optimum plan were tabled. The results from optimum plans were more superior.

The objective of this paper was to develop a LP model based on an economic efficiency criterion for determining optimal water allocation and crop combination. Specifically, there are three aims for this paper as follows: First, understanding the actual patterns of water allocation and crop production in Egypt. Second, examining the economic analysis of water use in crop production. Third, developing an optimization model to ensure the efficient allocation of water resources among the competing crops. This is to serve as a tool for policy makers of indicative planning in irrigation management in Egypt.

METHODOLOGY AND DATA

Model specification

Linear Programming (LP) is a mathematical technique well suited for such a study because of the following reasons: Many activities and constraints can be considered at the same time, secondary explicit and efficient optimum seeking procedure is provided, results from changing variables can easily be calculated once formulated (Hazell and Norton, 1986). Therefore, linear programming is used to make decisions ensuring optimal allocation of water.

The currently most used algorithm in Linear Programming LP software is the Simplex Method which was developed during the Second War in 1974 by a Northern American scientist staff, and has been published afterwards. However, breakthrough in terms of correlated algorithms efficiency only could be observed in the 1980 through developed studies. Nowadays, LP is broadly used around the world and can be applied for different objectives such as maximize net income or minimize costs, losses etc. The General Algebraic Modelling System (GAMS) modelling language is used to calculate the optimal solutions (Brooke et al., 1998). It is preferred for this study because of its flexibility, and it is easy to apply for modellers familiar with such language.

The objectives of the model tries to achieve are related to maximizing the net return and/or minimizing the use of water resources. The mathematical formulation of the applied model includes the following components.

The objective function

The model is to determine the optimum allocation of water resources among competing activities via optimal cropping pattern. It is assumed that the decision maker has perfect knowledge and that there is no risk. The model was applied in two possible future scenarios in accordance with objective function as the following:

Scenario 1: The model employed maximizes net return subject to a set of constraints on cultivated areas, water resources, and other constraints. The optimal number of *Feddans* of each crop depends on the total amount of water and the crop water requirement. The decision makers choose the optimal number of *Feddans* of each crop for which the optimal quantity of water will be applied. The maximization of net return per unit of area is equivalent to the maximization of net return per unit of water. Therefore, the objective function of LP is to maximize the net return per unit of land from all crops. This can be written as:

$$Max \quad \pi = \sum_{j=1}^{n} Z_j X_j$$

Scenario 2: LP model is formulated to suggest the optimal cropping pattern for minimizing the amount of irrigation water used. This scenario modeled a situation of water scarcity in time of drought or a reduction in water supply for agricultural expansion of new lands. This model is to inform decision makers about the impact of a reduction in water supply on crop production. In this case, the objective function is to minimize the total amount of water used for irrigation as follows:

$$Min \quad \pi = \sum_{j=1}^{n} W_j X_j$$

Where; π is the objective function value; Z_j is the net return per unit of land (*Feddan*), $Z_j = (P_j Y_j - C_j)$; N is the number of crops; P_j is the price of crop j (LE/Ton); Y_j is the yield per area unit (Ton/*Feddan*; C_j is total production costs per area unit (LE/*Feddan*), X_j is cultivated area under crop j (*Feddan*), decision variables (j = 1, 2,…n), and W_j is the amount of water needed for irrigation (m^3/*Feddan*).

The constraints

The previously stated objectives are subjected to sets of constraints that are to be satisfied within the model, which include the following.

Land area constraints

This implies that the sum of areas allocated to crops in a certain season must not exceed or equal the total land area available for that season. The mathematical illustration of the land resource constraints is presented as follows:

$$\sum_{j=1}^{n} \alpha_{sj} X_j \leq A_s$$

Where, α_{sj} = 1 if crop j is planted in a season s, otherwise α_{sj} = 0, A_S represents the total land areas available in season S of a year for different crops.

The total available cropped areas for the modeling was about 11540 thousand *Feddan*, representing about 90.06% of the total cropped area of the years (2009-2011) in Egypt. It is distributed over the 3 seasons of the year: 6272 thousand *Feddan* for winter season and 4858.63 thousand *Feddan* for summer and 409.56 thousands *Feddan* for *Nili* seasons, representing land restrictions. Due to the limitations of the data on fruits and other field crops that occupied an area of less than 1000 *Feddans*, they are excluded from this study.

Water constraints

The availability of water for irrigation from the *Nile* water source is limited. Since the amount of water available and water requirement of the crops are different in any month of the year, it is essential for water constraint to be monthly considered. So allocation of water must not exceed the available water in a month. Assuming that there is no recharge of *Nile* water during irrigation season, water constraints can be written as follows:

$$\sum_{j=1}^{n} W_{mj} X_j \leq W_m$$

Where W_{mj} represents a matrix of the water requirement in month m for crop j (m^3/*Feddan*).

W_m is a vector of the total irrigation water availability in month m . The total annual volume of water for the modeling amounted to about 36.7 billion m^3 accounting for 90% of the total irrigation water used at the field level as average of the years (2009-2011). It was distributed over the 12 months, representing monthly water restrictions, after excluding the quantity of water resources available for crops that are not included in the models.

Labor constraints

Labor demand per month for all crops should not exceed the total number of labor days available in that particular month; these constraints can be written as follows:

$$\sum_{j=1}^{n} L_{jm} X_j \leq L_m$$

Where L_{jm} represents a matrix of the labor requirement for crop j (man- day/*Feddan*) in month m . L_m is a vector of the total current number of labor days in month m .

The total annual number of labor days for the modeling amounted to about 790 million days. It was distributed over the 12 months, representing monthly Labor constraints.

Production cost constraints

The value of production cost for all crops should not exceed the total cost of production for the actual cropping pattern; this constraint can be represented for each input as follows:

$$\sum_{j=1}^{n} I_j X_j \leq \sum I$$

Where I_j is input-output coefficient that states the production cost to produce one *Feddan* of crop j , I represents the value of inputs quantities used in actual cropping pattern.

Organisation constraints

Management considerations restrict minimum and maximum value for areas under crops to ensure the supply of the minimum quantities of food commodities and avoid deficiencies and marketing problems. The lower limitations on corresponding acreage were based upon the minimum levels of historical cultivation over the five years (2007-2011) for each crop. On the other hand, to prevent one high value crop from dominating the maximum benefits maximum areas should be considered for each crop. These constraints can be expressed mathematically:

$$LC_j \leq X_j \leq UC_j$$

Where X_j is the area under j crop (*Feddan*), LC_j is the minimum area of crop j, UC_j is the maximum area of crop during the years (2007-2011).

Non- negativity constraints

$$X_j \geq 0$$

The constraint states that the algorithm must not allocate negative amounts of land use in order to optimize the objective function.

Data sources

This study is mainly based on published and unpublished statistical data from the Ministry of Agriculture and Land Reclamation (MALR, 2009, 2010 and 2011), the Ministry of Water Resources and

Irrigation (MWRI), and the Central Agency for Public Mobilization and Statistics (CAPMAS), Egypt. The technical coefficients that quantify resource requirements are determined as a weighted average for real values of the most recently available three years (2009-2011). Moreover, individual crops are subject to organization constraints, which are the upper and lower limitations.

RESULTS AND DISCUSSION

Egypt's water resources and demand

The main source of fresh water for Egypt is *Nile* River. Egypt relies on the availability of its annual share of *Nile* water that is stored in Lake Nasser. That is approximately 55.5 billion cubic meters annually by agreement between Egypt and Sudan in 1959.

The 1959 Agreement was based on the average flow of the *Nile* during the period 1900-1959, which was 84 billion m^3/year at Aswan. Average annual evaporation and other losses from the High Dam Lake were estimated to be 10 billion m^3/year, leaving a net usable annual flow of 74 billion m^3/year. It was agreed that 18.5 billion m^3/year is allocated to Sudan and 55.5 billion m^3/year to Egypt (International Water Law, Documents, http://www.internationalwaterlaw.org/documents/africa.html#*Nile*).

The groundwater aquifer underlying the agricultural lands of the *Nile* Valley and the Delta is entirely recharged and is dependent on deep percolation of irrigation water and seepage for the irrigation system. It cannot, therefore, be considered as an independent resource. And it cannot be added to the country water resources but rather be considered as a reservoir in the *Nile* River system. The total groundwater abstraction was estimated about 5.9 billion m^3/year as average of the period of 2009-2011. For the re-use of agricultural drainage water, the permitted total amount of the recycled water in the *Nile* Delta is about 7.5 billion m^3/year as average of the period of 2009-2011 (CAPMAS, 2009, 2010 and 2011). Sanitary drainage water is used in agriculture and tree planting after treating it to meet the specifications. Some amount of the treated water was about 1.3 billion m^3/year used in irrigation in specific locations outside the greater Cairo regions. Rainfall on the Mediterranean coastal strip decreases from 200 mm/year at Alexandria to 75 mm/year at Port Said. It also decreases inland to about 25 mm/year near Cairo. The average total amount of rainfall is about 1.30 billion m^3/year. This amount cannot be considered a reliable source of water due to a high spatial and temporal variability.

From the above, the actual water resources currently available for use in Egypt are 55.50 billion m^3/year and 1.3 billion m^3/year effective rainfall on the northern strip of the Delta, while water demands for different sectors are about 76 billion m^3, comprising of agriculture, industrial and municipal demand of 61.63, 1.33, and 10.43 billion m^3, respectively (CAPMAS, 2009, 2010 and 2011). The gap between the water supply and water needs is overcome by recycling.

Irrigation water requirements for actual cropping pattern

Allocation patterns of irrigation water use

Table 1 shows the irrigation water use for major crops in the current cropping pattern according to the season of cultivation at the field. The total water needs of summer crops was the highest, followed by winter crops and *Nili* crops, representing 62.59, 34.62 and 2.80% of the total irrigation demand, respectively, as shown in Figure 1; Winter Crops: Total area under winter crops was about 6.271 million *Feddan*. The corresponding water use reached about 13596 million m^3. Wheat and perennial clover are determined to be the most water consuming crops in winter season, as the irrigation requirements for these crops reached about 5377 and 4619 million m^3, respectively, representing 14.57 and 12.52% of the total irrigation requirements at the field level, respectively. Summer Crops: The area under summer crops amounted to 5.178 million *Feddan* and its annual water use reached about 21004 million m^3. Rice, maize, sugarcane, cotton, and sorghum are found to be the most water consuming crops in the summer season. The irrigation requirements for these crops amounted to 8879, 5363, 3065, 1154 and 1097 million m^3, respectively, representing about 24.06, 14.53, 8.31%, 3.13 and 2.97% of the total irrigation requirements, respectively. *Nili* Crops: Area under *Nili* crops was about 0.409 million *Feddan*. Maize is the most water consuming crop in the *Nili* season, consuming 748 million m^3. This represented about 2.03% of the total irrigation requirements.

Crop water requirements

Crop water requirements used in the study were directly taken from Government figures available and published by the Central Agency for Public Mobilization and Statistics (CAPMAS). The data on irrigation requirement are available as annual figures, and it is assumed that these annual requirements can be allocated over the months of plant growth-cycle. For modelling purposes, the computations of monthly irrigation water requirements were carried out by multiplying the theoretical monthly percentage crop consumptive water use by annual irrigation requirement. The theoretical consumptive water use is compiled by Water Management Research Institute (WMRI) in Egypt. Figures 2 and 3 show monthly water requirements of major winter and summer crops in Egypt, respectively.

Table 2 indicates that the average water consumption per irrigated *Feddan*. Perennial clover is considered the most water consuming crops in winter season, as the

Table 1. Actual cropping pattern and its water use for modelling in Egypt (2009- 2011).

Crop	Cropping area		Water used at field	
	Area (000 *Feddan**)	%	Quantity (MCM**)	%
Wheat	3055.15	25.76	5377.06	14.57
Barley	229.02	1.93	346.51	0.94
Broad bean	214.46	1.81	282.66	0.77
Fenugreek	10.36	0.09	12.75	0.03
Lentil	2.22	0.02	4.74	0.01
Lupine	3.40	0.03	5.16	0.01
Chickpeas	8.14	0.07	11.42	0.03
One-cut clover	354.88	2.99	391.79	1.06
Perennial clover	1583.61	13.35	4619.38	12.52
Flax	13.61	0.11	17.60	0.05
Onion	122.10	1.03	232.97	0.63
Sugar beet	302.65	2.55	667.04	71.81
Garlic	22.86	0.19	74.12	0.20
Winter tomatoes	229.18	1.93	481.05	1.30
Winter squash	29.05	0.24	60.97	0.17
Winter cabbage	36.31	0.31	76.22	0.21
Winter green peas	54.99	0.46	115.43	0.31
Total winter crops	**6271.98**	**52.88**	**12776.87**	**34.61**
Cotton	322.09	2.72	1154.00	3.13
Summer rice	1410.77	11.89	8879.41	24.06
Summer maize	1687.24	14.23	5363.73	14.53
Summer yellow maize	262.34	2.21	841.84	2.28
Summer Sorghum	341.91	2.88	1097.19	2.97
Soybean	24.65	0.21	78.19	0.21
Sesame	84.34	0.71	237.24	0.64
Peanut	152.33	1.28	631.08	1.71
Sunflower	31.38	0.26	78.45	0.21
Summer potatoes	125.54	1.06	388.05	1.05
Summer tomatoes	272.45	2.30	842.15	2.28
Summer squash	47.47	0.40	146.74	0.40
Summer eggplant	55.11	0.46	170.33	0.46
Summer cucumber	41.00	0.35	126.74	0.34
Sugar cane	320.21	2.70	3065.37	8.31
Total summer crops	**5178.83**	**43.67**	**23100.51**	**62.59**
Nili maize	287.99	2.43	748.49	2.03
Nili potatoes	52.13	0.44	121.67	0.33
Nili tomatoes	60.59	0.51	141.41	0.38
Nili cabbage	8.85	0.07	20.67	0.06
Total *Nili* crops	**409.56**	**3.45**	**1032.24**	**2.80**
Total crops	**11860.38**	**100.00**	**36.909.62**	**100.00**

Source: Data calculated from CAPMAS, irrigation and water resources bulletin, different issues. *Area in thousand *feddan*, **MCM, million cubic meters.

irrigation requirements for this crop was about 2917 m^3. Sugar cane and rice had the highest water consumption per *Feddan* in summer season. Their water requirements amounted to about 9573 and 6294 m^3, respectively.

Economic analysis of water use in crop production

Table 2 reports the profitability to scarce factors land in LE/*Feddan* and water in LE/1000 m^3, in which net return

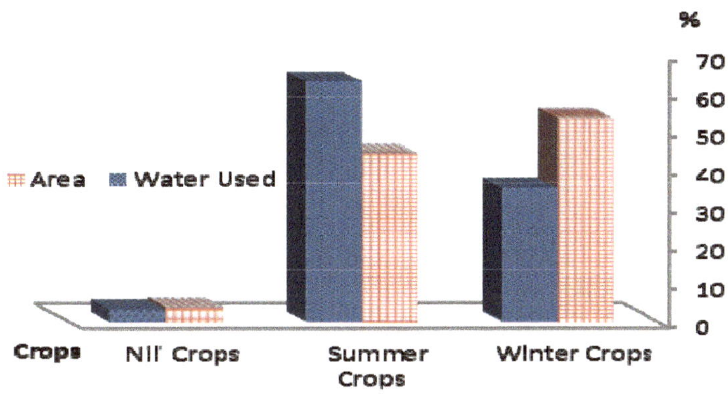

Figure 1. Actual cropping pattern and its water in Egypt (2009-2011).

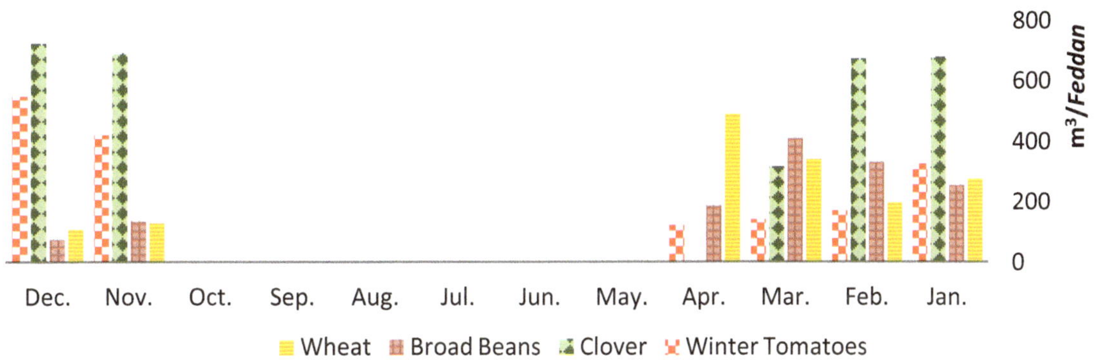

Figure 2. Monthly water requirements of major winter crops in Egypt.

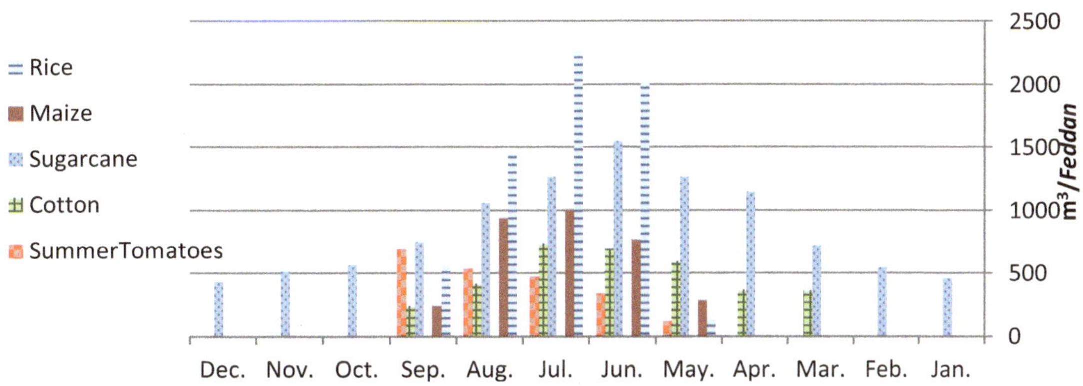

Figure 3. Monthly water requirements of major summer crops in Egypt

per unit of land and water are calculated based on the farm prices. The most profitable crops in winter season were winter tomatoes, garlic, onion, and perennial clover with net return of 9316, 11919, 6520, and 6191 LE/*Feddan*, respectively. For the summer crops: Summer tomatoes were also the most profitable, with net return of 9600 LE/*Feddan*. Summer potatoes cane was among the next most profitable crops, with net return of 6212 LE/*Feddan*. The net returns were 4522, 2716 and 1975 LE/*Feddan* for peanut, rice and cotton, respectively.

For net return per unit of water, it is observed that the most profitable crops in winter season were winter tomatoes and garlic with net return per unit of water of 4438 LE/1000 m^3 and 3675 LE/1000 m^3, respectively. For the summer crops: tomatoes and potatoes were also the most profitable, with net return of 3106 and 2010 LE/1000 m^3, respectively.

Virtual water represents the amount of water needed to raise a certain quantity of food (Allan, 1999). The "virtual water" concept can contribute to a change in water

Table 2. Average net return by unit of land and water for the most important crops in Egypt (2009- 2011).

Crops	Yield (Ton/*Feddan*)	Net return (LE*/*Feddan*)	Water requirement (M^3/*Feddan*)	Net return per unit of water (LE/1000M^3)	Virtual water (M^3/Ton)
Wheat	2.7	3109	1760	1766	651.85
Broad bean	1.4	2040	1318	1548	941.43
One-cut clover	12.8	2912	1104	2638	86.25
Perennial clover	29.4	6191	2917	2122	99.22
Sugar beet	20.5	3286	2204	1491	107.51
Onion	13.8	6520	1908	3417	1060.00
Garlic	9.7	11919	3243	3675	334.33
Winter tomatoes	19.4	9316	2099	4438	108.20
Cotton	1.0	1975	2985	662	2985.00
Summer rice	4.1	2716	6294	432	1535.12
Summer maize	3.3	1931	3179	607	963.33
Summer sorghum	2.3	1419	3209	442	1395.22
Soybean	1.4	1014	3172	320	2265.71
Sesame	0.5	1670	2813	594	5626.00
Peanut	1.3	4522	4143	1091	3186.92
Summer potatoes	12.0	6212	3091	2010	257.58
Summer tomatoes	16.3	9600	3091	3106	189.63
Nili maize	2.8	1489	2599	573	928.21
Nili potatoes	9.7	1789	2334	766	240.62
Nili tomatoes	16.9	1058	2334	453	138.11
Sugar cane	49.8	5748	9573	600	192.23

Source: Calculated from MALR and CAPMAS. *LE Egyptian Pound.

The basic linear programming model solutions

The LP model is used to determine the optimal allocation and crop pattern of the different crops. In order to calibrate the model, the actual cropping plan for the reference average of years (2009-2011) is compared with the results generated by the models. Table 3 compares the optimal values of the net return, land, water and crop areas in the two scenarios vs. the actual cropping pattern in 2011.

Scenario 1: Maximizing the net return

The outputs of maximization model are shown in terms of percentage change of actual values. In winter season, the area under wheat declined by approximately 8.43% below its actual cropped area. Similarly, area under one-cut clover and broad bean decreased by 12.68 and 11.20%, respectively. However, there was an increase in area under perennial clover by about 15.20% above the current area. Area under sugar beet and winter tomatoes would increase by 27.45 and 15.70% above its actual area because of their high profitability. In summer season, area under rice and summer maize decreased by 12.57 and 7.0%, respectively, below existing area, while summer potatoes and tomatoes would increase by 6.70 and 4.63%, respectively. *Nili* potatoes and *Nili* maize crops could decline, while *Nili* tomatoes and *Nili* cabbage crops would increase in the optimal plan. Because of its high profitability, sugar cane recorded an increase in the optimal solution at 4.60%.

The results showed the great potential to generate a net return equivalent to about 3.56% more than the actual total net returns. The optimized cropping pattern in Egypt has been coupled with about 3.24% saving in the water use and about 3.13% reduction in the production cost compared to the existing plan.

Scenario 2: Minimizing the amount of irrigation water used

The objective function of this scenario is to minimize the amount of irrigation water used taking into consideration the same specified constraints. This model is useful in informing water policy makers about the impact of water cuts on the crop production in old lands. The results show that the cropping pattern changes in favour of less water demanding crops. Also, cultivated winter and summer areas would decline by 10.42 and 12.05% below the basic level, respectively. This means that the fallowed lands appeared due to water becoming scarcer. Consequently, cropped area under most of the crops

Table 3. Comparison of cropping pattern under optimal plans with actual plan.

Indicators	Actual plan	Optimal plans	
		Δ_1 %	Δ_2 %
Total Net Return (M.LE)*	42963.11	3.56	-10.20
Resources Utilization:			
Total Water Used (M.CM)	36909.62	-3.24	-11.05
Winter Land Used (000*Feddan*)	6271.98	0.00	10.42
Summer Land Used (000 *Feddan*)	5178.83	0.00	12.05
Nili Land Used (000 *Feddan*)	409.56	0.00	14.37
Production Costs (M.LE)	39391.33	-3.13	-11.24
Crop Area (1000 Feddan)			
Wheat	3055.15	-8.43	-11.12
Barley	229.00	-20.50	-20.50
Broad Bean	214.33	-11.20	-11.20
Fenugreek	10.35	-23.40	-23.40
Lentil	2.12	-34.30	-34.30
Lupine	3.40	-8.94	-8.94
Chickpeas	8.14	-22.20	-22.20
One-cut Clover	354.88	-12.68	-12.68
Perennial Clover	1583.61	15.20	-4.10
Flax	13.61	-41.60	-41.60
Onion	122.10	9.80	-46.37
Sugar Beet	302.65	27.45	-44.71
Garlic	22.86	22.90	-25.57
Winter Tomatoes	229.18	15.70	-12.61
Winter Squash	29.05	3.75	-30.21
Winter Cabbage	36.31	4.20	-25.28
Winter Green Peas	54.99	9.20	-5.61
Cotton	322.09	43.62	-11.69
Summer Rice	1410.77	-12.57	-20.50
Summer Maize	1687.24	-7.00	-7.00
Summer yellow Maize	262.34	-46.00	-46.00
Summer Sorghum	341.91	-3.80	-3.80
Soybean	24.65	-30.80	-30.80
Sesame	84.34	-21.30	-21.30
Peanut	152.33	4.35	-13.29
Sunflower	31.38	-38.70	-38.70
Summer Potatoes	125.54	6.70	-37.02
Summer Tomatoes	272.45	4.63	-20.92
Summer Squash	47.47	-8.16	-8.16
Summer Eggplant	55.11	-3.54	-3.54
Summer Cucumber	41.00	-4.96	-4.96
Nili Maize	278.99	-0.38	-15.60
Nili Potatoes	52.13	-5.66	-5.66
Nili Tomatoes	60.59	21.60	-12.57
Nili Cabbage	8.86	15.80	-10.02
Sugar Cane	320.21	4.60	-1.10

Source: Mathematical programming models results based on CAPMAS, MWRI (NWRC) and MALR Data. MLE Million Egyptian Pound.

decreased. Area under wheat would decline by 11.12% below the current area allocated to wheat in winter season. Area under one-cut clover and perennial clover decreased by 12.68 and 4.10%, respectively. There was a decrease in area under sugar beet by about 44.71% below the actual area. Despite the high profitability of winter tomatoes, its area declined by 12.61% below the basic level due to its high water requirement. Area under cotton recorded also a decrease in the optimal solution at 11.69%. Similarly, in the summer season, area under rice and summer maize would decrease by 20.50 and 7.00% below the actual cropped areas, respectively. This may be attributed to more water consumption for these crops in relation to their net returns. Area under summer potatoes and tomatoes would decrease by 37.02 and 20.92% below the actual area, due to their high water requirements compared to other field crops. Area under Nili maize, Nili potatoes, and Nili tomatoes crops would decrease in by 15.60, 5.66 and 12.57%, respectively.

The results show that the total net returns decreased by 10.20% below the actual total net returns. The optimized cropping pattern in Egypt has been coupled with about 11.05% saving in the water use and about 11.24% reduction in the production cost use compared to the existing plan.

Conclusion

Regarding to the results of above analysis, the following conclusions can be summarized:

1. The model for maximizing net return produced an optimal allocation of water and cropping pattern that gives higher net return compared to the existing plan,
2. Land resource under maximization of net return, had an optimal use, where it is fully used,
3. There is a need for the governmental co-ordination in crop production ensuring the supply of food commodities and avoiding marketing problems.
4. Minimizing of irrigation water requirements is generally difficult. The policy had negative impacts on farm income, the irrigated area decreased and fallow lands appeared.

RECOMMENDATIONS

Several recommendations, based on the results can be made for the future water policies as follows:

1. In order to improve water allocation, farmers should be advised to follow the indicative optimal cropping pattern, which maximizes the net income return.
2. Cultivation of sugarcane in Egypt should be reduced based on minimum requirement of raw material for sugar processing plants.
3. Cultivation of rice should be restricted.
4. The Government of Egypt should encourage utilization of new irrigation technologies.
5. The applied model can be used to provide useful information to decision makers about likely optimal allocation policies for irrigation.

Conflict of Interest

The authors have not declared any conflict of interest.

REFERENCES

Abdelaziz HH, Abdalla AA, Abdellatif MA (2010). Optimizing the Cropping Pattern in North Darfur State. Sudan: A Case Study of Dar Elslam District. J. Appl. Sci. Res. 6(2):156-164.

AL-Weshah RA (2000). Optimal use of irrigation water in the Jordan Valley: A Case Study. Water Resourc. Manage. 14:327–338.

Amir I, Fisher FM (1999). Analysing agricultural demand for water with an optimising model. Agric. Syst. 61(1):45-56.

Brooke A, Kendrick D, Meeraus A, Raman R (1998). GAMS: A User's Guide, GAMS Development Corporation. Washington. D.C.

CAPMAS (2009). Central Agency for Public Mobilisation and Statistics. Annual Bulletin of Irrigation and Water Resources. Egypt (In Arabic).

CAPMAS (2010). Central Agency for Public Mobilisation and Statistics. Annual Bulletin of Irrigation and Water Resources. Egypt (In Arabic).

CAPMAS (2011). Central Agency for Public Mobilisation and Statistics. Annual Bulletin of Irrigation and Water Resources. Egypt (In Arabic).

Hazell PBR, Norton RD (1986). Mathematical Programming for Economic Analysis in Agriculture. MacMillian Publishing Company. New York.

Igwe KC, Onyenweaku CE, Nwaru S (2011). Application of Linear Programming To Semi-Commercial Arable And Fishery Enterprises in Abia State. Nigeria. Int. J. Econ. and Manage. Sc. 1 (1): pp.75-81.

International Water Law (Not dated). African regional documents, Addressing the future of water law and policy in the 21 st century. http://www.internationalwaterlaw.org/documents/africa.html#*Nile* River Basin.

Kheper SD, Chaturvedi MC (1982). Optimum Cropping and Groundwater Management. Water Resourc. Bull. 18(4):655-660.

Koundouri P, Nauges C, Tsouvelekas V (2006). Technology Adoption under Production Risk: Theory and Application to Irrigation Technology. Am. J. Agric. Econ. 88(3):657-670.

Mainuddin M, Gupta AD, Onta PR (1997). Optimal Cropping Planning Model for an Existing Groundwater Irrigation Project in Thailand. Agric. Water Manage. 33(1):43-62.

MALR (2009). Ministry of Agriculture and Land Reclamation. Economic Affairs Sector. Agricultural Statistics Bulletin. Egypt.

MALR (2010). Ministry of Agriculture and Land Reclamation. Economic Affairs Sector. Agricultural Statistics Bulletin. Egypt.

MALR (2011). Ministry of Agriculture and Land Reclamation. Economic Affairs Sector. Agricultural Statistics Bulletin. Egypt.

MWRI, NWRC (Unpublished). Ministry of Water Resources and Irrigation. Egypt. National Water Research Centre. Data.

Panda SN, Kaushal MP, Kheper SD (1983). Irrigation planning in a command area in a project: An application of deterministic linear programming. J. Agric. Eng. ISAE. 10(2):47-60.

Paudyal GN, Gupta AD (1990). Irrigation management by multilevel optimisation. J. Irrigat. Drain. Eng. ASCE. 116(2):273-291.

Salman AZ, Al–Karablieh EK, Fisher FM (2001). An inter–seasonal agricultural water allocation system (SAWAS). Agric. Syst. 68:233-252.

Singh DK, Jaiswal CS, Reddy KS, Singh RM, Bhandarkar DM (2001). Optimal cropping pattern in a canal command area. Agric. Water Manage. 50:1-8.

Tabieh MAS (2007). An optimal irrigation water allocation model: Management and Pricing Policy Implications for the Jordan Valley. Thesis for the Degree of Doctor of Philosophy Economics–Econometrics. University Sains Malaysia (USM).

World Bank (2002). Water priority for responsible growth and poverty reduction: An agenda for investment and policy change. Washington. D.C: World Bank.

Young RA (2005). Determining the Economic Value of Water: Concepts and Methods. Washington D.C: RFF Press.

Economic and technical efficiency of cassava production in Ika North East Local Government Area of Delta State, Nigeria

Eze, A. V. and Nwibo, S. U.

Department of Agricultural Economics Management and Extension, Ebonyi State University, Abakaliki, Nigeria.

Cassava production seems to be economically viable but there seems to exist no empirical documented evidence to this effect in Ika North East L.G.A of Delta State, Nigeria. In view of this, the study analyzed the economic and technical efficiency of cassava production in Ika North East Local Government Area of Delta State. A multistage random sampling was used to select a total of 120 respondents used for the study. Data used for the study was from primary source, which was collected using a well structured questionnaire. Both descriptive and inferential statistics were used to analyze the data based on the objective of the study. The result obtained showed that females (52.5%) are more than males. Majority (50%) of the respondents are married with an average household size of 6. The result further showed that the farmers were in their middle age (42 years) and had acquired reasonable years of farming experience of 10 years. More than half of the farmers had attended formal educational and earn average annual income of ₦180,000.00. The production systems practiced by the farmers was mono cropping. Cassava production was profitable in the area with a profit margin of ₦200,400.00 per a hectare. The Benefit Cost Ratio shows that in every ₦1.00k invested by farmers, ₦1.00k was realized as profit. The multiple regression result showed R^2 value of 0.833 or 83.3%. The coefficients of farm size, labour and cassava stem were positively signed. Farm size, labour, fertilizer and cassava cuttings were underutilized because their efficiency index was greater than one. The cassava farmers identified some of the factors that constrained their farming activities to include: lack of access to credit, high cost of transportation, poor extension services, among others. Based on the findings, it is therefore recommended that farmers should organize themselves into cooperative societies so as to access credit; viable extension service should be provided bridge the extension need of the farmers and effective transportation system to ease evacuation of cassava produce to urban centre where the demands are high.

Key words: Costs and returns, profitability, production system, constraints, input/output.

INTRODUCTION

Cassava (*Manihot* spp.) is important not only as food crop but moreso as a major source of income for rural households. According to Ogunniyi et al. (2012), cassava has some inherent characteristics, which makes it attractive especially to the smallholder farmer in Nigeria. Firstly, it is rich in carbohydrates, which make it useful in some industries and consequently has a multiplicity of end uses. Secondly, it is available all year round compared to other crops as it is more tolerant to low soil fertility and resistant to drought, pests and diseases. These attributes combined with other socio-economic considerations are therefore what IFAD has recognized in the crop as lending itself to a commodity-based approach to poverty alleviation (FAO/IC, 1995).

The comparative production advantage of cassava over other staples has made the government to encourage its cultivation even by the resource poor farmers. The crop production is generally thought to require less labour per unit of output than other major staples. It is a good staple whose cultivation if encouraged can provide the nationally required food security minimum of 2400 calories per person per day (FAO, 2000).

In 2002, cassava suddenly gained prominence in Nigeria following the pronouncement of a presidential initiative on the crop. The initiative was aimed at using cassava production as the engine of growth in Nigeria. In recent times, government has encouraged the use of the crop to produce a wide range of industrial products such as ethanol, glue, glucose syrup and bread. The Nigerian government has also promulgated a law, making it compulsory for bakers to use composite flour of 10% cassava and 90% wheat for bread production. The new regulation which came into effect January 2005, stipulated that the large flour mills that supply flour to bakeries and confectioneries must pre-mix cassava flour with wheat flour (Technical for Agricultural and Rural Cooperation (CTA), 2005).

Cassava has been reported as the chief source of dietary food energy for the majority of the people living in the lowland tropics, and much of the sub-humid tropics of West and Central Africa (Tsegia et al., 2002). Therefore, its production and utilization must be given prime attention in food policy. Even though farmers have not yet attained the desired technical efficiency in cassava production as a result of weak access to external inputs such as fertilizers and herbicides (Ezedinma et al, 2006), the wide scale adoption of high yielding varieties and the resulting increase in yield have shifted the problem of the cassava sector from supply (production) to demand issues, such as finding new uses and markets for it.

Nigeria produces more than 45 million metric tons (MT) of cassava, thus emerging as the world's largest producer (USAID, 2010). In spite of this volume, the full yield potential has not been realized since smallholder production rarely exceeds 11 MT per hectare as against 25 to 40 MT per hectare recommended by experts. This yield per hectare is indicative of the yields experienced in the south-south region of Nigeria including Ika North East L.G.A of Delta State. This region is one of the most productive in the country with respect to cassava. The national average is somewhat lower at 10.0 tonnes/ha. In contrast, Thailand national experienced yields of 17.1 tonnes/ha in 2002. Regional yields in countries such as India, Laos, Thailand and Barbados have been estimated as high as 25 to 40 tonnes/ha. Obviously, Nigeria's highest productivity yields fall short of these rates and this situation is due to a number of factors including small scale farming (on plots that are usually less than 1 ha), manual operation, little or no use of fertilizers and limited knowledge in the use of high yielding roots (Olomola, 2007). Farming at this level makes it difficult to achieve efficiency and economies of scale.

At the farm level, production costs for cassava are high relative to those in other countries. Production is not oriented towards commercial use; instead, farmers produce and process cassava as a subsistence crop. The Nigerian cassava system, is characterized by small-scale farmers/holdings cultivating less than 2 ha of cassava (average of 0.5 ha), primarily cultivated for the traditional food market, is subsistence in nature and not oriented to the industrial market. Any surplus cassava is either processed on the farm, or sold to local processors. The average production figures per hectare in Nigeria were 10.5 MT/Ha in the early 1970s, 11.5 MT/Ha in the 1980s, 10.5 MT/Ha by the end of 1980s, and 11.5 MT/Ha in the 1990s and up to 17.3 MT/Ha was achieved in Ondo State in 2004.

It is also important to note that cassava production is mostly done by rural smallholder farmers using low-level production techniques (Omonona, 2009; Oyegbami et al., 2010; Nweke et al., 2002). Though government at various levels has been trying in various ways to encourage rural farmers to adopt the modern cassava production technologies in order to increase the rural farmer's productivity (Frescro, 1993; Otoo, 1994), there are constraints to adoption in rural farming communities (Nweke et al., 2002; Teklewold et al., 2006). In some instances, farmers reject some of modern technology due to their cultural background and inhibitions due to perhaps illiteracy and religious beliefs. Nevertheless, credit constraint has been singled out as a major factor militating against adoption of modern cassava production techniques (Nweke et al., 2002). The technologies are herbicides application, use of hybrid cassava stake, use of insecticides, use of inorganic fertilizer, use of tractor, appropriate spacing, planting date and tillage practices. The adoption of modern cassava production technologies is an important route out of poverty and enhancing productivity for many in the developing world including Nigeria because of the major role cassava play in food security. Many studies have noted poor technology adoption in cassava production as a serious factor constraining outputs (Barham and Boucher, 1994; Ogboso, 2005).

Despite the importance of cassava as a means of livelihood of farmers in Nigeria, the dearth of empirically documented data on the economic and technical efficiency in the Ika North East L.G.A of Delta State necessitated this study. It is in view of the foregoing that the following specific objectives were addressed:

i) Describe the socio-economic characteristics of the farmers in the area;
ii) Identify the production system employed by the farmers;
iii) Determine the technical efficiency of production in the area;
iv) Analyze the costs and returns of production in the area; and
v) Analyze the constraints to production in the area.

METHODOLOGY

The study was carried out in Ika North East Local Government Area of Delta State, Nigeria. It has an area of 463 km² and a population of 183,657 (NPC, 2006). There are nine communities, namely; Owa, Ute-Ogbeje, Ute-Okpu, Umunede, Idumuesah, Igbodo, Otolokpo and Mbiri spread out into fourteen wards in the area. The Local Government Area has natural vegetation that supports agricultural activities such as crop production, fishing etc. thus; agriculture is the major activities of the people of this area. The principle crops grown in this area are: yam, cassava, melon, maize, tomatoes, plantain, among others.

A multiple-stage random sampling techniques was employed in selecting the respondents. This involves the random selection of four communities from the nine communities in the area. From the four randomly selected communities, three villages were randomly selected to give a total of 12 villages. Finally, ten cassava farmers were randomly selected from the 12 villages to give rise to 120 farmers. Thus, a total of 120 cassava farmers were randomly selected for the study. Primary data was used for the study. The data was collected through the use of structured questionnaire that was administered to the 120 randomly selected respondents. Data used for the study was analyzed using descriptive statistics such as mean, frequency distribution tables, percentages and inferential statistics. Descriptive statistics was used to analyze objective (i) and (ii); objective (iii) was analyzed using multiple regression analysis while objective (iv) was achieved using gross margin analysis and objective (v) was analyzed using mean score derived from 4 point likert scale.

Model specification

Multiple regression model

$Y =$ f $(X_1, X_2, X_3, X_4, X_5)$ - - - - - implicit form
$Y = a_0 + a_1X_1\ a_2X_2, + a_3X_3, + a_4X_4, X_4, +a_5X_5, +$ et ---- Explicit stochastic form

Where
Y=total output of cassava (tonnes)
X_1 = farm size (ha)
X_2 = labour used in man-days
X_3 = fertilizer used (kg)
X_4 = cassava cuttings (kg)
X_5 = herbicide used (litre)
et = Stochastic error term
$a_1 - a_5$ = Parameters estimate
a_0 = constant

Technical efficiency of each parameter was estimated using efficiency index Rxi = biPy/Pxi,

Where
pxi = unit price of input (N),
Py = unit price of output (N),
bi = marginal productivity of the input and
Rxi = Technical efficiency index of the input.

Model for gross margin

The model used for the estimation of the gross margin according to Olukosi and Ernabor (1988) is stated as:

GM = TR – TVC (GI – TVC)

Gross margin = Total revenue – Total variable cost

ϒ = GM – TFC

Profit = Gross margin – Total fixed cost

Where
GM = Gross Margin
TR = Total Revenue
GI = Gross income
TVC = Total variable cost
ϒ = Profit

Model for Likert scale

$$\overline{Xs} = \frac{\sum fn}{Nr}$$

Where:

$\overline{X}$s	=	mean
$\sum$	=	Summation
Fn	=	frequency of respondents responses
Nr	=	number of response of respondent

RESULTS AND DISCUSSION

Socio-economic characteristics of the respondents

Table 1 shows the age distribution of the famers. The result indicated that more than half (70%) of the respondents are between 31 to 50 years of age, which is regarded as economically active age according to FAO (1992). At this stage in life, Anyanwu et al. (2001) recognised that people are more likely to be energetic and have the capacity to use innovation. This justified the findings of Ebukiba (2010), who reported that 76% of the cassava farmers in Akwa Ibom State were aged between 31 to 50 years.

The results equally revealed that majority (52.5%) of the farmers are female while 47.5% are male. This implies that women participate more actively in cassava production than their men counterpart. This collaborate the findings of Ebukiba (2010), who reported that 60% of the cassava farmers in Akwa Ibom State were females.

Table 1. Percentage distribution of the farmers according to their socio-economic characteristics.

Characteristics	Description	Frequency (n=120)	Percentage	$\overline{X}$
Age (years)	20 - 30	24	20	42
	31 - 40	12	10	
	41 - 50	72	60	
	51 and above	12	10	
Gender	Male	57	47.5	
	Female	63	52.5	
Marital status	Single	27	22.5	
	Married	60	50	
	Separated	10	8.3	
	Divorced	8	6.7	
	Widowed	15	12.5	
Household size	1 - 4	48	40	5
	5 - 8	48	40	
	9 - 12	18	15	
	13 and above	6	5	
Educational level	Non-formal	15	12.5	
	Primary	26	21.7	
	Secondary	41	34.2	
	OND/NCE	27	22.5	
	HND/B.Sc	9	7.5	
	M.Sc	2	1.7	
Annual income	≤50,000	6	5	180,000
	50,001 - 100,000	36	30	
	100,001 - 150,000	30	25	
	150,001 - 200,000	12	10	
	200,001 - 250,000	6	5	
	250,001 - 300,000	18	15	
	300,001 and above	12	10	
Farming experience	1 - 5	24	20	10
	6 - 10	66	55	
	11 - 15	12	10	
	16 - 20	18	15	
Farm size	3 - 5	54	45	6
	6 and above	66	55	

Source: Field Survey (2012).

It was noted that most (60%) of the cassava farmers were married, 22.5% were single, 8.3% were separated while 6.7% were divorced and 12.5% were widowed. This is justified on the ground that the majority of respondents who engaged in cassava farming are married people. It also implies that cassava production is the means of livelihood for these households.

Household size is a very important factor especially in determining labour for farm work. A farmer with a large household size has the chance of using them as their farm labour. This will affect the size of land cultivated and enhance returns. From the result, it was observed that

Table 2. Percentage distribution of respondents according to production system.

Production system	Frequency (n=120)	Percentage
Mono cropping	93	77.5
Mixed cropping	15	12.5
Inter-cropping	12	10.0

Source: Field Survey (2012).

the farmer had an average household size of 6. This conforms to the findings of Oladeebo and Oluwaranti (2012), who reported average of 8 persons per cassava farmers in South Western, Nigeria.

Again most (34.2%) of the respondents had attended secondary school education, 21.7% of them had attended primary school and 31.7% of them had acquired post secondary school education, while a few (12.5%) of them did not acquire formal education. By implication, a reasonable number of farmers in the area should be able to understand the use of improved technologies and apply it to achieve increased production. Through education, the quality of labour is improved and with it the propensity to adopt new techniques (Tijani et al., 2006; Hyuha, 2006). Thus, cassava farmers in the study area would easily adopt new technologies which could improve their level of profit *ceteris paribus.*

The result reveal an average income of ₦180,000.00 per annual. The breakdown shows that most (45%) of the farmers earned an annual income of between ₦100,000 to ₦150,000, 12% earned between ₦150,001 to ₦200,000, 30% of them earned above ₦200,00 per annual income. Signifying that the respondents are low income earners and this will have a negative effect on the rate of adoption of improve cassava technologies in the area, since capital is needed to procure most of the modern cassava technologies.

The result equally showed that most of the farmers had been in the business of cassava farming for up to 10 years. This is an indication that majority of the farmers has taken into cassava farming for quite a while in the area. This is also in consonance with the findings of Oladeebo and Oluwaranti (2012), who reported average of 13 years farming experience for cassava farmers in South Western, Nigeria.

The result of the farm size as held by the farmers on average was 5 ha, while majority (45%) held a size of between 3 to 4 ha. This followed the study of Oladeebo and Oluwaranti (2012), who reported average of 4 ha farm size for cassava farmers in South Western, Nigeria.

Cassava production system

The result in Table 2 shows that majority (77.5%) of the cassava farmers practiced mono cropping production system, while few (12.5%) of them practiced inter-cropping and 10% practiced inter cropping production system in the area. It is justifiable to say that mono cropping system is the cassava production system practiced by the farmers in the area.

Relationship between inputs and outputs of cassava production

Table 3 shows the result of multiple regression analysis of the relationship between inputs used and outputs from cassava production in the study area. The multiple regression co-efficient (R) was 0.912 or 91.2%. The implication is that the included independent variables (farm size, labour, fertilizer used, cassava stem and herbicide used) were highly correlated with the farmers' outputs. Also the coefficients of multiple determination (R^2) was 0.833 or 83.3%, signifying that 83.3% of total variation in dependent variable (total outputs) was explained by the explanatory variables, that is, inputs (x_1-x_5) included in the model. The fitness of the model was confirmed by the low value of the overall standard error of the estimate (Std. error = 5.27849) and the Durbin-Watson value of 2.356, indicating absence of autocorrelation in the model.

Farm size (x_1): the coefficient of farm size was positively signed and statistically significant at 1%. This implies that increasing the farm size cultivated by the farmers will lead to proportionate increase in total cassava outputs. Again, the statistical significance indicated by farm size of the farmers signifies that farm size contribute to outputs of the farmers. This conforms to the *a priori* expectation.

Labour used (x_2): the coefficient of labour used in cassava production was negatively signed and statistically insignificant. This implies that a unit increase in labour used in cassava production will not contribute to total cassava output. This conforms to the a priori expectation, because increasing labour used in cassava production will add additional cost to total cost of production which will reduce the returns of the farmers.

Fertilizer used (x_3): by the farmers was negatively sign, but was statistically significant at 1%, indicating an

Table 3. Relationship between inputs and total outputs from cassava production in the area.

Variables	Coefficient	Standard error	t-value	Sig
Constant	-18963.514	2191.241	-8.654	*
Farm size (x_1)	0.580	0.200	2.904	*
Labour used (x_2)	-0.231	0.249	0.926	NS
Fertilizer used (x_3)	-5.145	0.378	-13.613	*
Cassava stem (x_4)	0.939	0.262	3.583	*
Herbicide used (x_5)	-5.022	0.883	-5.689	*
R	0.912			
R^2	0.833			
D.W	2.356			
F-statistics	113.432			
Standard error	5.27849			

Source: SPSS Analyzed Data (2012). NS = Non significant; NS = Non significant, *indicate significance at 1% level

Table 4. Technical efficiency of cassava production in the area.

Resource	MVP (N)	MFC (N)	Efficiency Index
Farm size (X_1)	16436.3	1524.5	10.8
Labour (X_2)	17945.4	2846.7	6.3
Fertilizer (X_3)	21050.0	16450.0	1.3
Cassava cuttings (X_4)	4360.0	1405.2	3.1
Herbicide (X_5)	1235.4	1846.1	0.07

Source: Computed Field Survey (2012).

inverse relationship between the fertilizer used and the total cassava output in the area. In other words, increasing fertilizer used for cassava production will lead to decreasing outputs. However, the statistical significance implies that fertilizer used contributes to total cassava outputs. This is in conformity to the *a priori* expectation, because the continued application of fertility to farm will lead to soil acidity and binding of certain important micro and macro nutrients which are needed for optimum crop growth. In addition to the cost it will impute to the overall production cost. Although, fertilizer is required in its optimal level for the improvement of soil fertility, its over-use is damaging to soil.

Cassava stem (x_4): used by the farmers was positively related to total output and statistically significant at 1%. This signifies that increasing use of cassava stem will result to a unit increase in total cassava output. Again, statistical significance indicated that the use of cassava stem is associated with outputs of farmers. Thus, the *a priori* expectation was met.

Herbicide used (x_5): the coefficient of herbicide used was negatively related to the total output but statistically significant at 1%. This implies that increasing the use of herbicides in cassava production will lead to decreasing cassava production output in the area. While the statistical significance signifies that herbicide application contributes to cassava outputs, this agrees with the *a priori* expectation because increasing herbicide used will add additional cost to the overall cost of production and decrease returns accruing to the farmer.

Technical efficiency

From the result in Table 4, it was observed that the farmers were not efficient in the utilization of all the specified resources as far as cassava production is concerned in the study area. Farm size had the highest efficiency index of 10.8, followed by labour (6.3), cassava cuttings (3.1), fertilizer (1.3) and herbicide (0.07). Farm size, labour, fertilizer and cassava cuttings were underutilized since the efficiency index was greater than one. This indicates that additional income can be made from the production of cassava by using more of these inputs efficiently by the farmers. There was over utilization of herbicide since the efficiency index is less than one. Therefore reducing the litres of herbicide used can lead to more income. It should be noted that the MVP's of all the inputs used were not negative, indicating that cassava farmers still use the resources within the

Table 5. Analysis of cost and returns of cassava production per hectare.

Items	Unit	Quantity	Unit Price	Total
A. Revenue				
Cassava tubes	tonnes	14.5	25,000	362,500.00
Cassava stems	tonnes	7	4,500	31,500.00
Total Revenue				**394,000.00**
B. Variable Cost Inputs				
Cassava stems (Cuttings)	tonnes	4	4,500	18,000.00
Fertilizers	Bag	3	5,000	15,000.00
Total Cost				**33,000.00**
C. Cost of Labour	**Mandays**			
Land preparation (Clearing, ploughing and harrowing)		18	750	15,000.00
Planting	mds	10	750	7,500.00
Weeding	mds	30	750	30,000.00
Harvesting	mds	10	750	7,500.00
Transportation				50,000.00
Miscellaneous cost				20,000.00
Total Cost				**130,000.00**
D. Total Variable Cost				**163,000.00**
Fixed Cost				
Depreciation on farm tools (hoes, matches) @ 10				5,600.00
Depreciation on land @ 5%				25,000.00
E. Total Fixed Cost				**30,600.00**
Total variable cost (TVC) = B + C				163,000.00
Gross margin = TR – TVC = A – D				231,000.00
Total cost = TFC + TVC = E + D				196,000.00
Benefit Cost Ration (TR/ TC)				2.0:1.0

Source: Computed From Field Survey (2012).

economically range even though they were not optimally used. This justifies the finding of Ogunniyi et al. (2012),who reported that cassava farmers in Atakunmosa Local Government Area of Osun State underutilized farm size labour, fertilizer and cassava cuttings, while herbicide, was over-utilized.

Cost and returns

From the result in Table 5, total cost of producing cassava per hectare was ₦196,000.00, the total revenue obtained was ₦394,000.00 and the gross margin was ₦231,000.00. The profit of ₦200,400.00 was actualised, this implies that cassava production in the area was profitable. Also the Benefit Cost Ratio was ₦2.00, indicating that for every ₦1.00k expended in cassava production, ₦1.00k was realized as a profit. This follows the findings of Ebukiba (2010) who reported BCR of N1.9:1.0 for cassava farmers in Akwa Ibom State.

Constraints militating against efficient cassava production

The farmers were constrained by the following factors: lack of access to credit facilities (3.8), lack of ready market (2.5), poor storage facilities (2.8), high cost of transportation (3.2), high cost of labour (3.4), inadequate supply of fertilizer (3.5), poor extension services (3.6), problems of pests and diseases (2.9) and poor road network (3.0). This follows the findings of Ebukiba (2010), who reported that cassava farmers in Akwa Ibom State face problems such as inadequate capital, lack technical, lack of government support, lack of improve cuttings and poor market, among others (Table 6).

Conclusion

The finding of this study shows that cassava production in the area is very lucrative, inspite of the inefficient use

Table 6. Mean score distribution of respondents according to constraints militating against cassava production.

Constraints	Mean score (xs)	Decision
Lack of ready market	2.5	Accepted
Lack of access to credit facilities	3.8	Accepted
Poor storage facilities	2.8	Accepted
High cost of transportation	3.2	Accepted
Lack/ inadequate improved varieties	2.0	Rejected
High cost of labour	3.4	Accepted
Inadequate supply of fertilizer	3.5	Accepted
Land fragmentation	2.2	Rejected
Poor extension services	3.6	Accepted
Problems of pests and diseases	2.9	Accepted
Poor road network	3.0	Accepted

Source: Field Survey (2012).

of certain factors of production by the farmers. Hence, the farmers should be encouraged through technical training on production techniques/ practices that will improve their productivity especially in those areas were the study identified inefficiency and underutilization of production resources. These are: farm size, fertilizer and cassava cutting. It is therefore recommended that more farmers should be encouraged to go into cassava farming since it is profitable and can tolerate soil with low fertility.

Conflict of Interests

The author(s) have not declared any conflict of interests.

REFERENCES

Anyanwu AC, Agwu AE, Musa AP (2001). Adoption of gender specific innovation by women in state. J. Agric. Ext. 5:64-72.

Barham B, Boucher S (1994). Credit constraints, credit unions, and small-scale producers in Guatemala. Mimeo.PMCid:PMC1238870

CBN (2004). Annual report, Central Bank of Nigeria, Abuja.

Ebukiba E (2010). Economic analysis of cassava production (farming) in Akwa Ibom State. Agric. Biol. J. North Am, pp. 2151-7517. http://www.scihub.org/ABJNA

Ezedinma CN, Nkang EI Simon (2006). Price transmission and market integration: A test of the central market hypothesis of geographical markets for cassava products in Nigeria. International Institute of Tropical, Collaborative Study of Cassava in Africa, International Institute of Tropical Agriculture, Ibadan, Nigeria.

FAO/IC (1995). Formulation reports root and tubers expansion programme. FAO, Rome, Italy.

Food and Agricultural Organisation (FAO) (2000). Agriculture Towards 2015/2030.Technical Interim Report. April, 2000. Rome

Frescro LO (1993). The dynamics of cassava in Africa, an outline of research issues. COSCA Working P. 9.

Hyuha TS (2006). Profit efficiency among rice producers in Eastern and Northern Uganda. PhD thesis Makerere University, Kampala, Uganda.

National Population Census (NPC) (2006). The Official Census Report of Ndokwe L.G.A.

Nweke FI, Spencer DSC, Lyman JK (2002). The Cassava Transformation Africa's Best-Keep Secret. East Lansing: Michigan State University Press.

Ogboso N (2005). Effect of improved technology on cassava production in bende Local Government Area of Abia State. An Unpublished B. Agric project, Dept. Agric. Econ, Univer. Niger. Nsukka.

Ogunniyi LT, Ajao AO, Olapade-Ogunwole F, Ganiyu MO (2012). Resource-use efficiency of cassava production in Atakunmosa Local Government Area of Osun State. Prime Journal of Social Science (PJSS), 1(2):27-30. Available @ www.primejournal.org/PJSS

Oladeebo JO, Oluwaranti AS (2012). Profit efficiency among cassava producers: Empirical evidence from South western Nig. J. Agric. Econ. Develop. 1(2):46-52.

Olomola AS (2010). Formal-Informal Institutional Linkages in the Nigerian Agribusiness sector and implications for pro-poor growth. IPPG Discussion Papers, Series 37, available @ www.ippg.org.uk

Omonona BT (2009). Efficiency of resource - use in cassava production in Kogi State, Nigeria: implications for food security and environmental degradation. Retrieved on 27th April, 2009 from http://www.cababstractsplus.org/abstracts/SearchResults.

Otoo JA (1994). Rapid multiplication of Cassava, IITA Training Guide P. 51. IITA, Ibadan Nigeria.

Oyegbami T, Oboh G, Omueti O (2010). Cassava processors awareness of occupation and environmental hazards associated with cassava processing in South Western Nigeria. Afr. J. Food. Agric. Nutr. Dev. 10(1):1982-2000.

Technical for Agricultural and Rural Cooperation (CTA), (2005). Cooperation for Agricultural Development in ACP Countries. SPORE August, 2005, P. 118.

Teklewold H, Dadi L, Yami A, Dana N (2006). Determinants of adoption of poultry technology: A double-hurdle approach, Livestock Res. Rural Develop. 18(3):2006.

Tell (2004). Agricultural Credit and Production efficiency: A Case Study of NISER Monograph Series No. 4, Nigeria Institute of Social Economics Research, Ibadan, Nigeria, P. 67.

Tijani AA, Alimi T, Adesiyan AT (2006). Profit efficiency among poultry egg farmers: a case study of Aiyetodo farm settlement, Nigeria. Res. J. Agric. Biol. Sci. 2(6):256-261.

USDA (2010). Forces shaping Global Food Demand and Agricultural Trade. Statistics. Rome, Italy.

Technical efficiency of maize production in Ogun State, Nigeria

Idris Akanbi Ayinde[1], Raheem Olatunji Aminu[1] and Shakirat Bolatito Ibrahim[2]

[1]Department of Agricultural Economics and Farm Management, Federal University of Agriculture Abeokuta, Nigeria.
[2]Agricultural and Food Economics Department School of Agriculture, Policy and Development, University of Reading, United Kingdom.

This study examined the technical efficiency of maize production in Ogun State, Nigeria. Primary data were collected for this study using a multistage sampling technique to select 100 maize farmers from the study area. The data were analysed using descriptive statistics, gross margin analysis and stochastic production frontier. The socio-economic characteristics of respondents as evident from the data analysis revealed that 73% were males with an average age of 50.1 years. Most (85%) of them were married with average household size of 6 people. Also, 89.0% had below secondary school education and 84.8% were full time farmers while 55.0% were engaged in inter-cropping activities. The mean total variable cost per hectare was estimated as ₦109,599.17 per year while the mean total revenue per hectare was ₦111,436.00. The gross margin per hectare was estimated as ₦1, 836.83. The significant variables affecting maize production were seeds (α0.05), herbicide (α0.10), labour (α0.01), and farm size (α0.05), while the factors affecting inefficiency were household size (α0.05) and educational level (α0.01). The study recommended that provision should be made by governments and other stakeholders in the agricultural sector to provide farmers with access to affordable inputs such as seed, herbicides as well as making provision for alternative source of family labour.

Key words: Technical efficiency, maize production, Ogun State, Nigeria.

INTRODUCTION

Maize and other cereals constitute important sources of carbohydrate, protein, vitamin B and mineral. It is a staple food crop for most sub-Saharan Africans including Nigeria (Zalkuwi et al., 2010). It is one of the most abundant food crops in Nigeria. According to FAO (2013), Maize (9,180,270 tonnes) has been rated as the second grown food crop in Nigeria after Cassava (52,403,455 tonnes) then followed by Sorghum (6,897,060 tonnes) and Rice (4,567,320 tonnes). Due to its high adaptability and productivity, the cultivation of maize spread rapidly around the globe and currently it is being produced in most countries of the world (Anupama et al., 2005). It provides food for man and feed for livestock. Maize is an important food crop grown on a large scale in Nigeria, Ghana and to a lesser extent in Sierra Leone (Oladipo et al., 2008). The global output of maize in 2011 was recorded to be 883,460,240 tonnes and Nigeria produced about 9,180,270 tonnes, which constituted about 1.04%

of the world's production. The world maize production ranking for year 2011 shows that Nigeria was ranked 14th position (FAO, 2013).

The need for improved technology has risen in recent years due to the geometrical increase in population rate and consequently, the forces of demand for maize relative to supply are evident in frequent rise in price of maize. This has a great effect on the food security status of Nigeria, given its importance as a staple crop. The production of maize is affected by the development of advanced technologies, as fertilizer, hybrid seeds, pesticides, herbicides and better management practices which remains a limiting factor for developing countries.

One of the strategies for increasing agricultural productivity is the use of improved technology. But no matter how productive the technology may be, optimal productivity can only be obtained when the technology is efficiently used. Due to inadequate knowledge about the optimum level of farm resources and their efficient utilization, high risk of uncertainties often characterize the entire process of production. Therefore, this research seeks to determine the profitability and technical efficiency of maize production in Ogun State, Nigeria.

Stochastic frontier production function

Ojo (2004), used stochastic frontier production function and confirmed the presence of technical inefficiency effect in yam production and suggested that production, productivity and technical efficiency would be improved if those variables with negative elasticities of production are improved upon. According to Battese and Coelli (1995) the stochastic frontier production function postulates the existence of technical inefficiencies of production of firms involved in producing a particular output. Since the Stochastic frontier production framework (SFPF) was developed by Aigner et al. (1997) and Meeusen and Broeck (1977), evaluating the efficiency of individual firm and industry has become popular with the increasing availability of firm level data and growing capacity of computer to process them. The most common approach to estimate stochastic frontier production function is to specify a deterministic production plus a random, symmetric, firm-specific error term. This frontier represents the largest production for individual firm. Associated with firms is a second, non-negative error term, denoted as the technical efficiency term. Total production for each firm is defined as the frontier minus the inefficiency (Dhawan and Jochumen, 2012). The stochastic frontier production postulates the existence of technical inefficiencies of production of firm involved in producing particular output. Battese and Coelli (1995) and Ajibefun (2002) stated that the stochastic frontier production function has the advantage in that it allows simultaneous estimation of individual technical efficiency of the respondent as well as determinants inefficiency.

The idea of frontier production can be illustrated with a farm using n inputs ($X_1, X_2 \ldots\ldots\ldots\ldots X_n$) to produce output Y. Efficient transformation of inputs into outputs is characterized by the production function f(X), which shows the maximum output obtainable from various input vectors (Oyewo, 2011). The Stochastic frontier production function, assumes the presence of technical inefficiency of production. Hence, the function is defined by:

$$Y_i = f(X_i, E) \exp(V_i - U_i) \; i = 1, 2, 3, \ldots\ldots\ldots\ldots n$$

Where V is a random error associated with random factors not under the control of the farmer. The model is such that the possible production Yi is bounded above by the stochastic quantity f (X_i, E) exp (V_i), hence the term stochastic frontier.

The random error Vi is assumed to be independently and identically distributed as N(0,Φ2V) random variables independent of Ui s. Technical efficiency of an individual farmer is defined in terms of the ratio of the observed output to the corresponding frontier output, given the available technology.

Technical Efficiency: (TE) =Yi/Yi*dfdfd

$$TE = f(X_i, E) \exp(V_i - U_i) / f(X_i, E) \exp(V_i)$$

$$TE = \exp(-Ui)$$

Where, Yi is the observed output and Yi*, the frontier output.

METHODOLOGY

This study was conducted in Ogun State, Nigeria. Ogun State is situated within the tropics, with a total land area of 16,762 square km which lies within latitude 6°20' South and 7° 58' North in the tropics and longitude 2° 40' West and 4° 35' East of the Greenwich Meridian, and has an estimated population of 4,054,272. The state borders Lagos state to the south, Oyo and Osun states to the North, Ondo state to the east and the republic of Benin to the west. A multistage random sampling technique was used for this study. The first stage involves purposive selection of one Local Government Area from each of the four ADPs zones in Ogun State (Abeokuta, Ilaro, Ijebu-ode and Ikenne). The second stage involves a random selection of one rural community in each of the Local Government Areas. Finally, a systematic random sampling technique was adopted to randomly select 25 respondents from each of the community. As a result one hundred farmers were used for the study. Primary data were collected from the selected farmers through a well structured questionnaire which was randomly administered to farmers. The data collected were subjected to descriptive analysis, gross margin analysis and Cobb Douglas stochastic frontier production functions.

Model specification

The stochastic Frontier Production Function proposed by Battese and Coelli (1995) which assumes the existence of technical inefficiency of different firms in production will be adopted for this study. This is depicted using the model below.

$Y_i = f(X_i, \beta_i) \exp (V_i - U_i); i = 1,2,......n$ (1)

The functional form of this model to be used in estimating the level of technical efficiency is the Cobb-Douglas type (Bravo-Ureta and Evenson, 1994):

$Ln\ Y_i = \beta_0 + \beta_1 LnX_1 + \beta_2 LnX_2 + \beta_3 LnX_3 + \beta_4 LnX_4$ +------Write the equation till the last variable written below. $V_i - U_i$ (2)

Where: ln = natural logarithm, i = 1, 2, 3, 4....6, Y = Maize output (kg), X_1= Seeds (kg), X_2 = Fertiliser (kg), X_3 = Insecticide (litres), X_4 = Herbicides (litres), X_5 = Labour (family + hired in man-days.), X_6 = Farm size (Ha), β_0 = Constant term, β_i = parameters to be estimated, V_i = Symmetric error associated with uncontrollable factors related to production process, U_i = Inefficiency component of error term.

Inefficiency model

$U_i = \delta_o + \delta_i Z_i$

Where: U_i = Technical inefficiency, $_o$ = Constant tern, δ_i= Coefficient to be estimated, Z_1 = Gender, Z_2 = Age (in years), Z_3 = Household size, Z_4 = Level of Education, Z_5 = Farming Experience, Z_6 = Age squared.

The value of U_i may be obtained from the observable value of $V_i - U_i$ with the assumption that the composed error $V_i - U_i$ is known and is the best predictor for technical efficiency. The prediction which is presented in Battese and Coelli (1995) is estimated at the maximum likelihood estimates of the parameters of the full frontier inefficiency model stated above.

RESULTS AND DISCUSSION

Tables 1 and 2 revealed that the socio-economic characteristics of respondents as evident from the data analysis revealed that 73.0% were males with an average age of 50.1 years. Most of them were married (85%), 63.0% had between 5 and 8 persons in their household, with an average of 6 people. Majority (89.0%) had below secondary school education and 84.8% were full-time farmers while 55.0% were engaged in inter-cropping activities. Table 3 shows the mean total variable cost per hectare is estimated as ₦109, 599.17 while the average total revenue is ₦111,436.00. The gross margin per hectare was estimated as ₦1, 836.84. This shows that profitability of maize farming was relatively low in the study area.

Estimated production function

The maximum likelihood estimate (MLE) of the stochastic frontier model of maize farmers is presented in Table 4. The sigma-square (δ^2) estimate of 1.43 ($\alpha_{0.01}$) attests to the good fit and correctness of the model. Also, the gamma (γ) estimate of 0.79 ($\alpha_{0.01}$) shows the amount of variation in output resulting from the technical inefficiencies of the farmers. This means that 79% of the variation in farmer's output was due to technical efficiency.

The results reveal that the variables as seeds, herbicide quantity, labour, farm size are factors which positively influence the quantity of outputs of maize. The seed variable had a positive sign which is statistically significant ($\alpha_{0.05}$). This indicated that a percentage increase in the quantity of seed planted would result in 0.21% increase in maize output. This finding corroborates with Shehu et al. (2007) and Oyewo (2011). The elasticity of seed equals 0.21 indicating inelasticity of seed in the production process; thus the importance of the input in maize production cannot be over-emphasised.

The estimated coefficient of herbicides which is another significant factor is at $\alpha_{0.10}$. This means that an increase in the quantity of herbicide used by the maize farmers will lead to increase in the quantity of output of maize produced by the farmers. The elasticity of herbicide equals 0.10 which also shows herbicide inelasticity indicating its invaluable nature as input in maize production. Labour is positively correlated and significant ($\alpha_{0.01}$), with an elasticity of 0.41 while farm size also have positive estimated elasticity of 0.19 implying that increase in these variables will also increase the quantity of maize produced. The mean technical efficiency of the farmers was estimated as 0.69 indicating relatively high efficiency of maize production within the ambit of production resources available in the study area.

Sources of inefficiencies

The sources of inefficiency were examined simultaneously and the results as specified by the maximum likelihood parameter estimates are presented in Table 4. The results of the inefficiency model show household size ($\alpha_{0.05}$) and educational level ($\alpha_{0.01}$) of the respondents are significant determinants of technical inefficiency. The sign of the variables in the inefficiency model is very important in explaining the observed level of technical efficiency of the farmers. A negative sign implied that the variable had the effect of reducing technical inefficiency, hence increasing farmers' production efficiency, while a positive coefficient indicate that the variable has the propensity of increasing inefficiency, thus reducing farmers' production efficiency. These indicate therefore, that increase in household size would significantly increase production efficiency. This juxtaposes the fact inherent in many literature that farmers usually rely on household labour to boost production given it availability, less cost and ease of manipulation to suit different farm activities. The fact that improvement in education reduces maize production efficiency leaves a worry as it does not conform to the *a priori* expectation. This may probably mean that non-formal education provided by extension officers, which directly impinges positively on the production process, would have been better captured in the model instead.

Table 1. Distribution of respondents by socio economic characteristics.

		Frequency	Percent	Mean	Standard deviation
Gender	Female	27	27		0.45
	Male	73	73		
Marital status	Single	8	8		0.71
	Married	85	85		
	Widow	4	4		
	Separate	3	3		
Age group(years)	30 or less	14	14	50.1	16.03
	31 to 40	23	23		
	41 to 50	15	15		
	51 to 60	17	17		
	61 to 70	19	19		
	Above 70	12	12		
Household size group	4 or less	22	22	6	2.43
	5 to 8	63	63		
	9 to 12	13	13		
	Above 12	2	2		
Farming exp group	10 or less	37	37	20.15	14.08
	11 to 20	23	23		
	21 to 30	19	19		
	41 to 50	15	15		
	Above 50	6	6		
	Total	100	100		
Level of Educational	No formal	21	21.0		1.17
	Primary	43	43.0		
	Secondary	25	25.0		
	NCE/OND	8	8.0		
	HND/BSC	3	3.0		
Total		**100**	**100**		

Source: Field survey 2013.

Conclusion

This study examined the technical efficiency of maize production in Ogun state, Nigeria with the aid of stochastic production frontier model. The estimated gross margin implies that profitability of maize farming in the study area was relatively low. The maximum likelihood estimate (MLE) revealed that efficiency of maize production in the study area is significantly influenced by seed; herbicide, labour and farm size. However, the mean technical efficiency was 0.693 which indicates that production can still be increased by 30.9% using available technology.

This means that substantial opportunities should be explored to increase productivity and income of such farmers through availability and efficient utilization of productive resources.

RECOMMENDATIONS

In order to meet up with the goal of improved productivity of maize production in Ogun state, this study recommends the following:

(i) For an effective improvement in the level of efficiency among the maize farmers, provision should be made by governments and other stakeholders in the agricultural sector to provide farmers with access to affordable inputs such as seed, herbicides as well as making provision for

Table 2. Distribution of respondents by production characteristics.

		Frequency	Percent	Mean	Standard deviation
Mode of farming	Full time	56	56		0.50
	Part time	44	44		
Fund source	Bank loan	1	1		0.61
	Esusu	6	6		
	Personal	88	88		
	Friend	1	1		
	Money lender	2	2		
	Others	2	2		
Cropping pattern	Sole cropping	45	45		0.50
	Inter cropping	55	55		
Variety	Local	36	36		0.65
	Improved	52	52		
	Both	12	12		
Maturity period	2.5	4	4	3.07	0.30
	3	86	86		
	3.5	2	2		
	4	8	8		
Farm size group(Ha)	0 through 1	55	55	1.63	2.67
	1.1 to 2	26	26		
	2.1 to 3	6	6		
	3.1 to 4	6	6		
	4.1 to 5	7	7		
Total		**100**	**100**		

Source: Field survey 2013.

Table 3. Gross margin per hectare for maize production in the study area.

Variable	Minimum (₦)	Maximum(₦)	Mean(₦)
Total Variable cost (TVC)	5,700.00	1,105,935.00	109,599.17
Total Revenue (TR)	10,270.00	500,000.00	111,436.00
Gross Margin (GM)			1836.83

Source: Computed from Field Survey, 2013.

Good Agricultural Practises (GAP). Also, the government should introduce the farmers to access non-formal education more, through extension education and establishment of demonstration farms to boost their productive capacity.

(ii) A land redistribution policy that will increase the farm size of farmers (since they are mainly small scale farmers) will boost maize production. And also policies that will encourage the expansion of existing farm lands that are not currently under cultivation should be formulated. This could be through the provision of support such as low interest funds for the purchase or development of such farm lands through the Nigerian Agricultural Bank.

Conflict of Interest

The authors have not declared any conflict of interest.

Table 4. Maximum likelihood estimates of the stochastic frontier production function.

Variable	Parameters	Coefficient	Standard-error	t-ratio
		Efficiency		
Constant	B	5.867892	0.44569592	13.16568500
Seeds	X_1	0.21037706**	0.1072168	1.9621651
Fertiliser	X_2	-0.030510525	0.033871828	-0.90076405
Insecticide	X_3	0.049002906	0.07446027	0.65810812
Herbicide qty	X_4	0.10201794*	0.05685250	1.7944321
Labour	X_5	0.41538462***	0.16329108	2.54382920
Farm size	X_6	0.19385539**	0.09890650	1.95998620
		Inefficiency		
Gender	Z_1	-0.322718	0.89506359	-0.3605526
Age	Z_2	0.003763	0.067568225	0.05569144
Household size	Z_3	-0.178588**	0.10595548	-1.68549870
Educational Level	Z_4	0.19328224***	0.061632006	3.13606930
Farming Experience	Z_5	0.042178	0.0299376	1.40887790
Age squared	Z_6	-0.001338	0.0013587638	-0.98505757
Diagnostic statistics				
Sigma square	δ^2	1.430479***	0.4556956	3.1391109
Gamma	γ	0.78534257***	0.11442896	6.8631451
Log likelihood		-100.75684		
Mean		0.69		

Source: Computed from Field Survey 2013, *** Parameters significant 1% probability level, ** Parameters significant at 5% probability level, *Parameters significant at 10% probability level.

REFERENCES

Aigner DJ, Knox Lovell C, Schmidt P (1997). Formulation and estimating of stochastic function models. J. Econ. 6:21-23.

Ajibefun AI (2002). "Analysis of policy issues on technical efficiency of small scale farmers using the stochastic frontier production function: With application to Nigerian farmers". Paper prepared for presentation at the International farm management association congress. Wageningen, Netherland.

Anupama J, Singh RP, Ranjit K (2005). Technical efficiency in maize production in Madhya Pradesh: Estimation and implications. Agric. Econ. Res. Rev. 18:305-315.

Battese GE, Coelli TJ (1995). "A model for technical inefficiency effects in a stochastic frontier production function for panel data". Empir. Econ. 20:325–332.

Bravo-Ureta BE, Evenson R (1994). Efficiency in Agricultural Production: The Case of Peasant Farmers in Eastern Paraguay. Agric. Econ. 10:27-37.

Dhawan R, Jochumen P (2012). Stochastic Frontier Production with Error in Variables. Internet resource downloaded, pp. 1-2.

FAO (2013). FAO STAT. Retrieved July 7, from FAO Statistics Division 2013: http://faostat3.fao.org/home/index.html#DOWNLOAD

Meeusen W, Broeck J (1977). Efficiency estimation from cobb-douglas production function with composed error. Int. Econ. Rev. 18:435-444.

Ojo SO (2004). Improving labour productivity and technical efficiency in food crop production. A panacea for poverty reduction in Nigeria. J. Food Agric. Environ. 2(2):227- 231.

Oladipo FO, Ayandiji A, Akande M (2008). The roles of Youths in Maize production in Surulere Local Government Area, Oyo. Int. J. Agric. Econ. Rural Dev. 1:1.

Oyewo IO (2011). "Technical efficiency of maize production in Oyo state". J. Econ. Int. Fin. 3(4):211-216.

Shehu JF, Mshelia SI, Tashikalma AK (2007). "Analysis of technical efficiency of small-scale rain-fed upland rice farmers in North-west Agricultural Zone of Adamawa State". Niger. J. Agric. Soc. Sci. 3:4.

Zalkuwi JW, Dia YZ, Dia RZ (2010). "Analysis of Economic Efficiency of Maize Production in Ganye local Government Area Adamawa state, Nigeria". Nassarawa: www.sciencepub.net/report

Determinants of the responsiveness of cooperative farmers to the cocoa renaissance programme in Osun State, Nigeria

Oluwasola O. , Ige A. O. and D. Omodara

Department of Agricultural Economics, Obafemi Awolowo University, Ile-Ife, Osun State, Nigeria.

A key policy instrument in the cocoa transformation agenda is to provide subsidized inputs for farmers to replace old grooves to increase cocoa production in Nigeria. The study examined the role of cocoa marketing cooperatives in enhancing the incomes of its member patrons in Osun State under the cocoa renaissance policy. Multi stage sampling method was used to select 100 respondents from four cocoa produce marketing unions in the State. Data obtained were analysed with descriptive statistics, budgetary analysis and the multiple regression technique. The study revealed that the average age of the respondents was 64.4 years and only 16% of the cocoa farms were developed after the commencement of the renaissance policy. The farmers operated an average of 5.4 ha farm sizes while inputs supplied respondents through cooperative societies were much cheaper compared to what obtained in the open market. The gross margin to enterprise was ₦387, 639 while the net income was ₦345, 282. The regression analyses revealed that while family size, age, level of education, farm size and cooperative experience were significant determinants of output, family size, farm size, cooperative experience and amount of cocoa marketed through cooperative societies were significant determinants of income realized in the cocoa enterprises. The study concluded that greater efforts should be put in place to attract younger and educated farmers to achieve policy objectives.

Key words: Renaissance, transformation agenda, marketing cooperatives, subsidize, agrochemicals, fertilizer, seedlings.

INTRODUCTION

The importance of cocoa to Nigeria's socioeconomic development cannot be overemphasized and has been documented in several empirical studies (Olayide, 1969; Olayemi, 1973; Folayan et al., 2006). Although its prime place has since been taking by petroleum production, cocoa remains the most important agricultural product as no other export commodity has earned more foreign exchange than it (Abang, 1984; Akinbola, 2001). It is a major employer of labour (Folayan et al., 2006) and also supplies raw materials to local industries. Unfortunately, the production of the commodity has continued to dwindle while the standing of the nation among producers of the product has continued to diminish. From a peak of 308,000 tonnes in early 1970s, output declined to 110,000 tonnes in 1981 and in spite of spirited efforts to increase output as engendered by the Structural Adjustment Programme (SAP), output only increased to 205, 000 tonnes in 2000 (Koekoek, 2003).

Major problems implicated in the dwindling fortunes of Nigeria in the production of cocoa included the old age of cocoa farmers (Amos, 2007) and the cocoa trees (Adegeye, 1977); incidence of pests and diseases; poor management of the economic deregulation system which affected product quality control (Ajayi and Okoruwa, 1996); depleted soil fertility, use of poor planting materials, poor maintenance of cocoa farms, defective method of harvesting, poor handling of post harvest process and inefficient agricultural extension service (Adenikinju et al., 1989). The adoption of SAP which liberalized the marketing of cocoa produce and the abolition of the Nigerian Cocoa Marketing Board led to the shipment of poor quality cocoa beans from Nigeria (Ajayi and Okoruwa, 1996) with serious implications for the demand and pricing of cocoa from Nigeria at the international commodity market. To worsen this, dwindling world market price of cocoa from US$3,000 per tonne in 2002 to US$1,860 in 2004 to US$1,580 in 2005 have according to Ogunleye and Oladeji (2007) crushed farmers' willingness to invest in measures to boost output.

In the attempt by the Nigerian government to diversify the nation's economic base, attention has shifted to the agricultural sector with cocoa production as key in the structural transformation of the economy. A key strategy is the efforts of government to rejuvenate cocoa production in the country through its Cocoa Renaissance Policy in 2000. The strategy adopted to achieve the policy objective were the provision of cocoa producing States with marketing grants to raise hybrids, disease resistant, high yielding and early maturing seedlings for distribution to the farmers at subsidized prices. The strategy adopted by government in supplying improved inputs and technology to the poor farmers follows the high input pay-off model. The model which posits that farmers are economically rational, productive and take wise economic decisions but have exhausted potentials available. Accessing new farm technologies and improved inputs to them will enhance their efficiency and hence output (Akinyosoye, 2005; Schultz, 1964) The success in achieving very high output from the development of new, high-yielding varieties of wheat in Mexico in the 1950's and rice in the Philippines in the 1960's lends credence to the relevance of the model in tackling the challenges confronting cocoa farm enterprises in Nigeria.

While farmers have responded to the new initiative of government, a major disincentive has been the dwindling prices of cocoa at the international commodity market and the size of this income that actually accrue to farmers as a large proportion of this are claimed by intermediaries (exporters, middlemen and assemblers) (Ogunleye and Oladeji, 2007). This affects their capacity to compete in the market place. However, the severity of the impact of the market price of the commodity as well as market imperfections will depend on national policies, production and marketing conditions as well as the existence and participation in socio-economic networks like the co-operative societies and farm organizations. Fortunately, the cocoa renaissance programme is taking place within the free market policy of government. While government is providing the necessary conducive policy environment and incentives, private commercial concerns like the cooperative societies should be able to tap into these initiatives to curb the unhealthy activities of market intermediaries and complement the government's goal of increasing output and incomes to cocoa producers.

Cooperatives are a voluntary association of persons or business enterprises owned by member patrons pursuing common goals and who contribute capital and business and is controlled by member patrons who run the affairs of the association along democratic lines (Gupta, 2012; Dogarawa, 2005; Adegeye and Dittoh, 1985). Key values of cooperatives include self-help, equality, equity, solidarity democratic control. Cooperatives harnesses the individual strengths and/or resources of members which on their own are too weak to be competitive in a free market system to meet their common economic, social and cultural needs. These are used to provide services or products at cost to members who share the net surplus as benefits. The close relations of members ensures trust and compliance with the rules and principles of the association (Aremu et al., 2013; Nkonya et al., 2010). These inherent advantages coupled with reduced costs in providing services to a large group of farmers in cohesive society is a veritable tool for government to explore to implement policy measures aimed at reaching a large group of atomized farmers in a free market system.

Cooperative societies according to Akinbola (2001) are out to promote fair trade as well as seek to give a higher share of the final consumer price directly to farmers through effective marketing system. Cooperative societies are able to buy farm inputs in bulk and sell to members (by cash or on credit) at cheaper prices at the beginning of the production year and pay higher prices to member- farmers through the maintenance of inventory and storage facilities. This enables the cooperative society to hold cocoa products when there is a glut and sell when market prices are more favourable. The incentives provided by the income raise is expected to provide an additional impetus for cocoa farmer members to take advantage of the policy initiatives to increase farm size, outputs and hence incomes. Since the introduction of the policy strategy in Osun state, little efforts have been made at examining its effect on the cocoa production sub-sector in general and the impact of cocoa marketing cooperatives in exploiting the opportunities offered under the programme for the benefit of its member patrons in particular. Hence, this study aims at identifying the role of cocoa marketing cooperatives in enhancing the outputs and incomes of its member patrons under the cocoa renaissance programme in Osun State, Nigeria. The specific objectives were to:

(i) Examine the extent to which the policy initiative has influenced the farming decisions of cocoa farmers,
(ii) Evaluate factors determining responsiveness of farmers to the cocoa renaissance policy initiative in Osun State, and,
(iii) Analyse factors determining output and incomes among cocoa farmers

METHODOLOGY

Sampling technique and data collection

The study was carried out in Osun State, Nigeria. The State covers an area of approximately 14,875 sq/km and lies between longitude 4° and 5°E and latitude 7° and 8°N. The State is divided into 30 Local Government Areas (LGAs) and experiences two main seasons: the rainy season that runs through April and October, and the dry season that covers the rest of the year (November to March). Mean annual rainfall averages 1,570 mm while the mean annual temperature is about 27°C. The State lies within the tropical rainforest belt of the western uplands (Agboola, 1979) where climatic and edaphic factors provide the ideal environment for cocoa cultivation.

The multi stage sampling technique was employed to obtain necessary data from the major cocoa producing areas of the State. First, five Cooperative Marketing Unions involved in cocoa marketing were purposively selected. Each of these Unions were based in five LGAs namely: Ife Central, Ife East, Ifelodun, Boluwaduro and Boripe LGAs. From each of the Marketing Unions, four cooperative societies were randomly selected. In all, 20 cooperative marketing societies were selected. Finally, from each of these marketing societies, five member participants were randomly selected for interview. In all, 100 respondents were interviewed for this study.

Data analysis

Descriptive statistics, budgetary analysis and the multiple regression analytical techniques were used to analyse information obtained from the respondents. Descriptive statistics, including frequency counts, means and percentages were used to describe the socio-economic characteristics (including age, education level, family size, farm size) of selected cooperative member participants in the study area. Budgetary analysis was employed to estimate costs and returns to cocoa production using the gross margin as stated in Equation (1):

$$\pi_i = P_i Q_i - TCi \quad (1)$$

Where, π_i = gross margin per tonne (₦/tonne), P_i = price per unit of cocoa produced (₦), Q_i = cocoa output (tonne), and, TCi = total costs of production (fixed cost {FC} plus variable cost {VC}) (₦) Variable costs (VC) included in the analysis were expenditures on labour, seedlings, fertilizers, agrochemicals and transportation. Items that could be used for more than a production cycle were classified as fixed costs (FC). These included cutlasses, sprayers and farm-bans.

Finally, two multiple regression models were used to estimate the socio-economic factors determining the production and marketing of cocoa through marketing cooperative channels as well as those determining the profitability of cocoa enterprises in the study area. The model on factors determining production and marketing of cocoa beans was specified as:

$$Q_1 = f(X_1, X_2, X_3, X_4, X_5, X_6, X_7, \varepsilon_i) \quad (2)$$

Q_1= output of cocoa (kg), X_1 = total cost of production (N), X_2= family size, X_3= age of respondents (years), X_4= educational level of respondents (years spent in formal schools), X_5= farm size (ha), X_6= age of cocoa farm (years), X7= experience in cooperation (years), ε_i = error term.

In terms of *a priori* expectations, X_1, X_4, X_5 and X_7 are expected to be positively correlated to farm output. X_2, could be either positively or negatively correlated depending on whether the family is a production or consumption unit. X_3 is expected to be positively correlated to farm output to a certain age where it starts to show a negative relationship as increasing age affects the productivity of farmers. X_6 is also expected to be negatively correlated to farm output as cocoa trees age beyond their prime productive years. The second model on factors determining the income realised from cocoa enterprises was also specified as:

$$Y_1 = f(X_1, X_2, X_3, X_4, X_5, X_6, X_7, X_8, X_9, \varepsilon_i) \quad (3)$$

Where, Y_1 = farm income (₦), X_8= output of cocoa produced and marketed through cooperative societies (kg), X_9= distance of farm to cooperative office (km). X_1, X_2, X_3, X_4, X_5, X_6, X_7, and ε_i are as defined earlier.

In terms of *a priori* expectations, X_1, X_4, X_5, $X_7$and X_8 are expected to be positively correlated to farm income while X_2, could be either positively or negatively correlated depending on whether it is a production or consumption unit. X_3 is expected to be positively correlated to farm income to a certain age where it starts to show a negative relationship as increasing age affects the productivity of farmers. X_6 and X_9 are expected to be negatively correlated to farm incomes. X_5 is expected to be positively correlated to farm income to a certain age where it starts to show a negative relationship as increasing age affects the productivity of farmers. Three functional forms of the regression models were fitted to the data namely the linear, semi-logarithm and the double logarithm models. The models that provided the best of fit were selected and discussed.

RESULTS AND DISCUSSION

Socioeconomic distribution of respondents

The age distribution of the respondents as shown in Table 1 indicated that 90% of the farmers were over 50 years old while only 4% aged less than 40 years. The mean age of the respondents was 64.4 years. This is slightly higher than that recorded among cocoa farmers by Adesiyan and Adesiyan (2012) and Idowu et al. (2007). A critical factor in the sustainability of the new

Table 1. Socio-economic characteristics of respondent farmers.

Variable	Frequency	Percentage	Cumulative percentage	Mean
Age (Years)				
21 – 30	3	3.0		
31 - 40	1	1.0	4.0	
41 - 50	6	6.0	10.0	64.4
>50	90	90.0	100.0	
Level of education				
Did not go to school	36	36.0		
Adult/Quaranic education	10	10.0	46.0	
Primary school education	19	19.0	65.0	
Secondary school education	26	26.0	91.0	
Technical/Teachers College	9	9.0	100.0	
Marital status				
Married	78	78.0	78.0	
Widowed	17	17.0	95.0	
Divorced	5	5.0	100.0	
Family size				
6 – 10	29	29.0		
11 – 15	43	43.0	72.0	13.1
16 – 20	28	28.0	100.0	
Farm Size (ha)				
≤ 2.5	8	8.0		
2.6 – 5.0	62	62.0	70.0	
5.1 - 7.5	20	20.0	90.0	5.4
7.6 – 10.0	5	5.0	95.0	
≥ 10.1	5	5.0	100.0	
Age of cocoa farms (years)				
≤ 10	16	16.0		
11 – 20	34	34.0	50.0	
21 – 30	46	46.0	96.0	20.8
31 – 40	2	2.0	98.0	
≥ 41	2	2.0	100.0	
Cooperative experience of farmers (years)				
≤ 10	9	9.0		
11 – 20	20	20.0	29.0	
21 – 30	52	52.0	81.0	24.5
31 - 40	7	7.0	88.0	
≥41	12	12.0	100.0	
Benefits derived from membership				
Access to inputs	96	96.0		
Access to credit	98	98.0		
Access to transportation of produce	77	77.0		
Marketing of produce	100	100.0		

Table 1. Contd.

Distance of farm to cooperatives (km)				
≤ 2.0	26	26.0		
2.1 – 4.0	55	55.0	81.0	3.1
4.1 – 6.0	16	16.0	97.0	
≥ 6.1	3	3.0	100.0	

Source: Field survey (2012).

Table 2. Input costs variations between cooperative societies and the open market.

Cocoa farm inputs	Input prices (₦) per unit at	
	Cooperative market outletsb	Open market outlets
Seedlings	5	10
Fertilizer	150	200
Ridomine	200	250
Copper sulphate	200	250
Harvesting equipment	290	317
Hoe	950	1,400
Cutlass/matchet	630	650

Source: Field survey (2012).

initiative will be the set of farmers that have keyed in to the new policy initiative. Clearly, most of the farmers were old, above their prime age of production and are most probably into cocoa farming as a way of life rather than the policy initiative. The policy has therefore not succeeded in attracting new and younger farmers to drive it. This is reflected in the age of the cocoa farms where only 16% of the respondents' farms were developed after the policy became operational. The remaining 84% were cultivated long before the policy came on board and the average age of the farms was 20.8 years (Table 1).

The level of education among the farmer respondents was very low as 36% did not even attend any school while 19% attended up to primary school. Only 35% of the respondents had either Grade II Teacher's College or Technical College education. The high level of illiteracy could not only hamper the farmers from participating in the new policy drive but also in acquiring new skills and accessing technical inputs (Ogundele and Okoruwa, 2006) necessary to modernize the cocoa production subsector. However, as members of cocoa produce marketing societies, they are key to the success of the cocoa renaissance policy initiative hence, the need for increased efforts to improve the knowledge base of these set of farmers. Most of the farmers had access to family labour as 78% of them were married while the remaining were either widowed or divorced. The mean family size as shown in Table 1 was about 13 which is typical of most traditional farming communities where family labour is critical to farm production system (Oluwasola and Alimi, 2007). While this is good to meet the labour requirements in the farm, large family sizes could result in high household expenditure that could in turn become a drain on family income, savings and farm capitalization. The mean farm size was 5.4 ha which is far higher than the 2 ha average farm size in Nigeria (NINCID, 2006, Idachaba, 1989) or the 2.2 ha found among cocoa farmers by Idowu et al. (2007) in the same region. This is quite understandable as cocoa grooves with their high density tree foliage requires less efforts to weed once established compared to food crop farms. It is also a major subsector of agriculture that has enjoyed considerable commercialization since the colonial times.

The respondents have been involved in cooperative marketing for an average of 24.5 years and the main benefits derived in the long association include the marketing of cocoa produce enjoyed by all members. Ninety six percent enjoyed the supply of inputs from the societies while 98% enjoyed credit facilities from the relationship. As much as 77% also enjoyed the benefit of transporting their cocoa products to the cooperative offices with the marketing unions vehicles. In terms of distance of farms operated by the respondents to the societies' purchasing offices, most farms were located within 3.1 km.

Table 2 reveals the advantages of buying farm inputs from the cooperative marketing unions as the unit price of inputs purchased by farmers for their cocoa farms were much cheaper compared to what obtained in the open market especially if farmers needed to buy in bulk. The

Table 3. Analysis of cost and returns to cocoa marketing.

Income/cost Items	Amount (N)	Percentage in cost category
Revenue	519,000	
Seedlings	10,230	8.36
Agrochemicals	35,405	28.94
Labour (i).Weeding	32,565	26.61
(ii) Harvesting	16,675	13.63
Transportation	27,486	22.46
Total variable costs	*122,361I*	
Gross margin	387,639	
Fixed costs items	41,105	
Depreciation	10,252	
Total fixed cost	*51,357II*	
Total Cost	173,718	
Net revenue	345,282	
Expense-Structure Ratio	0.41	
Benefit-Cost Ratio	2.99	
Rate of Return	1.99	

price advantage enjoyed by the cooperative organization comes from two main sources. First as a major organization of farmers targeted by the policy on cocoa renaissance, they could receive these inputs at subsidized rates. In addition, it could also buy the inputs in bulk and distribute to farmers as they needed thereby reducing marketing and handling costs.

Analysis of costs and returns to farm enterprise

The breakdown of the cost and return components of the enterprises are presented in Table 3. The table shows that on the average, the farmers spent only ₦10, 230 (US$63.93) or 8.36% of the total variable cost on seedlings in spite of the efforts of government to access the farmers with subsidized improved seedlings. Although the cooperative marketing unions have passed the benefits of the government assistance to cocoa farmers in terms of supplying subsidized inputs to farmers at cheaper prices as indicated in Table 2, the farmers have not responded maximally to this incentive. The replacement of old cocoa grooves which has been implicated as one of the major factors in the dwindling fortunes of Nigeria in cocoa production (Amos, 2007) by now and improved seedlings is very critical if the cocoa subsector is to be revitalized hence the need for the marketing unions to embark on aggressive drive to ensure members replaced old grooves or planted new farms. The age of the farmer-members of the union which on the average was 64.4 years is however, a major disincentive to this kind of efforts, hence the need to attract younger farmers into the cocoa production subsector. The mean expenditure on agrochemicals was ₦35, 405 (US$221.28) or 28.94% of the total variable cost. Labour cost constitute the major variable cost component on the cocoa farms as the farmers spent ₦32, 565(US$203.53) (26.61%) on weeding and ₦16, 675 (US$104.22) (13.63%) on harvesting. The two labour driven activities constituted 40.27% of the total variable cost. The gross margin to enterprise was ₦387, 639 (US$2,422.74) indicating that the enterprise was far able to meet its variable costs. The total variable cost component constitutes 70.44% of the total enterprise cost while the fixed cost components constitute only 29.56%. This is possible because the fixed cost component is recouped over a long period of time and as pointed out earlier, the mean age of the cocoa farms was 20.8 years. The table shows that the net revenue accruing to cocoa farmers was ₦345, 282 (US$2,158.01). This is equivalent to ₦28, 773.50 (US$179.83)/month which is far above the national minimum wage of ₦19, 000.00 (US$118.75) of the nation. The expense-structure ratio of 0.41 indicates that for every ₦100 spent on cocoa farm business ₦41 was spent on fixed inputs while the remaining ₦59 went on variable inputs. The relatively large expenditure incurred on fixed inputs indicates that farmers could find it difficult to adjust to vagaries in market conditions hence adverse market conditions could discourage cocoa farming. This is particularly so as the farmers have very little control of the cocoa export market. The subsidy provided on farm inputs is thus very significant in encouraging farmers to replace old grooves or plant new farms. The benefit-cost ratio of 2.99 indicates that cocoa farming is profitable as every ₦100 invested returned ₦199 while the rate of return of 1.99 suggests an increasing return to scale with every ₦100 invested returning ₦199. over and above the amount invested (Table 3).

Determinants of cocoa production and marketing

Equation (3) shows the factors determining the output of cocoa in the study area. The result shows that in conformity with *a priori* expectations, family size (X_2), educational level of respondents (X_4), farm size (X_5) and experience of respondents in cooperation (X_7) were positively signed while age of respondents (X_3) and age of cocoa farms (X_6) were negatively signed. Contrary to *a priori* expectations, total cost of production (X_1) was negatively related to cocoa output. All the variables, with the exception of cost of production (X_1) and age of cocoa farms (X_6) were statistically significant determinants of cocoa production. As shown, a unit increase in family size (X_2) would increase cocoa output by 28.7%. This indicates that the family sizes are productive and an increase in the number of family sizes will provide additional family labour that can boost output of cocoa. This is particularly important as labour is the most important variable input in smallholder agriculture (Oluwasola, 2012). On the other hand, a unit increase in the age of farmers (X_3) would decrease output by 14.1%. As indicated earlier, the mean age of farmers was 64.4 years. Clearly the farmers were very old and as they age, their strength and drive will decrease with a consequent negative effect on output. Furthermore a unit increase in the educational level of respondents (X_4) would increase cocoa output by 43.8%. Increased education will help farmers understand the policy drive of government as well as access the key elements of the policy as related to the inputs. In addition, it will enable the farmers understand the instructions on the usage of these inputs. A unit increase in farm size (X_5) will also increase output by 78.2%, indicating that in smallholder farms where minimal inputs are used, farm sizes tend to be strongly correlated with output and income. Finally, the result also indicated that a unit increase in the years of experience of farmers in cooperative activities (X_7) will increase output by 18.6%. This is important in that cooperative societies provide ready market for farm produce and inputs as well as transportation of produce to cooperative stores. In addition, they also offer opportunities for agricultural extension which could enhance cocoa output. Cost of production (X_1) and age of cocoa farms (X_6) were not statistically significant. The adjusted coefficient of determination of 0.950 indicates that about 95% of the variability in cocoa output is explained by the variables specified in the model.

$$Y_1 = \underset{}{0.572} - \underset{(0.133)}{0.196\ln X_1} + \underset{(0.194)}{0.287\ln X_2^*} - \underset{(0.202)}{0.141\ln X_3^*} + \underset{(0.139)}{0.438\ln X_4^{**}} + \underset{(0.237)}{0.782\ln X_5^{**}} - \underset{(0.114)}{0.09\ln X_6} + \underset{(0.216)}{0.186\ln X_7^{**}} \quad (4)$$

R2 = 0.967
R̄2 = 0.950
F-value= 56.3

N.B. *significant at 1% level; ** significant at 5% level

Determinants of income from cocoa enterprises

Equation (4) presents the regression analysis of factors affecting income from cocoa farms. As indicated, total cost of production (X_1) and respondents level of education (X_4) were negatively signed contrary to *a priori* expectations. Age of cocoa farms and (X_6) and distance of farm to cooperative offices (X_9) were also negatively signed but in conformity with *a priori* expectations. The remaining variables, family size (X_2), age of farmers (X_3), farm size (X_5), experience of farmers in cooperative activities (X_7) and quantity of cocoa marketed through cooperative societies (X_9) were positively signed. Four of these independent variables were statistically significant. Family size (X_2) was positively and significantly related to income and as shown, a unit increase in family size will increase cocoa income by 26.3% while a unit increase in farm size (X_5) will also increase income by 42.9%. The result further shows that a unit increase in the years of experience of farmers in cooperative activities (X_7) would increase income by18.7% while a unit increase in the quantity of cocoa products marketed through cooperative societies (X_8) will increase income by 13.4%. Although not statistically significant, there was a negative correlation between total cost of production (X_1), educational level of farmers (X_4), age of cocoa farms (X_6) and distance of farms to cooperative offices (X_9). The adjusted coefficient of determination of 0.734 indicates that about 73.4% of the variability in income derived from cocoa farms is explained by the variables specified in the model.

$$Y_2 = -6.357 - 0.422\ln X_1 + 0.263\ln X_2^{**} + 0.363\ln X_3 - 0.383\ln X_4 + 0.429\ln X_5^{**} - 0.383\ln X_6 + 0.187\ln X_7^{**} + 0.134\ln X_8^{**} - 0.383\ln X_9$$

$$(0.500) \quad (0.690) \quad (0.607) \quad (0.628) \quad (0.207) \quad (0.391) \quad (0.694) \quad (0.421) \quad (0.578) \qquad (5)$$

R^2 = 0.857
$\bar{R}^2$ = 0.734

F-value = 53.8
N.B. *significant at 1% level; ** significant at 5% level

CONCLUSION AND RECOMMENDATION

In spite of the fact that Nigeria's fortune in cocoa has considerably declined, the study has shown that the enterprise is still very profitable. Although the efforts of government to transform cocoa production through the Cocoa Renaissance Policy is laudable, the farmers involved are very old and very insignificant efforts have been put into developing new cocoa farms since the policy was introduced. There is thus the need to explore strategies that will attract young and educated Nigerians into the sector if the objectives of the policy is to be realized. The use of cooperative societies to target farmers is very important as the inputs obtained through this medium were cheaper than obtained in the open market. The cooperative societies also offer more services including credit and transportation than could be obtained even through public agricultural outlets. They also tend to be closer to the farmers, hence, greater attention should be focused on them to reach more farmers.

The major determinants of output and hence the quantum of cocoa available for sale through cooperative societies were family size, age of respondents, educational level of respondents, farm size, and experience of respondents in cooperation. Although large family sizes could significantly increase output through the supply of cheap family labour as revealed in this study, the large family expenditure that will be incurred in the process could be a major financial drain that could hinder ploughing back farm income hence, the need for enlightenment campaign among farm households to reduce family sizes. This should be complemented by accessing farmers with farm implements that can substantially reduce labour requirements especially for weeding and harvesting. As farmers age, production will decline hence the need to attract younger, able bodied and educated farmers to the sector. Farm size among the cocoa farmers was above national average and is very significant in determining output. Policy efforts should also focus on increasing the yield of cocoa per hectare to further enhance output.

Conflict of Interest

The authors have not declared any conflict of interest.

REFERENCES

Abang SO (1984). "Stabilization Policy; An Economic Analysis and Evaluation of its importance for Nigerian Cocoa Farmers". Unpublished Ph.D Thesis, Oklahoma State University, Stillwater, U.S.A.

Adegeye AJ (1977). "Production and Marketing of Cocoa in Nigeria: Problems and Solutions". pp. 1-10

Adegeye AJ, Dittoh JS (1985). "Essentials of Agricultural Economics". Impact Publishers, Ibadan. Pp. 183 - 195.

Adenikinju SA, Esan EB, Adeyemi AA (1989). "Nursery Techniques in Propagation and Management of Cocoa, Kola, Coffee, Cashew and Tea. Progress in Tree Crop Research" (2nd edition). A commemorative book to mark the 25th Anniversary of CRIN.

Adesiyan AT, Adesiyan OF (2012). "Technical Efficiency of Cocoa Farmers in the Rain Forest Agro-Ecological zone of Osun State" in Idowu E.O., Ayanwale A.B., Bamire A.S. and A.O. Adejobi (eds) Agriculture in the National Transformation Agenda: The Policy Mix. Proceedings of 2012 Annual Conference of the Nigerian Association of Agricultural Economists (NAAE) pp. 613-620.

Agboola SA(1979). "An Agricultural Atlas of Nigeria". Oxford University Press, Ibadan.

Ajayi SI, Okoruwa A (1996). Managing of Uncertainties and Risks in Cocoa Production and Marketing in Nigeria. NCMB paper pp. 12-18.

Akinbola CA (2001) "International Project on Cocoa Marketing and Trade in Nigeria". Manual on Cocoa Quality and Training Manual for Extension Workers. pp. 10-24.

Akinyosoye VO (2005). "Government and Agriculture in Nigeria: analysis of policies, programmes and administration". Macmillan Publishers, Ibadan, Nigeria.

Amos T (2007). "An Analysis of Productivity and Technical Efficiency of Smallholder Cocoa Farmer in Nigeria". J. Soc. Sci. 15(2):127-133.

Aremu MA, Ekpo IC, Mustapha AM (2013). "Determinants of Banks Profitability in a Developing Economy: Evidence from Nigerian Banking Industry". Int. J. Contem. Res. Bus. 4(9):155-181.

Dogarawa AH (2005). "The Role of Cooperative Societies in Economic Development". MPRA Paper No. 23161.

Folayan JA, Daramola GA, Ogunrade AE (2006). "Structure and Performance Evaluation of Cocoa Marketing Institutions in Southwestern Nigeria: An Economic Analysis". J. Food Agric. Environ. 4(2):123-128.

Gupta PK (2012). "Agricultural Economics". Vrinda Publications Ltd. p. 29.

Idachaba FS (1989). "The Implications of Forest Fires and Bush Burning for Nigerian Food Production Strategies". Lecture delivered on the occasion of World Environment Day, Lagos, FEPA Special Publication. 2:7-21.

Idowu EO, Osuntogun DA, Oluwasola O (2007). "Effects of Market Deregulation on Cocoa (Theobroma cacao) Production in Southwest Nigeria". Afr. J. Agric. Res. 2(9):429-434.

Koekoek FJ (2003). The Organic Cocoa Marketing Europe Market Study. Export Promotion of Organic Product from Africa (EPOPA).

Nigerian National Committee on Irrigation and Drainage (NINCID) (2006). Annual Report 2006. Abuja, Nigeria.

Nkonya E, Pender J, Kato E, Oni O, Phillip D, Ehui S (2010). "Options for Enhancing Agricultural Productivity in Nigeria". The Nigeria Strategy Support Programme (NSSP) Background Paper 11. IFPRI.

32 p.
Ogundele OO, Okoruwa VO (2006). "Technical Efficiency Differentials in Rice Production Technologies in Nigeria". AERC Research Paper 154. African Research Consortium, Nairobi. pp. 1-37.
Ogunleye KT, Oladeji JO (2007). "Choice of Cocoa Market Channel Among Cocoa Farmers in Ila Local Government of Osun State, Nigeria". Middle-East J. Sci. Res. 2(1):14-20.
Olayemi JK (1973). Some Economic characteristics of Peasant Agriculture in the Cocoa belt of Western Nigeria". Bull. Rural Econ. Sociol. 1:24-30.
Olayide SO (1969). "Some estimates of supply and demand elasticity for selected Commodities in Nigeria's Foreign Trade". J. Bus. Soc. Stud. 1(a):176-193.
Oluwasola O (2012). "Integrating smallholder food crop farmers into the National Policy for Commercialization and large scale agriculture in Nigeria: a Case study of Ekiti State". Int. J. Agric. For. 2(5):247-256.
Oluwasola O, Alimi T (2007). "Financial Intermediation in Agriculture in Nigeria: Emerging role of Non-Governmental Organizations (NGOs)". Agrosearch 9(1-2):43-51.
Schultz TW (1964). "Transforming Traditional Agriculture". New Haven, CN: Yale University Press.

23

Technical and allocative efficiency gains from integrated soil fertility management in the maize farming system of Kenya

Morgan C. Mutoko[1], Cecilia N. Ritho[2], James K. Benhin[3] and Oliver L. Mbatia[2]

[1]Kenya Agricultural and Livestock Research Organization (KALRO), Food Crops Research Institute, P. O. Box 450-30200, Kitale, Kenya.
[2]Department of Agricultural Economics, University of Nairobi, P. O. Box 29053, Nairobi, Kenya.
[3]Plymouth Business School, University of Plymouth, Drake Circus, Plymouth, Devon PL4 8AA, United Kingdom.

Declining land productivity and *per capita* food availability poses challenges to overcoming land degradation and poverty in sub-Saharan Africa. There is a need to identify ways of improving land productivity particularly among smallholders. This study investigated the contribution of integrated soil fertility management (ISFM) practices to both technical and allocative efficiencies in the maize farming system of Kenya. To determine efficiency gains from ISFM, we compared efficiencies of two groups of smallholders: those within the contact areas and their counterfactuals. We estimated Cobb-Douglas stochastic functions based on maize production data collected from a stratified sample of 373 farmers. The results indicate that farmers who applied ISFM were more efficient both technically and allocatively than those who did not. Application of ISFM practices increased technical and allocative efficiencies by 26 and 30%, respectively. However, other favourable factors are required for farmers to realize maximum efficiency gains from maize farming activity. They included farming experience, extension contacts, off-farm income and market access. Therefore, policies and practices aimed at enhancing farming efficiency in smallholder agriculture should address these factors. We recommend increased dissemination of ISFM technologies to the wider farming community through effective and participatory approaches to increase efficiency and enhance farm returns.

Key words: Maize, land husbandry, productivity, small-scale, stochastic frontier.

INTRODUCTION

Sub-Saharan Africa (SSA) is the only region in the world where land productivity and *per capita* food availability continues to fall over time (Clover, 2003; Lambin et al., 2003). Declining soil fertility and high cost of purchasable inputs are the main contributory factors to low agricultural productivity among farming communities in SSA (CGIAR, 2002). Soil fertility loss is viewed as a key source of land degradation and environmental damage in the long-term

(Henao and Baanante, 2006). This is because continuous farming without adequate replenishment of nutrients drains the productive potential of the soil. The soil becomes less fertile when the nutrients and trace elements are constantly used without taking proper care of the mass balance of the soil (Bojö, 1996). This problem is more profound among resource-poor, small-holder farming households because they lack knowledge on better soil management options. They also have low capacity to invest in soil nutrient replenishment—especially using chemical fertilizers—and have less ability to bear risk and wait for future payoffs from such investments (Jayne et al., 2010). According to Todaro and Smith (2008), widespread abject poverty can precipitate over-use and destruction of the natural resources where short-term survival goals and practices are pursued with little regard to long-term sustainability concerns.

Kenya, like many SSA countries, grapples with the twin problems of increasing poverty incidence and land degradation, especially in rural areas. Declining soil fertility in high agricultural potential areas of the country has raised concerns regarding the sustainability of the smallholder maize production system (Mureithi et al., 2002). For example, the resource-poor smallholders in Western Kenya hardly invest in farming activities due to liquidity constraints, experience more than twice the erosion rates and achieve less than one-third of potential maize yields (Mureithi et al., 2002). This raises food security concerns as smallholder farmers are the major producers of maize in the country; hence, there is a need for them to increase their farm productivity in order to satisfy the increasing food needs in Kenya. The low maize productivity attributable to both insufficient farm resources and inefficient allocation of available farm inputs, hinders progress in this direction (Seyoum et al., 1998). To bridge the resource insufficiency gap, low-cost, integrated soil fertility management (ISFM) technologies have been availed through participatory approaches such as farmer participatory research and farmer field schools, to tackle soil fertility loss and boost productivity in smallholder farming system of North-western Kenya (Nyambati et al., 2003). The promoted ISFM technologies included the application of organic residues and animal manure; inorganic fertilizers; integration of leguminous crops e.g. soya beans, groundnuts, pigeon peas, *Mucuna pruriens* and *Crotalaria spp;* and agro-forestry practices such as incorporation of *Tithonia diversifolia* residues. Others included integrated pest management using extracts from neem, hot pepper and tephrosia plants and low-cost soil conservation methods such as grass strips. However, knowledge about the efficiency contribution of the ISFM technologies within the maize farming system of Kenya remains unknown.

There is an increasing interest in determination of productive efficiency in various fields since the pioneering work by Farrell (1957), and analytical advancements that followed (Aigner et al., 1977; Battese and Broca, 1997; Coelli, 1996; Meeusen and van Den Broeck, 1977). Determination of actual efficiency levels is essential in effective policy-making and practical implementation of various economic activities. Therefore, many researchers have empirically investigated whether economic units such as farms, firms, and organizations, were utilizing the scarce resources to produce maximum quantities of goods and services.

Efficiency studies in SSA have reported varied technical efficiencies ranging from 46% in Nigeria (Olowa and Olowa, 2010), 56% in Ethiopia (Seyoum et al., 1998) to between 64 and 76% among two groups of farmers in Lesotho (Mochebelele and Winter-Nelson, 2000). Two studies conducted in Kenya reported technical efficiency of 49% (Kibaara 2005) and 71% (Liu, 2006) in maize production, while in Malawi Tchale and Sauer (2007), found on average 87% technical efficiency among small-holder maize farmers. These empirical findings clearly show that SSA farming system generally is not efficient and produces less output than the possible potential. This suggests therefore that inefficiency is one of the principal causes of low productivity of agriculture in SSA. Consequently, there is a need to establish whether the application of ISFM practices contribute to efficiency in maize farming system and which factors are key to maximizing the efficiency benefits from ISFM practices. This is important because the greatest challenge to adoption of sustainable production practices is not only liquidity constraints but also a lack of knowledge on efficient production plans (Place et al., 2002). In fact, it is not only the lack of credit and poor farm revenue but also the absence of information that often prevent the poor from making the best resource-augmenting investments important for improving farm productivity (Todaro and Smith, 2008). As observed by Bationo et al. (2004), tackling poor soil fertility and low farm productivity requires both a long-term perspective and an all-inclusive approach to which this study aims to contribute.

The specific objective of this study was to estimate the prevailing technical and allocative efficiencies and examine their determinants in two maize producing systems of North-western Kenya. Unlike many efficiency studies conducted in SSA, which focus on technical efficiency alone (Mochebelele and Winter-Nelson, 2000; Olowa and Olowa, 2010; Seyoum et al., 1998; Sherlund et al., 2002; Tchale and Sauer, 2007), we concurrently estimate both technical and allocative efficiencies and evaluate farming efficiency gains from the ISFM technologies availed to smallholder farmers. Providing information on ways to enhance efficiency in maize production is essential in improving *per capita* output and farmers' incomes to re-invest in soil fertility improvement, including the use of available ISFM technologies in Kenya. This is in line with the Kenya government's vision that sustainable and efficient production practices within the smallholder agriculture is key to ameliorating the negative environmental

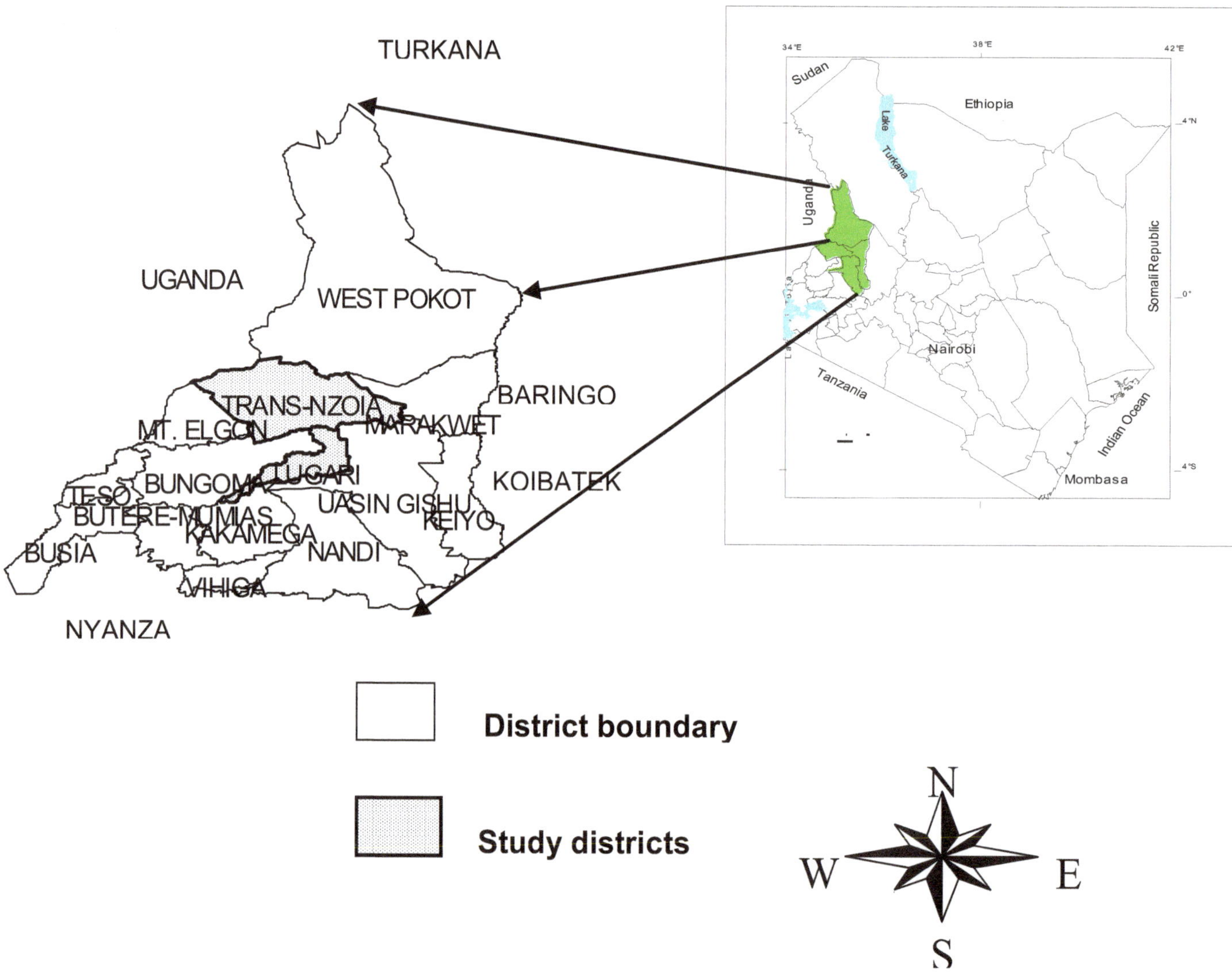

Figure 1. Map showing study districts (dotted) in North-western Kenya.

effects of poverty and improving better livelihoods in rural areas (Government of Kenya, 2004). As noted in the policy document, the greatest emphasis should be given to improving efficiency among smallholders so as to simultaneously mitigate poverty-related land degradation and raise agricultural productivity (Government of Kenya, 2005). This study therefore provides useful information on policy options and best practices to improve maize yields to enhance food security and sustainable land management not only in Kenya but also in other similar SSA countries.

METHODOLOGY

Study area

The North-western Kenya is a high agricultural potential region and accounts for about 90% of total maize output in the country. Trans Nzoia and Lugari Districts are located in this region (Figure 1). Trans Nzoia District dubbed, 'Kenya's granary' remains the major maize producer in the country (Wangia et al., 2002). Trans Nzoia District was selected for the study because yields have been declining in recent years. As a result, the district was targeted with ISFM options aimed to address low yields. On the other hand, we chose Lugari District because it has comparative maize farming system like that in Trans Nzoia, but was not covered by the soil management project.

The two districts receive between 1000 to 2100 mm of bimodal rainfall pattern. Rainfall received is considered reliable for agricultural activities. The elevation ranges from 1300 to 1900 m above sea level, with Upper Midlands (UM_4) being the predominant agro-ecological zone accounting for 94 and 47% of all land area in Lugari and Trans Nzoia Districts, respectively. This zone is the area of intensive maize cultivation in the study districts (Government of Kenya, 2006; Jaetzold et al., 2007). The major soil type is humic Acrisols, which is deep and well-drained. Soil fertility is moderate given that poor soil fertility is one of the most limiting factors to agricultural productivity in the study area (Government of Kenya, 2006; Nyambati et al., 2003). Farm sizes are on average 2.5 ha in

Lugari and 3.6 ha in Trans Nzoia. The common farming system is mixed crop-livestock production. Maize-bean production takes about 90% of total cropped land (Government of Kenya, 2006). Maize yields have been declining in both districts to as low as just 2 tha^{-1}. The declining trend has been blamed on among other factors, the soil fertility loss due to continuous monoculture cropping.

Population density is 328 people/km^2 for Trans Nzoia District and 437 people/km^2 in Lugari District (KNBS 2010). Poverty incidence is 50.2% for Trans Nzoia and 47% for Lugari District (Government of Kenya, 2011). In both districts, smallholders form the bulk of maize producers amid the waning importance of large-scale maize production due to continuous land sub-division (Mose et al., 2006).

Sampling techniques

Stratified sampling, based on agro-ecological zonation and concentration of smallholders, was applied to select two localities in each district for the study. These were: Kaplamai (UM_4) and Kiminini (LH_{3-4}) in Trans Nzoia District, as the contact areas where ISFM technologies was promoted for about a decade (Nyambati et al., 2003), and Mautuma (UM_4) and Matete (LM_3) in Lugari District, as the matching counterfactual areas.

The optimum sample size was chosen in a two-step process (Rangaswamy, 1995). First, a total sample size of 373 farmers was derived based on the number of strata, total farming households and variance of maize yields (calculated from data reported in Mose (2007), in each district. The total sample was made up of 154 farmers for Trans Nzoia District and 219 farmers for Lugari District. Second, we used the Neyman allocation method to distribute the total sample across the four study strata. For each stratum, we developed updated sampling frames with the assistance of frontline agricultural extension staff and local leaders. We used randomly generated numbers in MS Excel computer program to select individual farming households for interviews.

Data collection

We conducted face-to-face interviews at each of the selected households using a detailed and pre-tested, semi-structured questionnaire. The questionnaire was administered to the household head or member knowledgeable about farm and off-farm activities. We obtained data on physical quantities and monetary value of farm inputs (that is, fertilizers, manures, labour, seeds, and land) and maize output. We also collected farm level data on the ISFM practices that they applied in maize production. In addition, we collected socio-economic data on farmer's age, number of years in farming, family members and their level of education. Farmers also provided information on the cost of market access, distance and condition of the main roads as well as access to credit and the number of contacts they had with agricultural extension agents during the year. Finally, we collected data on planting date, maize varieties grown, weeding frequency and pest control. These data were analysed applying the analytical procedures specified next.

Analytical framework

The economic theory of production provided the analytical framework for this efficiency study (Debertin, 1986). The fundamental idea underlying the measurement of technical efficiency is that of attaining maximum possible output from a set of physical inputs. A farmer is considered technically inefficient if little output is produced from a given bundle of inputs (Ogundari et al., 2006). Allocative efficiency on the other hand, reflects the ability of the farmer to use inputs they have in optimal combinations given their relative prices (Coelli, 1996). A farmer is deemed allocatively inefficient if excessive cost is incurred to achieve the same level of output.

Following Ogundari et al. (2006), two self-dual stochastic functions were estimated from production data to generate technical and allocative efficiency values. Stochastic functions used in this study attribute part of the inefficiencies to external factors and are suitable when analysing the role of measurable socio-economic factors in observed efficiency differences (Coelli, 1996). This made it possible to establish the effects of farmers' responsiveness to the incentive structure and technologies that defines their production environment. This was important in this study because efficiency gains from ISFM interventions had to be estimated taking into consideration all possible relationships (Tchale and Sauer, 2007).

We estimated a self-dual, stochastic Cobb-Douglas production (Equation 1) and cost function (Equation 2) to generate technical and allocative efficiency values, respectively.

$$\ln(y_i) = \ln\beta_0 + \sum_{i=1}^{4}\beta_i \ln(x_i) + \beta_5 dist + (v_i - u_i) \qquad [1]$$

Where y_i is maize output (tha^{-1}), and x_i are physical inputs (fertilizer, seeds, total labour and manure per ha). A binary variable *dist* (1=Trans Nzoia; 0=Lugari) accounted for the difference in physical attributes important for farm production such as natural soil fertility and rainfall. β_0 is a parameter common to all farms while β_i and β_5 are unknown coefficients estimated in the model. v_i is the ordinary two-sided error term assumed to be normally, identically and independently distributed and u_i is the one-sided error term assumed half-normal that captured technical inefficiency.

$$\ln(c_i) = \ln\alpha_0 + \sum_{i=1}^{4}\alpha_i \ln(r_i / w_i) + \alpha_5 \ln(y_i) + (v_i + u_i) \qquad [2]$$

Where c_i is the total variable cost of maize production (KES/ha), r_i are the unit prices for fertilizer, seed, ploughing, w_i is the labour wage and y_i is maize yield (tha^{-1}). α_0 is the intercept taking care of the fixed costs in maize production, while α_i is a vector of coefficients estimated for the prices of fertilizer, seed, ploughing, labour and yield. μ_i is the half-normal error term that measured allocative inefficiency.

We applied a one-step maximum likelihood estimation procedure (Wang and Schmidt, 2002) to estimate each of the above equations simultaneously with those determinants of technical and allocative efficiency in maize production, specified in Equation 3.

$$\mu_i = \delta_0 + \delta_1 EXP_i + \delta_2 EDU_i + \delta_3 CRAC_i + \delta_4 HSIZ_i + \delta_5 OFIN_i + \delta_6 SFM_i + \delta_7 EXT_i + \delta_8 MAC_i + \delta_9 AEZ_i + \varepsilon \qquad [3]$$

Where μ_i is the inefficiency (technical or allocative) score; *EXP* is farming experience of the farmer (years); *EDU* is formal education level of the decision-maker (years of schooling); *HSIZ* is the number of household members (those living and eating in the same household); *OFIN* is a binary variable for off-farm income earning (1 = for household with positive earnings; 0 = otherwise). *CRAC* is a binary variable for credit access (1 = for households that obtained credit; 0 = otherwise); *SFM* is binary variable for soil fertility management practice (1 = for ISFM practices; 0 = fertilizer alone). *EXT* is the number of extension contacts during the year; *MAC* is market access (transport cost/bag of maize in KES). *AEZ* is a binary variable for agro-ecological zone (1 = Upper Midland; 0 = otherwise) controlling for the influence of natural soil fertility, rainfall and temperature and ε is the error term. The selection of these variables was based on past studies that found their significant influence on various efficiency measures (Mochebelele and Winter-Nelson, 2000; Mutoko et al., 2014; Ogundari et al., 2006; Olowa and Olowa, 2010; Seyoum et al., 1998; Sherlund et al., 2002; Tchale and Sauer, 2007). We used FRONTIER 4.1(c) for efficiency estimations (Coelli, 1996).

Before estimations, we tested for the violations of classical assumptions of OLS commonly expected in cross-sectional data used in this study, such as heteroscedasticity, multi-collinearity and endogeneity (Gujarati, 2005). The Breusch-Pagan-Godfrey test did not show evidence of heteroscedasticity in the data; hence, the parameter estimates were unbiased, consistent and efficient. The endogeneity test ensured that the error term μ_i and the explanatory variables do not co-vary. Since the estimation of the stochastic production and cost frontiers is based on the distribution of this error term, this independence is critical for two reasons. First, the variables describing the inputs in the stochastic frontier functions need to be independent from the socio-economic variables explaining inefficiency effects. Second, the stochastic frontier functions and the equation explaining inefficiency have to be estimated simultaneously. If the independence condition were not satisfied, the parameter estimates from both functions would be biased and inconsistent (Verbeek, 2008). The procedure to establish independence between the error term and the explanatory variables involved a regression of each variable against the others in the set and assessing the strength of the R^2 (Verbeek, 2008). Those R^2 values greater than 0.5 indicate high dependence and therefore such a variable is endogenous (Gujarati, 2005). Soil fertility management choice had the highest value (R^2=0.4) whereas all the other variables in Equation 3 had lower values (R^2≤0.2). Given that all R^2 values were less than 0.5, we concluded that none of the explanatory variables was endogenous. Following Maddala (2001), we confirmed the presence of multi-collinearity based on the high degree of variance inflation factors for seed and fertilizer. Gujarati (2005), recommends expressing variables as deviations from the mean as one practical ways of reducing the effect of multi-collinearity in estimations. We followed this suggestion with the implication of the transformation being that the results had to be interpreted at the mean values. Independent samples t-tests were used to determine statistical difference on key variables between the two main study districts.

RESULTS AND DISCUSSION

Description of the maize production system

Survey results indicate that maize farmers were on average 48 years old, decision-makers were mostly male (70%) with eight years of formal schooling. The average family had six members out of which two had attained secondary education level. This result implies that each household had appreciable levels of both physical and technical aspects of human capital. However, the low active participation of the youth and women may influence the degree of interest in and implementation of new technologies including ISFM practices.

Off-farm earnings averaged only KES 2,400 per month per household, mainly from casual employment and remittances. Only 6% of the farmers obtained agricultural credit, mainly from informal sources including 'merry-go-rounds', input stores, family members and neighbours. Most farmers blamed low access to credit on the lack of information about credit providers and lack of land title deeds (by 62%) that would serve as collateral for the loan. Some farmers cited the main deterrent as stringent requirements imposed by formal credit institutions and the perceived risks in case they defaulted re-payment. This finding indicates poor injection of liquidity into the farming system from external financial sources thereby limiting farmers' affordability of essential inputs. Most of the maize production costs (61%) were financed from farm income demonstrating the need to improve farm returns to guarantee considerable investments in maize production.

Agricultural extension contacts with the farmers were low (only 27%), on average just one visit per year. This is because farmers were yet to embrace the new demand-driven extension delivery system. In the earlier system, the extension agents were entirely responsible for making visits to individual farmers or organizing group trainings in order to provide them with better agricultural knowledge and skills.

The average cost structure of maize production included expenditure on chemical fertilizers (34%), land preparation (20%) and seed (12%). These were the major costs taking about two-thirds of all variable costs incurred in maize production. To enhance smallholders' access to these inputs, there is need for appropriate policy intervention to minimize transaction costs thereby making their acquisition more affordable. All other expenditures on labour input accounted for 34% of total production costs, indicating that labour was not a limiting resource in the study area.

Forty per cent of the sampled farmers used some components of low-cost ISFM options, which included incorporation of maize crop residues (30%), use of farmyard (24%) and compost manures (22%) as well as integration of crotalaria (10%) and groundnuts (6%). The preference of these ISFM practices was due to the availability of the manures or the bonus benefits to the household from the legumes. Within such farming environment therefore, we hypothesized that the average maize yield of 2.6 tha^{-1} was below the technically feasible and allocatively efficient levels.

Status of efficiency in maize production

Results in Table 1 indicate that overall farmers on the average achieved 64% technical efficiency. Therefore, it is possible to improve yields by an additional 36% through adoption of better farm practices such as improved soil fertility management, early land preparation, timely planting; proper spacing, use of hybrid maize varieties and effective weed control. The significant gamma (γ) estimate indicates that 65% of the technical inefficiencies can be explained jointly by the socio-economic variables in the technical inefficiency equation. The coefficients for chemical fertilizer and seed are statistically significant. This means that inorganic fertilizer

Table 1. Stochastic production function estimated using maximum likelihood method to determine technical efficiency in maize production.

Variable	Parameter	Coefficient	SE[a]
Production frontier function			
Intercept	β_0	0.50*	0.30
Fertilizer	β_1	0.19**	0.05
Seed	β_2	0.20*	0.13
Labour	β_3	-4.26***	1.68
Manure	β_4	0.03	0.07
District	β_5	0.06***	0.01
Efficiency measures			
Sigma-squared, $\sigma_\mu^2+\sigma_v^2$	σ^2	0.31**	0.08
Gamma, $\sigma_\mu^2/(\sigma_\mu^2+\sigma_v^2)$	γ	0.65**	0.18
Mean technical efficiency[b]	TE	64%	

[a]SE is standard error of the estimate, [b]Technical efficiency estimates a farmer's actual yield in relation to the optimal yield, given a production technology. The maximum possible technical efficiency level is 100%. Significant at the following levels: *10%; **5%; ***1%.

and seed are the main limiting inputs in maize production because as shown by positive coefficients, their use beyond the current levels will increase yields. The practised seed rate of 24 $kgha^{-1}$ was closer to the recommended rate of 25 $kgha^{-1}$; hence, yield increases can only be realized by planting improved varieties. Therefore, the results demonstrate that the current stage of production is inefficient (Debertin, 1986). The expectation in this study was that when efficiency is improved in the use of available inputs, farmers are more likely to expand their scale of production, since most of them are constrained by lack of finances to invest in farming consistent with Jayne et al. (2010).

The coefficient for labour is negative indicating that at the mean, increased labour use has a decreasing effect on maize yields because the current level is beyond the optimal amount required for efficient production. Since most of the labour (67%) was from own family, it was likely under-valued and over-used. The result is consistent with past findings (Seyoum et al., 1998; Tchale and Sauer, 2007) that associated negative marginal product for labour with production systems that relied on cheap family labour and usually employed it beyond the economically optimal level.

The significant coefficient for district showed that on average farmers in Trans Nzoia realized higher maize yields than those in Lugari. This is due to the relatively large farm sizes of better quality and favourable climatic conditions over there, confirming the considerable role of conducive environmental conditions in farm productivity (Sherlund et al., 2002).

The estimates of the cost frontier showed that maize farmers on average exceeded the minimum cost of production by 34% (Table 2). We calculated allocative efficiency score as the inverse of allocative inefficiency value. This translated to allocative efficiency level of 75% and meant that there was opportunity to enhance efficiency by up to 25% through better allocation of scarce financial resources in maize production.

The significant estimate of the intercept indicates that there were considerable fixed costs in maize production. When farmers do not engage in any maize farming activity (and total variable costs are zero), they still incur significant opportunity cost of land. Cost of ploughing, price of seed and labour wage have positive and significant coefficients indicating that a marginal increase in their unit prices has sizeable effect on the production cost of maize. This implies that pricing of these inputs was beyond reach of many resource-poor smallholders. Therefore, efforts targeted at reducing cost of purchasable inputs will go a long way in enhancing affordability and access by majority of the resource-poor farmers. Similarly, an increase in yields would raise total cost of production, an indication that farmers were operating in the inefficient stage I of production (Debertin, 1986). This implies that there existed scope for increasing the scale of production without necessarily raising production costs so much by improving technical efficiency to benefit from economies of scale.

Assessment of differences in farming efficiencies

Farmers in the project area where they were exposed to ISFM practices achieved higher technical efficiency and lower allocative inefficiency compared to those in the counterfactual area (Table 3). The difference in estimated efficiency levels between farmers within the project area and those outside the project area represents ISFM contribution to technical and allocative efficiencies in

Table 2. Stochastic cost function estimated using maximum likelihood method to determine allocative efficiency in maize production.

Variable	Parameter	Coefficient	SE
Cost frontier function			
Intercept	α_0	6.771***	0.476
Fertilizer price	α_1	-0.026	0.077
Seed price	α_2	0.116**	0.058
Labour wage	α_3	0.109**	0.026
Ploughing cost	α_4	0.282***	0.037
Yield	α_5	0.118**	0.030
Efficiency measures			
Sigma-squared, $\sigma_\mu^2+\sigma_v^2$	σ^2	0.074**	0.013
Gamma, $\sigma_\mu^2/(\sigma_\mu^2+\sigma_v^2)$	γ	0.351***	0.029
Mean allocative inefficiency[c]	AE	34%	

[c]Allocative efficiency measures by how much the farmer exceeded the minimum feasible cost of production for a given level of output. We subtract 100 from the allocative inefficiency percentage to estimate the excess costs incurred by the farmer or group of farmers above the minimum efficient cost. This computation is implicit in all interpretations of differences in allocative inefficiency. Significant at the following levels: *10%; **5%; ***1%.

Table 3. Differences in technical and allocative efficiencies between farmers within and outside the project area.

Efficiency by site	Technical efficiency (%)		Allocative efficiency (%)	
	Mean	SD	Mean	SD
Within project area	84***	11	110***	16
Outside project area	58***	19	140***	17
Efficiency gain	26		30	

SD is standard deviation; *** Significantly different at 1% level.

maize production.

As presented in Table 3, significant differences (p=0.001) between the two sites (i.e. within and outside the project areas) demonstrate that adoption of ISFM practices has potential to narrow the yield gap by 26%, (84% less 58%), which is comparable to 30% reported in Tchale and Sauer (2007) and reduce cost incurred in maize production by 30% (140% less 110%). This clearly indicates that there is room to increase yields through more use of ISFM options to improve returns for smallholder farmers who cannot afford recommended rates of chemical fertilizers.

Factors influencing technical and allocative efficiencies

Table 4 shows the influence of the factors identified to contribute to farming efficiencies in maize production. They include farming experience, education level of the household head, household size (proxy for family labour), extension contacts and soil fertility management option. Others were credit access, off-farm income, market access and agro-ecological zone. We reversed the signs for all coefficients to enable direct inferences in relation to efficiency gains as opposed to inefficiency effects.

The coefficient for farming experience is significant and negative indicating that technical efficiency decreased with every year spent in farming (Table 4), in contradiction with previous findings (Külekçi, 2010; Seyoum et al., 1998). Although we expected higher efficiency among farmers with longer experience, the knowledge and skills gained over time may become less relevant with new technologies and constraints. However, farming experience enhanced allocative efficiency, supporting the view that the ability to acquire and process useful financial information increases with time, in line with Ogundari et al. (2006). The finding indicates that most experienced farmers gain various cost-saving strategies over time, which they apply in maize production. For instance, experience must have taught the seasoned maize producers to purchase key inputs such as fertilizers and seeds, and plough maize fields before the peak planting period when costs rise rapidly.

Formal education was found to increase technical efficiency, consistent with Külekçi (2010). The result points to

Table 4. Maximum likelihood estimates of factors affecting technical and allocative efficiency in smallholder maize production.

Factor	Coefficient	SE	Coefficient	SE
	Technical efficiency		Allocative efficiency	
Constant	0.54	1.55	0.34	0.56
Farming experience	-0.03**	0.01	0.01*	0.001
Education level	0.03*	0.02	-0.01*	0.007
Household size	-0.04*	0.03	0.01	0.01
Extension contacts	0.05*	0.04	0.03***	0.003
Soil fertility management	0.01***	0.001	0.02**	0.005
Credit access	-0.22	0.20	-0.05	0.09
Off-farm income earning	-0.61	0.55	-0.09*	0.08
Market access	-0.004**	0.001	-0.01**	0.001
Agro-ecological zone	0.23*	0.13	-0.08	0.15

Significant at the following levels: *10%; **5%; ***1%.

the importance of human capital in making and implementing informed and timely farming decisions. This means that most educated farmers have the capacity to source for, interpret and apply technical information well than the less educated ones. Moreover, better adoption of complex production technologies may call for technical knowledge and skills. Therefore, it is possible that these decisions and skills certainly benefit from some level of formal education. However, we found that higher education reduced allocative efficiency, consistent with Ogundari et al. (2006). This was surprising because it contradicts the view that the higher the number of years of schooling, the better the ability of farmers to match input use to their relative costs. Nevertheless, higher education level is likely to give farmers other off-farm income generating alternatives, which compete with maize production for management attention.

Agricultural extension contacts were associated with relatively higher technical and allocative efficiencies. The result demonstrated the value of providing farmers with skills and modern production techniques to improve yields and minimize production constraints. This finding is in agreement with other studies (Seyoum et al., 1998; Tchale and Sauer, 2007), which established that the farmers that regularly received extension information recorded higher technical efficiency compared to their counterparts. In fact, this study indicates that farmers who applied ISFM practices operated closer to their efficient frontiers. Therefore, promoting these practices through an effective extension approach will lead to greater efficiency gains in the entire farming system.

The application of ISFM practices in maize production contributed to both technical and allocative efficiencies than the use of chemical fertilizers alone. This observation point to the beneficial role played by organic nutrient sources in improving the productive capacity of the soil (Nyambati et al., 2003; Zingore et al., 2008). Further analysis already presented clearly show that those farmers who implemented some ISFM practices incurred on average 30% less costs of production at the same level of maize output. This confirms findings in other studies that have advocated for combination of inorganic and organic nutrient sources in different farming systems (Ranamukhaarachchi et al., 2005; Tchale and Sauer, 2007; Wanyama et al., 2010).

Off-farm income generation had an impact on allocative efficiency but not on technical efficiency contrary to the finding by Mochebelele and Winter-Nelson (2000). This is probably due to the possibility that farmers who earn more income away from the farm engage hired labour to carry out most activities in maize farming. However, hired labourers are less thorough in implementing agronomic activities. Moreover, owing to high demand for labour during peak periods, the implementation of critical agronomic activities such as planting, weeding and top-dressing may be untimely and this may eventually lead to low yields. The significant contribution to allocative efficiency was due to better financial capacity that enabled households that had off-farm income sources to acquire farm inputs timely before prices rose rapidly.

Higher cost of accessing the input-output markets led to lower technical and allocative efficiencies in maize production. This result can be associated with low use of purchasable inputs such as chemical fertilizers and hybrid seeds because of poor roads and costly transport system. The cost of accessing markets adds an extra financial burden to farmers located in remote areas characterized by poor roads network. The finding indicates the importance of enhancing access to input and output markets in order to improve farm productivity as also found by Tchale and Sauer (2007).

CONCLUSIONS AND RECOMMENDATIONS

Determining and overcoming possible constraints to

efficiency in smallholder farming system can contribute to sustainable use of farm resources in sub-Saharan Africa (SSA). Efforts aimed at enhancing overall efficiency among smallholders aims to improve maize productivity, net farm returns and soil fertility management in the maize farming system of Kenya. We investigated whether the availed integrated soil fertility management (ISFM) technologies have had impacts on both technical and allocative efficiencies in maize production.

We found that overall farmers achieved only 64% technical efficiency and 75% allocative efficiency. Farmers who applied ISFM practices were closer to their efficient frontiers compared to those who did not. We established that ISFM contributed about 26% to technical efficiency and 30% to allocative efficiency. Other factors that were found to determine efficiency gains included farming experience, provision of extension services, market access and off-farm income. These factors indicate the need for farming knowledge and profitability, and that farmers were responsive to policy-induced incentives. Therefore, we recommend the promotion of ISFM technologies through farmer groups and participatory extension system, in order to achieve greater efficiency gains in maize production in the country. We suggest the integration of efficiency considerations in agricultural research and policy formulation processes to ensure continued use of improved technologies and for enhanced food availability and incomes among the rural poor in similar SSA countries.

Conflict of Interest

The authors have not declared any conflict of interests.

ACKNOWLEDGEMENTS

The authors are grateful to Director General, Kenya Agricultural and Livestock Research Organization and the World Bank for KAPAP funding of this study. We sincerely appreciate logistic facilitation during fieldwork from Centre Director, KALRO Kitale. We appreciate the assistance of Noel Makete (KALRO, Kakamega), Hildah Wambani (KALRO Kitale), Joseph Muguide and the team of committed enumerators who conducted the interviews. We deeply thank the small-scale farmers in Trans Nzoia and Lugari districts who freely provided the data for this study. We appreciate two anonymous reviewers whose useful comments improved the quality of this paper.

REFERENCES

Aigner DJ, Lovell CAK, Schmidt P (1977). Formulation and estimation of frontier stochastic production function models. J. Econom. 6:21-37.

Bationo AJ, Kimetu S, Ikerra S, Kimani D, Mugendi M, Odendo M, Silver M, Swift J, Sanginga N (2004). The African Network for Soil Biology and Fertility: New Challenges and Opportunities. In: A Bationo (ed.) Managing Nutrient Cycles to Sustain Soil Fertility in Sub-Saharan Africa. Academy Science Publishers, Nairobi.

Battese GE, Broca SS (1997). Functional Forms of Stochastic Frontier Production Functions and Models for Technical Inefficiency Effects: A Comparative Study for Wheat Farmers in Pakistan. J. Prod. Anal. 8:395-414.

Bojö J (1996). The costs of land degradation in Sub-Saharan Africa. Ecol. Econ. 16:161-173.

CGIAR (2002). Combating soil fertility degradation in Sub-Saharan Africa In: CGoIA Research (ed) Proceedings of the Annual General Meeting, October 30-31,2002, Washington, D.C.

Clover J (2003). Food security in Sub-Saharan Africa. Afr. Security Rev. 12:5-15.

Coelli T (1996). A guide to FRONTIER version 4.1: A computer program for stochastic frontier production and cost function estimation. Centre for Efficiency and Productivity Analysis. CEPA Working paper No.96/07. University of New England, Armidale, New South Wales, Australia.

Debertin DL (1986). Agricultural Production Economics. Macmillan Publishing, New York.

Farrell MJ (1957). The measurement of productive efficiency. J. Royal Stat. Soc. 120(3):253-290.

Government of Kenya (2011). Kenya Open Data. Kenya Open Data Project.

Government of Kenya (2004). Investment Programme for the Economic Recovery Strategy for Wealth and Employment Creation (2003-2007). Government Printers, Nairobi.

Government of Kenya (2005). Strategy for Revitalizing Agriculture 2004-20014. Government Printers, Nairobi.

Government of Kenya (2006). District Agricultural Annual Reports for Lugari and Trans Nzoia Districts. In: Ministry of Agriculture (ed). Ministry of Agriculture, Nairobi.

Gujarati DN (2005). Basic Econometrics. fourth edn. Tata McGraw-Hill Publishing Company, New Delhi

Henao J, Baanante C (2006). Agricultural production and soil nutrient mining in Africa, excerpts from technical bulletin Agricultural Production and Soil Nutrient Mining in Africa- Implications for Resource Conservation and Policy Development. International Centre for Soil Fertility and Agricultural Development, IFDC, Muscle Shoals.

Jaetzold R, Hornetz B, Shisanya CA (2007). Farm Management Handbook of Kenya Vol II: Natural Conditions and Farm Management Information, Part B, Western Kenya. Ministry of Agriculture, Nairobi.

Jayne TS, Mather D, Mghenyi E (2010). Principal challenges confronting smallholder agriculture in Sub-Saharan Africa. World Dev. 38:1384-1398.

Kibaara BW (2005). Technical Efficiency in Kenyan's Maize Production: An Application of the Stochastic Frontier Approach. Colorado State University, Colorado.

KNBS (2010). 2009 Kenya Population and Housing Census. Population Distribution by Administrative Units. Kenya National Bureau of Statistics (KNBS), Nairobi p. 217.

Külekçi M (2010). Technical efficiency analysis for oilseed sunflower farms: a case study in Erzurum, Turkey. J. Sci. Food Agric. 90:1508-1512.

Lambin EF, Geist HJ, Lepers E (2003). Dynamics of land-use and land -cover change in tropical regions. Ann. Rev. Environ. Resourc. 28:205-241.

Liu Y (2006). Model selection in stochastic frontier analysis: Maize production in Kenya. Michigan State University, East Lansing 36 p.

Maddala GS (2001). Introduction to Econometrics. third edn. John Wiley and Sons, Chichester.

Meeusen W, van Den Broeck J (1977). Efficiency estimation from Cobb-Douglas production functions with composed error. Int. Econ. Rev. 18(2):435-444.

Mochebelele MT, Winter-Nelson A (2000). Migrant Labor and Farm Technical Efficiency in Lesotho. World Dev. 28:143-153.

Mose LO (2007). Who gains, who loses? The impact of market liberalisation on rural households in northwestern Kenya Social Sciences. Wageningen University, Wageningen p. 207.

Mose LO, Burger K, Kuvyenhoven A (2006). The role of traders in fertilizer, seed, maize and milk marketing in Kenya after market liberalization. Wageningen University.

Mureithi JG, Gachene CKK, Muyekho FN, Onyango M, Mose LO, Magenya O (2002). Participatory Technology Development for Soil Management by Smallholders in Kenya. KARI, Nairobi.

Mutoko MC, Hein L, Shisanya CA (2014). Farm diversity, resource use efficiency and sustainable land management in the western highlands of Kenya. J. Rural Stud. 36(2014):108-120.

Nyambati EM, Mureithi JG, Wamuongo JW (2003). Soil Fertility Improvement Technologies for Western Kenya, Synthesis Report of the Soil Management Project (1994-2002). KARI, Nairobi.

Ogundari K, Ojo SO, Ajibefun IA (2006) Economies of scale and cost efficiency in small scale maize production: empirical evidence from Nigeria. J. Soc. Sci. 13:131-136.

Olowa OW, Olowa OA (2010). Sources of technical efficiency among smallholder maize farmers in Osun State of Nigeria. Res. J. Appl. Sci. 5:115-122.

Place F, Franzel S, Dewolf J, Rommelse R, Kwesiga F, Niang A, Jaba B (2002). Agro forestry for soil fertility replenishment: Evidence on adoption processes in Kenya and Zambia. In: CB Barret, F Place, AA Abdillah (eds) Understanding and Improving Current Practices. CABI, Wallingford.

Ranamukhaarachchi SL, Rahman MM, Begum SN (2005) Soil fertility and land productivity under different cropping systems in highlands and medium highlands of Chandina Sub-district, Bangladesh. Asia-Pac. J. Rural Dev. 15:63-76.

Rangaswamy R (1995). A Text Book of Agricultural Statistics. first edn. New Age International (P) Ltd, New Delhi.

Seyoum ET, Battese GE, Fleming EM (1998). Technical efficiency and productivity of maize producers in eastern Ethiopia: a study of farmers within and outside the Sasakawa-Global 2000 project. Agric. Econ. 19:341-348.

Sherlund SM, Barrett CB, Adesina AA (2002). Smallholder technical efficiency controlling for environmental production conditions. J. Dev. Econ. 69:85-101.

Tchale H, Sauer J (2007). The efficiency of maize farming in Malawi. A bootstrapped translog frontier. Cahiers d'économie et sociologie rurales 82-83:33-56.

Todaro MP, Smith SC (2008) Economic Development. Longman Group, Addison-Wesley.

Verbeek M (2008). A Guide to Modern Econometrics. 3rd edn. John Wiley and Sons Ltd, Chichester.

Wang HJ, Schmidt P (2002). One-step and two-step estimation of the effects of exogenous variables on technical efficiency levels. J. Prod. Anal. 18:129-144.

Wangia C, Wangia S, De Groote H (2002). Review of maize marketing in Kenya: Implementation and impact of liberalization, 1989-1999. In: DK Friesen, AFE Palmer (eds) Integrated Approaches to Higher Maize Productivity in the New Millennium Proceedings of the 7th Eastern and Southern Africa Regional Maize Conference. CIMMYT, Nairobi.

Wanyama JM, Nyambati EM, Mose LO, Mutoko MC, W. M. Wanyonyi, Wanjekeche E, Rono SC (2010) Assessing Impact of soil management technologies on smallholder farmers' livelihoods in north western Kenya. Afr. J. Agric. Res. 5: 2899-2908.

Zingore S, Delve R, Nyamangara J, Giller K (2008). Multiple benefits of manure: The key to maintenance of soil fertility and restoration of depleted sandy soils on African smallholder farms. Nutr. Cycl. Agroecosyst. 80:267-282.

Economic analysis of factors influencing adoption of motor pumps in Ethiopia

Gebrehaweria Gebregziabher[1], Meredith A. Giordano[2], Simon Langan J.[1] and Regassa E. Namara[3]

[1]International Water Management Institute, East Africa Office, Addis Ababa, Ethiopia.
[2]International Water Management Institute, Colombo, Sri Lanka
[3]International Water Management Institute, West Africa Office, Accra, Ghana

The Ethiopian economy depends heavily on smallholder agriculture, and this sector directly affects the country's economic development, food security and poverty alleviation efforts. The adoption of smallholder irrigation technologies as a means to tackle these challenges has become an important policy issue in the development agenda of the country. The lack of access to low-cost irrigation technologies is, however, one of the major bottlenecks to increase smallholder irrigation. This paper examines the factors influencing farmers' decisions to adopt low-cost small motor pumps. The analysis is based on a survey of 800 farm households in four regions of Ethiopia. We use a combination of econometric techniques to find comparable households among adopter and non-adopter sample households. First, we employ a multivariate probit model to check whether a correlation exists between motor pumps and other water lifting technologies (that is, bucket, treadle and electric pumps). A non-parametric matching method is used to identify a counterfactual (control group) among the non-adopter sample households. Finally, a probit model is adopted to model the determinants of farmers' motor pump adoption decisions. Our analysis reveals that gender; age; ownership of oxen; access to extension; access to surface and shallow ground water; social capital and regional differences captured by a regional dummy, all influence farmers' decision of motor pump adoption.

Key words: Smallholder, irrigation technology, propensity score matching, probit.

INTRODUCTION

Investment in irrigation, particularly in small-scale and household level irrigation, has been identified as a core strategy in Ethiopia to reduce the strength of the link between agricultural production from rainfall and climate risk to improve crop production (Hagos et al., 2009). Irrigation also requires the use of modern inputs (such as, fertilizers and improved seeds), which further enhance agricultural productivity (World Bank, 2006; MoFED, 2006; Diao et al., 2010; Gebregziabher and Holden, 2011). To alleviate poverty, the financial gains from irrigation need to be geographically scaled out to widen the access and participation of many poor farmers. Small-scale private irrigation using small pumps has a greater chance to reach and involve many smallholders than command and scheme level irrigation approaches (Rydzewski, 1990; Fan and Hazell, 2001; Shitundu and Luvanga, 1998). Experience from Sub-Saharan Africa shows that investments in scheme level irrigation in the 1970s and 1980s did not meet targets for food production because economic problems, such as high capital investment and management costs, impeded the performance of large scale irrigation (Adams, 1991). Similarly, Lam (1996) shows that in Asia, small-scale schemes perform better than large-scale systems partly due to constraints by government bureaucracy on the latter and has promoted a shift to small-scale irrigation. Furthermore, D'Souza and Ikerd (1996) argued that from a sustainability perspective, small-scale farms are more effective and competitive compared with large-scale farms. Likewise, Ofosu et al. (2010) documented that in the Volta basin, irrigation technologies are frequently better managed by farmers and consequently result in higher productivity and good profit margins. Ofosu et al. (2010) also suggest that as compared to scheme level irrigation, small-scale irrigation technologies are more profitable and financially sustainable than large-scale irrigation, because they provide income opportunities to the wider society in terms of employment and participation of women. Moreover, experience from India suggests that given the right conditions, the use of small pumps and other micro-irrigation technologies commonly used in water scarce areas is an efficient use of irrigation water that can improve the productivity of water; generate income and financial benefits; and enhance food security of farm households (IWMI, 2006).

Likewise, D'Souza and Ikerd (1996) and Lam (1996) argue that smallholder and household level irrigation technology is more likely to bring higher returns per hectare than large-scale irrigation schemes. However, FAO (2005) has documented that only 13% of the irrigation potential of Sub-Saharan Africa is currently developed, largely due to past experience in irrigation development in the region emphasizing large-scale irrigation, which in most cases is constrained by high cost and management complexity. In the 1980s and 1990s, emphasis began to shift to smallholder irrigation using simple technologies, such as small and inexpensive pumps (Abric et al., 2011; Kay, 2001). For example Perry (1997) recommended low-cost manual and/or mechanized irrigation technologies as promising interventions in Sub-Saharan Africa while de Lange (1997) concluded that small-scale irrigation developed from farmers' initiative in Sub-Saharan Africa is more successful than government initiated large-scale irrigation.

Ethiopia has substantial surface and groundwater potential (Makombe et al., 2007; Awulachew, 2010; Awulachew et al., 2006; Cherre, 2006), although to date, farmers have not accessed this at a large enough scale to produce enough food to remove issues of regional poverty and food insecurity. Whilst irrigation has the potential to increase cereal yields by up to 40% (Diao et al., 2010), agricultural producers in Ethiopia have used only about 5 to 6% of the country's irrigation potential (Awulachew et al., 2007), mainly through large- and small-scale community irrigation schemes.

For the purpose of this paper, 'small' motor pumps are between 1 to 10 horsepower and costs between US$200 to US$1,000. Smallholder farmers usually use their own financing mechanisms to purchase these pumps to irrigate less than 5 ha of land to produce cash crops. The pumps are owned and managed individually or by small informal groups of farmers to pump water from rivers, lakes, reservoirs and shallow aquifer.

Data on private smallholder irrigation and the use of small pumps are not readily available and national estimates vary considerably. For example, Kay (2001) report that Ethiopia's potential irrigated area is approximately 670,000 ha, of which in 1992 about 82,000 ha and 5,000 ha were irrigated using large-scale and small-scale irrigation, respectively. More recently, Awulachew et al., (2005) report that the aggregated maximum irrigation potential in Ethiopia (including small, medium and large-scale) is about 3.7 million ha, of which only about 197,000 ha, or 5.3%, is irrigated. Furthermore, Santini et al. (2011) suggest that the potential for small private motor pump irrigation in Ethiopia is between 1.4 and 2.8 million ha, from which about 9 to 18 million people could benefit. However, except for some indicative government statistics, information on the current status of motor pump adoption in Ethiopia is largely unavailable.

The premise of this paper is that smallholder farmers can play a significant role in Ethiopia's irrigation development provided they have access to appropriate low-cost water lifting technologies. Ofosu et al. (2010) defined irrigation technology as "a method and techniques for diverting and/or pumping, storing, transporting and distributing ground, surface and rainwater to agricultural crops", and Perry (1997) has characterized motor pumps as "low-cost irrigation technologies". Based on unpublished reports in the regional bureaus of water resources, motorized small pumps are among the emerging private irrigation technologies in rural Ethiopia. The spread of small pumps occurs through the regional bureaus of water resources mainly distributed on credit and through direct purchase with farmers' own resources. According to Ethiopian government statistics (Ethiopian Revenue and Customs Authority), about 800,000 motor pumps have been imported between August 2004 and December 2010, while unpublished reports from the regions show that at the end of 2009, the regional bureaus of water resources have distributed 19,338 pumps in Oromia, 20,916 in

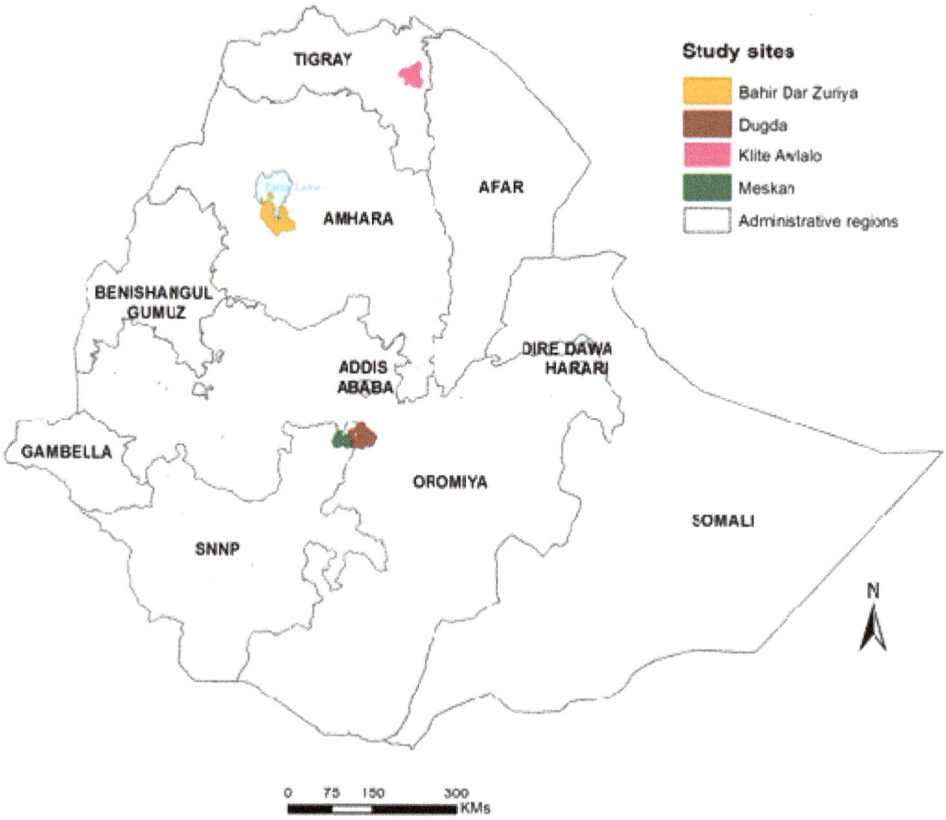

Figure 1. Study areas of Region and Districts

Amhara, and 18,348 in Tigray. Given that only 5 to 6% of Ethiopia's irrigation potential is being used, it is likely that private small-scale irrigation using pumps would benefit smallholder farmers. However, information on factors that influence the adoption of smallholder water lifting irrigation technologies is scant. The main objective of this paper is, therefore, to study the factors that affect smallholder farmers' adoption of small motor pumps in rural Ethiopia. The paper aims to contribute to the growing literature on adoption of smallholder irrigation and informing policy making in a country that has put irrigation at the heart of its agricultural development strategy.

METHODOLOGY

Data collection

In this study, we utilise data from a household survey collected during January-April, 2010 from four districts of the four main regions (Amhara, Oromia, SNNP and Tigray) of Ethiopia (Figure 1). Primary data were collected from 800 randomly selected farm households, using a multi-stage stratified random sampling method. In the first stage, we used information from the regional bureaus of agriculture to identify *wereda* (Districts) with a high concentration of smallholder irrigation technologies, such as buckets, treadle pumps, motorized pumps, and electric pumps. In the second stage, information from agricultural offices of the selected *weredas* was used to select *Kebeles* (communities) that have high adoption rates of these technologies. In the third stage, a list of farm households in the selected communities was used to disaggregate them into adopter and non-adopter households. Finally, we used a proportional random sampling technique to select our sample households. Of the total sample households, 266 were classified as adopters of motor pumps (Table 1).

Empirical analysis

This paper uses both descriptive and econometric analysis techniques. We assume that the production system represents a multi-crop agricultural production unit where land holding is fixed, but the allocation of land into crop type and irrigation is possibly endogenous. The adoption decision of irrigation technology is a discrete outcome where the farmer faces a dichotomous decision to adopt or not to adopt a motor pump. In our context, motor pump adopters are those farmers who were using motor pumps (petrol or diesel, rented or purchased) during the data collection to irrigate all or part of their land, while the rest are non-adopters.

Among the sample households, some of them have adopted a combination of technologies, such as motor pumps, bucket, treadle pumps, electric pumps and other water lifting technologies. These households may have adopted these technologies as substitutes or complements as they may have faced interdependent/correlated choices of technologies in their adoption decisions. Moyo and Veeman (2004); Marenya and Barrett (2007); Nhemachena and Hassan (2007); Yu et al. (2008) and Kassie et al. (2009) argue that farmers usually consider a set of possible technologies and try to

Table 1. Sample households and type of technologies by region.

Type of technology	Region				Total
	Amhara	Oromia	SNNPR	Tigray	
Purely rain-fed cultivators (Non-adopters)	115	118	120	146	499
Bucket	0	0	5	0	5
Treadle pump	3	1	5	0	9
Motor (petrol/diesel) pump[a]	66	68	73	59	266
Electric pump	0	21	0	0	21
Other type of technology	21	6	0	0	27
Total number of sample household[b]	200	200	200	200	800

a, Other type of technology includes rope and washer, wind mill, solar pumps, etc; **b,** the sum exceeds the total sample size, because some households have adopted more than one technology

Table 2. Correlation coefficients between irrigation technologies.

Irrigation technology	Motor pump	Bucket	Treadle pump
Bucket	ρ_{21} 0.436***(0.072)		
Treadle pump	ρ_{31} 0.298***(0.092)	ρ_{32} 0.193(0.270)	
Electric pump	ρ_{41} 0.222* (0.119)	ρ_{42} 0.198(0.184)	ρ_{43} -0.024(0.148)

χ^2 (6) = 27.483; probability > χ^2 = 0.000.

Table 3. Sample households that adopt bucket, treadle pump, electric pump and other irrigation technologies by region.

Technology	Region				Total
	Amhara	Oromia	SNNPR	Tigray	
Bucket	7(7)	14(13)	9(3)	2(2)	32(25)
Treadle pump	5(5)	7(4)	9(7)	0	21(16)
Electric pump	1(1)	23(14)	0(0)	0(0)	24(15)
Other	21(2)	7(3)	0(0)	0(0)	28(5)
Total	34(15)	51(34)	19(10)	2(2)	105(61)

1) Figures in parenthesis shows number households who also adopt motor pump; 2) Other types of technology includes rope and washer, wind mill, solar pumps, etc.

adopt a mix of technologies they assume can maximize their expected utility. While the adoption decision is inherently multivariate, recent studies on technology adoption (Tsefay, 2011; Nata and Bheemalingeswara, 2010; Deressa et al., 2009; Amha, 2006) assume a single technology without addressing the correlation and interdependence between the technologies. When a multitude of technology option exists, like in our case, a household may have equal opportunity, given their financing options, to choose from the set of technologies. In this situation farmers may well consider some combination of technologies as complementarity or competing. Hence, failure to capture such correlation/interdependence is likely to mask the reality that decision-makers face in their adoption decision. Consequently the results will be potentially biased and inefficient leading to underestimate or overestimate the influences of various factors in the adoption decisions.

Therefore, to identify the possible correlation that may exist between the irrigation technologies, we adopt and estimate a multivariate probit (MVP) model, which establishes a positive correlation between motor pumps and the other three (bucket, treadle pump and electric pump) irrigation technologies. This implies that a household's decision to adopt one of these technologies is likely to influence motor pump adoption (Table 2). A likelihood ratio test [χ 2 (6) = 27.48 and probability > χ 2 = 0.000] indicates significant joint correlations between the irrigation technologies under discussion implying the error term in the adoption of motor pumps is not independent of the other irrigation technologies. Both the correlation and likelihood ratio test justify that the estimation of the multivariate as opposed to separate univariate model is appropriate.

Given the very small number of bucket, treadle pump and electric pump adopters (Table 3); it may not be possible to generate illustrative regression results in relation to these technologies. Moreover, information on the adoption of bucket, treadle pump and electric pump is missing from the dataset of some study areas. For example, the data that we have from Tigray is only on motor pumps. Although, unless due to problems attributed to data collection, this does not mean bucket and treadle pumps are not used in the region. Furthermore, electric pumps are only an option in Oromia, where there is a large-scale electric pumped scheme at Lake Zeway. In general, due to practical reasons, our dataset on adoption of irrigation technologies is dominated by motor pumps, hence, even though small in number, treating the households that adopt bucket, treadle pump or electric pump as non-adopters in the analysis is likely to result in data contamination and inefficient results. To control for this potential problem, we omitted the 94 sample households who adopted irrigation technologies other than motor pumps. This left a sample of 706 (212 motor pump adopters and 494 are purely non-adopters) households. The reason the number of omitted sample households is less than the total in Table 3 (that is, 105) is because 11 households adopted more than two technologies and were double counted in the summary.

Since the adoption of motor pumps is not random, a selection bias is still a potential problem, as the adoption of motor pumps can be related to a number of factors (such as: unobserved household characteristics; proximity to water source; access to information and others). In addition, the remaining non-adopter sample households may not properly approximate the adopting sample households to serve as a counterfactual (control group). Hence, comparing adopters with the non-adopters without matching may still result in biased and inconsistent results.

In controlling the potential problem of selection bias, a propensity score matching method was used to identify 'real' comparable (counterfactual) sample households (Cobb-Clark and Crossley, 2003; Heckman et al., 1998; Ravallion, 2005). The basic assumption of using propensity score matching is that the matched non-adopter sample households approximate the adopters if they had not adopted. Given the control variables, this implies that the counterfactual outcome for the adopter group is the same as the observed outcomes for the non-adopting group (Heckman et al., 1998). In some cases, however, matching of adopting and non-adopting households based on observable characteristics may not be feasible, especially when the dimension of control variables is large. To overcome this problem, we employ the propensity score, $p\ (X\ )$ method that summarizes the multi-dimensional variables (Rosenbaum and Rubin, 1985). Propensity score is a conditional probability that household i has adopted a motor pump given the conditioning variables, written as:

$$p(x) = pr(MP = 1)|x \qquad (1)$$

Where p (the propensity score) represents the probability of motor pump adoption given unobservable household characteristics (x) and $MP\ (motor\quad pump\ )$ is equal to 1 for adopters and 0 for non-adopters. For the propensity score to be valid, the balancing properties need to be satisfied, implying that households with the same probability of adoption will be placed in the treated (adopter) and untreated (non-adopter) samples in equal proportions. Once the propensity score (pscore) is estimated, the data is split into equally spaced pscore intervals, implying that within each of these intervals, the mean pscore of each conditioning variable is equal for the adopter and non-adopter (control) households, known as the balancing property (Cobb-Clark and Crossley, 2003).

In line with this, the adopter and non-adopter households were matched based on observable characteristics (such as, household head's gender, off-farm participation, family size, access to credit, access to extension service, social capital in the form of household's membership in farmer associations, household's leadership role, farm size and tenure arrangement). Finaly, we found that out of the 494 non-adopter sample households, 420 of them have satisfied the balancing property implying that they can be used as counterfactuals (control group) in the adoption analysis. Concern about endogeneity is quite high in the adoption decision, because only households with access to water sources might consider adopting motor pumps. In an effort to account for such structural issue, we used WU-Hausman for the endogeneity test and found insignificant *F-test* coefficients [*F (1,669) =2.330 and P=0.128*], implying that the suspected variable (that is, access to water sources) is not endogenous in the adoption equation.

After identifying counterfactuals (control households) in the adoption analysis and validating that the suspected variable is not endogenous, we employed a binary outcome (probit) adoption model to estimate factors that influence households' adoption of motor pumps using the matched sample households. The probit model assumes that while we only observe the values of 0 for non-adopters and 1 for adopters for the outcome variable (Y), there is a latent unobserved, continuous variable that determines the value of (y^*). The probit model is specified as:

$$\begin{aligned} & y^* = \alpha + \Sigma X_i\beta + \varepsilon, \varepsilon \approx N(0,1) \\ & If\ \ y^* > 0, \quad Y = 1 \\ & if\ \ y^* \le 0, \quad Y = 0 \end{aligned} \qquad (2)$$

Where y^* is the outcome variable (adoption of motor pump) equal to 1 if household i adopted motor pump and 0 otherwise. X_i is a vector of values for the ith observation, β is a vector of parameters to be estimated and ε_i is the error term.

Explanation of variables and hypotheses

Following the adoption literature (e.g., Kassie et al., 2012; Pender and Gebremedhin, 2007; Marenya and Barrett, 2007; Bandiera and Rasul, 2006; Lee, 2005), the explanatory variables included in our regression analysis and their hypothesized effect on adoption of motor pumps are discussed below.

Human capital

Household characteristics, such as education, age, family size and gender may affect a households' decision to adopt irrigation technologies. Households with more educated members may have greater access to non-farm income and are able to finance the purchase of irrigation technologies. Furthermore, better educated farmers are likely to be more informed about the benefits of modern technologies and may have a greater ability to translate information and analyse the importance of technologies (Pender and Gebremedhin, 2007; Kassie et al., 2011). On the other hand, educated farmers are able to earn higher returns on their labour and capital if they are used in other activities (Pender and Gebremedhin, 2007).

Similarly, age may capture farming experience and exposure to production technologies implying an ability to respond to unforeseen events/shocks. It may also imply that older farmers have a life time accumulation of physical and social capital suggesting greater respect in their community. On the other hand, age can be associated with loss of energy, short-planning horizons and being risk averse. Thus, the impact of age on technology adoption is ambiguous prior to being empirically tested. It has been argued that women have less access to crucial farm resources (land, labour, and cash) and are generally discriminated in terms of access to external inputs and information (De Groote and N'Golo, 1998; Quisumbing et al., 1995). In Sub-Saharan Africa, there are gender specific constraints, such as women's poorer access to education, land and production assets (Ndiritu et al., 2011). It is obvious that these constraints have direct effects on technology adoption including irrigation technologies where women are usually less likely to adopt. In this paper, gender is specified as dummy variable equal to 1 for male and 0 for female.

Access to market

Access to markets can influence farmers' decision making in various ways, such as availability of technology, the use of output and input markets, and access to information and support organizations, for example, credit institutions (Jansen et al., 2006; Wollni et al., 2010; Pender and Gebremedhin, 2007). It can also increase the amount of labour and/or capital intensity by rising output to input price ratios. The hypothesis here is that the further away a village or farming household is from a market, the less likely it is to adopt new technology.

Physical capital

This variable is represented by livestock ownership and farm size as proxies of household wealth. Wealthier households are better able to bear risk associated with the adoption of motor pumps and to finance purchase of motor pumps. Furthermore, as mixed farming (crop-livestock farming) production system is common practice in the Ethiopian context, livestock may serve as source of manure and draft power. In such a situation irrigated crop production may generate fodder for livestock; hence, the linkage between crop and livestock production systems may encourage adoption of irrigation technologies.

Off-farm participation

Economic incentives play an important role in the adoption of technologies, although their effects may be complex and subtle (Lee, 2005). Household access to alternative sources of employment and return from such activities are likely to influence the adoption of motor pumps, but in different directions. For example, households that have alternative sources of income may have greater capacity to pay and adopt the technologies. On the other hand, off-farm activities may divert time and labour from agricultural activities, reducing investments in irrigation technologies and the availability of labour that can be used in irrigation. In this paper, off-farm participation is defined as equal to 1, if the household has participated in off-farm activity and 0 otherwise. The hypothesized effect of off-farm participation on the adoption of irrigation technologies is, therefore, ambiguous.

Land tenure

A number of studies have demonstrated that security of land ownership has a substantial effect on the agricultural performance of farmers (Besley, 1995; Kassie and Holden, 2007; Deininger et al., 2009). In this paper, tenure security is indicated by land tenure (1=owned by the farmer, 0=otherwise) and we assume that households who produce on their own land have better tenure security and are more likely to invest in irrigation technologies.

Social networks

This represents a combination of variables, such as membership in farmer groups or associations and number of traders that a respondent knows as a proxy of market network. Isham (2007) and Bandiera and Rasul (2006) suggest the positive effects of social networks and personal relationships on technology adoption. With scarce or inadequate information and imperfect market, a social network allows and facilitates the exchange of information, enables farmers to access inputs and overcome credit constraints. Social networks also reduce transaction costs and increase farmers' bargaining power, helping farmers to earn higher returns when marketing their products that can also affect technology adoption (Wollni et al., 2010; Lee, 2005). Farmers who do not have contacts with extension agents may still find out about new technologies from their networks, as they share information and learn from each other. Membership in farmers' groups or associations is therefore hypothesized to be positively associated with adoption of motor pumps.

Biophysical characteristics

Agricultural production in Sub-Saharan Africa is characterized by wide variability of agro-ecological and biophysical factors. We asked our respondents whether they have year round access to surface and shallow groundwater. Two dummy variables (access to ground and surface water) are included in the regression. The assumption is that those households have access to surface and/or ground water are more likely but not certain to adopt motor pumps. Moreover, other biophysical (e.g., rainfall, topography, soil type) and socioeconomic characteristics (e.g., population, production risk) may influence the adoption of water lifting technologies. For example, in the Ethiopian highlands, topography follows a gradient from flat lowlands to mountainous area (Pfeifer et al., 2012). The same report indicates that most of the western Ethiopian highlands are dominated by Nitisols that are stable and relatively less prone to erosion, while the eastern part and highland plateau of the Blue Nile Basin are dominated by leptosols and vertisols, respectively. Leptosols are relatively shallow and prone to erosion while vertisols are low drainage heavy clay soils, implying that topographical and soil characteristics may influence the recharge and availability of groundwater, suitability of irrigation technologies. However, due to lack of site specific biophysical and socio-economic data, we use region dummies to capture unobserved site specific biophysical and socioeconomic differences.

RESULTS

Descriptive results

The definition and summary statistics of variables used in the analysis are presented in Table 4. About 34% of the total matched sample households have adopted motor pumps. In many parts of the Sub-Saharan African countries, male farmers dominate the farming system and technology adoption, which our data also show. Male headed households constitute about 92% of the total sample households and about 97 and 89% of the adopter and non-adopter sample households, respectively.

Table 4. Definition of variables and descriptive statistic.

Dependent variable			Mean		SD		
Adoption of Motor pump (1 = yes, 0 = no)			0.335		0.473		T-test/significance of difference
Independent variables	Total sample households		Non-adopters		Adopters		
	Mean	SD	Mean	SD	Mean	SD	
Household head's gender (1 = male, 0 = female)	0.916	0.277	0.890	0.313	0.967	0.179	0.001***
Household head age (years)	44.323	14.024	45.464	14.043	42.061	13.739	0.004***
Ownership of oxen in tropical units (TLU)	1.212	1.116	1.069	1.049	1.495	1.190	0.000***
Ownership of non-oxen livestock in tropical units (TLU)	2.747	3.006	2.451	2.827	3.333	3.261	0.000***
Adult household member (number)	3.036	1.564	2.983	1.453	3.142	1.763	0.230
Educated household member (number)	2.723	2.096	2.590	2.034	2.986	2.195	0.025**
Access to extension (1 = yes, 0 = otherwise)	0.560	0.497	0.524	0.500	0.632	0.483	0.010**
Access to credit (1 = yes, 0 = otherwise)	0.237	0.426	0.224	0.417	0.264	0.442	0.261
Household has market network (1 = yes, 0 = no)	0.071	0.257	0.057	0.232	0.099	0.299	0.053*
Household membership in farmer association (1 = yes, 0 = otherwise)	0.650	0.477	0.626	0.484	0.698	0.460	0.074*
Farm size (ha)	2.323	1.540	2.184	1.397	2.600	1.761	0.001***
Land tenure (1 = owned, 0 = leased in)	0.992	0.089	0.990	0.097	0.995	0.069	0.520
Availability of surface water (1 = yes, 0=otherwise)	0.381	0.486	0.236	0.425	0.670	0.471	0.000***
Availability of ground water (1 = yes, 0=otherwise)	0.166	0.373	0.100	0.300	0.297	0.458	0.000***
Region dummies							
Amhara (1 = yes, 0 = otherwise)	0.245		0.245		0.245		
Oromia (1 = yes, 0 = otherwise)	0.223		0.240		0.189		
SNNPR (1 = yes, 0 = otherwise)	0.264		0.248		0.297		
Tigray (1 = yes, 0 = otherwise)	0.267		0.267		0.269		

The summary statistics also show that younger farmers are more likely to adopt motor pumps as compared to older farmers. This is consistent with Ahmed et al. (2002) that older farmers are risk averse and usually stick to traditional farming systems. Physical assets (proxied by ownership of livestock) are significantly different from those who own more physical assets being in a better position to finance the purchase of new technologies, especially when credit is a constraint. Farm size as a proxy of physical assets is also significantly higher for adopters as compared to non-adopters. Education and access to extension are positively related to adoption. Finally, the summary statistics reveal that those who have a positive perception about the availability of surface and shallow ground water are more likely to adopt motor pumps. One may argue that these households are located in more favorable settings, so that they have better access to a source of irrigation water leading to a high adoption rate. However, the fact that we use matched sample households in the analysis possibly invalidates such an argument.

Assessment of market prices of motor pumps is an integral part of this study. A motor pump of 3.5 HP[*] that can irrigate about 2 ha costs about US$1,087 (equivalent to 12,500 Ethiopian Birr). Data from the revenue and customs authority of Ethiopia also show that the average cost of a motor pump is estimated at US$565 of which government taxes account for about 37% of the costs[†]. Furthermore, since motor pumps do not stand alone, the cost of accessories and other irrigation infrastructure are important in the motor pump adoption process. Information from our survey suggests that the average cost of motor pump accessories, maintenance and construction of wells is in the order of US$165 (Table 5), which makes the investment more expensive.

Results from the regression analysis

Here, we discuss the results obtained from the probit model. Table 6 presents regression results of the adoption (probit) model. The data suggest that household, socioeconomic and biophysical characteristics all affect households' motor pump adoption decisions. For example, male headed households are more likely to adopt motor pumps as compared to female headed households indicating that female headed households are less likely to benefit from motor pump adoption than male headed households.

The negative association between adoption of motor pumps and age imply that older farmers are less likely to adopt as compared to younger farmers. This can be associated with short planning and risk averse behavior

[*] HP represents horsepower.

[†] The Ethiopian Birr was devalued by about 20% in September 2010, significantly increasing the price of imports, including motor pumps, so that the current price of pumps is likely much higher.

Table 5. Average cost and tax rate of imported water pumps.

Cost Component	Average
Average CIF value of water pump (Birr)	4668
Average tax per unit of water pump (Birr)	1832
Average purchase price/water pump (CIF+Tax) (Birr)	6500
Tax rate	36%

Source: Summarized Based on Data from Ethiopian Customs and Revenue Authority

Table 6. Regression results of the adoption (Probit) model.

Variable description	Coefficient	Robust Std. Err.
Household head's gender (1 = male, 0 = female)	0.664***	0.249
Household head age (years)	-0.015**	0.006
Ownership of oxen in tropical units (TLU)	0.145*	0.080
Ownership of non-oxen livestock in tropical units (TLU)	0.008	0.031
Adult household member (number)	0.051	0.057
Educated household member (number)	-0.024	0.043
Access to extension (1 = yes, 0 = otherwise)	0.319**	0.140
Access to credit (1 = yes, 0 = otherwise)	-0.112	0.187
Household has market network (1 = yes, 0 = no)	0.365	0.262
Household membership in farmer association (1 = yes, 0 = otherwise)	0.292*	0.151
Farm size (ha.)	0.067	0.045
Land tenure (1 = owned, 0 = leased in)	*0.614*	*0.484*
Availability of surface water (1 = yes, 0 = otherwise)	*1.767****	*0.169*
Availability of ground water (1 = yes, 0 = otherwise)	0.936***	0.183
Region dummies (control region is Tigray)		
Amhara (1 = yes, 0 = otherwise)	-1.391***	0.278
Oromia (1 = yes, 0 = otherwise)	-0.918***	0.272
SNNPR (1 = yes, 0 = otherwise)	-0.641***	0.240
Constant	-2.103***	0.591
Number of observation	632	
Log pseudo likelihood	-278.493	
Wald $\chi^2(17)$	155.190	
Prob > chi2	0.000	
Pseudo R^2	0.31	

*, **, *** are levels of significance at 10, 5 and 1%, respectively. Figures in parentheses are robust standard errors.

of older farmers and supports the findings of previous research (Kassie et al., 2012; He et al., 2007). In terms of wealth factors, ownership of oxen is positively related to the adoption of motor pumps, suggesting that wealthier farmers are more likely to take risk as compared to poor farmers. This is consistent with the findings of research carried out in Egypt (Mourshed, 1995), which found that risk causes anxiety towards new innovations and unfamiliar techniques can produce uncertain yields. As a result, farmers with limited incomes or assets are reluctant to adopt unproven/unfamiliar technologies.

Access to extension is also positively related with the adoption of motor pumps as farmers' awareness and skills to efficiently use of the technology is expected to increase. For example, Mourshed (1995) documents that Egyptian small desert farmers adopt drip irrigation after witnessing the success of nearby large farmers. This may hint to the importance of strengthening extension service. This can be achieved, for example, by organizing formal and informal experience-sharing tours and farm 'field-days' to learn from nearby better performing model farmers and from that scaling up best practices in technology adoption.

As expected, farmers' perception about the availability of surface and shallow ground water has both a positive and significant effect (both at 1% level of significance) on the probability of motor pump adoption. This suggests that scientific evidence about the potential of ground/surface availability to increase farmers' confidence and willingness to adopt irrigation technologies is an important factor. Furthermore, the adoption of motor pumps varies by region. This variation is most likely due to region specific socio-economic and biophysical characteristics differences, such as rainfall, topography, erosion, and soil and water conservation. The negative coefficients for Amhara, Oromia and SNNPR dummies for adoption of motor pumps suggest a lower probability of adoption of motor pumps in these regions as compared in Tigray. This probably reflects the effect of unobservable spatial differences (such as rainfall, land degradation and land fertility) as well as the difference in soil/water conservation and watershed management activities between the regions. For example, since the 1970s, there have been intensive and relatively successful soil/water conservation and watershed management activities in Tigray (Woldearegay, 2012), which has led to increased infiltration and groundwater recharge and in turn an increased adoption of household level private irrigation technologies. On the other hand, previous research (Pender et al., 2006; Aiayi, 2007; Kassie et al., 2012) stated that several biophysical and socioeconomic factors have been identified as limiting factors for increasing food production for most smallholder farmers in Sub-Saharan Africa. Furthermore, Kassie et al. (2012) argues that such unfavourable biophysical factors are likely to encourage farmers to adopt production enhancing technologies as a coping mechanism.

DISCUSSION AND CONCLUSIONS

In Ethiopia, agriculture is the main sector that substantially influences economic development, food security and poverty alleviation. The sector is dominated by smallholder farmers. However, low and high variability of rainfall combined with low levels of technology adoption characterize the performance of agriculture. As a result, in a country where there is substantial surface and groundwater potential, farmers are unable to access it to produce enough food. Moreover, information on private smallholder irrigation and the use of smallholder irrigation technologies is not readily available and lacks consistency. Recent studies indicate that the potential for small private motor pump irrigation in Ethiopia is in the order of 1.4 up to 2.8 million ha (Santini et al., 2011) and can benefit between 9 to 18 million smallholder farmers.

Regression results show that there is heterogeneity with regard to the factors that influence the adoption of motor pumps. It underscores the importance of gender; age; ownership of oxen; access to extension; social capital in the form of farmers' membership in farmer associations; access to surface and shallow ground water and region specific socio-economic and biophysical differences. There is a need for more research to identify site specific socio-economic and biophysical factors in the adoption and dissemination of smallholder water lifting technologies and then targeting these technologies where they perform well.

Our results also suggests that the probability of motor pump adoption increases with farmers' participation in farmer associations implying local rural institutions can assist farmers in providing information, credit, experience sharing and market outlets. The positive effect of access to extension on motor pump adoption emphasizes the need to improve the extension system. Finally the adoption of motor pumps is influenced by farmers' gender, age and wealth. The policy implication of this result is that targeting women's groups to address their constraints to actively participate in the adoption of irrigation technologies and rural economic activities in general can have a significant impact on the adoption of smallholder water lifting irrigation technologies and improved livelihoods.

Beyond the regression results, our survey data also show that the cost of motor pumps is high and prices continue to increase. Government taxes account for about 37% of the prices. The cost of accessories and irrigation infrastructures are also quite high for resource poor farmers. The supply of agricultural inputs, fuel and maintenance service is a critical problem. The output market is highly fragmented where informal brokers have un-proportional power to set market prices, usually against the interest of farmers. Frequent mechanical breakdowns are widespread due to farmers' lack of skills, while the supply of spare parts and maintenance services are lacking in the rural areas. Knowledge about environmental risk of motor pump use (that is, risk of groundwater depletion) is seldom. Hence, we suggest that further studies need to understand welfare and environmental implication of motor pump adoption and policies to support the dissemination of motor pumps for smallholder irrigation as a poverty reduction strategy.

Conflict of Interest

The authors have not declared any conflict of interest.

REFERENCES

Abric S, Sonou M, Augeard B, Onimus F, Durlin D, Soumaila A, Ga dElle F (2011). Lessons Learned in the Development of Smallholder Private Irrigation for High-Value Crops in West Africa. The International Bank for Reconstruction and Development /The World Bank.

Adams WM (1991). Large scale irrigation in Northern Nigeria: Performance and Ideology. Transactions of the Institute of British Geographers. New Series 16(3):287-300.

Amha R (2006). Impact assessment of rainwater harvesting ponds: The Case of Alaba Woreda, Ethiopia. Faculty of Business and Economics, Addis Ababa University (Thesis).

Ahmed MM, Gebremedhin B, Benin S, Ehui S (2002). Measurement and sources of technical efficiency of land tenure contracts in Ethiopia, Environ. Dev. Econ. 7:507–527.

Aiayi OC (2007). User acceptability of sustainable soil fertility technologies: lessons from farmers' knowledge, attitude, and practice in southern Africa. J. Sustain. Agric. 30(3):21–40.

Awulachew SB (2010). Irrigation potential in Ethiopia: Constraints and opportunities for enhancing the system, Int. Water Manage. Institute

Awulachew SB, Yilma AD, Loulseged M, Loiskandl W, Ayana M, Alamirew T (2007). Water Resources and Irrigation Development in Ethiopia. Work. Int. Water Manage. Inst. P. 123.

Awulachew SB, Menker M, Abesha D, Atnafe T, Wondimkun Y (2006). Background: About the Symposium and Exhibition. Best practices and technologies for small-scale agricultural water management in Ethiopia: MoARD/MoWR/USAID/IWMI symposium and exhibition, held at Ghion Hotel, Addis Ababa 7-9 March: International Water Management Institute (IWMI).

Awulachew SB, Merrey DJ, Kamara AB., Van Koppen B, Penning de Vries F, Boelee F (2005). Experiences and opportunities for promoting small-scale/micro irrigation and rainwater harvesting for food security in Ethiopia. IWMI Working P. 98. Colombo: IWMI.

Bandiera O, Rasul I (2006). Social networks and technology adoption in Northern Mozambique. Econ. J. 116(514):869-902.

Besley T (1995). Property rights and investment incentives: theory and evidence from Ghana. J. Pol. Econ. 103:903–937.

Cherre S (2006). Irrigation policies, strategies and institutional support conditions in Ethiopia. In Awulachew SB, Menker M, Abesha D, Atnafe T, Wondimkun Y (2006). Proceedings of Symposium on Best Practices and Technologies for small-scale agricultural water management in Ethiopia: MoARD/MoWR/USAID/IWMI symposium and exhibition, held at Ghion Hotel, Addis Ababa 7-9 March: International Water Management Institute (IWMI).

Cobb-Clark DA, Crossley T (2003). Econometrics for evaluations: An introduction to recent developments. Econ. Record 79(247):491-511.

De Lange M (1997). Promotion of low-cost and water saving technologies for small-scale irrigation. in FAO (1997). Irrigation Technology Transfer in Support of Food Security (Water Reports - 14). Proceedings of a subregional workshop, Harare, Zimbabwe, 14-17 April 1997. Available at: http://www.fao.org/docrep/w7314e/w7314e00.HTM

Deininger K, Ali DA, Alemu T (2009). Impacts of Land Certification on Tenure Security, Investment, and Land Markets, EfD Discussion Paper no. 09–11, Environment for Development and Resources for the Future, Gothenburg, Sweden, and Washington, DC.

De Groote H, N'Golo C (1998). Gender and generation: an intra-household analysis on access to resources in southern Mali. Afr. Crop Sci. J. 6(1):79-95.

Deressa T, Hassan RM, Ringler C, Alemu T, Yesuf M (2009). Determinants of farmers' choice of adaptation methods to climate change in the Nile Basin of Ethiopia. Glob. Environ. Change 19:248-255.

Diao X, Seyoum AT, Yu B, Nin Pratt A (2010). Economic importance of agriculture for sustainable development and poverty reduction: The Case Study of Ethiopia: Global Forum on Agriculture, 29-30 November 2010, Policies for Agricultural Development, Poverty Reduction and Food Security, OECD Headquarters, Paris.

D'Souza G, Ikerd J (1996). Small farms and sustainable development: Is small more sustainable. J. Agric. Appl. Econ. 28(1):73-83

FAO (2005). Irrigation in Africa in figures. FAO Water Report 29. Rome

Fan S, Hazell P (2001). Returns to Public Investments in the Less-Favored Areas of India and China. Am. J. Agric. Econ. 83(5):1217–1222.

Gebregziabher G, Holden S (2011). Does Irrigation enhance and food deficits discourage fertilizer adoption in a risky environment, Evidence from Tigray, Ethiopia. J. Dev. Agric. Econ. 3(10):514-528.

Hagos F, Makombe G, Namara RE, Awulachew SB (2009). Importance of Irrigated Agriculture to the Ethiopian Economy: Capturing the Direct Net Benefits of Irrigation. Colombo, Sri Lanka. Int. Water Manage. Institute. IWMI Research Report 37:128.

He XF, Cao H, Li FM (2007). Economic analysis of the determinants of adoption of rainwater harvesting and supplementary irrigation technology in the semiarid Loess Plateau of China. Agric. Water Manage. 89:243-250

Heckman JJ, Ichimura H, Todd, P (1998). Matching as an econometric evaluation estimator. Rev. Econ. Stud. 65(2):261-294.

IWMI (2006). Promoting micro-irrigation technologies that reduce poverty. Water Policy Brief. Issue P. 23.

Isham J (2007). The effect of social capital on fertilizer adoption: evidence from rural Tanzania. J. Afr. Econ. 11:39–60.

Jansen HGP, Pender J, Damon A and Chippers R (2006). Land Management Decisions and Agricultural Productivity in the Hillsides of Honduras, in: Paper presented at the International Association of Agricultural Economists Conference, Gold Coast, Australia, August 12–18.

Kassie M, Jaleta M, Shiferaw B, Mmbando F, Mekuria M (2012). Adoption of interrelated sustainable agricultural practices in smallholder systems: Evidence from rural Tanzania. Technol. Forecast. Soc. Change 80(3):525–540.

Kassie M, Shiferaw B, Geoffrey M (2011). Agricultural technology, crop income, and poverty alleviation in Uganda. World Dev. 39(10):1784-1795.

Kassie M, Zikhali P, Manjur K, Edwards S (2009). Adoption of organic farming technologies: Evidence from Semi-Arid Regions of Ethiopia. Natural Resour. Forum (33):189–198.

Kassie M, Holden ST (2007). Sharecropping efficiency in Ethiopia: threats of eviction and kinship. Agric. Econ. 37:179–188.

Kay M (2001). Smallholder irrigation technology: Prospects for Sub-Saharan Africa: International Programme for Technology and Research in Irrigation and Drainage: Knowledge Synthesis Report Food and Agriculture Organization of the United Nations Rome P. 3.

Lam WF (1996). Improving the Performance of Small-Scale Irrigation Systems: The Effects of Technological Investments and Governance Structure on Irrigation Performance in Nepal. World Dev. 24(8):1301-1315.

Lee DR (2005). Agricultural sustainability and technology adoption: Issues and Policies for Developing Countries. Am. J. Agric. Econ. 87(5):1325-1334.

Makombe G Kelemework D, Aredo D (2007). A comparative analysis of rainfed and irrigated agriculture production in Ethiopia. Irrigation and Drainage Systems: Published Online 21:31-44.

Marenya PP, Barrett CB (2007). Household-level determinants of adoption of improved natural resources management practices among smallholder farmers in western Kenya. Food Policy 32:515–536.

MoFED (Ministry of Finance and Economic Development) (2006). Ethiopia: Building on Progress: A Plan for Accelerated and Sustained Development to End Poverty 2005/6 –2009/10, September 2006.

Mourshed M (1995). Rethinking Irrigation Technology Adoption: Lessons from the Egiptian Desert: Working No. 23, Program in Sciency, Technology and Society, Massachusetts Institute of Technology Moyo.

Nata T, Bheemalingeswara K (20100. Prospects and constraints of household irrigation practices, Hayelom Watershed, Tigray, Northern Ethiopia. Momona Ethiop. J. Sci. 2(2):87-109.

Ndiritu SW, Kassie M, Shiferaw B, Ouma J, Odendo M (2011). Adoption of agricultural technologies in Kenya: How Does Gender Matter?, CIMMYT, Nairobi. PMCid:PMC3163847

Nhemachena C, Hassan R (2007). Micro-Level Analysis of Farmers' Adaptation to Climate Change in Southern Africa, IFPRI Discussion Paper No. 00714. International Food Policy Research Institute. Washington DC.

Ofosu EA, van der Zaag P, van de Giesen NC, Odai SN (2010). Productivity of irrigation technologies in the White Volta basin. Phys.

Chem. Earth 35(13-14):706-716.

Pender J, Gebremedhin B (2007). Determinants of agricultural and land management practices and impacts on crop production and household income in the highlands of Tigray. Ethiopia. J. Afr. Econ. 17:395–450.

Pender J, Place F, Ehui S (2006). Strategies for Sustainable Land Management in the East African Highlands, International Food Policy Research Institute, Washington, DC, 2006.

Perry E (1997). Low-cost irrigation technologies for food security in sub-Saharan Africa, in FAO. 1997. (Proceeding), Irrigation Technology Transfer in Support of Food Security. (Water Reports – 14) at: http://www.fao.org/docrep/W7314E/W7314E00.htm accessed on 26/12/2012

Pfeifer C, Notenbaert A, Omolo A (2012). Similarity analysis for the Blue Nile Basin in the Ethiopian highlands. Nairobi, Kenya: ILRI. Available at: http://r4d.dfid.gov.uk/PDF/Outputs/WaterfoodCP/NBDCTechnicalReport3.pdf

Quisumbing AR, Brown L, Hillary R, Feldsten S, Haddad L, Pena C (1995). Women: The Key to Food Security, Food Policy Statement, No. 21, International Food Policy Research Institute, Washington, DC.

Ravallion M (2005). Evaluating Anti-Poverty Programs. World Bank Policy Research, Development Research Group. Working Paper 3625.

Rosenbaum PR, Rubin DB (1985). Constructing a control group using multivariate matched sampling methods that incorporate the propensity score. Am. Stat. 39:33–38.

Rydzewski JR (1990). Irrigation: A viable development strategy. Geographical J. 156(2):175-180.

Santini G, Peiser L, Faurès JM, Neves B, Vallée D (2011). Planning smart investments in agricultural water management through a livelihood mapping approach: the case of Ethiopia (Draft).

Shitundu J, Luvanga N (1998). The Use of Labor-Intensive Irrigation Technologies in Alleviating Poverty in Majengo, Mbeya Rural District, REPOA Research Report 98:3.

Tesfay G (2011). On-farm water harvesting for rainfed agriculture development and food security in Tigray, Northern Ethiopia: Investigation of technical and socioeconomic issues. Drylands Coordination Group Report P. 61. Available at http://www.drylands-group.org

Veeman M (2004). Analysis of joint and endogenous technology choice for protein supplementation by smallholder dairy farmers in Zimbabwe, Agrofor. Syst. 60:199–209.

Woldearegay K (2012). Regreening Tigray - Upscaling 3R Catchment Management in Ethiopia. World Water Week, Stockholm. at: http://www.worldwaterweek.org/documents/WWW_PDF/2012/Tue/Investment-options/Kifle-Woldearegay-Regreening.pdf, accessed on January, 11, 2013.

Wollni M, Lee DR, Thies JE (2010). Conservation agriculture, organic marketing, and collective action in the Honduran hillsides. Agric. Econ. 41:373–384.

World Bank (2006). Ethiopia: Managing Water Resources to Maximize Sustainable Growth: A World Bank Water Resources Assistance Strategy for Ethiopia. Washington DC, USA.

Yu L, Hurley T, Kliebenstein J, Orazen P (2008). Testing for Complementarity and Substitutability among Multiple Technologies: The Case of U.S. Hog Farms, Working Paper No. 08026. Ames, IA, USA: Iowa State University, Department of Economics.

Factors determining allocation of land for improved wheat variety by smallholder farmers of northern Ethiopia

Leake Gebresilassie[1] and Adam Bekele [2]

[1]College of Dray Land Agriculture and Natural Resources, Mekelle University, Ethiopia.
[2]Melekasa Research Institute, Melekasa, Oromia, Ethiopia.

This study was conducted in Northern Ethiopia, Adwa district. The main objective of the study was to examine factors influencing allocation of land for improved wheat variety by smallholder farmers of the study area. Descriptive, inferential and econometric methods were used to analyze data. Results of descriptive and inferential analyses showed that; adopters had high family size in adult-equivalent, high number of tropical livestock unit, large land size, high frequency of extension contact, access to credit service, they were followed formal schooling, and they were nearest to main road and market as compared to non-adopters. Tobit model was used to analyze factors influencing adoption of improved wheat technology econometrically. A total of thirteen explanatory variables were included in the model. From the tested variables only eight variables (education level of household head, family size, tropical livestock unit, distance from main road and nearest market, access to credit service, extension contact and perception of household towards cost of the technology) were found to be the significant factors affecting adoption of improved wheat variety. Implication of results of this study is that any development intervention through improved wheat technologies should consider the aforementioned socioeconomic characteristics and determinants of adoption for success.

Key words: Adoption of Improved wheat variety, Adwa, smallholder farmers, Topit model.

INTRODUCTION

Agriculture is the mainstay of the Ethiopian economy. It employs 80% of the population and contributes about 41% of GDP and 86% of exports (Bingxin et al., 2011). Besides its contribution as the main income-generating sector for the majority of the rural population, it serves as the main source of household food consumption (Samia, 2002).

The agricultural sector in Ethiopia is dominated by subsistence, low input, low output and rain-fed farming system. The use of improved seeds is quite limited despite government efforts to encourage the adoption of modern agricultural system and intensive agricultural practices. Therefore, improving the productivity, profitability, and sustainability of smallholder farming is the main pathway out of poverty in using agriculture for development (World Bank, 2008). One important way to increase agricultural productivity is through the introduction of improved agricultural technologies and management systems. Adoption of new agricultural technology such as high yielding varieties stimulates the transition from low productivity subsistence agriculture to a high productivity agro-industrial economy (World Bank, 2008).

Cereals dominate Ethiopian agriculture; accounting for about 70% of agricultural GDP. Wheat is one of the major cereal crops grown in Ethiopia (Hailu, 2003). It is grown by smallholder farmers in the highlands and mid highland areas of the country (Bingxin et al., 2011). The productivity of the crop has been low and a number of yield improving technologies like seeds of improved varieties have been recommended for use by wheat producing smallholder farmers in the country. However, the level of adoption of the technologies is not as expected. Farmers of the study area faces problem of low productivity of the crop due to use of traditional method of farming system and use of low productive inputs.

Studies conducted in Yelmana Densa and Farta Districts of Northwestern Ethiopia (Tesfaye et al., 2001) indicate that socioeconomic, institutional and technical factors are accountable for determining technology adoption. However, these recommendations are location specific and would justify the need for research elsewhere. It is expected that geographical and climatic differences would affect the adoption decision of farmers and studies done elsewhere may not be of direct relevance to address the problems and opportunities of the present study area. Therefore it is relevant to examine the specific factors that affect the adoption of improved wheat variety by farmers of study area. This information is expected to make easy the distribution of the improved wheat technologies in the study area and suggest interventions that may help improve the efficiency of agricultural research and extension in promoting smallholder, climate risk-prone agriculture in wheat production.

The main objective of this study was to analyze factors affecting allocation of land for improved wheat variety by smallholder farmers of the study area.

METHODOLOGY

Description of the study area

This study was conducted in Northern Ethiopia, rural Adwa district. Adwa is found about 1006 kilometers from Addis Ababa and 223 kilometers away from Mekelle. The district has total area coverage of 66,618 ha of which 13,714 ha is cultivated land. The geographical structure of the district is both low land and semi-low land. About 32.2 and 67.8% of the cultivated land is found in the low land and semi-low land respectively. The district has a total household of 24,692 and has a total population of 108,647, out of which 54,659 were females and the rest of 53,988 were males. The average temperature of the area is 27°C and average annual rainfall ranges from 600 to 850 mm. The main economic activity of the study area includes both crop and livestock production. Some of the major crops grown in the area include teff, wheat, barley, finger millet, sorghum and maize and the major livestock production includes cattle, sheep, goat, donkey and poultry.

Data collection

The study uses both primary and secondary sources of data. The primary data was collected through individual interviews of the selected respondents whereas the secondary data was gathered from annual and monthly report of district agriculture Office and reports from the center statistical agency. During sampling process two-stage sampling procedure was used to select sample farmers that were included in the study. In the first stage, out of the total 18 *peasant associations* of the district four *peasant associations* were selected purposively based on their wheat production performance. In the second stage, from the selected *peasant associations,* 160 respondents were identified based on probability proportional to size of households of each *peasant associations* and the subsequent application of random sampling technique. After the sampling process was completed data were collected by using formal and informal survey methods of data collection.

Data analysis

In this study both descriptive statistics and econometric models were utilized to assess the relationship between explanatory and dependent variables. Descriptive statistics involving mean, percentage and standard deviations was used to assess the socio-economic characteristics of the sample households and farmer's response for adoption of improved wheat technologies and the type and distribution of improved wheat variety among the farmers of the study area. Also, t-test and χ^2-test were employed to assess the relationship among the variables of interest. For the econometrics model Tobit model was used to analyze factors affecting the farmer's decision to allocate land for improved wheat variety and the intensity of adoption by farmers. In Tobit model, decisions whether to adopt or not and how much to adopt are assumed to be made jointly and hence the factors affecting the two level decisions were taken simultaneously (Solomon et al., 2010). As stated in Gujarati (2004) the Tobit model to estimate the factor affecting the adoption was defined as:

$$\begin{cases} y_i = y* = X_i\beta + u_i \quad \text{if} \quad \mathrm{y_i}^* > 0 \ \mathrm{u_i} \approx \mathrm{n}(0, \sigma^2) \\ \mathrm{y_i} = 0 \ \text{otherwise} \end{cases} \tag{1}$$

Where: y_i = land size allocated for improved wheat variety at a given level of Xi; y^* = unobserved latent variable, n = number of observations; Xi = vector of explanatory variables; β = vector of unknown coefficients (parameter to be estimated); and U_i = independently and normally distributed error term with zero mean and constant variance σ^2.

The model parameter was estimated by maximizing the Tobit likelihood function of the following:

$$L = \prod_{Y_i^*>0} \frac{1}{\sigma} f\left(\frac{Y_i - \beta_i X_i}{\sigma}\right) \prod_{Y_i^*\leq 0} F\left(\frac{-\beta_i X_i}{\sigma}\right) \tag{2}$$

Where; f and F are respectively, the density function and

Table 1. Description of independent variables.

Variables	Nature of the variable	Unit of measurement	Expected sign
Age of household head	Continuous	Years	+
Education level of household heads	Continuous	Year of formal schooling	+
Sex of household head	Dummy	Male/female	Male adopt more than female
Sizes of land holding of household	Continuous	Hectare	+
Frequency of contact with extension agents	Continuous	Number of visit farmer's land by development agents per month	+
Access to credit facility	Dummy	Yes/Not	+
Distance from market	Continuous	Kilometer	-
Distance to the main road	Continuous	Kilometer	-
Family size in adult-equivalent	Continuous	Number of adult-equivalent	+
Livestock holding (TLU):	Continuous	Number of TLU	+
Perception of farmers about cost of technology	Dummy	Ordinal variable	-
Perception of farmers about yield of improved wheat technologies	Dummy	Ordinal variable	+
Participation of the household head in leadership position	Dummy	Yes/no	+

cumulative distribution function (Maddala, 2005).

The marginal effect of an explanatory variable on the expected value (mean proportion) of the dependent variable was estimate by:

$$\frac{\partial E(Y_i)}{\partial Xi} = F(z)\beta_i \quad (3)$$

Where z is defined by:

$$\frac{\beta ixi}{\sigma}$$

The change in the probability of adopting improved wheat technology as independent variable X_i changes was estimate by:

$$\frac{\partial F(z)}{\partial X_i} = f(z)\frac{\beta_i}{\sigma} \quad (4)$$

Where, $z = X\frac{\beta}{\sigma}$, F (z) is the cumulative distribution function, f (z) is the value of derivative of the normal curve at a given point, z is the Z-score for the area under normal curve, ß is a vector of Tobit maximum likelihood estimates and σ is the standard error of the error terms. Similarly, the change in intensity of adoption with respect to change in an explanatory variable among adopters was estimated by:

$$\frac{\partial E(Y|Y_i^* > 0)}{\partial X_i} = \beta\left[1 - z\frac{f(z)}{F(z)} - \left(\frac{f(z)}{F(z)}\right)^2\right] \quad (5)$$

In this study the dependent variable was the land size allocated for the production of improved wheat varieties. Whereas the independent variables that were expected to affect the dependent variable with their unit of measurement and expected sign are presented in Table 1.

RESULTS AND DISCUSSION

Socio-economic characteristics of sample households

The descriptive statistics of some selected socio-economic characteristics of sample farmers examined in this study are presented in Tables 2 and 3. Table 2 presents for continuous variables whereas Table 3 presents for dummy variable. Out of the total sample respondents 118 were adopters and 42 were non-adopters.

Table 2 shows the result of descriptive statistics for continuous variables. As shown from the table, t-value was computed for all continuous variables and it was found to be statistically significant for family size in adult equivalent, education level, TLU, Average extension contact per month and farm size at 1% level of significance. This implies that there was significant difference in all these variables between the two categories (adopters and non-adopters).

Table 3 shows the result of descriptive statistics for dummy variables. The chi-square test was computed for the dummy variables and it was found to be statistically significant for credit access, Perception of farmers towards yield and cost of the variety at significance level

Table 2. Descriptive statistics of some selected continuous variables.

Variables	Adopters		Non adopters		Total		t-value
	Mean	Std.dev	Mean	Std.dev	Mean	Std.dev	
Age	48.89	9.93	47.45	11.66	48.51	10.40	0.769(NS)
Family size in adult equivalent	3.89	0.99	2.54	0.63	3.54	1.09	8.282***
Education level of household head in year	2.96	2.59	1.4	2.64	2.29	2.64	5.937***
Farm experience in year	26.5	10.69	24.7	12.9	26	11.29	0.883(NS)
TLU	4.44	0.64	3.21	0.48	4.13	0.81	11.453***
Average extension contact per month	2.5	0.95	1	0.74	2	1.18	10.669***
Land holding	0.71	0.3	0.52	0.17	0.65	0.30	5.53***

Source; Own computed result, 2011. ***, Significant at 1%; **, significant at 5%; NS, not significant.

Table 3. Descriptive statistics of some selected dummy variables.

Variables		Adopters		Non adopters		Total		χ^2
Sex		No.	%	No.	%	No.	%	
Sex	Male	96	81.4	35	83.3	131	81.9	4.738(NS)
	Female	22	18.6	7	16.7	29	18.1	
Participation in leader ship activity	Yes	50	42.37	15	35.71	65	40.63	5.134(NS)
	No	68	57.63	27	64.29	95	59.37	
Access to credit service	Yes	110	93.2	9	2.38	119	74.38	12.79***
	No	8	6.8	33	97.62	41	25.62	
Perception towards yield	Low	1	0.8	2	4.8	3	1.9	34.435***
	Medium	15	12.7	22	52.4	37	23.1	
	High	102	86.4	18	42.9	120	75	
Perception towards cost	Cheap	30	25.4	-	-	30	18.8	16.93***
	Medium	46	39	-	-	46	28.8	
	Expensive	42	35.6	42	100	84	52.4	

Source; Own computational result, 2011; NS, Not significant; ***, significant at 1%.

Table 4. Category of respondents based on type of improved wheat Varieties used and total land allocated for improved wheat.

Description	Number of respondents	%
Variety type		
HAR1685	79	66.95
HAR1686	32	27.12
HAR2501	7	5.93
Total	118	100.0
Land size in hectare		
<0.25	76	64.4
0.25	39	33.1
>0.25	3	2.5

Sources: Own computational result, 2011.

of 1%. This indicated that there was systematic difference in these variables between the two categories.

Category of respondents based on type of improved seed they used and allocation of land for improved seed

The types of improved wheat varieties distributed to farmers of the study area were HAR1685, HAR2501 and HAR 1686. From Table 4; out of the total adopters 79(66.95%) of them were user of HAR1685, 32 (27.12%) were used HAR1686 and the remaining 7 (5.93%) were user of the variety type of HAR2501. Since most farmers of the study area had problem of land shortage; the proportion of land allocated for improved variety was very small. From Table 4, 64.4% of adopters allocated less than 0.25 ha, 33.1% of them allocated 0.25 ha and 2.5% of them allocated greater than 0.25 ha of land to the improved wheat variety.

Table 5. Maximum likelihood estimates of Tobit model of adoption of improved wheat variety.

Land	Coefficient	Robust Std. Err.	t-Value
Sex of house hold head	0.087	0.008	1.13
Educational level of household head	0.064	0.001	5.22***
Participation of HH head in leadership activity	0.065	0.006	1.12
Farming experience in year	1.94e-06	0.000	0.01
Family size in adult-equivalent	0.128	0.005	2.65***
Sizes of land holding of HH	0.132	0.012	1.13
Distance to main road	-0.082	0.003	-2.64***
TLU	0.304	0.007	4.42***
Access to credit facility	0.563	0.011	4.90***
Frequency of contact with extension agents/month	0.012	0.004	3.38***
Perception of HH about yield of the variety	0.012	0.009	1.34
Perception of HH about cost of the technology	-0.015	0.005	-3.07***
Distance from market	-0.033	0.002	-2.44**
CONSTANT	-0.108	0.044	-2.45**
Numbers considered	160		
Log likelihood function	226.81403		
Lift censored	0		
Right censored	+infinity		

Source: Computed from the field survey data 2011; TLU, Tropical Livestock Unit, HH, household. ***,** significant at 1 and 5% respectively.

Table 6. Marginal effect of explanatory variables on use of improved wheat variety.

Variable	Change in probabilities as independent variable changes $\frac{\partial F(z)}{\partial X_i}$	Change among entire sample $\frac{\partial E(Y_i)}{\partial Xi}$	Change among adopters $\frac{\partial E(Y \vert Y_i^* > 0)}{\partial X_i}$
Educational level of household head	0.117	0.051	0.049
Family size in adult equivalent	0.023	0.13	0.098
Distance from main road market	-0.015	-0.064	-0.063
TLU	0.056	0.025	0.021
Access to credit facility	0.103	0.513	0.425
Frequency of contact with extension agents/month	0.022	0.126	0.092
Perception of HH about cost technology	-0.027	-0.014	-0.012
Distance from market	-0.060	-0.028	-0.025

Source: Computed from the field survey data 2011; TLU, Tropical livestock unit; HH, household.

Determinants of adoption of improved wheat variety

It is well known that adoption of improved technologies depends on different socio-economic, demographic and institutional factors. Different variables are important across different space and time in explaining adoption of new technologies. For this study thirteen (5 discrete and 8 continuous) variables were hypothesized to influence the adoption of improved wheat variety in the study area. Among these, eight of the explanatory variables were found to be statistically significant in explaining the status and intensity of adoption of improved wheat technology in the study area. The estimated results of the Tobit model in Tables 5 and 6 showed that the significant variables affecting use of improved wheat variety in the study area include; educational level of household head, family size in adult-equivalent, distance to main road, tropical livestock units, access to credit facility, frequency of contact with extension agents, perception of household towards cost of the technology and distance from the nearest market.

Educational level of household head

As expected educational level of household head was affected the adoption decision of farmer of the study area positively and significantly at less than one percent significance level (Table 5). This result is similar with studies by Nzomoi et al. (2007) education of household head affects positively and significantly adoptions of production of horticultural export produce. Also the studies by Ozor and Madukwe (2005), Motuma et al. (2010) and Isaiah et al. (2007) confirmed similar results. Results of analysis of marginal effect show that an increase in the level of education by one year increases the probability of being an adopter by 11.7% and it increases the level of adoption by 0.049 and by 0.051 among the adopters and the total sample in that order (Table 6). This implies that having high formal year of education increases the level of adoption of new agricultural technology by farmers. Farmers who have higher formal year of education are expected to analyze information and adopt earlier than the uneducated persons; because farmers with higher education level are eager to grasp new ideas and to try the technology by allocating some proportion of their land. Hence education level and adoption have positive relationship.

Family size in adult-equivalent

As expected family size in adult-equivalent affects the adoption decision of farmers of the study area positively and significantly at 1% (t=2.65) significance level (Table 5). A unit increase in family size increases the probability of adoption by 2.34% whereas it increases the level of adoption among adopters and the total sample by 0.098 and 0.13 respectively (Table 6). This result is consistent with the study on adoption of improved maize seed by Motuma et al. (2010). From this; household with high number of family size in adult-equivalent adopts more agricultural technology (improved wheat technology) than households with low number of family size. This could be because households with high number of family size can undertake the agricultural activity in time and effectively manage the wheat fields. On the other hand, the increase in number of family members would urge the families to look for high productivity and return options to meet the demand for food and expenditure. These scenarios would increase the adoption of improved wheat technologies providing better options to meet the pressing demand.

Distance from the main road

This variable affects adoption decision of farmers negatively and significantly at 1% (t = -2.64) (Table 4). This result agrees with the study by Isaiah et al. (2007); according to his study; accesses to means of transportation affect positively adoption of improved barley varieties. A unit increase in distance from home to main road in kilometer decreases the probability of adoption in favour of adopters by 1.5% and it increases level of adoption by 0.063 among adopters and by 0.064 among the whole sample (Table 5). This implies that farmers near the main road adopts more than farmers away from the main road. Farmers near the road can get transportation facility easily and they can transport the improved wheat variety easily and at low cost than the other farmers.

Tropical livestock unite (TLU)

As expected TLU affects the adoption level of farmers positively and significantly at 1% (t = 4.42) level of significance (Table 4). A unit increase in TLU increases the probability of adoption by 5.56% and increases level of adoption by 0.021 and 0.025 among adopters and among the total sample respectively (Table 5). This implies that being owner of more livestock increase the level of adoption of improved agricultural technology. Livestock increases household income from sale of animals and farmers can finance their agricultural requirement easily from their livestock income. The study by Solomon et al. (2011) confirms this result. According to his study TLU affects adoption of agricultural technology positively and significantly.

Access to formal credit facility

As expected this institutional factor affected adoption level of improved wheat variety by farmers of the study area positively and significantly at significance level of 1% (t = 4.9) (Table 5). Results of analysis of marginal effects; show that having access to credit service increases the probability of being an adopter by 10.3% and it increases level of adoption by 0.42 and 0.51 among adopters and the total sample respectively (Table 6).

The reason behind is that most farmers of the study area suffers from shortage of money to purchase improved agricultural inputs and it force them to use the input what they have on hand; which is the local one. But having access to credit facility solves such type of problem and farmers can purchase the improved input. According to Namwata et al. (2010) access to credit facility affect adoption of improved agricultural technology for Irish potatoes positively and significantly. Also studies by Isaiah et al. (2007), Motuma et al. (2010) and Odoemenem and Obinne (2010) confirmed similar results.

Frequency of contact with extension agents

This variable represents the number in which extension agents visit farmer's field of production per month. As expected this institutional factor affects adoption of improved wheat variety of farmers of the study area positively and significantly at 1% significance level (t=3.38) (Table 5). According to Namwata et al. (2010) extension contact was affected adoption of improved agricultural technology for Irish potatoes positively and significantly. And also according to the study by Isaiah et al. (2007), Solomon et al. (2011), Ayinde et al. (2010), Odoemenem and Obinne (2010) and Matata et al. (2010) frequency of contact with extension agent affect positively and significantly adoption decision of farmers for improved agricultural technology. From the analysis of marginal effects a unit increase in frequency of contact with extension agent increases the probability of being an adopter by 2.18% and it increases level of adoption by 0.092 and 0.126 among adopters and the entire sample respectively (Table 6). This implies that contact with extension agent increases availability of information about the improved technologies to farmers. Farmers can learn more about the technology. Hence farmers with more contact with extension agents adopt more than farmers with less contact.

Perception of households about the cost of the technology

As expected this variable affects the adoption decision of farmers negatively and it was statistically significant at 1% (t = -3.07) level of significance (Table 5). From the analysis of marginal effect perceiving the cost of technology in high as compared with the local one decreases the probability of adoption of the improved wheat variety by 2.73% and it decreases adoption level of the technology by 0.0123 and 0.0145 for adopters and for the entire sample respectively (Table 5).

Distance from nearest market

This variable affects adoption decision of farmers negatively and significantly at 5% level of significance (Table 4). The study by Solomon et al. (2011) was consistent with this result; distance from nearest market affects adoption of improved agricultural technology negatively and significantly. And also the studies by Isaiah et al. (2007) and Mesfin (2005) were consistent with this result. Table 5 shows a unit increase in distance of the nearest market from farmers home decreases the probability of adoption by 6% and it decreases the level of adoption by 0.025 among adopters and by 0.028 among the entire sample. This implies that farmers nearest to market can get and buy the technology without any difficulties and it decreases transportation and marketing cost, in addition to this farmers nearest to market are nearest to any market information than farmers away from the market and they have updated market information.

CONCLUSION AND POLICY IMPLICATION

The use of improved variety is considered as the most important input for the achievement of increased agricultural productivity and food security status of farm households in Ethiopia. However, adoption of improved variety remains very low, especially among small-scale farmers of the country. The results of this study showed that variables like access to credit facility, family size in adult-equivalent, TLU, extension contact and education level of household head affect adoption of improved wheat variety positively and significantly. Whereas variables like distance from nearest market and main road and perception of households about cost of the technology affects adoption of improved wheat variety negatively and significantly.

The fact that access to extension service affect adoption of improved wheat variety positively and significantly; implies the important role the extension personnel played in order to impact farmers' attitude and enhance farmers' awareness on the benefit of improved wheat technology. This in turn implies the need for advancing farmers perception on the use and advantage of improved wheat technology to increase the sustainable food production. Therefore, the government and other stakeholders should encourage access to extension agents to enhance dissemination of improved wheat varieties among the farmers through workshops, seminars, trainings and pertinent demonstration activities.

Formal credit service had been found as one of the important factors affecting the adoption of improved wheat variety. As credit service should provide better ground for making improved decision to access improved inputs particularly those unaffordable to smallholder farmers through its effect of reducing the existing cash constraint for undertaking agricultural decisions and accessing high value inputs. Therefore it is recommended that credit service should be made available to farmers at an affordable rate to increase better and wider adoption of improved wheat technologies.

Distance from farmers' home to main road was an important variable which affects adoption of improved wheat variety negatively and significantly. Hence attention should be given to expand the road infrastructure in the rural area to increase farmer's access to transportation facility and decrease transaction cost to get better access to improved agricultural technologies. Perception of household towards cost of technology was significant variable which affect the adoption of improved wheat variety. Information about the benefits of new technology should be given for farmers to increase farmer's awareness about the technology and to develop farmer's attitude towards the technology. However, since in most cases cost is associated with the existing scenario of lack of capacity to buy the technologies, any agricultural development effort advocating the adoption of improved technologies should consider an enabling environment such as access to credit.

Conflict of Interest

The authors have not declared any conflict of interest.

ACKNOWLEDGEMENTS

The authors express their sincere thanks to Wolaita Sodo University (minister of education Ethiopian) for their financial support in conducting the field survey. They are also grateful to the farmers of the study area for kindly providing the necessary information used in the study.

REFERENCES

Ayinde OE, Adewumi MO, Olatunji GB, Babalola OA (2010). Determinants of Adoption of Downy Mildew Resistant Maize by Small-Scale Farmers in Kwara State, Nigeria. Glob. J. Sci. Front. Res. 10(1):32-35.

Bingxin Y, José F, Sinafikeh A (2011). Cereal Production and Technology Adoption in Ethiopia. Development Strategy and Governance Division. Int. Food Policy Res. Inst. 36p.

Gujarati DN (2004). Basic Econometrics, 4rd edition. McGraw-Hill, Companies, New York.1002p.

Hailu G (2003). Wheat Production and Research in Ethiopia. Addis Ababa, Ethiopia. 12p.

Isaiah KO, Fred UN, Washington WO (2007). Socio-Economic Determinants of Adoption of Improved Sorghum Varieties and Technologies Among Smallholder Farmers in Western Kenya. 18p.

Maddala GS (2005). Introduction to Econometrics.3rd edition. Formerly of Ohio State University Press, New York.

Matata PZ, Ajay OC, Oduol PA, Agumya A (2010). Socio-Economic Factors Influencing Adoption of Improved Fallow Practices Among Smallholder Farmers in Western Tanzania. Afr. J. Agric. Res. 5(8):818-823.

Mesfin A (2005). Analysis of Factors Influencing Adoption of Triticale (*X -Triticosecale Wittmack*) and Its Impact: The Case of Farta District. An M.Sc. Thesis Presented to the School of Graduate Studies of Haramaya University 112p.

Motuma T, Dejene A, Wondwossen T, Roberto LR, Girma TMW, Germano M (2010). Adoption and Continued Use of Improved Maize Seeds: Case Study of Central Ethiopia. Afr. J. Agric. Res. 5(17):2350-2358.

Namwata BML, Lwelamira J, Mzirai OB (2010). Adoption of Improved Agricultural Technologies for Irish Potatoes (Solanum Tuberosum) Among Farmers in Mbeya Rural District, Tanzania: a Case of Ilungu Ward. J. Anim. Plant Sci. 8(1):927-935.

Nzomoi JN, Byaruhanga JK, Maritim HK, Omboto PI (2007). Determinants of Technology Adoption in the Production of Horticultural Export Produce in Kenya. Afr. J. Bus. Manage. 1(5):129-135.

Odoemenem IU, Obinne CPO (2010). Assessing the factors Influencing the Utilization of Improved Cereal Crop Production Technologies by Smallscale Farmers in Nigeria. "Agriculture in Nigerian. Indian Society for Education and Environment. http://www.indjst.org Indian J. Sci. Technol. 3:1.

Ozor N, Madukwe MC (2005). Obstacles to the Adoption of Improved Rabbit Technologies by Small Scale Farmers in Nsukka Local Government Area of Enugu State. J. Agric. Food Environ. Ext. 4(1):70-73.

Samia Z (2002). Innovative and Successful Technical Experience in the Production of Agricultural Statistics and Food Security of Ethiopia. Contributing Paper Presented at a Seminar on a New Partnership to Strengthen Agricultural and Rural Statistics in Africa for Poverty Reduction and Food Security, Paris, France 16-17. Central Statistics Authority. Addis Ababa, Ethiopia.

Solomon A, Bekele S, Franklin S (2010). Does Technology Adoption Promote Commercialization? Evidence from Chickpea Technologies in Ethiopia. 27p.

Solomon A, Bekele S, Franklin S, Mekbib G (2011). Agricultural Technology Adoption, Seed Access Constraints and Commercialization in Ethiopia. J. Dev. Agric. Econ. 3(9):436-447.

Tesfaye Z, Girma T, Tanner D, Verkuijl H, Aklilu A, Wilfred M (2001). Adoption of Improved Bread Wheat Varieties and Inorganic Fertilizer by Small-scale Farmers in Yelmana Densa and Farta Districts of Northwestern Ethiopia. Ethiopian Agric. Res. Org. 42p.

World Bank (2008). Agriculture for Development, World Development Report, Washington, DC. www.worldbank.org

26

Bio-economic modelling of the influence of family planning, land consolidation and soil erosion on farm production and food security in Rwanda

J. C. Bidogeza[1,2], P. B. M. Berentsen[2], J. De Graaff[3] and A. G. J. M. Oude Lansink[2]

[1]AVRDC – The World Vegetable Center, West and Central Africa, Yaounde – Messa, 2008, Cameroon.
[2]Business Economics / Wageningen University, Netherlands.
[3]Land Degradation and Development / Wageningen University, Netherlands.

Rwandan agriculture is not able to meet its population's food needs from its own production, which results in food insecurity. Land degradation is a serious problem which contributes to a low and declining agricultural productivity and consequently to food insecurity. The objective of this paper is to develop a bio-economic model capable of analysing the impacts of soil erosion, family planning and land consolidation policies on food security in Rwanda. The results of the bio-economic model show that a higher availability of good farm land would increase the farm income. Additionally, preserving soils against erosion and reducing risk would allow for releasing more marginal land which would increase food production for home consumption and for the market. Increasing the opportunities for off-farm employment can also increase farm household income. The outcomes of the model support the Rwanda policy on family planning, while the policy on land consolidation is not endorsed.

Key words: Rwanda, land degradation, food security, bioeconomic model, family planning policy, land consolidation policy.

INTRODUCTION

Agricultural statistics indicate that per capita food production in Rwanda is declining (Minecofin, 2003a; RADA, 2005; NISR, 2008). This trend is putting at stake the food security of the rural and urban poor. Rwandan agriculture is not able to meet its population's food needs with the national production.

Land degradation is a serious problem which contributes to the low and declining agricultural productivity and consequently to food insecurity. Land degradation can be defined in terms of loss of actual or potential productivity as a result of natural or human factors (Anecksamphant et al., 1999). Soil erosion and soil mining are believed to be the most important causes of land degradation in Rwanda with a soil loss of 50 to 400 tons per hectare per year depending on location (Mugabo, 2005). Some slopes are totally degraded by erosion and no production is possible without restoring fertility. In addition, Rwandan soils have a very low organic matter content and weak soil fertility potential except for the marshy and volcanic soils (Gecad, 2004). Furthermore, land scarcity due to the high population density is limiting the option to extend agricultural land size. In Rwanda, the biophysical causes of land degradation are relatively well known, but less is known about the economic impact of land degradation on farming activities. Very little modelling analysis exists at farm level on the economic consequences of land degradation (Byiringiro and Reardon, 1996; Clay et al., 1998; Musahara, 2006).

Rwanda's population, which is made up mostly of subsistence farmers, has quadrupled during the last 50 years. At present, Rwanda has 9.3 million inhabitants with a density of 380 inhabitants/km^2. The average size of a family farm is 0.76 ha (Minagri, 2004). If the human reproduction rates are not slowing down, the population will double by 2030 (Kinzer, 2007), with dramatic consequences for natural resources and food security. Thus, it is important to balance the increasing population with the limited available land, and ensure food security.

The new land law put in place by the Rwandese government stipulates that, under its article 20, landholdings less than one hectare (ha) are deemed insufficient for effective and efficient agricultural exploitation (Minerena, 2005). Therefore, the Rwanda government prepared to use the land law as one of the drivers of agricultural reform, notably through the provision on land consolidation and minimum land holdings. The farm households whose land is less that 1 ha would have difficulties to register their land (Huggins, 2012). The land law and land policy tend to stimulate farm households whose landholdings are less than one hectare to consolidate their land, but those who are reluctant to comply to the land law and land policy are vulnerable to confiscation of their land (Huggins, 2012; Pottier, 2006). This ruling follows a recommendation made by the Poverty Reduction Strategy paper (Minecofin, 2003b): "households will be encouraged to consolidate plots in order to ensure that each holding is not less than 1 ha. This will be achieved by the family cultivating in common rather than fragmenting the plot through inheritance".

Decisions on land use are basically made by heads of farm households. As in many other developing countries, a farm household system in Rwanda concerns production (of crops and livestock), off-farm activities and consumption (of food, other basic needs and some leisure). A major characteristic is the non-separability of production and consumption decisions. The allocation of productive resources and the choice of activities could affect land degradation and subsequently food security. It is assumed that farm households are rational in pursuing certain meaningful objectives which guide their behaviour (Upton, 1996; Anderson, 2002; Woelcke, 2006; Laborte et al. 2007; Laborte et al., 2009). However, the decision-making process is restricted by the range of possible alternative activities that can be undertaken by farm households and constraints imposed by limited resources availability and other external conditions like agricultural and/or environment policies (Senthilkumar et. al, 2011).

To understand the complex relations at farm level between technical, ecological and economic components, there is a need to combine information from biophysical and social sciences (Kruseman, 2000). Bio-economic modelling is at the interface of biophysical and social sciences, enabling the accommodation of biophysical data in economic analysis (Kanellopoulos et al., 2010; Louhichi et al., 2010).

In developing countries, many studies have made use of bio-economic farm models and there is growing interest for its application (Jansen and Van Ittersum, 2007). However, little modelling analysis at farm household has been conducted in subsistence or semi-subsistence farming. Barbier (1990), Cárcamo et al. (1994), Barbier and Bergeron (1999) and Louhichi et al. (1999) evaluated the economic nature of land degradation and estimated net returns from erosion control. Van Keulen et al. (1998), Kruseman and Bade (1998), Kuyvenhoven et al. (1998), Ruben et al. (1998), Struif Bontkes and Van Keulen (2003) assessed different sustainable technologies to improve farm household income and soil fertility. Dorward (1999) investigated the conditions under which peasant farm household models may need to allow embedded risk. Anderson (2002), Mudhara et al. (2002), Thangata et al. (2002) examined the options for improving household food security for small-scale farms.

Modelling farm households might bring some insights into the ongoing debate on land and family planning reforms and the potential impacts of soil erosion. So far no modelling studies in sub-Saharan countries have incorporated at the same time soil erosion, soil fertility, soil quality and food consumption in terms of energy and proteins, risk, labour, land, cash and credit availability in their economic evaluation of crop production for farms.

The objectives of this paper are:i) to develop a general bio-economic model capable of analysing the impacts of family planning, land consolidation and soil erosion on farm production and food security in Rwanda; ii) to apply the bio-economic model for a typical farm in Rwanda.

The remainder of this paper is structured as follows. The next section describes the study area and the farm household model. Next, data and application of the model for a typical farm are presented. This typical farm household has available resources that are the average of farm types distinguished in (Bidogeza et al., 2009).

This is followed by the presentation of the modelling results regarding food security, technical and economic results for the typical farm. The outcomes of the farm household model are compared with observed farm household data; and the effects of family and land size changes on food security, income and soil loss results are determined and discussed. Thereafter follow the conclusions.

MATERIALS AND METHODS

Area of study and typical farm

The area of study is in Umutara, a former province located in the eastern part of Rwanda, approximately 180 km from Kigali along the main tarmac road between Kigali and Kagitumba (border with Uganda). It has a border with two countries, Uganda in the north, and Tanzania in the southeast. The tarmac road and the geographical position of Umutara imply that the market access is fairly good.

Most inhabitants of Umutara are former refugees who arrived from Tanzania and Uganda after the genocide which ended in 1994. When they returned to Rwanda, Umutara was chosen for their resettlement. The increasing population puts a high pressure on natural resources of the province, and different land uses often compete for the same piece of land.

Umutara province belongs almost entirely to the agro-climatic zone of the Central Bugesera and the Savannahs of the East, which is the driest agro-climatic region of Rwanda. The annual precipitation is quite variable in the region and is on average lower than 1000 mm (Sirven et al., 1974). The irregularity of the precipitation is a frequently stated problem for Umutara. The climate of Umutara is bimodal (Fleskens, 2007), with two growing seasons annually. The agricultural activities for one season referred to as B last from January to June, and agricultural activities for the other season referred to as A take place from July to December.

The pedology of Umutara is quite diverse, notwithstanding that it is only a small area. Two types of soils are dominant in Umutara: Inceptisols and Oxisols (USDA, 1999), mostly located on gentle (2-6%) and moderate (6-13%) slopes, respectively. These land types are covering 60% of the total soil in Umutara province, respectively 40% for Oxisols and 20% for Inceptisols (GhentUniversity, 2002). The chemical fertility of Oxisols is poor; weathered minerals and cations retention by mineral soil fraction is weak, while Inceptisols have a satisfactory chemical fertility and contain at least some weathered minerals in silt and sand fraction (FAO, 2001). Despite of the low fertility of the soils, small-scale farmers maintain soil fertility and reduce soil erosion by using low input systems such as crop rotations, organic fertilisers and few of them also use some chemical fertilisers. However, these land management strategies are not suficient for a sustainable farming.

With respect to the importance of the different crops cultivated in the region: 33% of the cultivated land is occupied by cereals, followed by tubers (29%), leguminous crops (21%) and bananas (15%) (Minagri, 2002).

The farm household analysed in this paper is typical for the province. Important socio-economic variables used to characterise the typical farm household were average farm data at regional or national level derived from the literature and field survey (Kinzer, 2007; Loveridge et al., 2007; Strode et al., 2007; Ansoms and McKay, 2010).

Model specification and data used

General structure

The basic structure of the bioeconomic farm household model is shown in Equation (1). It has the mathematical form of a quadratic programming model (Hazell and Norton, 1986):

$$\text{Maximise } \{Z = c'x - \text{Ø}\,\sigma\}$$
$$\text{Subject to } Ax \leq b$$
$$\text{and } x \geq 0 \qquad (1)$$

where: Z = expected utility; c = vector of gross margins, costs or revenues per unit of activity; x = vector of activities; A = matrix of technical coefficients; b = vector of resource availabilities; Ø = risk aversion coefficient (Ø >0); σ = standard deviation of totalgross margin.

The model presented here is a quadratic programming model with a time span of one year (two seasons). The expected utility is the objective function and this is maximized. The farmer is assumed to maximise expected utility which is defined as discretionary income minus the risk premium. Discretionary income is defined as income available for spending after essential expenses have been made (Castano, 2001; Laborte et al., 2009). The most important essentials include clothes, taxes, medication, school fees, kitchen ustensils and food ingredients.

Activities include crop production for home consumption, crop production for sale, off-farm activities, hiring labour, family expenditures, borrowing credit. Major constraints include land, labour in three different periods per season, rotations, available cash, maximum credit, food consumption requirements, soil loss and soil organic matter.

The major activities and constraints are summarized bythe Equations (2) to (14). For the description of the indices, coefficients and variables see Tables 1, 2 and 3, respectively.

$$c'x = \begin{array}{l} \sum_{c,se,lu}(Y_{c,se,lu} * vL^{m}_{c,se,lu}) * Pr_{c,se} - \sum_{c,se,lu}[(Cs_{c,se,lu} * vL^{m}_{c,se,lu}) + (Cs_{c,se,lu} * vL^{c}_{c,se,lu})] - \\ (\sum_{pe,se} vHlab_{pe,se} * Wage) + (\sum_{pe,se} vOfflab_{pe,se} * Wage) - \sum_{pe,se}(Exp_{pe,se}) \end{array} - vI \qquad (2)$$

The discretionary income per year is defined as returns from the sale of crops production ($\sum_{c,se,lu}(Y_{c,se,lu} * vL^{m}_{c,se,lu}) * Pr_{c,se}$) plus wages from off-farm activities ($\sum_{pe,se} vOfflab_{pe,se} * Wage$) minus costs of seeds/establishment costs ($\sum_{c,se,lu}[(Cs_{c,se,lu} * vL^{m}_{c,se,lu}) + (Cs_{c,se,lu} * vL^{c}_{c,se,lu})]$) and costs of hired labour ($\sum_{pe,se} vHlab_{pe,se} * Wage$) and expenditures. The standard deviation for total gross margin is calculated from the variance/covariance matrix of gross margins for the crops per

Table 1. Indices used in the farm household model.

Index	Description	Elements
C	Crop	Banana, beans, cassava, groundnut, maize, sorghum, sweet potato
Leg	Leguminous	Beans, groundnut
Len	Non leguminous	Cassava, maize, sorghum, sweet potato
Lu	Land type	Inceptisols, Oxisols
Pe	Period	Periods 1, 2, 3 (in each season)
Se	Season	Season A, season B

Table 2. Coefficients used in the farm household model.

Coefficient	Description	Dimension
AVL	Available land	ha
AVlab	Available labour	man-day
Credilim	Credit limit	fr.rw
Cs	Cost of seed/establishment costs	fr.rw ha^{-1}
En	Energy content per crop	Kcal kg^{-1}
Enreq	Energy requirement	Kcal $season^{-1}$
Exp	Expenditure	fr.rw
Labreq	Labour requirement	man-day ha^{-1}
MaxOfflab	Maximum off farm labour	man-day
Pr	Price products	fr.rw kg^{-1}
Prot	Protein content per crop	g kg^{-1}
Protreq	Protein requirement	g $season^{-1}$
Ri	Rate of interest	%
Soc	Soil organic matter	t ha^{-1} $season^{-1}$
Socav	Soil organic matter available	t ha^{-1} $year^{-1}$
Soill	Soil loss	t ha^{-1} $season^{-1}$
Soilltol	Soil loss tolerance	t ha^{-1} $year^{-1}$
Totcostse	Total cost of seeds/establishment costs	fr.rw ha^{-1}
Totrev	Total returns from crop sales	fr.rw ha^{-1}
$Varcovar^{c}$	Variance /covariance matrix of Gross Margins of crops for home consumption (using constant product prices)	-
$Varcovar^{m}$	Variance /covariance matrix of Gross Margins of marketed crops.	-
Wage	Wage	fr.rw day^{-1}
Y	Yield	Kg ha^{-1}

($\sum_{pe,se}(Exp_{pe,se})$) and total interest (*vI*).

$$\sigma = \sqrt{\sum_{c,se,lu}[varcovar^{m}_{c,se} * (vL^{m}_{c,se,lu})^2 + varcovar^{c}_{c,se} * (vL^{c}_{c,se,lu})^2]} \quad (3)$$

season and the area of crops per season for consumption and for marketing, respectively.

Land constraint (for each season and land type)

$$\sum_c(vL^{m}_{c,se,lu} + vL^{c}_{c,se,lu}) \leq AVL_{se,lu} \quad (4)$$

Labour constraint (for each season and each period)

Table 3. Variables used in the farm household model.

Variable	Description	Dimension
vCach	Cash	fr.rw
vCred	Credit required	fr.rw
vHlab	Hired labour	man-day
vI	Total interest	fr.rw $year^{-1}$
vL^c	Land allocated to crop for consumption	ha
vL^m	Land allocated to crop for market	ha
vNewcred	Credit added each period	fr.rw
vOfflab	Days allocated to off farm activities	man-day
vRepay	Repayment	fr.rw
σ	Standard deviation of income	

$$C'X \sum_{c,lu}[(Labreq_{c,pe,se,lu} * vL^m_{c,se,lu}) + (Labreq_{c,pe,se,lu} * vL^c_{c,se,lu}) + vOfflab_{pe,se}] \leq vHlab_{pe,se} + AVlab_{pe,se} \quad (5)$$

$$vOfflab_{pe,se} \leq MaxOfflab_{pe,se} \quad (6)$$

Rotations constraint (for each season and each land type)

$$vL^m_{leg,se,lu} + vL^c_{leg,se,lu} = vL^m_{len,se,lu} + vL^c_{len,se,lu} \quad (7)$$

Minimum food consumption constraints (for each season)

$$\sum_{c,lu}[(Y_{c,se,lu} * vL^c_{c,se,lu}) * En_c] \geq Enreq_{se} \quad (8)$$

$$\sum_{c,lu}[(Y_{c,se,lu} * vL^c_{c,se,lu}) * Prot_c] \geq Protreq_{se} \quad (9)$$

Cash constraints (for each season and each period)

$$vCash_{pe,se} = vCred_{pe,se} + vOfflab_{pe,se} * Wage + \sum_c(Totrev_{c,se}) - vHlab_{pe,se} * Wage - Exp_{pe,se} - \sum_c(Totcostse_{c,se}) \quad (10)$$

Required credit (for each season and each period)

$$vCred_{pe,se} = vCred_{pe-1,se} * (1 + ri) - vRepay_{pe-1,se} + vNewcred_{pe,se} \quad (11)$$

Credit constraint (for each season and period)

$$\sum_{pe} vCred_{pe,se} \leq Credlim_{se} \quad (12)$$

Soil loss constraint (per year for each land type)

$$\sum_{c,se}[(Soill_{c,se,lu} * vL^c_{c,se,lu}) + (Soill_{c,se,lu} * vL^m_{c,se,lu})] \leq Soilltol_{lu} \quad (13)$$

Soil fertility constraint (per year for each land type)

$$\sum_{c,se}[(Soc_{c,se,lu} * vL^c_{c,se,lu}) + (Soc_{c,se,lu} * vL^m_{c,se,lu})] \leq Socav_{lu} \quad (14)$$

The software used for optimization of the quadratic programming farm household model is General Algebraic Modelling System, version 22.6 (GAMS) with the solver CONOPT.

Sources of data used

In 2004 and 2005 data were collected in Umutara province by the National Institute of Statistics of Rwanda, in the framework of a national agricultural farm survey held twice annually. This farm survey database can be obtained from the authors upon request. In addition, a small survey was conducted in October, November and December 2007 through interviews in order to collect information supplementary to the national farm survey. For the latter survey, farm households were asked questions about family expenditure and income, crops and rotations, production costs and output prices, labour use and costs, market availability. Supplementary information related to coefficients of the current farming were estimated from literature (MCDF, 1984; Birasa et al., 1990; Minagri, 1991; Ghent university, 2002; CPR, 2002; Minagri, 2002; Zaongo et al., 2002; Van Ranst, 2003; CIRAD, 2004 and Minagri, 2006). These coefficients are estimated under low input systems. Low inputs are defined as no significant use of purchased inputs such as artificial fertilizers, improved seeds, pesticides or equipment. Input and output prices in the region were derived from the database on the market prices list provided by the Minagri (2007). Data to generate many of the coefficients for soil characteristics of the region were obtained from the natural resource database hosted by the "Carte Pedologique" Unit at the Ministry of Agriculture (Birasa et al., 1990).

Activities

Farm household activities consist mainly of crop production, off-farm activities and hiring in labour or working as farm labour on other farms. Livestock is not a major activity for the farm type

Table 4. Input-output information on crops for Oxisols.

Season		A	A	A	A	A	A&B	A&B
Crop activities	**Unit**	**Beans**	**Groundnuts**	**Maize**	**Sorghum**	**Sweet potatoes**	**Cassava**	**Banana**
Inputs-output								
Yield[1]	kg ha^{-1}	670	500	990	1,050	6,300	6,000	15,000
Price[2]	fr.rw kg^{-1}	141	419	100	106	90	177	83 (98)[a]
Revenues	fr.rw ha^{-1}	94,470	209,500	99,000	111,300	567,000	1,062,000	1,357,500
Cost of seed/establishment cost	fr.rw ha^{-1}	8,500	23,000	2,000	800	23,000	50,000	93,200*
Gross margin	fr.rw ha^{-1}	85,970	186,000	97,000	110,500	544,000	1,012,000	1,264,300
Labour[3]	man-days ha^{-1}	104	98	116	108	94	129	264
Total energy[4]	10^3kcal ha^{-1}	2,231	2,835	3,237	3,664	5,720	6,378	5,190[b]
Total Protein[4]	g ha^{-1}	158,120	129,000	75,240	112,350	86,310	48,000	56,250[b]
Soil loss[5]	t ha^{-1}	11.8	19.9	21.8	24.9	14.3	16.2[b]	2.5[b]
Soil Organic carbon[6]	t ha^{-1}	3020	1887	3020	1887	3775	1887[b]	3020[b]

[1]Minagri (1991) and Minagari and INSR (2006); [2]Minagri (2007a); [3] MCDF (1984); CIRAD (2004) and White et al. (2005); [4]WHO (1985); FAO/WHO (2000) and own calculations; [5] Roose (1994), Wischmeier (1995), Roose and Ndayizigiye (1997) , CPR (2002), Lufafa et al. (2003), Fleskens (2007), USDA (2009), and own calculations; [6] Sys et al. (1993) and own calculations.; [a] price per Kg of banana, respectively in season A and B. [b] These values are concerning season A. *The costs and revenues of Banana (multiyear crop) are based on annuities.

considered. Major food crops in Umutara include beans, groundnut, maize, sorghum, cassava, sweet potatoes, and banana. Crop activities in the model are production for sale and production for home consumption, since farm households consume a large part of their own products and sell what remains. Thus, we have assumed in the farm household model that any production above subsistence requirements will be sold.

All crop activities are defined at the level of annual cropping systems except banana and cassava, which are perennial crops. Subsequently, each of the perennial crops is assumed to have equal land area in the two growing seasons.

Table 4 presents a summary of input-output information for the different crop activities for season A and for Oxisols. Input-output information for the other season and the other soil type is provided in Appendix A.

Off-farm activities are important for the household systems in Rwanda. Off-farm activities represent an alternative source of income which must be taken into account when maximizing the farm income. Available off-farm activities concern informal sector work and include activities such as running small businesses, hiring out labour or working as vendor in the market etc. The family labour that can be devoted to off-farm activities depends on the available labour of the head of the household since he is the one who is mostly involved in these activities. However, off-farm opportunities are scarce in these rural areas. In our farm household model, we have assumed that a head of household can devote at most 50% of the available time for labour to off-farm activities. The daily wage received by the head of household for participating in off-farm activities is 400 fr.rw. This is the average daily wage for agricultural and informal non agricultural labour in eastern region (Strode et al., 2007).

Hired labour can be used in addition to farm household labour when cash is available. Hence, hired labour and farm household labour may be regarded as perfect substitutes. The wage of hired labour amounts to 400 fr.rw per man-day (Strode et al., 2007).

Borrowing can be seen as an option to supplement insufficient cash in order to finance seeds, hired labour, school fees, etc. Financial institutions could be important sources of credit facilities. However, in practice farmers find it quite difficult to acquire credit from these institutions due to the lack of collateral. Instead, credit can be obtained from informal sources like "credit club", the primary source of credit in Umutara. Loans plus interest must be repaid at the end of the cropping season. Interest is paid at the rate of 10% per month depending on the "credit club" to which a farm household has subscribed. Although that credit can be available, it is constrained by a credit limit .

Constraints

Land available for the representatative farm household is based on the average farm size in the eastern region, which is 0.7 ha (Loveridge et al., 2007). Crops may be grown on two soil types Oxisols and Inceptisols.

Labour requirements for crop activities (Table 4) vary depending on crop development stage. Most of the field operations on crops (land preparation, planting/sowing, crop maintenance, hand weeding and harvesting) have to be performed during a particular period of the season. Thus, each season is divided into three periods of two

Table 5. Energy and proteins recommended by World Health Organization (1985).

Age	Energy/day (kcal)		Proteins/day (gr)	
	Male	Female	Male	Female
0-11 months	679.8	628.3	11.9	11
1 to 3 years	1123	1057.3	12.8	12.2
4 to 6	1454.4	1408.5	16.7	16.9
7 to 9	1758	1570	22.7	22.8
10 to 12	1984.4	1805.1	28.6	30
13 to 14	2177.3	1942.6	37.8	38
15 to 16	2435.7	2055.1	46.8	44.1
17 to 18	2657.2	2113.0	51.9	42.2
19 to 29	3324.8	2315.3	44.3	39.6
30 to 60	3285.6	2344.8	44.3	39.6
60+	2287	1886.7	44.3	39.6

months. Small-scale farm households typically use family labour. Composition of the household determines labour capacity. The labour capacity of an adult farm household member is 100%, while children (10-18 years) and adults over 65 years of age are assumed to have 50% working availability. The available farm family labour may be subject to fluctuations over the year.

In fact, for school-going adolescents, labour contributions vary, depending on whether they live at home during school year. Additionally, children also contribute to the farm labour force during their vacations in April, July, November and December. We assume that available labour that can be allocated to activities is equivalent to 5 days per week per adult. However, 1 day per week per adult is substracted since farm households allocate labour to other necessary activities such social and household activities (e.g. firewood and water collection). The total labour requirements for crop production should be met by farm household labour and hired labour.

Rotation restrictions are set for individual crops for agronomic reasons. Crop rotations can be very important for pest and disease control, for maintaining soil fertility and reducing soil erosion. Seasonal crop rotation practices are widely adopted by farmers throughout the country. Crop rotations are incorporated in the model as strict equality constraints and imply that areas of the crops in the rotation are equal. The most frequently adopted rotations for the region are cereals-leguminous (that is, maize and sorghum with beans and groundnut) and tubers-leguminous (that is, sweet potatoes with beans and groundnut).

Cash is required to finance expenses of crop production during each cropping season and is a major constraint for small-scale farm households. These expenses include family expenditures, purchase of seeds and hiring labour. Cash is also needed for family expenditures. Cash is available from farm household's own savings made in the previous harvesting season. Moreover, cash may come from off-farm activities and credit. Credit limits set a limit to the amount of credit to be lent to a farmer. The limit varies from 5,000 Fr. Rw to 50,000 Fr. Rw depending on the wealth of the farmer. In the model, we assumed a credit limit of 10,000 fr.rw (Bidogeza et al., 2009).

Food consumption constraints in the model reflect the need of the household to first secure the household food requirements since the primary objective of small-scale farmers in Rwanda is to provide their families with adequate food . Food purchases have not been considered in model since the food consumption is mainly from the farm's food production. Small-scale farmers can hardly buy food. Consumption constraints are specified to guarantee minimum energy (in kilocalories) and proteins (in grams) per season. The minimum food requirements are obtained from the World Health Organization (WHO) recommendation level of energy and proteins per person (Table 5).

Soil organic carbon (SOC) is one of the key factors that affect agricultural production, nutrient availability and soil stability (Tang et al., 2006), particularly in highly weathered Rwanda soils where organic matter is the major source of nutrients. SOC is a dynamic property of soil, not a static one (Cooperband, 2002). The crop requirements for SOC are derived from Sys et al. (1993). The right hand side of the SOC constraint specifies its tolerance value below which yields begin to decrease (Barbier, 1998). Arshad and Martin (2002) suggested that for SOC a decrease of 15% over the average or the baseline value seems reasonable to use as critical value.The baseline SOC values considered are the organic carbon content of the two soil types for a soil depth of 1m (Ghent University, 2002).

Soil loss above certain limits will lead to the degeneration of soil reserve and soil fertility resulting in the destruction of the usable agricultural land. The farm household model takes soil loss into consideration as a constraint. Soil loss values are required for each crop activity. These values are incorporated into a soil loss constraint for each of two land types, respectively Inceptisols and Oxisols. The Wischmeier's model (Universal Soil Loss Equation) is used to calculate the soil loss coefficients (Wischmeier, 1995). The model predicts gross soil loss per unit of land as:

$$A = R*K*L*S*C*P \quad (15)$$

where A is the estimated soil loss in tons per hectare. R is the rainfall erosivity calculated based on the total kinetic energy of the rainfall and the maximum rainfall intensity over a continuous 30 min period. It represents the potential erosive risks for a particular region. R values have been derived from Equation (16) and are obtained from measurements in a region of Uganda which has close similarities with Umutara (Lufafa et al., 2003).

$$R = 47.5+0.38*Pr \quad (16)$$

In formula (16) Pr is the seasonal precipitation (mm). K is soil erodibility and represents soil resistance. K is a function of texture,

Table 6. Variance/Covariance Matrix of GM with variable prices (VP) and constants prices (CP) for Season A.

Parameter	Sorghum		Maize		Beans		Peanuts		Banana		Sweet potatoes		Cassava	
	VP	CP	VP	CP	VP	CP	VP	CP	VP	CP	VP	CP	VP	CP
Sorghum	$9.6.10^{8}$	5.10^{8}	$4.5.10^{7}$	$2.2.10^{8}$	$9.8.10^{8}$	$3.4.10^{8}$	$2.6.10^{8}$	$1.01.10^{9}$	$5.1.10^{9}$	$1.2.10^{9}$	$2.9.10^{9}$	$1.06.10^{9}$	$2.5.10^{9}$	$1.9.10^{8}$
Maize	$4.5.10^{7}$	$2.2.10^{8}$	$4.7.10^{8}$	$3.9.10^{8}$	$4.1.10^{7}$	$-1.6.10^{7}$	$5.6.10^{8}$	$1.1.10^{9}$	$6.8.10^{8}$	$2.5.10^{8}$	$9.5.10^{8}$	$1.4.10^{9}$	$2.1.10^{9}$	$6.3.10^{8}$
Beans	$9.8.10^{8}$	$3.4.10^{8}$	$4.1.10^{7}$	$-1.6.10^{7}$	$1.4.10^{9}$	$5.6.10^{8}$	$8.7.10^{7}$	$2.5.10^{8}$	$7.4.10^{9}$	$1.3.10^{9}$	$3.8.10^{9}$	$-5.9.10^{6}$	$1.5.10^{9}$	$-2.3.10^{8}$
Peanuts	$2.6.10^{8}$	$1.01.10^{9}$	$5.6.10^{8}$	$1.1.10^{9}$	$8.7.10^{7}$	$2.5.10^{8}$	$4.5.10^{9}$	$6.4.10^{9}$	$-4.1.10^{9}$	$1.8.10^{9}$	$1.2.10^{9}$	$5.9.10^{9}$	$1.6.10^{8}$	$1.7.10^{9}$
Banana	$5.1.10^{9}$	$1.2.10^{9}$	$6.8.10^{8}$	$2.5.10^{8}$	$7.4.10^{9}$	$1.3.10^{9}$	$-4.1.10^{8}$	$1.8.10^{9}$	4.10^{10}	$5.4.10^{9}$	2.7.1010	$1.6.10^{9}$	$2.3.10^{10}$	$2.3.10^{8}$
Sweet potatoes	$2.9.10^{9}$	$1.06.10^{9}$	$9.5.10^{8}$	$1.4.10^{9}$	$8.4.10^{9}$	$-5.9.10^{7}$	$1.2.10^{9}$	$5.9.10^{9}$	$2.7.10^{10}$	$1.6.10^{9}$	$1.4.10^{10}$	$8.9.10^{9}$	$2.5.10^{10}$	$2.92.10^{9}$
Cassava	$2.5.10^{9}$	$1.9.10^{8}$	$2.1.10^{8}$	$6.3.10^{8}$	$1.5.10^{9}$	$-2.3.10^{8}$	$1.6.10^{8}$	$1.7.10^{9}$	$2.3.10^{10}$	$2.3.10^{8}$	$2.5.10^{10}$	$2.9.10^{9}$	$4.9.10^{10}$	$2.95.10^{9}$

organic matter, permeability and soil structure. K values for Inceptisols and Oxisols are respectively 0.20 and 0.25 (Roose and Ndayizigiye, 1997; Henao and Baanante, 2006; Fleskens, 2007). L*S represent hillslope length and steepness, and reflects the effect of topography on soil loss rates at a particular site. Values used for Inceptisols with slope of 4% and Oxisols with slope of 9% are respectively 0.42 and 1.3 (Roose, 1994). C is the land use and land cover factor and expresses effects of surface cover and roughness, soil biomass, soil-disturbing activities on rates of soil loss at particular sites. Values used are obtained from Lewis (1988; cited by Fleskens, 2007). Banana has the lowest C-value of 0.04, while sorghum has the highest C-value of 0.45. P is management practice and expresses the effects of supporting conservation practices, such as contouring, buffer strips, terracing, etc. on soil loss at a particular site. When no erosion control practice is used, P equals 1. Planting crops with dispersed trees could be attributed a P value of 0.6, use of grass strips lowers this to 0.4 and grass strips with hedgerows P to 0.1. Thick mulching also has a P of 0.1 (Fleskens, 2007 and Roose, 1994).

The right hand side of the soil loss constraint specifies the soil loss tolerance. The concept of soil loss tolerance is defined as the maximum acceptable soil loss from an area which will allow a high productivity to be maintained for a long period of time. In the model, soil tolerance values used for Oxisols and Inceptisols were derived from Pretorius and Cook (2002), 12 t ha^{-1} yr^{-1} and 16 t ha^{-1} yr^{-1}, respectively. Pretorius and Cook (2002) have assigned soil tolerance values to soils depending on their root penetration depth. The Oxisols have generally steeper slopes and lower soil depth, while Inceptisols are on gentle solpes and deeper soils.

Inclusion of risk in the farm household model

It is important to account for risk in any agricultural productive activity (Hardaker et al., 2004; Anderson and Dillon, 1992). Risk is defined as a measure of the effect of uncertainty on the decision-maker (Upton, 1996). Farm households in Rwanda are facing an unstable income from season to season due to unpredictable rainfall and fluctuations of market prices. Most small farmers typically behave in risk-averse ways, they are willing to forgo some expected income for a reduction in risk (Acs et al., 2009). Ignoring risk-averse behaviour in farm household models may lead to results that are unacceptable to the farmer, or that have little relation to the decisions he actually makes.

From Equation (1) risk is explicitly incorporated in the farm household model. The risk is calculated following a quadratic programming approach (Hazell and Norton, 1986). This method computes the standard deviation from the variance-covariance matrix and the level of the stochastic activities. Since seasonal fluctuations in farm prices and rainfall have a large effect on farmers' income, risk has been calculated for two types of production activities: home consumption and market. To compute risk for home consumption, we use gross margin with constant prices, while gross margin with variable prices is used to compute market risk. Data from six years are used to determine the variance and covariance matrix. This is refered to in Table 6 for Season A. The variance and covariance matrix table for Season B is shown in Appendix B.

Given the difficulty to objectively assign a risk aversion parameter for typical farm, we have assumed that small-scale farm households are somewhat risk averse (Ø = 1.0). This value is derived from Anderson and Dillon (1992) as they grouped relative risk aversion as follow: Ø = 0, risk neutral; Ø = 0.5, hardly risk averse at all; Ø = 1.0, somewhat risk averse (normal); Ø = 2.0, rather risk averse; Ø = 3.0, very risk averse; Ø = 4.0, extremely risk averse.

The risk aversion coefficient of 1 is within the range of values reported by Senkondo (2000), that is, -0.98 and 2.64 with an average value of 0.774 for the situation when the farmer has inadequate food stocks, which is a situation quite similar for the typical farm household in Rwanda.

Set up of the calculations

Calculations are made for a typical farm household on Oxisoils and Inceptisoils for the two growing seasons of a year. Table 7 shows some specific farm characteristics for the representatative farm household considered in the model. The farm household is composed of one adult male, one adult female, two kids under 10 years old and four children are of age 10-18. This family size follows from

Table 7. Characteristics of the typical farm household used as input in the model.

Parameter	Unit	Farm household
Total farm size	ha	0.7
Inceptisols (slope of 4%)	ha	0.28
Oxisols (slope of 9%)	Ha	0.42
Family size	Person	8
Available Labour	man-day	
Season A		
Period 1		104
Period 2		64
Period 3		144
Season B		
Period 1		64
Period 2		104
Period 3		64
Wage off-farm income	fr.rw/day	400
Available cash at the start of the year	fr.rw	5,000
Credit limit per season	fr.rw	10,000
Rate of interest per month	%	10
Family expenditure	fr.rw.	128,000
Energy requirement (Kcal/Household)		
Season A	10^3 kcal	3,067
Season B	10^3 kcal	3,067
Proteins requirement (Grams/Household)		
Season A	10^3 gr	49
Season B	10^3 gr	49

Note: Average exchange rate in 2007: US$1 = 550 Fr.Rw.

the average national rate of birth with six child per woman (Kinzer, 2007). The household is supposed to benefit of the labour from the children while they have vacation. Consequently, the available labour within the household fluctuates within the year as it can be seen from the Table 7. Average yearly expenditures of the typical farm household are estimated on the basis of national value representing the consumption poverty line per adult equivalent per year. That value is estimated at 64,000 Rwandese francs per adult equivalent per year (Ansoms and McKay, 2010). The farm household is assumed to have two adults (the head of household and his wife). The children are added to this adult equivalent. For the cash availability, we assume that the farm houshold has a cash of 5,000 Fw.Fr at the beginning of the year (Bidogeza et al., 2009).

Subsequently, the results from the typical farm household model are compared with actually observed values. Lastly, additional calculations are made with the model to examine the effects of the land area and family size on food security , income and soil loss results. Therefore, the farm household model is optimized with nine different combinations of land area and family size. Three households with a family size of five, eight, and ten persons are combined each, with a land area of 0.5, 0.7 and 1 ha, respectively. The household size of five, eight and ten reflect respectively: the Government's policy on family planning which encourages families to have at most 4 children per woman (Solo, 2008); the current average family size (about 8) and a rather high household size, also often encountered in Rwanda. The land areas embody, respectively the possible future, the actual, and the minimum recommended land size.

RESULTS AND DISCUSSION

Calculations have been made first to determine the optimal farm plan for the typical farm.

Technical results

The optimal cropping plan for the typical farm is presented in Table 8. A large proportion of land is allocated to banana, beans, sweet potatoes and sorghum which reflects the food habits in Umutara province. Banana and sweet potato have higher calories per hectare while beans have the highest level of proteins per hectare. Banana covers a much larger proportion (47%) of the land in the optimal farm plan than other crops because of its high calories per hectare. In addition, banana protects well the soil since it causes less soil loss. Sweet potato also has high yield of calories per hectare but, because of the high soil loss rate compared

Table 8. Optimal cropping plan for season A and B.

	Area (ha)			
	Season A		Season B	
Land type	**Inceptisols**	**Oxisols**	**Inceptisols**	**Oxisols**
Crops for home consumption				
Banana	0.067	0.209	0.067	0.209
Beans	0.035	0.161	0.056	0
Cassava	0	0	0	0
Groundnut	0.015	0	0.006	0
Maize	0	0	0	0
Sorghum	0.134	0	0.004	0
Sweet potatoes	0.019	0.032	0.125	0.082
Crops for sale				
Banana	0.010	0.018	0.010	0.018
Sweet potatoes	0.0006	0	0	0
Unused land	0	0	0	0.112
Total	0.28	0.42	0.28	0.42

to banana, a smaller land area is allocated to sweet potato than banana. Beans are produced to a relatively large extent (20%) because of its highest level of proteins. A small proportion of the available land is allocated to sorghum and groundnut to supply additional calories and proteins and secure the nutritional requirements of the farm household.

From the model results, both nutritional requirements and soil loss are binding constraints. However, soil loss is restricting only on marginal land (Oxisols). Banana and beans cause less erosion compared to other crops. This explains why they are grown mostly on marginal land (70%).

Cassava is not considered in the optimal farm production although it has the highest yield of calories per hectare. The model considers that an optimal plan including cassava is too risky since it has a higher variability of production and prices compared to other crops.

Technical results for fixed resources, specifically land, on-farm labour and off-farm labour are shown in Table 9. The area under Inceptisols is fully used in both seasons, whereas the model leaves 0.112 ha of the area under Oxisols unexploited in season B. This is because of constraining soil loss and SOC. A total of 172 man-days and 106 of man-days remain available, for in seasons A and B, respectively. In both seasons, labour allocated to the off-farm activity is at its maximum level.

In our farm household model we have differentiated the crop production for home consumption from crop production for sale. The model results reveal that 88% of the land is allocated to crop production for home consumption, while 8% remains unused and 4% of land is used for crop production for sale. A large proportion of land for home consumption is needed to secure the World Health Organisation's (WHO) nutritional requirements, that is, to maintain the food security status. The model results identify soil loss and risk as the major explanations why some land remains idle while a small portion of land is allocated to crop production for sale. At relatively low extent, SOC has some influence on the optimal farm production.

From the model results, crops which contribute mostly to secure calories for the representatative farm household throughout the year are banana and sweet potatoes providing respectively 48% and 26% of total energy, respectively. Beans is the major supplier of proteins with 48% of total proteins required.

Economic results

The farm income can come from off-farm activities and crop production for sale. Although there is sale of crops, revenues from crop production for sale are small since the model has allocated major portion of land to crop production for home consumption. Therefore, the major contributor of farm income is from off-farm activities with 55%, while sale of crops production contributes 45%. Net farm Income equals to 18,680 fr.rw, yearly. Net farm income is the cash income after substrating the cash expenditures. Banana is almost the only cash crop, because of its high gross margin per hectare. The model has shown that risk and soil loss are playing a role to maintain this subsistence trait. The restricting food requirements explain why the typical farm in our model is

Table 9. Optimal seasonal resource use and constraint and their shadow prices or slack values activities.

Parameter	Unit	Season A			Season B		
		Level of activity	Shadow price (fr.rw/ha)	Slack value	Level of activity	Shadow price (fr.rw/ha.)	Slack value
Land type	**Ha**						
Inceptisols (slope: 4%)		0.28	63,845	0	0.28	42,515	0
Oxisols (9%)		0.42	16,354	0	0.308	0	0.112
Soil loss*	$t\ ha^{-1}$						
Inceptisols (slope: 4%)		4.48	0	2.2			
Oxisols (9%)		5.04	1,785				
SOC*	$kg\ ha^{-1}$						
Inceptisols (slope: 4%)		1,960	0	726			
Oxisols (9%)		2,286	163	0			
On-farm labour Use in:	man-day						
Period 1		26	0	58	26	0	18
Period 2		25	0	19	17	0	66
Period 3		29	0	95	22	0	22
Off-farm labour use for the head of household	man-day						
Period 1		20	400	0	20	400	0
Period 2		20	400	0	20	400	0
Period 3		20	400	0	20	400	0
Credit	fr.rw	10,000	0	10,000	10, 000	0	10,000
Nutrition requirements							
Calories	10^3 kcal	3067	-12.81	0	3067	-12.80	0
Proteins	$10^3 g$	81	0	32	55	0	6

*Values of soil loss and SOC are for a year. Note: Average exchange rate in 2007: US$1 = 550 Fr.Rw.

willing to forego some land or prefers to grow subsistence crops in order to avoid risk.

The farm household model reports the shadow prices for the fixed resources and constraints that are fully used. A shadow price indicates the maximum amount by which the model's objective function could be increased if an additional unit of the resource were to become available (Hazell and Norton, 1986). For example, in case of the land constraint expressed in ha, a shadow price of 1.5 indicates that the value for the objective function would increase by 1.5 if the availability of land would increase by one 1 ha. Table 9 presents shadow prices of some of the fixed resources and constraints. Off-farm activities are extremely important for the typical farm. One man day labour allocated to off-farm activities would increase farm income with 400 fr.rw. Scarcity of employment opportunities refrain farm households from hiring out labour. In the case of land: the maximum rent a farmer should be willing to pay for one additional hectare of land type Inceptisols would be 63,845 fr.rw and 42,515 fr.rw, respectively in season A and B. Land with Oxisols is only fully used in season A with a shadow price of 16,354 fr.rw.

The farm household model calculates the shadow prices for levels of soil loss for the two types of soil. In the case of soil loss, shadow prices represent the amount by which the objective function would change if the constraint on soil loss were increased by one unit. They represent the maximum allowable cost of erosion reductions (Carcamo et al, 1994). Thus, allowing 1 $t\ ha^{-1}$ more soil loss can increase farm income with 1,785 fr.rw for Oxisols. The shadow price of soil loss for Inceptisols is zero. Likewise for SOC the shadow price for the Inceptisols is zero, while for Oxisols, it is restricting. This implies that soil loss on Inceptisols and SOC do not entail negative economic consequences. However, in the long run, an acceptable solution from both economic and environmental perspective should be found, i.e. less erosive solution which generates at the same time an acceptable level of profitability.

Comparison of the household model results with observed household data

The model results are compared with information from literature and farm surveys. With regards to crop allocation the farm model results indicate that banana occupy a large proportion of the land (43%), followed by beans (20%), sweet potatoes (20%) and sorghum (10%). These results are relatively consistent with the information from the farm survey done in the region , which affirms that the most cultivated crops are beans (95% of the farmers), banana (85%), maize (75%), sweet potatoes (72%), sorghum (70%) and cassava (60%) (Minagri and INSR, 2006).

Banana and sweet potatoes are known to have less calories and proteins per kg compared to other crops, but are favoured in the model and in the real farming since they have high calories per hectare. Additionally, the two crops tend to produce even when other crops fail completely; they also produce during the nutritionally critical pre harvest period such April-May and November-December (Kangasniemi, 1999). Moreover, banana is causing less soil loss.

Despite its high energy yield per hectare, the model hasn't selected cassava due to its high production and price variance. The cassava production is varying over years because of the recurrent virus of African mosaic which quite often damages the crop (Mukakamanzi, 2004).

The model indicates that a major proportion of crop production is self-consumed to secure nutritional requirements of the typical farm household, a small proportion is sold. The food security status is maintained at the expense of getting cash from the crops. This fact is widely observed in Rwanda where farming is mostly subsistence oriented.

However, the model has attributed a small portion of banana production for sale. This is consistent with the findings from Kangasniemi (1999) and Okech et al. (2001), expressing that in regions where traditional cash crops are missing (coffee and tea), bananas are by far the most remunerative cash crop for Rwandan farmers.

The farm model reveals that the shadow prices of the good land (Inceptisols) are very high compared to the cost of renting one hectare of land per year in southern and eastern regions of Rwanda, which is 22,600 fr.rw. as reported by Takeuchi and Marara (2007). However, these shadow prices are more close to the cost of renting one hectare of land per year in the northern region of Rwanda, which is 50,000 fr.rw as reported by Fané et al. (2004). The shadow prices of marginal land are small or zero. Therefore, the model has left out a portion of marginal land where we would expect the farm to fully exploit his farm due to its small size. The cultivation of marginal land causes much more soil loss than cultivation on the good soils, which may explain why the model abandons some of the marginal land because of much soil loss, which may prevent their profitability. Barbier and Bergeron in Honduras (1999) also found that farmers were likely to crop less on erodible fields. Furthermore, we have observed from the farm survey (Minagri and INSR, 2006) that despite of the small size of the farms, 25% of the farmers prefer to put some land on fallow to enrich the soil or because they don't see any profitability to farm the whole farm once not all land is needed for their subsistence.

With regard to labour, the model shows that there is much on-farm labour available since the shadow price is zero, while off-farm activities are used to the maximum. This corresponds with the current situation in Rwanda where off-farm employment is already an important source of income for rural households (Loveridge et al., 2007). However, this option is limited by low availability of off-farm activities. Therefore, availability of off-farm employment would improve the income of farm households.

The results from the bio-economic model of the typical farm provide a valid and acceptable approximation of the reality. Hence, we use the model to test for different policy simulations for the typical farm and also for other farm types.

Effects of household size and land area changes on food security, income and soil loss results

Table 10 indicates the effects of household and land size on food security, income and soil loss results. According to the model, for the majority of farm households, it is possible to meet the WHO nutritional requirements. However, households with 8 members and a farm size of 0.5 ha and household of 10 members with farm size of 0.5 ha and 0.7 ha are not able to secure the WHO energy requirements. Therefore, calorie requirements were lowered (Table 10) until a feasible solution was reached. However, from Table 10 it can be seen that a household with 5 members and a farm size of either 0.5, 0.7 or 1 ha can obtain a high income and that soil loss has relatively little economic impact. This is in accordance with the family planning policy of Rwanda Government which promotes a fertility rate less than 4 children per woman. Indeed, for a household of 5 members even with the lowest farm size (0.5 ha) considered, it is possible to secure the WHO' s recommended level of calories and proteins, and additionally get a relatively high income. Table 10 highlights the fact that with more people having less land food security cannot be achieved and soil loss has a high economic impact at least for the marginal land with Oxisols. This finding seems to contradict the conclusion made by Tiffen et al. (1994). In their study conducted, in Machakos region in Kenya, they asserted that population growth has a positive impact on the

Table 10. Effects of household size and land area changes on food security, income and soil loss results.

Parameter	Land area (Ha)	Food requirements met		Income in Rwandese francs	Soil loss Shadow prices in Rwandese francs	
Household size (members)		Energy (%)	Proteins (%)		Inceptisols	Oxisols
	0.5	100	100	33,891	0	1,964
5	0.7	100	100	40,295	0	603
	1	100	100	41,545	0	0
	0.5	78	81	17,118	0	3,703
8	0.7	100	100	18,680	0	1,785
	1	100	100	32,095	0	1,052
	0.5	62	63	17,118	0	3,703
10	0.7	87	92	4,728	0	3,988
	1	100	100	17,855	0	2,062

economic development. Contrary to the findings of Tiffen et al. (1994): rather than saying "More people, less erosion", our findings indicate that fewer people leads to little economic impact of soil erosion and enough food for each household. However, these differences have to be distinguished by keeping in mind that the farm household model is a yearly based model and under low farming inputs while Tiffen et al. (1994) examined interactions of people and environment over a period of sixty years in association with intensive farming systems.

GENERAL DISCUSSION

In this article, a bio-economic farm model has been presented that can be applied for a typical farm household and be used to simulate the impact of family size, farm size, and soil erosion on farm production and food security. The bio-economic farm household model was developed by using a mathematical modelling approach. Here, some of the important underlying assumptions are discussed.

Capturing subsistence farming in the model

In this paper, the authors did not consider the option of purchasing food. Considering the option of purchasing food for the current typical farm with very low inputs and a farming fully focussed on subsistence would not represent the reality of livelihoods of farmers in the east region of Rwanda. However, this option may be appropriate for the livestock farms (they are large farms of more than 3 ha) who are also found in the region, but are less important in terms of total population in the province. Castaño (2001) and Laborte et al. (2009) have considered the option of purchasing food in their respective farm household models in the contexts of semi-subsistence and subsistence farming. Livestock activities, however, have been considered in their models. These activities are missing for the typical farm household considered in our model. It is known that livestock activities may constitute another source of income and a form of savings, which may then allow farmers to purchase food when necessary.

Subsistence farmers used the food produced from their own farms to feed their families. However, during the period of starvation, subsistence farmers may consider the option of purchasing food. In this article, the year considered in the model is assumed to be 'normal' where farmers do not have to face starvation due to droughts or inundation. Moreover, although that we did not program the option of purchasing food in our model, which would give more flexibility to household, we have dealt this in a flexible way in the sense that we relaxed the food constraints at the moment when the model was not able to produce enough food (Table 10).

Risks consideration in the model

In this article, the authors have used the method

of standard deviation of the gross margin to compute risk instead of a safety-first approach, including Target-MOTAD. The model somehow makes already use of the safety-first approach principle in the sense that food requirements are explicitly formulated as constraints.3.5.

Maximizing the objective function in the model

In this paper, we have assumed that the farmer is pursuing one objective that is to maximize the expected utility. Thus, the expected utility is the objective function and this is maximized. Subsistence farming characterizes most of the agricultural production of rural developing countries. Mishev et al. (2002) have stated that subsistence farmers are prone to maximize utility functions. Castaño (2001) and Laborte et al. (2009) have conducted empirical studies wherein the objective function was to maximize utility defined as discretionary income, in Andean hillside farms of Columbia and northern Philippines, respectively. Discretionary income is defined as income available for spending after essential expenses have been made. The farmer is assumed to maximise one objective function, which is the expected utility defined as discretionary income minus the risk premium. However, subsistence farmers may, also pursuit several objectives as Berkhout et al. (2010) have shown that there is heterogeneity in the farmer goals and preferences, in relation to the role of farm enterprise. Therefore, not considering all objectives of the farmer in the modelling approach, may lead to the results that differ from the reality. Given that the different approaches to capture the objective (s) of the farmer have their own limitations, the results should be analysed with respect to the particular farming system (Van Calker, 2004).

Conclusions

In this paper, a bio-economic model was developed to analyse the impacts of family planning, land consolidation and soil erosion on farm production and food security on a typical farm in Rwanda and on other farm types.

The results of the model show that a higher availability of good land increases farm income, whereas a higher availability of marginal land has slight impact on income. Considering that soil erosion is a restricting factor on marginal land, preserving soils against erosion would release more marginal land and increase food production. Farm household income would also benefit from better off-farm employment opportunities.

Household size and land area changes have a large impact on food security, income and soil loss. Our model results suggest that most farm households can satisfy the WHO minimum nutritional requirements. However, with more people and less land, it is difficult to fulfill the WHO's energy and proteins requirements. Households with a large family size and small land area cannot ensure their food security. The model results show that a household with 8 members and a farm size of 0.5 ha and a household of 10 members with farm size of 0.5 ha and 0.7 ha are not able to secure the WHO energy requirements. Also, results show that soil loss has in those situations a relatively high economic impact. However, households with the lowest person: land ratio easily secure their food security and soil loss has relatively little economic impact for those households.

The outcome of the model supports the Rwanda policy on family planning which intends to encourage every woman to have a human reproduction rate below 4. However, the land policy to encourage farmers with a total land area below 1 ha either to consolidate their land or to quit farming is not supported by the results. Our results show that a household of 5 members with a farm size of at least 0.5 ha is able to comply with the minimum food security requirements and to get a relatively high income; additionnally, the soil loss has little economic impact. In the context of Rwanda with a rapidly growing population, a minimum area of 0.5 ha instead of 1 ha should be considered (for the time being).

Moreover, policy makers should target adoption of technologies that reduce land degradation and risks to further improve food security.

Conflict of Interest

The authors have not declared any conflict of interest.

ACKNOWLEDGEMENTS

The authors are grateful to the reviewer and the editor who have greatly contributed to improve of this paper.

REFERENCES

Acs S, Berentsen PMB, Huirne RBM (2009). Effect of yield and price risk on conversion from conventional to organic farming. Austr. J. Agric. Resourc. Econ. 3:393–411.

Ansoms A, McKay A (2010). A quantitative analysis of poverty and livelihood profiles: The case of rural Rwanda. Food Policy 35:584-598.

Anderson AS (2002). The effect of cash cropping, credit, and household composition on household food security in southern Malawi. Afr. Urban Q. 6:161-186.

Anderson J , Dillon J (1992): Risk analysis in dry land farming systems. FAO. Farm systems management Series No.2, Rome.

Anecksamphant C, Charoenchamratcheep C, Vearasilp T, Eswaran H (1999). 2nd International Conference on Land degradation: Meeting the Challenges of Land Degradation in the 21st Century. Conference report. Bangkok, Thailand.

Arshad MA, Martin S (2002). Identifying critical limits for soil quality indicators in agro-systems. Agric. Ecosyst. Environ. 88:153-160.

Barbier EB (1990). The farm-level economics of soil conservation: the uplands of Java. Land Econ. 66:199-211.

Barbier B, Bergeron G (1999): Impact of policy interventions on land management in Honduras: results of a bioeconomic model. Agric. Syst. 60:1-16.

Barbier B (1998). Induced innovation and land degradation: Results from a bio-economic model of a village in Western Africa. Agric. Econ. 19:15-25.

Berkhout ED, Robert A, Schipper RA, Kuyvenhoven A, Coulibaly O (2010).: Does heterogeneity in goals and preferences affect efficiency? A case study of farm households in northern Nigeria. Agric. Econ. 41:265-273.

Bidogeza JC, Berentsen PBM, De Graaff J, Oude Lansink A (2009). A farm typology for Umutara province in Rwanda. J. Food Security: The Science, Sociology, Economics of Food Production and Access to Food 1:321-335.

Birasa EC, Bizimana I, Bouckaert W, Deflandre A, Chapelle J, Gallez A, Maesschalck G, Vercruysse J (1990). Les sols du Rwanda: methodologie, legende et classification. CPR et Minagri, Kigali.

Byiringiro F, Reardon T (1996). Farm productivity in Rwanda: effects of farm size, erosion, and soil conservation investments. Agric. Econ. 15:127-136.

Cárcamo AJ, Alwang J, Norton WG (1994). On-site economic evaluation of soil conservation practices in Honduras. Agric. Econ. 11:27-269.

Castaño J (2001). Agricultural marketing systems and sustainability: study of small Andean hillside farms. PhD thesis, WageningenUniversity, Wageningen, The Netherlands.

CIRAD, GRET, Ministere francais des affaires etrangeres (2004). Momento de l'agronome. Paris, France.

Clay D, Reardon T, Kangasniemi J (1998). Sustainable intensification in the highland tropics: rwandan farmers' investments in land conservation and soil fertility. Econ. Dev. Cultural Change 46:351-377.

Cooperband L (2002). Building soil organic matter with organic amendments. A resource for urban and rural gardeners, small farmers, turfgrass managers and large-scale producers. Unpublished report. Center for integrated agricultural systems, University of Wisconsin-Madison, USA.

CPR (2002). Descriptions des profils. Minagri, Kigali, Rwanda.

Dorward A (1999). Modelling embedded risk in peasant agriculture: methodological insights from northern Malawi. Agric. Econ. 21:191-203.

Fané I, Kribes R, Ndimurwango P, Nsengiyumva V, Nzang Oyono C (2004). Les systèmes de production de la pomme de terre au Rwanda. Propositions d'actions de recherche et de développement dans les provinces de Ruhengeri et Gisenyi. Série de documents de travail N° 122. Centre International pour la recherche Agricole orientée vers le développement (ICRA), Réseau des Organisations paysannes du Rwanda (ROPARWA), Institut des Sciences Agronomiques du Rwanda (ISAR), Montpellier, France.

FAO (2001). Lecture notes on major soil of the world. Edited by WageningenUniversity, International Institute for Aerospace Survey and Earth Sciences (ITC), Catholic University of Leuven, International Soil Reference and Information Centre and FAO. Rome, Italia.

FAO/WHO (2000). Preliminary report on recommended nutrient intakes. Joint FA/WHO Expert Consultation on Human Vitamin and Mineral requirements, FAO, Bangkok, Thailand, September 21-30, 1998, revised July 13, 2000. Food and Agricultural Organization of the United Nations Rome, Italy and World health Organization, Geneva, Switzerland.

Fleskens L (2007). Prioritizing rural public works interventions in support of agricultural intensification. International center for soil fertility & agricultural development (IFDC), Kigali, Rwanda.

Ghent University (2002). Rwanda soil database. Department of soil science. Laboratory of soil. Ghent, Belgium.

Groupe d'Expertise, de Conseil et d'Appui au Développement (GECAD) (2004). Opérationnalisation de la politique agricole nationale, ébauches des stratégies sectorielles et sous sectorielles. Kigali, Rwanda.

Hardaker JB, Huirne RBM Anderson RJ, Lien G (2004). Coping with risk in agriculture. CABI Publishing.

Hazell PBR, Norton RD (1986). Mathematical programming for economic analysis in agriculture. Macmillan Publishing Company, New York, USA.

Henao J, Baanante C (2006). Agricultural production and soil nutrient mining in Africa: implications for resource conservation and policy development. Technical bulletin IFDC T-72. IFDC – An InternationalCenter for Soil Fertility and Agricultural Development., Muscle Shoals, Alabama, USA.

Jansen S , Van Ittersum KM (2007): Assessing farm innovations and responses to policies: a review of bio-economic farm models. Agric. Syst. 94:622-636.

Huggins C (2012). Consolidating land, consolidating control: What future for smallholder Farming in Rwanda's 'Green Revolution'? International Conference on Global Land Grabbling II, October 17-19, 2012.

Kanellopoulos A, Berentsen PBM, Heckelei T, Ittersum Van M, Oude Lansink A (2010). Assessing the forecasting performance of a generic bio-economic farm model calibrated with two different PMP variants. J. Agric. Econ. 61:274-294.

Kangasniemi J (1999). People and banana on steep slopes: agricultural intensification and food security under demographic pressure and environmental degradation in Rwanda. PhD thesis. Department of Agricultural Economics, MichiganState University, USA.

Kinzer S (2007). Rwanda plans to control population growth, New York Times. In http://www.news-medical.net, accessed on 15/04/2009.

Kruseman G, Bade J (1998). Agrarian policies for sustainable land use: Bio-economic modeling to assess the effectiveness of policy instruments. Agric. Syst. 58:465-481.

Kruseman G (2000). Bio-economic household modeling for agricultural intensification. Mansholt Studies 20, Wageningen University, Wageningen, The Netherlands.

Kuyvenhoven A, Ruben R, Kruseman G (1998). Technology, market policies and institutional reform for sustainable land use in southern Mali. Agric. Econ. 19:3-62.

Laborte AG, Van Ittersum, MK, Van den Berg MM (2007). Multi-scale analysis of agricultural development: A modelling approach for IlocosNorte, Philippines, Agricultural Systems 94:862-873.

Laborte AG; Schipper RA; Van Ittersum MK, Van Den Berg MM, Van Keulen H, Prins AG, Hossain M (2009). Farmers' welfare, food production and the environment: a model-based assessment of the effects of new technologies in the northern Philippines. NJAS - Wageningen J. Life Sci. 56:345-373.

Louhichi K, Flichman G, Zekri S (1999). Un modele bio-economique pour analyser l'impact de la politique de conservation des eaux et du sol. Le cas d'une exploitation agricole tunisienne. Econ. Rurale 252:55-64.

Louhichi K, Kanellopoulos A, Janssen S, Flichman G, Blanco M, Hengsdijk H, Heckelei T, Berentsen PBM, Oude Lansink A , Ittersum Van M (2010). FSSIM, a Bio-Economic farm model for simulating the response of EU Farming Systems to agricultural and environment policies. Agric. Syst. 103:585-597.

Loveridge S, Orr A, Murekezi A (2007). Agriculture and poverty in Rwanda. A comparative analysis of the EICV1, EICV2, and LRSS surveys. NISR, Kigali, Rwanda.

Lufafa A, Tenywa MM, Isabirye Majaliwa GJM , Woomer LP (2003). Prediction of soil erosion in a Lake Victoria basin catchment using a GIS-based Universal Soil Loss model. Agric. Syst. 76:883–894.

MCDF (1984). Momento de l'agronome, troisieme edition. Collection "techniques rurales en Afrique. Paris, France.

Minagri (1991). Enquete national agricole 1989: production, superficie, rendement, elevage et leur evolution 1984-89. Kigali, Rwanda.

Minagri (2002). Statistiques agricoles: Production Agricole, Elevage, Superficies, et Utilisation des terres . Division des Statistiques Agricoles (DSA) - Annee Agricole 2002. Kigali, Rwanda.

Minagri (2004). Strategic plan for agricultural transformation in Rwanda. Kigali, Rwanda

Minagri, INSR (2006). Enquete agricole 2005: Methodologie, Production Agricole, Elevage, Superficies, et Utilisation des terres. Kigali, Rwanda.

Minagri (2007). Market price list database 1997-2007. Kigali, Rwanda.

Minecofin (2003a). Indicateurs de développement du Rwanda. Kigali, Rwanda.

Minecofin (2003b). Poverty reduction strategy paper. Kigali, Rwanda.

Minerena (2005). OrganicLaw determining the use and management of land in Rwanda. Law N°08/2005of14/07/2005. http://www.minirena.gov.rw/IMG/pdf/LOI_ORGANIQUE_FONCIERE.pdf. Accessed on 10th December 2010.

Mishev P, Ivanova N, Kostov P (2002). Agricultural policy options distinguishing a subsistence sub-sector in Bulgaria. In in G. Mergos (ed.) (2002) EU Enlargement and the CAP: A Quantitative Analysis for Slovenia and Bulgaria, Kiel: WissenschaftsverlagVauk, P. 210.

Mudhara M, Hildebrand EP, Gladwin HC (2002). Gender-sensitive LP models in soil fertility research for smallholder farmers: Reaching de Jure female headed households in Zimbabwe. Afr. Urban Q. 6:1-2.

Mugabo JB (2005). Etude de la qualite et de la degradation des sols dans le bassin de l'Akagera. Akagera Transboundry Agro-ecosystem management. FAO Project. Kigali, Rwanda.

Mukakamanzi MC (2004). Contribution à une comprehension de la commercialisation du manioc dans la région de Mayaga, province de Gitarama. Mémoire présente en vue de l'obtention du grade d'ingénieur technicien. ISAE. Ruhengeri, Rwanda.

Musahara H (2006). Economic Analysis of natural resource management in Rwanda. UNDP, UNEP and REMA. Kigali, Rwanda.

National Institute of Statistics of Rwanda (NISR) (2008). Rwanda Development Indicators 2006. Kigali, Rwanda.

Okech SHO, Gaidashova SV, Gold CS, Nyagahungu I, Musumbu JT (2005). The influence of socio-economic and marketing factors on banana production in Rwanda: Results from a Participatory Rural Appraisal. The International J. Sustain. Dev. World Ecol. 12:149-160.

Pottier J (2006). Land reform for peace? Rwanda's 2005 land law in context. J. Agrarian Change 6(4):509-537.

Pretorius RJ, Cook J (1989). Soil loss tolerance limits: an environmental management tools. Geo J. 19:67-75.

RADA (2005). Business plan 2006-2008. Kigali, Rwanda.

Roose E , Ndayizigiye F (1997). Agroforestry, water and soil fertility management to fight erosion in tropical mountains of Rwanda. Soil Technol. 11:109-119.

Roose E (1994). Introduction a la gestion conservatoire de l'eau, de la biomasse et de la fertilite des sols. (GCES). Bulletin Pedologique de la FAO 70. Rome, Italie.

Ruben R, Moll H , Kuyvenhoven A (1998): Integrating Agricultural research and policy analysis: analytical framework and policy applications for bio-economic modelling. Agric. Syst. 58:331-349.

Senkondo EMM (2000). Risk attitude and risk perception in agroforestry decisions: case of Babati, Tanzania. PhD thesis, Wageningen University, Wageningen, The Netherlands.

Senthilkumar K, Lubbers MTMH, De Ridder N, Bindraban PS Thiyagarajan TM, Giller KE (2011). Policies to support economic and environmental goals at farm and regional scales: outcomes for rice farmers is Southern India depend on their resource endowment. Agric. Syst. 104:82-93.

Sirven P, Gotanegre JF, Prioul C (1974). Geographie du Rwanda. Eds. De Boeck, Bruxelles; Bruxelles; Eds. Rwandaises, Kigali, Rwanda

Solo J (2008). Family planning in Rwanda, how a taboo topic became priority number one. IntraHealth International, North Carolina, USA.

Strode M, Wylde E, Murangwa Y (2007). Labour Market and economic activity trends in Rwanda, analysis of the EICV2 survey. Final Report, National Institute of Statistics of Rwanda.

Struif BT , Van Keulen H (2003): Modelling the dynamics of Agricultural development at farm and regional level. Agric. Syst. 76.379-396.

Sys C, Van Ranst E, Debaveye J, Beernaert F (1993). Land evaluation. PartIII. Crop requirements. Agricultural publications no 7. General administration for development cooperation. Brussels, Belgium.

Takeuchi S, Marara J (2007). Regional differences regarding land tenancy in rural Rwanda with special reference to sharecropping in coffee production area. (2007). The Center for African Area Studies, KyotoUniversity. Afr. Stud. Monogr. 35:111-138.

Tang H, Qiu J, VanRanst E, Li C (2006). Estimations of soil organic carbon storage in cropland of China base don DNDC model. Geoderma 134:200-206.

Thangata HP, Hildebrand EP, Gladwin HC (2002). Modeling agroforestry adoption and household decision making in Malawi. Afr. Urban Q. P. 6.

Tiffen M, Mortimore M, Gichuki F (1994). More people, less erosion: environmental recovery in Kenya. John Wiley and Sons, New-York, USA.

Upton M (1996). The economics of tropical farming systems. Cambridge University Press.

USDA (1999). Soil Taxonomy. A basic system of soil classification for making and interpreting soil surveys. Soil Survey Agricultural Handbook No 436. Second Edition. WashinghtonDC.

USDA (2009). General Guidelines for Assigning Soil Loss Tolerance "T". Inhttp://soils.usda.gov/technical/handbook/contents/part618ex.html. Consulted on 06/05/2009.

Van Calker KJ, Berentsen PBM, De Boer IJM, Giesen GWJ, Huirne RBM (2004). An LP-model to analyse economic and ecological sustainability on Dutch dairy farms: model presentation and application for experimental farm "de Marke". Agric. Syst. 82:139-160.

Van Keulen H, Kuyvenhoven A, Ruben R (1998). Sustainable land use and food security in developing countries: DLV's approach to policy support. Agric. Syst. 58:285-307.

Van Ranst E (2003). Tropical soils. Lecture notes. University of Ghent, Ghent, Belgium.

White SD, Labarta AR, Leguí JE (2005). Technology adoption by resource-poor farmers: considering the implications of peak-season labor costs. Agric. Syst. 85:183-201.

Wischmeier WH, Smith DD (1995). Predicting rainfall erosion losses – A guide to conservation planning. USDA Agricltural Handbook 537, Washington, DC.

Woelcke J (2006). Technological and policy options for sustainable agricultural intensification in eastern Uganda. Agric. Econ. 34:129-139.

World Health Organisation (WHO) (1985). Energy and protein requirement. Report of a joint FAO/WHO/UNU expert Consultation. World Health Organisation Technical Report Series 724. Geneva, Swizteland .

Zaongo C, Nabahungu NL, Ruganzu MV, Bijula M, Habamenshi D (2002). Etude des sols du Rwanda. Volume II. Pedologie, aptitudes, proprietes physiques, chimiques, hydrologiques et leurs dynamiques. Isar/Icraf/Unr/ADR/Kanguka. Butare, Rwanda.

Appendix A

Table A1. Input-output information on crops for Oxisols in Season B.

Season		B	B	B	B	B	B&A	A&B
Crop activities	**Unit**	**Beans**	**Groundnuts**	**Maize**	**Sorghum**	**Sweet potatoes**	**Cassava**	**Banana**
Inputs-output								
Yield	kg ha^{-1}	690	450	940	1,050	6,300	6,000	15,000
Price	fr.rw kg^{-1}	167	538	127	118	93	177	83 (98)[a]
Revenues	fr.rw ha^{-1}	115,230	242,100	119,380	123,900	585,900	1,062,000	1,357,500*
Cost of seed/establishment cost	fr.rw ha^{-1}	10,000	30,000	2500	900	30,000	50,000	93,200*
Gross margin	fr.rw ha^{-1}	105,230	212,100	116,880	123,000	555,900	1,012,000	1,264,300
Labour	man-day ha^{-1}	104	98	116	108	94	129	264
Total energy	10^3kcal ha^{-1}	2,297	2,551	3,073	3,664	5,720	6,378	5,190[b]
Total protein	g ha^{-1}	162,840	116,100	71,440	112,350	86,310	48,000	56,250[b]
Soil loss	t ha^{-1}	14.4	30.4	26.6	30.4	17.5	19.7[b]	3[b]
Soil organic carbon	t ha^{-1}	3020	1887	3020	1887	3775	1887[b]	3020[b]

[a] Price per Kg of banana, respectively in season A and B. [b]These values are concerning Season B. *The costs and revenues during establishment of Banana (multiyear crop) are based on annuities.

Table A2. Input-output information on crops for Inceptisols in Season A.

Season		A	A	A	A	A	A&B	A&B
Crop activities	**Unit**	**Beans**	**Groundnuts**	**Maize**	**Sorghum**	**Sweet potatoes**	**Cassava**	**Banana**
Inputs-output								
Yield	kg ha^{-1}	1,000	760	1,490	1,580	7,300	7,000	17,000
Price	fr.rw kg^{-1}	141	419	100	106	90	177	83 (98)[a]
Revenues	fr.rw ha^{-1}	141,000	318,440	149,000	167,480	657,000	1,239,000	1,538,000
Cost of seed/ establishment cost	fr.rw ha^{-1}	8,500	18,000	2,000	800	30,000	50,000	93,200*
Gross margin	fr.rw ha^{-1}	132,500	300,440	147,000	166,680	627,000	1,189,000	1,445,300
Labour	man-day ha^{-1}	95	91	110	106	85	91	264
Total energy	10^3kcal ha^{-1}	3,330	4,309	4,872	5,514	6,628	7,440	5,882[b]
Total protein	g ha^{-1}	236,000	196,080	113,240	169,060	100,010	56,000	63,750[b]
Soil loss	t ha^{-1}	3.1	1450	5.6	6.4	3.7	4.2[b]	0.7[b]
Soil organic carbon	t ha^{-1}	2320	1450	2320	1450	2900	1450[b]	2320[b]

[a] Price per Kg of banana, respectively in Season A and B. [b] These values are concerning Season A. *The costs and revenues of Banana (multiyear crop) are based on annuities.

Table A3. Input-output informat on on crops for Inceptisols in Season B.

Season		B	B	B	B	B	B&A	B&A
Crop activities	**Unit**	**Beans**	**Groundnuts**	**Maize**	**Sorghum**	**Sweet potatoes**	**Cassava**	**Banana**
Inputs-output								
Yield	kg ha^{-1}	1,050	700	1,410	1,580	7,300	7,000	17,000
Price	fr.rw kg^{-1}	167	538	127	118	93	177	83 (98)[a]
Revenues	fr.rw ha^{-1}	175,350	376,600	179,070	186,440	678,900	1,239,000	1,538,000
Cost of seed/ establishment cost	fr.rw ha^{-1}	10,000	30,000	2500	900	30,000	50,000	93,200*
Gross margin	fr.rw ha^{-1}	165,350	346,600	176,570	185,540	648,900	1,189,000	1,445,300
Labour	man-day ha^{-1}	95	91	110	106	85	91	264
Total energy	10^3kcal ha^{-1}	3,496	3,969	4,610	5,514	6,628	7,440	5,882
Total protein	g ha^{-1}	247,800	180,600	107,160	169,060	100,010	56,000	63,750[b]
Soil loss	t ha^{-1}	3.7	7.8	6.8	7.8	4.5	5.11[b]	0.8[b]
Soil organic carbon	t ha^{-1}	2320	1450	2320	1450	2900	1450[b]	2320[b]

[a] Price per Kg of banana, respectively in season A and B. [b] These values are concerning season B. *The costs and revenues of Banana (multiyear crop) are based on annuities

Appendix B

Table B1. Variance/Covariance Matrix of GM with variable prices (VP) and constants prices (CP) for Season B.

Parameter	Sorghum		Maize		Beans		Peanuts		Banana		Sweet potatoes		Cassava	
	VP	CP	VP	CP	VP	CP	VP	CP	VP	CP	VP	CP	VP	CP
Sorghum	$1.7.10^9$	$2.2.10^8$	$9.6.10^8$	-3.2.107	$1.6.10^9$	-3.9.106	$2.1.10^9$	$6.6.10^8$	$1.1.10^{10}$	$3.2.10^8$	$4.5.10^9$	-2.10^8	$1.7.10^{10}$	$5.3.10^8$
Maize	$9.6.10^8$	$-3.2.10^7$	$2.5.10^9$	$6.6.10^8$	$4.3.10^8$	$-1.8.10^8$	$1.5.10^9$	$-3.4.10^8$	$4.5.10^9$	$-3.4.10^8$	$2.2.10^9$	$-5.7.10^8$	$1.1.10^{10}$	7.3.107
Beans	$1.6.10^9$	$-3.9.10^6$	$4.3.10^8$	$-1.8.10^8$	3.10^9	$4.7.10^8$	$2.3.10^9$	$1.8.10^7$	$1.5.10^{10}$	2.6.108	$6.8.10^9$	-1.10^8	$2.5.10^{10}$	$2.2.10^9$
Peanuts	$2.1.10^9$	$6.6.10^8$	$1.5.10^9$	$-3.4.10^8$	$2.3.10^9$	$1.8.10^7$	$4.7.10^9$	$3.1.10^9$	$1.4.10^{10}$	$1.2.10^9$	$7.4.10^9$	-6.10^7	$2.6.10^{10}$	$1.9.10^9$
Banana	$1.1.10^{10}$	$3.2.10^8$	$4.5.10^9$	$-3.4.10^8$	$1.5.10^{10}$	$2.6.10^8$	$1.4.10^{10}$	$1.2.10^9$	$1.4.10^{11}$	$1.4.10^9$	$5.1.10^{10}$	$-4.9.10^8$	$1.4.10^{11}$	$1.6.10^9$
Sweet P.	$4.5.10^9$	-2.10^8	$2.2.10^9$	$-5.7.10^8$	$6.8.10^9$	-1.10^8	$7.4.10^9$	-6.10^7	$5.1.10^{10}$	$-4.9.10^8$	3.10^{10}	$2.9.10^9$	$6.9.10^{10}$	$-1.3.10^9$
Cassava	$1.7.10^{10}$	$5.3.10^8$	$1.1.10^{10}$	$7.3.10^7$	$2.5.10^{10}$	$2.2.10^9$	$2.6.10^{10}$	$1.9.10^9$	$1.4.10^{11}$	1.6.109	$6.9.10^{10}$	$-1.3.10^9$	$3.5.10^{11}$	$2.3.10^{10}$

27

Comparative economic analysis of tomato (*Lycopersicon esculenta*) under irrigation and rainfed systems in selected local government areas of Kogi and Benue States, Nigeria

Josephine Bosede Ayoola

Institute of Food Security, University of Agriculture, Makurdi, P. M. B. 2373, Makurdi, Benue State, Nigeria.

The study compared the economic performance of tomato (*Lycopersicon esculenta)* under irrigation and rain fed systems in Bassa and Makurdi Local Government Areas of Kogi and Benue States of Nigeria, with the aim of assessing the determinants of its profitability. Primary data obtained from a sample of 120 farmers by stratified and multi-staged random sampling from four villages were analyzed using percentages, means, gross margin, net profit, Shepherd-future coefficient and exponential regression model of combined profit function. Results revealed gender inequality; all respondents under irrigation system were male, compared to 71.7% female participation under rain fed system. Average net profits were ₦128,750 and ₦57,050; and economic efficiencies were 1.380 and 0.986 for irrigated and rain-fed systems respectively. Results also showed that farm size, planting material and herbicide were significant at one and five percent levels, and positively correlated with farmers' profit; while age and costs of fertilizer and labor were negative. The study concludes that tomato is more profitable and economically efficient under irrigation; and that increased access to land, herbicides, and improved seeds will promote profitability of the crop in the study area.

Key words: Irrigation, rain-fed, tomato (*Lycopersicon esculenta*), profitability, economic efficiency.

INTRODUCTION

Tomato (*Lycopersicon esculenta*) is an important vegetable crop grown in many parts of the world, contributing significantly to income security and the nutritive diets of many households. According to Mofeke et al. (2003) vegetable crops constitute 30 to 50% of iron and vitamin A in resource poor diet. Vegetable crops including tomatoes are widely cultivated in most parts of Sub Sahara Africa, particularly by small scale farmers in most states of Nigeria (Adeolu and Taiwo 2009; Giroh et al., 2010). Global production of fruits and vegetables tripled from 396 million MT in 1961 to 1.34 billion MT in 2003 (International Institute of Tropical Agriculture, 2005) and Nigeria ranked 16th on the global tomato production scale, accounting for 10.79% of Africa's and 1.2% of total world production of tomatoes (Weinberger and Lumpkin, 2007).

Denton and Swarup (1983) observed that tomato production in the Northern States as in other parts of the country is done during the dry season, while its production is scarce during the rainy season because of high disease incidence associated with growing tomatoes and preference of tomato producers for grain food crops during rainy season. Nigeria is unable to meet its growing domestic requirements for vegetables, fruits, floriculture, herbs and spices, dried nuts and pulses. Between 2009 and 2010, Nigeria imported a total of 105,000 metric tons of tomato paste valued at over 16 billion Naira to bridge the deficit gap between supply and demand in the country (Food and Agriculture Organization, 2006). Kalu (2013) attributed this situation to socio-economic constraints surrounding the key actors in the tomato value chain, institutional weaknesses and declining agricultural research.

Irrigation farming relatively low in Nigeria and Africa as a whole, with irrigated area estimated at only 6% of total cultivated area, compared with 37% for Asia and 14% for Latin America (FAOSTAT, 2009). Svendsen and Sangi (2009) observed that more than two-third of existing irrigated area is concentrated in five countries namely Egypt, Madagascar, Morocco, Aouth Africa and Sudan. Given that irrigated crop yields are more than double of rainfed yields in Africa (Liangzhi et al., 2010), it is important to invest on irrigation development with particular focus on locations and technologies with greatest potential for irrigation. The efforts of the Federal Government of Nigeria, with the support of the World Bank and the African Development Fund to develop irrigation systems in the country started with the approval of the implementation of the National Fadama Development Project in 1992 (World Bank, 1992), followed by second National Fadama Development Project between 2004 and 2010, and the on-going third National Fadama Development Project (2008-2016). Small scale irrigation systems have gone a long way to support dry season farming of crops all over the country. Dry season production of vegetables is common along the banks of the rivers Niger and Benue that cut across cities and towns in Kogi and Benue States, respectively.

Tomato crop is cultivated in traditional small holdings in Nigeria and specifically in the study area. Denton and Swarup (1983), observed that tomato had ceased to be the main crop during rainy season in Northern Nigeria. In cognizance of the characteristic competition among major food crops for the limited resources of the farmers, one then wonders if the capital investment on irrigation for dry season farming of tomato is worthwhile relative to rainy season production; and what factors could further promote the profitability of tomato under irrigation or rain-fed system? Within this context, the general objective of the study was to assess the relative performance of tomato crop under irrigation and rain-fed systems in Bassa and Makurdi Local Government Areas of Kogi and Benue States of Nigeria respectively. Specifically, the study assessed the relative profitability and economic efficiency of tomato crop under irrigation and rain-fed systems, and identified the determinants of its profitability in the study area. It was thus, hypothesized that production of tomato crop under irrigation was not more economically efficient and profitable than those grown under rain-fed system; and that the socio-economic characteristics of the farmers did not affect profitable production in the area. It is expected that the results of the study would contribute to agricultural transformation policies and promote food security in Nigeria.

CONCEPTUAL AND ANALYTICAL FRAMEWORK

The conceptual framework is within the context of relative efficiency and profitability of investments in the production of tomato under irrigation and rain-fed systems. Thus, the concepts of gross margin and net profit were employed to compare the profitability, while Shepherd-future model was used to compare the economic efficiency of tomato crop under irrigation and rain-fed systems. Gross margin was measured in terms of the amount in Naira that is contributed to the enterprise after paying for direct variable unit costs, while the net profit accounts for the direct fixed costs in addition. The total variable costs comprised the expenses incurred on variable inputs such as fertilizers, seeds, and labor; while fixed costs comprised expenses on rent on land, depreciation of capital assets such as irrigation pumps, hoes, cutlasses, and wheel barrow. Shepherd-future coefficient was used as a measure of how well each naira return on the enterprise is utilized to cover the operational and overhead expenses. Shepherd-future model was expressed as the ratio of gross margin obtained from the production of tomato crop to the total cost of production (Shepherd, 1962). Greater positive net profit and Sheperd-futre coefficient would indicate higher profitability and efficiency of tomato crop production.

METHODOLOGY

Study area

The study was carried out in North Central Nigeria, covering Sharia and Gboloko villages in Bassa LGA of Kogi State, and Ugondo and Mbayong villages in Makurdi LGA of Benue State, where Fadama land is very prominent. The area is located in Southern Guinea Savanna zone within latitude 14°N and 16°N and longitude 12°E and 13°E, respectively for Kogi and Benue States; with annual rainfall of between 1100 and 1600 mm and an average temperature of 35°C (National Population Commission, NPC 2006). Both states are bound on the North and West by Rivers Niger and Benue.

Data sampling and collection methods

A total sample of 120 farmers used for the study was selected using stratified and six-staged random techniques. The sample was stratified into two; 60 farmers under one rrigation system and 60 farmers under rain-fed system. The six-stage random sampling firstly comprised, of the two States (Kogi and Benue) randomly selected from among six States in the Southern Guinea agro-ecological zone (Kwara, Niger, Benue, Kogi, Nasarawa and Taraba States). Secondly, one Local Government Area (LGA) was selected from each State (Bassa LGA from Kogi state and Makurdi LGA from Benue state). Thirdly, two villages were selected in each LGA (including Sharia and Gboloko in Bassa LGA of Kogi State, and Udongo and Mbayong in Makurdi LGA of Benue State). Fourthly, 3 Fadama Associations (FAs) were selected in each of the four villages making 12 FAs; and fifthly, 5 farmers were randomly selected from each of the FAs making a total of 60 farmers (15 farmers per village) under irrigation system. Lastly, 15 farmers that cultivated tomato crop under rain-fed system were purposively selected from each of the four villages to make a total of 60 farmers under rain-fed system.

Structured questionnaire was used to obtain primary data about farmers' socio-economic characteristics; such as sex (male or female), family size (number of persons in the household), age (years), and educational level (number of years in school). Also, data were obtained on production variables such as farm size (ha), farm output (₦); variable costs (₦) such as costs of fertilizer, labour, planting materials, pesticide and herbicide; and fixed costs (₦) such as rent on land, as well as depreciation of capital assets such as irrigation pumps, hoes, cutlasses, and wheel barrow.

Analytical methods

Descriptive and inferential statistics including frequency, percentage, gross margin and net profit analyses were used to describe the socio-economic characteristics of farmers and level of profitability of tomato crop production. Multiple regression was also used to analyze the socio-economic determinants and coefficients of profitability, while Shepherd-future coefficient was used to determine the economic efficiency of tomato crop under irrigation and rain-fed systems.

Models specification

(1) The functional form of the multiple regression function is:

$$Y = \beta o + \beta_1X_1 + \beta_2X_2 + \beta_3X_3 + \beta_4X_4 + \beta_5X_5 + \beta_6X_6 + e_1$$

Where: Y = Profit (₦); Bo = Intercept (₦); β = Marginal effect of Xs on Y; X_1 = Farm size (ha); X_2 = Educational level (number of years in school); X_3 = Age of the farmer (years); X_4 = Cost of fertilizer (₦); X_5 = Cost of labour (₦); X_6 = Cost of planting materials (₦); X_7 = Cost of herbicide (₦); e_1 = Error term,
(2) Gross Margin: GM = TR – TVC; where GM = Gross Margin (₦); TR = Total Revenue (₦); TVC = Total Variable Cost (₦),
(3) Net Profit: NP = TR – TC; where: NP = Net Profit (₦); TR = Total Revenue (₦); TC = Total Cost (₦) = Total Variable Cost (TVC) + Total Fixed Cost (TFC),
(4) Shepherd-future: S.F = (GMTC) x 100; Where: S.F. = Shepherd-future.

RESULTS AND DISCUSSION

Socio-economic characteristics of tomato farmers

The results revealed that all the farmers under irrigation s ystem were male while about 72% of farmers under rain-fed system were female farmers (Table 1); indicating gender inequality in access of farmers to irrigation facilities, and insensitivity of the irrigation programme to the Millennium Development Goal of gender equality and women empowerment (United Nations Development Programme, 2002, and International Food Policy Research Institute, 2006). About 62% of farmers under irrigation system and 22% of farmers under rain-fed system were above 50 years of age; indicating that majority of the farmers were getting advanced in age and may lack sufficient vigor for large scale and efficient production of vegetable crops. About 73% of farmers under irrigation system and 68% of farmers under rain-fed system have no formal education; indicating that low literacy rate among the farmers might hinder adoption of innovations since education has been reported to influence the level of technology adoption (Chinaka et al., 1995).

Despite the family size that was above six for about 94% of the farmers, more than 80% of farmers under irrigation system employed hired labor for farm operations; indicating that family labor was not a preferred option for reducing the cost of labor. About 98% of farmers under irrigation system have more than 10 years of farming experience; suggesting the possession of necessary farming and irrigation skills for increased productivity and efficiency. About 53% of farmers under irrigation system were land owners while about 47% obtained their farm land on lease from government; land ownership structure might be a decision factor with respect to investment on irrigation facilities as farmers that owned land could guarantee continuous access to the use of irrigated land. Majority of the farmers (58% under irrigation system and 98.4% under rain-fed system) cultivated less than 2 ha of farm size; indicating that tomato crop farming is generally small scale and corroborating Kalu (2013) that tomato is produced on small holdings in Northern Nigeria.

Relative profitability and economic efficiency of tomato crop

Table 2, results showed that gross margins obtained per hectare of tomato crop under irrigation and rain-fed systems were ₦153,500 and ₦68,000, respectively; and average net profits per hectare were ₦128,750 and ₦57,050. Shepherd-future coefficients were 1.380 and 0.986t, indicating that tomato crop was economically efficient under both irrigation and rain-fed systems. The results indicated that investment on tomato crop production under irrigation system was worthwhile, as it yielded greater revenue in excess of operational and overhead expenses in comparison with that of rain-fed system. These results corroborate Hussain and Wijerathna (2004) linking irrigation and poverty alleviation in developing countries. Adewumi et al. (2005) that

Table 1. Socio-economic characteristics of farmers under irrigation and rain-fed systems.

Socio-economic variables	Irrigation system		Rain-fed system		Total	
	Frequency	%	Frequency	%	Frequency	%
Sex: Male	60	100	17	28.3	77	64.17
Female	-	-	43	71.7	43	35.83
Age (years): <40	9	15	21	35	30	25
41-50	14	23.33	26	43.3	40	33.33
>50	37	61.67	13	21.7	50	41.67
Educational status: No formal	44	73.3	37	61.7	81	67.5
Primary (1-6 years)	16	26.7	22	36.7	38	31.67
Secondary (7-12 years)	-	-	1	1.6	1	0.83
Family size: <6	2	3.3	5	8.3	7	5.83
6-10	32	53.3	34	56.7	66	55
>10	26	43.4	21	35	47	39.17
Farming experience (years): <10	1	1.7	30	50.0	31	25.83
10-20	28	46.6	26	43.33	54	45.00
>20	31	51.7	4	6.67	35	29.17
Land status: Bought	3	5.0	26	43.0	29	24.17
Inheritance	29	48.3	24	40.0	53	44.17
Rent / Leased	28	46.7	10	16.7	38	31.66
Farm size (Ha): <1	2	3.1	34	56.7	36	30
1<X<2	33	55.0	25	41.7	58	48.33
2-3	23	38.4	1	1.6	24	20
>3	2	3.3	-	-	2	1.67
Type of labor: Family	6	10.0	41	68.3	47	39.17
Hired	48	80.0	19	31.7	67	55.83
Both	6	10.0	-	-	6	5

tomato farming under small scale irrigation systems was profitable in Sokoto State; where an average gross margin of N87,543.00/ha and average net income of N77,559.80/ha, with a rate of return to investment greater than 1 were obtained. Denton and Swarup (1983) had earlier observed that tomato crop has ceased to be the main crop during the raining season in Northern States, because farmers sustained greater loss during the rainy season due to diseases, nematodes, insect pests and high flower drops; resulting in lower yield and poor quality fruits (Sabo and Dia, 2009). Kalu (2013) later observed that farmers in Northern Nigeria engaged in the production of other crops during raining seasons, while they planted tomatoes in the dry season using the irrigation system, as a strategy for reducing losses incurred on tomato farming. Gani and Omonona (2009) also confirmed greater profitability and economic efficiency for maize production under irrigation system relative to rain-fed system. These results indicated that investments in irrigation facilities for tomato crop production would promote higher income among small scale farmers, and thus contribute to poverty alleviation in the study areas. Similarly, irrigation agriculture has been linked to poverty reduction in six Asian Countries (Hussain, 2007).

Determinants of profitability of tomato production

The exponential regression model of combined profit function gave the coefficient of multiple determination (R^2) value of 0.91 (Table 3), implying that 91% of the variation in farmers' profit is explained by the independent variables while the remaining 9% could be accounted for by the error term. The parameter estimates of combined profit function for tomato crop showed that age and planting material as well as farm size and herbicide were significant at one percent and five percent respectively. Farm size, level of education, cost of herbicide and planting material have positive correlation with farmers' profit; while age and costs of fertilizer and labor were negatively correlated to farmers' profit. These results implied that an increase in farm size, level of education, planting material and cost of herbicide would lead to an increase in income from tomato. Formal education could aid managerial ability of farmers and enable them achieve greater efficiency in tomato crop production. Age of farmers was significantly negative;

Table 2. Costs and Returns per hectare on tomato crop under irrigation and rain-fed systems.

Cost and return	Life span (years)	Unit price (₦)	Irrigation system		Unit price (₦)	Rain-fed system	
			Quantity	Total cost (₦)		Quantity	Total cost (₦)
Fixed costs items							
Cutlass	2	1000	5	2,500*	1000	3	1500*
Irrigation equipment	2	25000	1	12,500*			
Wheel barrow	2	5000	2	5,000*	5000	2	5,000*
Land (rent)	-		-	4,000			4,000
Hoe	6	900	5	750*	900	3	450*
Total fixed cost				**24,750**			**10,950**
Variable costs items							
Labor (mandays/ha)	-	1000	20	20,000	1000	10	10,000
Pesticide (litres/ha)	-	1000	5	5,000	1000	2.5	2,500
Herbicide (litres/ha)	-	1200	5	6,000	1200	5	6,000
Fertilizer (kg/ha)	-	4500	4	18,000	4500	2	9,000
Seed (kg/ha)	-	1500	15	22,500	1500	15	22,500
Transportation	-			15,000			8,000
Total variable cost				**86,500**			**58,000**
Total production cost				**111,250**			**68,950**
Total Revenue		2000	120	240,000	1200	105	126,000
Average Gross Margin per ha				**153,500**			**68,000**
Average Net profit/ha				**128,750**			**57,050**
Economic Efficiency				**1.380**			**0.986**

Note: Asterisked (*) figures are depreciated costs for the respective fixed capital items.

Table 3. Parameter estimates of exponential regression model of combined profit function for tomato production.

Variables	Coefficient	Standard error	t-value	p>/t/
Farm size	72105.89	29675.94	2.43	0.017**
Education	3564.318	4589.122	0.78	0.439
Age	-6951.96	1494.602	-4.65	0.000***
Fertilizer	-0.9568392	0.6603224	-1.45	0.150
Labor	-0.2029115	1.674137	-0.12	0.904
Planting material	3.411511	0.5859465	-5.82	0.000***
Herbicide	5.885283	2.618464	2.25	0.027**
Constant	202212.5	77502.95	2.61	0.010

Note: R^2 = 0.91, F-value = 126, ** = 5% significance, *** = 1% significance.

implying that older farmers tend to be less efficient. The cost of fertilizer was also negatively correlated with farm income and was not significant; suggesting that achieving greater efficiency in fertilizer utilization would not likely lead to significant increase in profit obtained on tomato crop production. Result also showed that cost of labor was negative while cost of herbicide was positively correlated with profit; meaning that an increase in the amount spent on labor will lead to reduced profit while an increase in the level of herbicide cost will lead to increase in profit on tomato production. This suggests that substitution of herbicide for labor may reduce cost of weeding and thus, enhance profitability of tomato production. The cost of planting material was also positively correlated with profit and significant at one percent; indicating that more use of improved planting materials might enhance the profitability of tomato crop production. These results confirmed the findings of Ayanwale and Abiola (2008) that farm size, education level and capital inputs are critical determinants of efficiency of vegetable production under tropical conditions.

SUMMARY AND CONCLUSIONS

The study aimed at assessing the extent to which investment in small scale irrigation have contributed to profitability of tomato crop farming in the North Central Southern Guinea Savanna agro-ecology of Nigeria. Specifically, the study estimated the relative profitability and efficiency of tomato crop production under irrigation and rain-fed systems, and identified the socio-economic determinants of profitability of tomato crop. Primary data obtained from a sample of 120 farmers were analyzed using descriptive statistics, gross margin, net profit, as well as Shepherd-future and multiple regression models.

Results showed that all the respondents under irrigation system were male farmers, indicating gender inequality in farmers' access to irrigation facilities. Majority of the farmers were above 50 years of age, with low literacy and average family size of 8 per household. Results also showed that the gross margins per hectare under irrigation and rain-fed systems were ₦153,500 and ₦68,000 respectively; and that average net profits per hectare were ₦128,750 and ₦57,050; indicating profitability of tomato crop under both systems. Shepherd-future coefficients were 137.98 and 98.62% respectively for irrigation and rain-fed systems; indicating that tomato crop was more economically efficient under irrigation system than rain-fed system. Parameter estimates from the combined profit function also revealed that farm size, level of education, cost of herbicide and planting material have positive correlation; while age of farmers, costs of fertilizer and labor were negatively related to farmers' profit. Age of farmers and planting materials were significant at one percent, while farm size and costs of herbicide were significant at five percent.

Thus, it was concluded that tomato crop was more profitable and economically efficient under irrigation; and that increased access to land, herbicides, and improved seeds would promote profitability of the crop in Bassa and Makurdi Local Government Areas of Kogi and Benue States, Nigeria. Land policies that increase access of farmers (with special consideration for women farmers) to adequate land, irrigation facilities, herbicides and good quality improved planting material would promote gender equality in irrigation farming and profitability of tomato production; thus contributing to food security in Nigeria and enhancing the income potential of farmers.

Conflict of interest

The authors have not declared any conflict of interest.

REFERENCES

Adewumi MO, Omotesho OA, Tsoho B (2005). Economics of irrigation farming in Nigeria: a case study of tomato based farming in Sokoto State, Nigeria. Int. J. Agric. Rural Dev. 6:103-110.

Adeolu BA, Taiwo A (2009). Potential and Opportunities for Sustainable Production and Utilization of Horticultural Crops in Nigeria. Horticultural EOS magazine, September 2009. P. 9.

Ayanwale AB, Abiola MO (2008). Efficiency of Fluted Pumpkin Production under Tropical Conditions. Int. J. Veget. Sci. 13(3):35-49.

Chinaka CC, Chikwendu DO, Asumugha GN (1995). A Socio-economic Study of the Adoption and Sustainability of Improved Cassava Varieties Among Resource Poor Farmers in South-East Zone of Nigeria. In Root crops and poverty alleviation. International Society for Tropical Crops African Branch, Malawi. pp. 333-336.

Denton T, Swarup V (1983). Tomato Cultivation and its Potential in Nigeria. Acta Hort. (ISHS) 123:257-272. http://www.actahort.org/books/123/123_23.htm)

Food and Agriculture Organization (2006). State of Food Insecurity in the World. Food and Agricultural Organization, FAO. Rome.

FAOSTAT (2009). Food and Agriculture Organization, FAO AQUASTAT: http://www.fao.org/nr/water/aquastat/dbase/index.sfm> Accessed May 2009.

Gani BS, Omonona BT (2009). Resource Use Efficiency Among Small Scale Irrigated Maize Producers in Northern Taraba State of Nigeria. J. Hum. Ecol. 28(2):113-119.

Giroh DY, Waizah Y, Umar HY (2010). Technical Efficiency and Cost of Production Among Gum Arabic Farmers in Jigawa State. Report Opin. 2(1):52-57.

Hussain I (2007). Irrigation and Poverty Alleviation: Pro-poor Intervention Strategies in Irrigated Agriculture in Asia. Paper presented in the 4th Asian Regional conference and 10th International Seminar on Participatory irrigation Management. 2-5 May, 2007, Tehran, Iran. http://fadama.org/taraba.html

Hussain I, Wijerathna D (2004). Irrigation and Income-poverty Alleviation: A comparative analysis of irrigation systems in developing Asia. International Water Management Institute (IWMI) February 2004.

International Food Policy research Institute (2006). Agriculture and Achieving the Millenium Development Goals. Agriculture and Rural development Department. *Report No. 32729 GLB.* Pp. 10

International Institute of Tropical Agriculture, IITA (2005). Annual Report 2005, IITA, Ibadan, Nigeria. pp. 29-30.

Kalu S (2013). The great tomato waste in Nigeria. http://kalusam.wordpress.com/2013/09/11

Liangzhi Y, Claudia R, Gerald N, Ulrike W-S, Richard R, Stanley W, Zhe G, Tingju Z, Yau S (2010). What is the Irrigation Potential for Africa? A combined Biophysical and Socio-economic Approach. IFPPRI Discussio Paper 00993. Environment and Production Technology Division. June 2010.

Mofeke ALE, Ahmada A, Mudiane OJ (2003). Relationship Between Yield and Seasonal Water Use for Tomatoes, Onions, and Potatoes Grown under Fadama Irrigation. Asset Series A. 3:35-46.

Sabo E, Dia YZ (2009). Awareness and Effectiveness of Vegetable Technology Information Packages by Vegetable Farmers in Adamawa State, Nigeria. Afr. J. Agric. Res. 4(2):065-070, February, 2009

Shepherd GS (1962). Marketing Farm Products Economic Analysis. 4th Edition. Ames Iowa, Iowa State University Press.

Svendsen E, Sangi M (2009). Measuring Irrigation Performance in Africa: IFFPRI Discussion Paper 894, Washington D.C., IFFPRI

United Nations Development Programme (UNDP) (2002). Human Development Report 2002. Available at: www.undp.org/hdr2002/

Weinberger K, Lumpkin TA (2007). Diversification into Horticulture and poverty Reduction: A research agenda. World Dev. 35(8):1464-1480.

World Bank (1992). National Fadama Development Project (NFDP), Staff Appraisal Report (SAR). Washington D.C. World Bank, pp. 55-72.

Permissions

All chapters in this book were first published in JDAE, by Academic Journals; hereby published with permission under the Creative Commons Attribution License or equivalent. Every chapter published in this book has been scrutinized by our experts. Their significance has been extensively debated. The topics covered herein carry significant findings which will fuel the growth of the discipline. They may even be implemented as practical applications or may be referred to as a beginning point for another development.

The contributors of this book come from diverse backgrounds, making this book a truly international effort. This book will bring forth new frontiers with its revolutionizing research information and detailed analysis of the nascent developments around the world.

We would like to thank all the contributing authors for lending their expertise to make the book truly unique. They have played a crucial role in the development of this book. Without their invaluable contributions this book wouldn't have been possible. They have made vital efforts to compile up to date information on the varied aspects of this subject to make this book a valuable addition to the collection of many professionals and students.

This book was conceptualized with the vision of imparting up-to-date information and advanced data in this field. To ensure the same, a matchless editorial board was set up. Every individual on the board went through rigorous rounds of assessment to prove their worth. After which they invested a large part of their time researching and compiling the most relevant data for our readers.

The editorial board has been involved in producing this book since its inception. They have spent rigorous hours researching and exploring the diverse topics which have resulted in the successful publishing of this book. They have passed on their knowledge of decades through this book. To expedite this challenging task, the publisher supported the team at every step. A small team of assistant editors was also appointed to further simplify the editing procedure and attain best results for the readers.

Apart from the editorial board, the designing team has also invested a significant amount of their time in understanding the subject and creating the most relevant covers. They scrutinized every image to scout for the most suitable representation of the subject and create an appropriate cover for the book.

The publishing team has been an ardent support to the editorial, designing and production team. Their endless efforts to recruit the best for this project, has resulted in the accomplishment of this book. They are a veteran in the field of academics and their pool of knowledge is as vast as their experience in printing. Their expertise and guidance has proved useful at every step. Their uncompromising quality standards have made this book an exceptional effort. Their encouragement from time to time has been an inspiration for everyone.

The publisher and the editorial board hope that this book will prove to be a valuable piece of knowledge for researchers, students, practitioners and scholars across the globe.

List of Contributors

Maxwell Darko Asante
West African Centre for Crop Improvement, College of Agriculture and Consumer Sciences, University of Ghana, P. M. B. LG 30, Legon, Ghana
CSIR-Crops Research Institute, P. O. Box 3785, Kumasi, Ghana

Bright Owusu Asante
CSIR-Crops Research Institute, P. O. Box 3785, Kumasi, Ghana
Deparment of Agricultural and Resource Economics, Faculty of the Professions, University of New England, Armidale, NSW, 2350, Australia

George Kwasi Acheampong
CSIR-Crops Research Institute, P. O. Box 3785, Kumasi, Ghana

Alexander Nimo Wiredu
CSIR-Savanna Agricultural Research Institute, P. O. Box TL 52, Tamale, Ghana

Samuel Kwame Offei
West African Centre for Crop Improvement, College of Agriculture and Consumer Sciences, University of Ghana, P. M. B. LG 30, Legon, Ghana

Vernon Gracen
West African Centre for Crop Improvement, College of Agriculture and Consumer Sciences, University of Ghana, P. M. B. LG 30, Legon, Ghana

Hans Adu-Dapaah
CSIR-Crops Research Institute, P. O. Box 3785, Kumasi, Ghana

Eric Yirenkyi Danquah
West African Centre for Crop Improvement, College of Agriculture and Consumer Sciences, University of Ghana, P. M. B. LG 30, Legon, Ghana

Arega D. Alene
International Institute of Tropical Agriculture (IITA), Lilongwe, Malawi

R. Khataza
Bunda College of Agriculture, University of Malawi, Lilongwe, Malawi

C. Chibwana
United States Agency for International Development (USAID), Lilongwe, Malawi

P. Ntawuruhunga
International Institute of Tropical Agriculture (IITA), Lilongwe, Malawi

C. Moyo
International Institute of Tropical Agriculture (IITA), Lilongwe, Malawi

G. M. Aloyce
Mikocheni Agricultural Research Institute, P. O. BOX 6226, Dar es Salaam, Tanzania

D. M. Gabagambi
Department of Agricultural Economics and Agribusiness, Sokoine University of Agriculture, P. O. Box 3007, Chuo Kikuu, Morogoro, Tanzania

J. P. Hella
Department of Agricultural Economics and Agribusiness, Sokoine University of Agriculture, P. O. Box 3007, Chuo Kikuu, Morogoro, Tanzania

N. C. Ehirim
Department of Agricultural Economics, Federal University of Technology Owerri, Imo State, Nigeria

R. A. Okere
Nigerian Institute for Oil Palm Research, Benin City, Edo State, Nigeria

S. U. O. Onyeagocha
Department of Agricultural Economics, Federal University of Technology Owerri, Imo State, Nigeria

M. A. Y Rahji
Department of Agricultural Economics, University of Ibadan, Ibadan Oyo State, Nigeria

T. T. Awoyemi
Department of Agricultural Economics, University of Ibadan, Ibadan Oyo State, Nigeria

B. I. Oluwatayo
Department of Agricultural Economics, University of Ibadan, Ibadan Oyo State, Nigeria

K. K. Salman
Department of Agricultural Economics, University of Ibadan, Ibadan Oyo State, Nigeria

Amsalu Bedemo
Department of Economics, Faculty of Business and Economics, Assosa University, P. O. Box 018,Assosa, Ethiopia

Kindie Getnet
International Water Management Institute (IWMI), P. O. Box, 5689, Addis Ababa, Ethiopia

Belay Kassa
School of Agricultural Economics and Agribusiness, Haramaya University, P. O. Box 138, Ethiopia

S. P. R. Chaurasia
School of Agricultural Economics and Agribusiness, Haramaya University, P. O. Box 138, Ethiopia

Abel Kwaku K. Bruce,
Department of Agricultural and Resource Economics, Faculty of Agribusiness and Communication Sciences, University for Development Studies, Tamale, Ghana

Samuel A. Donkoh
Department of Agricultural and Resource Economics, Faculty of Agribusiness and Communication Sciences, University for Development Studies, Tamale, Ghana

Michael Ayamga
Department of Agricultural and Resource Economics, Faculty of Agribusiness and Communication Sciences, University for Development Studies, Tamale, Ghana

Christian Crentsil
School of Economics, Keynes College, University of Kent, Canterbury, CT2 7NP, UK

Felix Larry Essilfie
Departments of Agricultural Economics and Agribusiness, College of Agriculture and Consumer Sciences, University of Ghana, Legon, Accra, Ghana

Dadirayi Manyumwa
Department of Agricultural Economics, Education and Extension, Faculty of Agriculture and Environmental Sciences, Bindura University of Science Education, Bindura, Zimbabwe

Never Mafuse
Department of Agricultural Economics, Education and Extension, Faculty of Agriculture and Environmental Sciences, Bindura University of Science Education, Bindura, Zimbabwe

Munyaradzi Matovore
Department of Agricultural Economics, Education and Extension, Faculty of Agriculture and Environmental Sciences, Bindura University of Science Education, Bindura, Zimbabwe

Joseph Musara
Department of Agricultural Economics, Education and Extension, Faculty of Agriculture and Environmental Sciences, Bindura University of Science Education, Bindura, Zimbabwe

Vincent Tinashe Munyati
Department of Agricultural Economics, Education and Extension, Faculty of Agriculture and Environmental Sciences, Bindura University of Science Education, Bindura, Zimbabwe

Joseph Chimvuramabwe
Department of Agricultural Economics, Education and Extension, Faculty of Agriculture and Environmental Sciences, Bindura University of Science Education, Bindura, Zimbabwe

Godfrey Chagwiza
Department of Agricultural Economics, Education and Extension, Faculty of Agriculture and Environmental Sciences, Bindura University of Science Education, Bindura, Zimbabwe

Emmanuel Zivenge
Department of Agricultural Economics, Education and Extension, Faculty of Agriculture and Environmental Sciences, Bindura University of Science Education, Bindura, Zimbabwe

Violet Dudu
Department of Agricultural Economics, Education and Extension, Faculty of Agriculture and Environmental Sciences, Bindura University of Science Education, Bindura, Zimbabwe

Bingxin Yu
International Food Policy Research Institute, United States

Alejandro Nin-Pratt
International Food Policy Research Institute, United States

ThankGod Peter Ojimba
Department of Agricultural Science, Ignatius Ajuru University of Education,Ndele Campus, P.M.B. 5047,Port Harcourt, Nigeria

Jacob Akintola
Department of Agricultural Economics, Bowen University,Iwo, Osun State, Nigeria

Sixtus O. Anyanwu
Department of Agricultural Science, Ignatius Ajuru University of Education,Ndele Campus, P.M.B. 5047, Port Harcourt, Nigeria

Hubbert A. Manilla
Department of Agricultural Science, Ignatius Ajuru University of Education,Ndele Campus, P.M.B. 5047,Port Harcourt, Nigeria

T. Abdoulaye
International Institute of Tropical Agriculture (IITA), Ibadan, Nigeria

A. Abass
International Institute of Tropical Agriculture (IITA), Ibadan, Nigeria

B. Maziya-Dixon
International Institute of Tropical Agriculture (IITA), Ibadan, Nigeria

G. Tarawali
International Institute of Tropical Agriculture (IITA), Ibadan, Nigeria

R. Okechukwu
International Institute of Tropical Agriculture (IITA), Ibadan, Nigeria

J. Rusike
International Institute of Tropical Agriculture (IITA), Ibadan, Nigeria

A. Alene
International Institute of Tropical Agriculture (IITA), Ibadan, Nigeria

V. Manyong
International Institute of Tropical Agriculture (IITA), Ibadan, Nigeria

B. Ayedun
International Institute of Tropical Agriculture (IITA), Ibadan, Nigeria

Daniel P. KABORE
Centre d'Analyse des Politiques Economiques et Sociales, 01 BP 1919 Ouagadougou 01, Burkina Faso

Mogos Teweldemedhin
Polytechnic of Namibia, School of Natural Resources and Spatial Sciences, Department of Agriculture and Natural Resources Sciences, Namibia

Namyenya Angella
Department of Agribusiness and Natural Resource Economics, School of Agricultural Sciences, Makerere University, P. O. Box 7062, Kampala Uganda

Sserunkuuma Dick
Department of Agribusiness and Natural Resource Economics, School of Agricultural Sciences, Makerere University, P. O. Box 7062, Kampala Uganda

Bagamba Fred
Department of Agribusiness and Natural Resource Economics, School of Agricultural Sciences, Makerere University, P. O. Box 7062, Kampala Uganda

B. A. Awotide
Department of Agricultural Economics, University of Ibadan, Nigeria

T. Abdoulaye
International Institute of Tropical Agriculture (IITA), Malawi

A. Alene
International Institute of Tropical Agriculture (IITA), Malawi

V. M. Manyong
International Institute of Tropical Agriculture (IITA), Tanzania

T. O. Adedeji
Department of Agricultural Production and Management Science, Tai Solarin University of Education, Ijebu-Ode, Ogun State, Nigeria

M. O. Nosiru
Department of Agricultural Production and Management Science, Tai Solarin University of Education, Ijebu-Ode, Ogun State, Nigeria

A. A. Akinsulu
Department of Agricultural Production and Management Science, Tai Solarin University of Education, Ijebu-Ode, Ogun State, Nigeria

I. O. Ewebiyi
Department of Agricultural Production and Management Science, Tai Solarin University of Education, Ijebu-Ode, Ogun State, Nigeria

B. G. Abiona
Department of Agricultural Extension and Rural Sociology, University of Agriculture, Abeokuta, Ogun State, Nigeria

T. S. Jimoh
Department of Agricultural Production and Management Science, Tai Solarin University of Education, Ijebu-Ode, Ogun State, Nigeria

Orachos Napasintuwong
Department of Agricultural and Resource Economics, Faculty of Economics, Kasetsart University, Bangkok, Thailand

Carl Pray
Department of Agricultural, Food, and Resource Economics, Rutgers: the State University of New Jersey, New Brunswick USA

Leone Iga Mawa
Department of Markets, Gender and Livelihoods, International Livestock Research Institute (ILRI), P. O. Box 30709-00100, Nairobi, Kenya

Mutuku Muendo Kavoi
Faculty of Agriculture, Jomo Kenyatta University of Agriculture and Technology, P. O. Box 62000-00200,Nairobi, Kenya

Isabelle Baltenweck
Department of Markets, Gender and Livelihoods, International Livestock Research Institute (ILRI),P. O. Box 30709-00100, Nairobi, Kenya

Jane Poole
Research Methods Group, International Livestock Research Institute, P. O. Box 30709-00100, Nairobi, Kenya

Dawoud
Department of Agricultural Economics, Faculty of Agriculture, Damietta University, Egypt

D. Z. Seham
Department of Agricultural Economics, Faculty of Agriculture, Damietta University, Egypt

A. V. Eze
Department of Agricultural Economics Management and Extension, Ebonyi State University, Abakaliki, Nigeria

S. U. Nwibo
Department of Agricultural Economics Management and Extension, Ebonyi State University, Abakaliki, Nigeria

Idris Akanbi Ayinde
Department of Agricultural Economics and Farm Management, Federal University of Agriculture Abeokuta, Nigeria

Raheem Olatunji Aminu
Department of Agricultural Economics and Farm Management, Federal University of Agriculture Abeokuta, Nigeria

Shakirat Bolatito Ibrahim
Agricultural and Food Economics Department School of Agriculture, Policy and Development, University of Reading, United Kingdom

O. Oluwasola
Department of Agricultural Economics, Obafemi Awolowo University, Ile-Ife, Osun State, Nigeria

A. O. Ige
Department of Agricultural Economics, Obafemi Awolowo University, Ile-Ife, Osun State, Nigeria

D. Omodara
Department of Agricultural Economics, Obafemi Awolowo University, Ile-Ife, Osun State, Nigeria

Morgan C. Mutoko
Kenya Agricultural and Livestock Research Organization (KALRO), Food Crops Research Institute,P. O. Box 450-30200, Kitale, Kenya

Cecilia N. Ritho
Department of Agricultural Economics, University of Nairobi, P. O. Box 29053, Nairobi, Kenya

James K. Benhin
Plymouth Business School, University of Plymouth, Drake Circus, Plymouth, Devon PL4 8AA, United Kingdom

Oliver L. Mbatia
Department of Agricultural Economics, University of Nairobi, P. O. Box 29053, Nairobi, Kenya

Gebrehaweria Gebregziabher
International Water Management Institute, East Africa Office, Addis Ababa, Ethiopia
International Water Management Institute, Colombo, Sri Lanka

Meredith A. Giordano
International Water Management Institute, Colombo, Sri Lanka

Simon Langan J.
International Water Management Institute, East Africa Office, Addis Ababa, Ethiopia

Regassa E. Namara
International Water Management Institute, West Africa Office, Accra, Ghana

Leake Gebresilassie
College of Dray Land Agriculture and Natural Resources, Mekelle University, Ethiopia

Adam Bekele
Melekasa Research Institute, Melekasa, Oromia, Ethiopia

J. C. Bidogeza
AVRDC - The World Vegetable Center, West and Central Africa, Yaounde - Messa, 2008, Cameroon
Business Economics / Wageningen University, Netherlands

P. B. M. Berentsen
Business Economics / Wageningen University, Netherlands

J. De Graaff
Land Degradation and Development / Wageningen University, Netherlands

A. G. J. M. Oude Lansink
Business Economics / Wageningen University, Netherlands

Josephine Bosede Ayoola
Institute of Food Security, University of Agriculture, Makurdi, P. M. B. 2373, Makurdi, Benue State, Nigeria